Modelling Pedestrian Movement

ice
Institution of Civil Engineers
Publishing

Modelling Pedestrian Movement

Urban Space – AECOM

Published by ICE Publishing, One Great George Street, Westminster, London SW1P 3AA.

Full details of ICE Publishing representatives and distributors can be found at: www.icebookshop.com/bookshop_contact.aspx

Other titles by ICE Publishing:
Pedestrian Facilities: Geometric design for safety and mobility, Second edition
John G. Schoon. ISBN 978-0-7277-6309-9
Designing for Cycle Traffic: International principles and practice
John Parkin ISBN: 978-0-7277-6349-5
Practical Road Safety Auditing, 3rd edition
Martin Belcher, Steve Proctor, and Phil Cook ISBN: 978-0-7277-6016-6

www.icebookshop.com

A catalogue record for this book is available from the British Library

ISBN 978-0-7277-6381-5

ICE Publishing is a division of Thomas Telford Ltd, a wholly-owned subsidiary of the Institution of Civil Engineers (ICE).

Cover photo: Crowd-modelling visualisation of the 2016 Olympic Park in Rio De Janeiro

Commissioning Editor: Michael Fenton
Production Editor: Madhubanti Bhattacharyya
Marketing Specialist: April Asta Brodie

Typeset by Sunrise Setting Ltd
Index created by Sunrise Setting Ltd
Printed and bound by in Great Britain by Bell and Bain, Glasgow

Contents

Preface

An attempt to explore an abstract but vital science

Every traveller becomes a pedestrian at a point during their journey. However, the portion of the journey spent walking has historically been overlooked by those planning and designing cities or infrastructure as it is often viewed as a part of everyday life.

Global population increase and a shift in economic demands have driven people to cities, which in turn has led to a need to ensure public safety in crowded stations, airports and urban environments. Physically constrained, older cities have struggled to renew and improve outdated public transport networks, often working incrementally to meet the ever increasing demand from their growing populations. Planning for the smooth movement and safety of large crowds has also moved into other spheres, such as sports stadia, major events, large buildings, urban regeneration, amusement parks and popular tourist attractions.

Though it is a nascent subject – only 30 or so years old – the science and techniques of pedestrian modelling have moved from the periphery of the planning process to become an integral part, aided by advances in micro-simulation software and hardware. Pedestrian modelling, which was once an afterthought, is now routinely included in most major transport planning projects which involve the movement of people.

When the Institute of Civil Engineers approached me two years ago with a request to write a book on pedestrian modelling, it occurred to me that for an emerging and important discipline there are remarkably few textbooks which can provide a guide to the subject for university students, researchers, event planners, public transport operators, transport authorities, architects and urban designers. This book is a modest attempt to be that guide – a source of reference for interested readers, students and practitioners around the globe. Keeping this objective in mind, the book is composed of two parts: the first outlines the theories, principles, techniques and tools for pedestrian modelling, and the second describes case studies and the lessons learnt through delivering them.

While contemplating how best to address this challenging proposition, I felt it prudent to draw on collective knowledge rather than be constrained by a singular perspective. A decision was made to involve AECOM's Urban Space team, which has conducted pedestrian modelling on a wide range and scale of projects over the last 15 years across the world and has rightfully established its position as a 'centre of excellence' on the subject. While the company and the team extended its full support throughout the writing of this book, it would be unjust if I didn't specifically thank two senior directors of AECOM, Kate Morris and Keyvan Rahmatabadi. Without their wholehearted backing and continuous encouragement this book would not have been produced.

I must also acknowledge the inputs of the two main contributors to this book, Emmet Ruxton and Martynas Didzys. They have undertaken the majority of the research, collating the relevant information and producing the text so that it flows in the form of a book. A brief biography of both contributors is provided at the end of this preface. Finally, it is important to recognise the input of a few other contributors during their association with the Urban Space team, and, while they are not mentioned by name, their contributions are appreciated in bringing this book together.

If this book can generate interest in the oldest and most fundamental mode of travel and help understand how, through modelling pedestrian movement, we can build and operate sustainable and value-engineered infrastructural assets in the future, we will consider our effort and hard work over the last two years more than rewarded.

Samya Ghosh

About the authors

Samya Ghosh BArch, MPlan (Transport Planning), CMILT
Head of Urban Space, Regional Director, Transportation
AECOM

Samya is a Regional Director in the Transportation division and leads the Urban Space team of AECOM. He is responsible for providing technical leadership, project direction, mentoring of staff and identifying directions for new research and development. He has 25 years' experience undertaking projects on varied aspects of transport planning, pedestrian and vehicle modelling and economics. Samya contributes to various international conferences, steering committees, the development of standards and guidance for micro-simulation modelling.

Emmet Ruxton BEng, PgCert, CMILT, MCIHT
Associate Director, Transportation
AECOM

Emmet has more than 15 years' experience undertaking projects in transport modelling, traffic engineering, data analysis and wide-reaching transport studies in both the public and private sector. He has extensive knowledge and expertise in pedestrian modelling, primarily focused on simulation, design, audit, impact assessment and specialist advice for a range of authorities. He previously helped produce best-practice guidance for Transport for London and continues to undertake research into pedestrian movement.

Martynas Didzys BD(A), BConstrMgt, MSc
Consultant
AECOM

Martynas is a pedestrian modeller with a background in architecture and urban planning. During his three years in the Urban Space team at AECOM, Martynas has undertaken micro-simulation modelling on a wide range of transport-related and other crowd simulation projects within the UK and abroad. Martynas' area of expertise extends from data analysis and strategic transport planning to micro-simulation modelling. He is proficient in several modelling software packages.

Urban Space – AECOM
ISBN: 978-0-7277-6381-5
https://doi.org/10.1680/mpm.63815.001

Chapter 1
Why model pedestrian movement?

1.1. Introduction

As the population of the planet increases at a steady rate, greater importance is being placed on the effective use of space. The proportion of the world living in urban areas has grown rapidly from 751 million in 1950 to 4.2 billion in 2018. Today, 55% of the world's population lives in urban areas, and this is expected to increase to 68% by 2050. Some forecasts show that the gradual shift in residence of the human population from rural to urban areas combined with the overall growth of the world's population could add another 2.5 billion people to urban areas by 2050.

The nature of work is also rapidly changing. Online working, flexible hours, the rise of self-employment and outsourcing are altering conventional employment patterns. Major urban areas are experiencing a period of accelerating technological and social change. Demographics, the capabilities of smart devices, the needs of citizens and even the climate are changing. An ageing society and shifting working and living arrangements will fundamentally influence expectations of how people travel.

The demands of a growing, changing population combined with a need for cost-effective design and technological advances have driven research into pedestrian movement and crowd behaviour at both an academic and industry level. Increased understanding of crowd behaviour and the development of tools to assess pedestrian comfort is important in the design and operation of transport, the infrastructure used by pedestrians and in other safety-critical environments. Using these tools to develop models – for example, to validate design – comes with several challenges, typically in terms of time, skills and data.

This book provides readers with an understanding of the fundamental research undertaken in relation to pedestrian movement, which is the basis for many of the modelling software packages available. It provides guidance on data collection, the uses of both macroscopic- and microscopic-simulation model development, analysis of model outputs and practical experience gained in projects.

This book targets two audiences. First, it is a textbook for university students undertaking courses related to transport planning, transport modelling, urban design or other relevant fields. No prior knowledge of pedestrian modelling is required, since the book introduces all the fundamental research. Second, it can be used as a reference text for fellow transport planners and modellers. As the book does not require prior knowledge of pedestrian dynamics, it can be used as a detailed introduction to the field of pedestrian modelling.

The book is divided into two main parts. Part I, 'Data and techniques', discusses the fundamental research and methods of pedestrian modelling. Part II, 'Case studies', provides examples of projects in which pedestrian modelling has helped make design or safety decisions in the real world. The two parts of the book are introduced in more detail in the following sections.

1.2. Part I: Data and techniques

Pedestrian movement is commonly understood through metrics such as speed, flow and density. These metrics have been tested, measured and illustrated in fundamental diagrams which visualise their relationship in different environments and conditions. Chapter 2 introduces these fundamental diagrams, explaining the patterns of pedestrian movements that they show. A range of other factors, such as trip purpose, age, and gender, are also shown to affect pedestrian behaviour along with the collective effects of crowd behaviour. Although the fundamental properties of pedestrian flows and Level of Service (LoS) methods for assessing capacity are well established, it has been difficult to develop microscopic models of pedestrian flows that replicate these fundamental properties. The chapter concludes by discussing the main mathematical models that form the basis of pedestrian modelling software platforms.

The next chapter, 'Data collection', focuses on the information needed to develop a pedestrian model. The process of planning a data-collection exercise is explained along with the most common types of data and how they are collected. The importance of appropriate geographic and temporal coverage is described before a discussion of other elements that should be considered prior to collecting data, such as intrusiveness, safety and weather. The common techniques for collecting data are then described, along with their benefits and drawbacks. The limitations of data collection are then explained so that readers can consider and minimise their impact.

Spatial analysis, macroscopic modelling and pedestrian routing are discussed in Chapter 4. Much can be gained from understanding how space can, and is expected to, be used. The outcome of spatial analysis can identify the most appropriate location for new infrastructure or desirable routes for specific journeys. Spatial network analysis and the use of mathematics to provide information about space and the relationship between the objects and different areas in space are explained with key terms such as centrality and connectivity. Macroscopic simulation is described in the next section. Macroscopic models focus on the aggregate representation of pedestrian movements in a crowd through speed, flow and density relationships. Static analysis, a subset of macroscopic modelling, is often undertaken early in a design or master planning process to give a high-level indication of capacity requirements or any issues likely to be encountered. Finally, pedestrian routing is discussed. Routing is one of the most important elements of modelling, but the variations of pedestrian route choice make it a challenging task. Pedestrian behaviour theory is explained, in which choices are categorised into *strategic*, *tactical* and *operational* levels. Various techniques to model pedestrian route choice are described, from sophisticated route choice logit-based models to straightforward shortest- or fastest-route models.

Microscopic movement is discussed in Chapter 5, expanding on the types of mathematical models discussed at the end of Chapter 2. The principles of pedestrian models are outlined, as are their focus on the behaviour of pedestrians in crowded conditions. The various definitions of a crowd are described along with the key components of their behaviour, and then the dynamic modelling process is explained. A summary of the most common commercial software platforms is then provided to give readers an understanding of what is typically used in the industry. The chapter concludes with a discussion of uncertainty through the modelling process to provide readers with an understanding of common issues during the life of a pedestrian modelling study.

Chapter 6 provides an in-depth explanation of the model-development process. It describes the order in which to tackle the development of a pedestrian model, from establishing the scope through to analysis. First, it focuses on the information required to develop a model. Guidance is given on what to observe during site visits, how to develop reasonable and defensible assumptions and the pedestrian demand and routing information required to feed the model. The next section focuses on model development and gives readers a background on the type of behaviour which can be coded into most models to control pedestrian behaviour. Finally, the most common techniques for analysing pedestrian models and how this information is used are described. Some relevant guidance documents are also identified to give readers an understanding of the standards for different infrastructure environments.

Part I of the book ends with a look at the future of pedestrian modelling, in Chapter 7. The industry commonly focuses on the traditional morning and evening peak movements when considering movement; however, this is likely to change in the future. Greater work flexibility, including working from home, is changing travel patterns. The data available to modellers will increase because of the popularity of mobile devices, but this brings further challenges. Finally, advanced transportation systems and the way they will be accessed in the future are discussed.

1.3. Part II: Case studies

Part II of the book provides an in-depth overview of pedestrian modelling approaches and nuances for various infrastructure types commonly modelled as part of design or operational projects. This is provided through a series of discussion topics and case studies grouped into chapters based on the type of infrastructure being assessed.

The first chapter in Part II of the book, Chapter 8, acknowledges the growing importance of rail systems as the dominant mode for transporting people within urban areas. As the globe becomes ever more urbanised, facilitating this movement of people in congested urban areas is becoming increasingly valued by policy makers and planners. Legacy rail networks around the world are continually being upgraded to increase both reliability and capacity as demand continues to grow. Meanwhile, new rail transit systems are rapidly being constructed, particularly in developing countries, where societal consciousness of environmental issues is increasing. The result of this growing investment in rail infrastructure is that pedestrian modelling is being increasingly used for the planning and operation of rail interchanges to optimise the efficiency and safety of the stations and interchanges that form the key nodes of all rail networks. A major underground interchange station in the Middle East is presented as a case study to highlight the advantages of using pedestrian modelling for this purpose. The chapter concludes with a discussion related to pedestrian modelling in degraded operations scenarios, such as periods of service disruption and delay. Pedestrian modelling allows station crowd-management plans to be developed for use in these scenarios.

Some of the largest crowds occur at major sporting and entertainment venues during major events, so the safety of the occupants of these venues is critically important. Chapter 9 discusses pedestrian modelling undertaken for various kinds of events and the venues and precincts that host them. Case studies are provided to highlight how both static and dynamic modelling have been undertaken covering the scenarios that experience the greatest congestion, such as ingress and egress, or which pose the greatest safety concerns, such as emergency evacuation situations. The chapter includes a discussion on modelling major sport and entertainment venues and provides a case study of one such stadium, for which pedestrian modelling was successfully undertaken. Modelling outputs provided the evidence base for not implementing a proposed move of seating location for away supporters within the stadium. A second case study concludes the chapter with an outline of the use of static and dynamic analysis to anticipate crowding levels and visitor comfort at a major global multi-day event precinct.

While the initial chapters in Part II of the book concentrate on the buildings and infrastructure that draw crowds, the focus in Chapter 10 shifts to the public spaces surrounding these buildings – that is, to 'public realm' in our cities. The companion term 'urban realm' can take on a variety of forms, including pedestrianised urban areas, such as busy shopping streets, plazas and the spaces surrounding transit interchanges that facilitate heavy flows of people to and from transport services. It can also refer to places in our cities where high interactions between pedestrians and other traffic occurs, which adds an additional layer of complexity from a pedestrian modelling perspective. Urban-realm assessments can range in scale from small models of isolated spaces, such as a city square or junction, right up to an entire town or city, and everything in between. An example of a qualitative city-wide assessment of transportation and walking infrastructure undertaken as part of a larger regeneration project in a major North American city is presented as the first case study in this chapter. It exemplifies how pedestrian analysis can be used to inform potential multi-model improvements to mobility on a city-wide scale.

Chapter 10's second case study is of a busy high street in suburban London, highlighting a completely different set of considerations for pedestrian modelling. As part of an urban-realm improvement scheme, pedestrian modelling was used in conjunction with traffic micro-simulation to test proposed improvements to pedestrian safety, connectivity and congestion along the busy retail strip, and the case study emphasises the challenges posed by the interaction between pedestrians and vehicles. The chapter concludes with a discussion of the value of pedestrian modelling in planning for disruptions. Disruption in cities, whether planned or unexpected, can be mitigated with measures to ensure that the quality of the urban environment, as well as accessibility and safety for pedestrians, is not compromised. Pedestrian modelling allows alternate mitigation options to be tested and optimised prior to any disruptions occurring.

Chapter 11 provides an overview of modelling pedestrian movement in buildings. Buildings are comprised of series of interconnected enclosed spaces with limited capacities. Facilitating the movement of people into, around and out of these spaces can often be problematic, especially with public buildings to and from which large numbers of people arrive and leave at similar times. Vertical-circulation capacity is often one of the major considerations in pedestrian analyses of multi-storey buildings. The chapter presents two case studies on this topic: a major exhibition building and a multi-storey school, where optimisation of vertical circulation was assessed to facilitate the peak demands of occupants entering, circulating and leaving the buildings. A final case study details the pedestrian micro-simulation of a multilevel department store in Paris. Outputs from the modelling led to changes to the entrance layout that reduced congestion at this critical pinch-point. The case studies in this chapter exemplify the value in pedestrian modelling in understanding peak occupant capacities and movements within buildings and identifying areas where crowd management may be beneficial in optimising pedestrian flows.

The final chapter of the book discusses the use of pedestrian modelling for airports. Growing worldwide demand for air travel has driven airport operators to expand to accommodate the rising numbers of people moving through their facilities. Cost-effective design requires pedestrian modelling of terminal buildings and arrival spaces, such as forecourts, car parks and public transport terminals. The movement of people through airports is often more complex than for other building typologies, as passengers follow routes and undertake several procedures at set locations along the way. Guidelines covering acceptable comfort levels for pedestrians inside airports also differ from those for other building types. These relate to both spatial capacity and waiting times. Pedestrian modelling provides a valuable measure of these parameters that allows spaces within airports to be designed to allow efficient movement of people while maintaining comfort and safety.

The chapter presents two case studies. The first looks at the interface at the forecourt of a Middle East airport, where vehicular pick-up and drop-off occur. Pedestrian modelling assessed various parking-allocation scenarios, identifying a preferred option that minimised journey times for both pedestrians and vehicles. In the second case study, the departures terminal of a major hub airport in Africa was modelled to analyse pedestrian flows, with a proposed terminal layout after expansion of the terminal. The modelling was able to highlight areas in the terminal experiencing unnecessary cross-flows leading to congestion and delays for passengers, allowing the design to be altered to minimise the specific movements causing these opposing pedestrian flows.

Each chapter of the book concludes with a useful set of summary points that recap its content. These summary points provide readers with a reference of key considerations when approaching pedestrian modelling projects for each of the infrastructure types covered in the case studies.

Part I

Data and techniques

Urban Space – AECOM
ISBN: 978-0-7277-6381-5
https://doi.org/10.1680/mpm.63815.009

Chapter 2
Modelling concepts

2.1. Overview of the chapter

This chapter first provides an overview of existing research on pedestrian flow and fundamental diagrams as the basis for analysing pedestrian flow (Section 2.2). Fundamental diagrams depict the relationship between speed, flow and density in pedestrian facilities. They are tools to be used by engineers, planners and designers to estimate the capacity of pedestrian facilities. Fundamental diagrams have also been used to validate pedestrian models and micro-simulation. The common definitions for speed, flow and density have been provided in this chapter. This gives readers an introduction to fundamental diagrams available in the literature. Pedestrian walking speeds and factors influencing them, such as gender, age, culture and structural facilities, are discussed in Section 2.2.3 to give readers an understanding of how pedestrian behaviour can vary within a sample of the population.

Pedestrian behaviour is discussed in Section 2.3, as is flow in a range of scenarios that have direct application to everyday situations. Three types of flow are considered: unidirectional, multidirectional and bottlenecks.

Movement in crowds is discussed in Section 2.4. Different phenomena have been observed in crowds, all of which have been extensively researched and documented in research studies. Examples of these phenomena are lane formation, jamming, density waves, oscillations and emergency situations.

Approaches to pedestrian modelling is the chapter's final topic, discussed in Section 2.5. It discusses the different types of models, inspired by a wide range of sources, including fluid mechanics, probability functions and discrete choice models. Five commonly used models for pedestrian modelling are explained with relevant formulas provided. Schematic examples are also given to help readers understand these modelling approaches. The pedestrian modelling approaches discussed in this section are continuum, social force, optimisation, grid-based and discrete choice.

2.2. Pedestrian flow

This section explains the behaviour of pedestrians in uni and bidirectional flows in pedestrian structural facilities. Understanding pedestrian behaviour is important for designing and planning for pedestrians in public places, such as railway stations, shopping malls and stadia, to provide a safe, comfortable, efficient and enjoyable experience. Pedestrian behaviour in large crowds can be made unpredictable by interpersonal interactions.

Fundamental diagrams, which show the relationships between observable parameters such as speed, flow and density, are essential for understanding the characteristics of pedestrian movement and can be used to validate pedestrian movement models. After introducing commonly used observables in the fundamental diagrams, the fundamental diagrams for different pedestrian flows and various pedestrian structural facilities are discussed.

2.2.1 Observables and parameters

Fundamental diagrams provide a graphical representation of macroscopic relationships between observables. Pedestrian movement is often defined by terms such as 'speed', 'flow' and 'density', which have varying definitions in the literature. The most common definitions are provided in this section.

Density, ρ, is defined as the number of pedestrians, P, in an area at a given moment. The area is usually a square with a width of $1m$ and hence the unit of density is $\frac{P}{m^2}$.

Speed, v, can be calculated as the average speed over time or over space. The average speed over time is the average speed of pedestrians passing a specific cross-section over a given period of time. The average speed over space is the average instantaneous speed of pedestrians in an area at a given moment [5].

Pedestrian flow can be thought of in the same way as water passing through a pipe. Pedestrian flow, J, is calculated based on the average time interval, $\langle \Delta t \rangle$, between consecutive pedestrians that pass a cross-section. For this purpose, a cross-section can be defined as a walkway or corridor of known width and infinite length. Flow rates are often defined as the number of pedestrians per minute per metre width. The formula to calculate pedestrian flow using this definition is shown in (2.1).

$$J = \frac{1}{\langle \Delta t \rangle}, \quad \left(\langle \Delta t \rangle = \frac{1}{N} \sum_{i=1}^{N} (t_{i+1} - t_i) = \frac{t_{N+1} - t_1}{N} \right) \tag{2.1}$$

Where t_i is the time that pedestrian i passes the cross-section and N is the total number of pedestrians.

Another commonly used approach to calculate pedestrian flow is borrowed from fluid dynamics, using density, ρ, pedestrian average speed, v, and the width of the facility, b. The formula to calculate pedestrian flow using this definition is shown in (2.2).

$$J = \rho v b = J_s b \tag{2.2}$$

Specific flow, J_s, is also used as a measure to show the flow in fundamental diagrams. Specific flow shows the pedestrian flow in a facility with a width of one unit. This is also called *hydrodynamic relation* in the literature [28].

2.2.2 Fundamental diagrams

The primary objective of fundamental diagrams is to provide characteristics and patterns of pedestrian movements in different structural facilities by showing the relationship between speed, flow and density of pedestrians. These fundamental diagrams are usually produced using empirical studies to accurately reflect real cases and are essential for engineers, planners and designers to estimate the accurate capacity of pedestrian facilities. Therefore, they are appropriate tools to validate pedestrian dynamic flow models and pedestrian micro-simulation models.

This section briefly explains fundamental diagrams for common pedestrian structural facilities, such as stairs and corridors, for different flow types, such as uni and bidirectional flows.

In Figure 2.1, the *Stable* condition exists at low densities and a linear relationship between flow and density is observed. In practical terms, this Stable condition would see pedestrians moving at free speed for increasing flow rates until the *Free flow* limit is reached and it moves to the *Unstable* condition.

Figure 2.1 Explanation of flow–density relationship in fundamental diagrams [5]

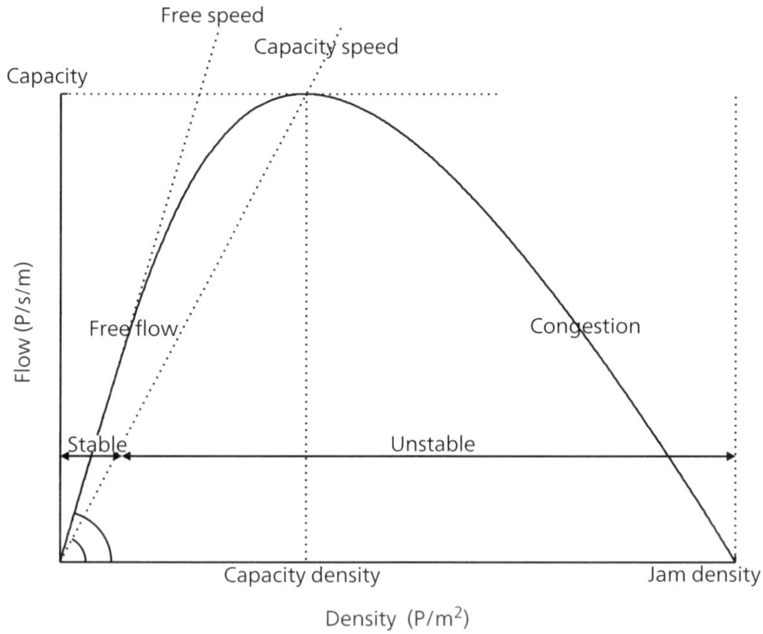

Density increases at a greater rate than flow until *Capacity density* is reached, which is the maximum flow rate for the cross-section. A further increase in density corresponds to a reduction in flow rate and *Congestion* until density becomes so great that *Jam density* is reached and flow becomes zero [5]. In reality, this would look like a crush situation with pedestrians unable to move.

2.2.2.1 Fundamental diagrams by flow types

Uni and bidirectional pedestrian movements exist in most pedestrian structural facilities, such as corridors, stairs, transport terminals and shopping malls. Experiments have shown that pedestrian flow in uni and bidirectional facilities could be different. Zhang and Seyfried [43] have compared fundamental diagrams in different studies, guidelines and handbooks. In terms of guidelines and handbooks, they considered WM [37], PM [25], FN [10] and HD [15] and identified discrepancies between their diagrams in terms of maximum flow values, density and specific flow. One reason for this discrepancy is the lack of distinction between fundamental diagrams for corridors, ramps, aisles and doorways, as pedestrian flow in each of these facilities could be different. Only PM [25] made this distinction in the equations. Table 2.1 shows the comparisons provided in [43]. In this table, *Free speed* should be considered as the limit of unimpeded movement.

Figure 2.2 depicts this comparison for uni and bidirectional flow from different empirical studies.

In this figure, the speed–density relationship diagram (Figure 2.2(a)) and speed–density relationship diagram (Figure 2.2(b)) show similar patterns for uni and bidirectional pedestrian flow. This comparison was originally provided by Zhang and Seyfried [43], who concluded there is not enough evidence to clearly show that the fundamental diagrams for uni and bidirectional flow are different. The authors argued that in

Table 2.1 Comparison of design variables from different handbooks [43]

	WM [37]	PM [25]	FN [10]	HD [15]
Maximum density (m^{-2})	5.4	8.14	5.4	3.8
Free speed (m/s)	1.34	0.95	1.27	1.19
J_{sm} ($(m.s)^{-1}$)	1.22	1.49	1.43	1.30
ρ (m^{-2})	1.75	6.64	2.16	1.88
v (m/s)	0.7	0.2	0.7	0.7

Figure 2.2 Fundamental diagrams in uni and bidirectional facilities [43]

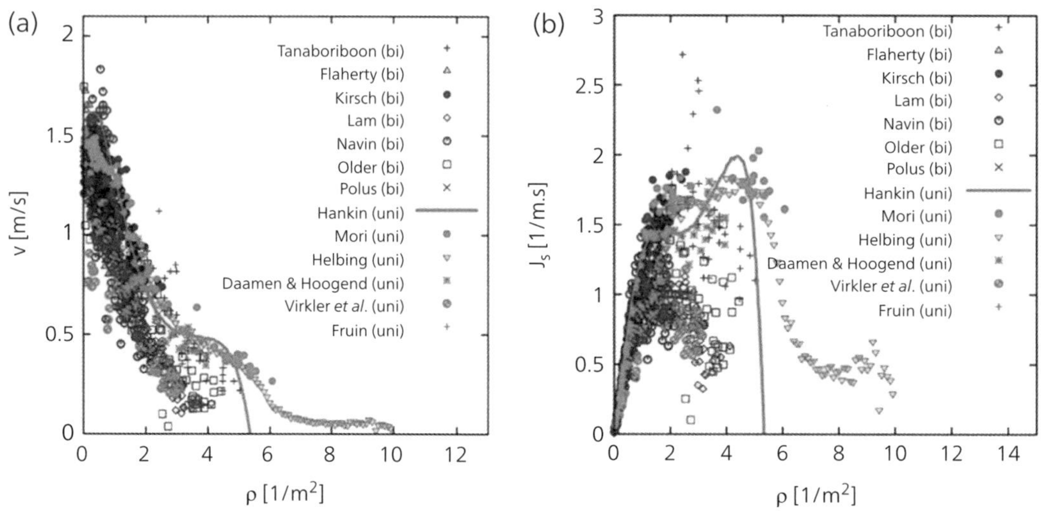

some research (e.g. [25] and [37]) differences between the fundamental diagrams for uni and bidirectional flow were not found to be significant. Other studies (e.g. [22], [26] and [18]), however, showed that the flow depends on directional imbalance and hence the flow in one direction would be reduced if the density of the opposite direction is considerably higher. Figure 2.2(b) also shows that the average pedestrian speed in unidirectional situations is higher than that in bidirectional situations when $\rho > 1.0m^{-2}$.

It appears there is no consensus on the characteristics of different types of pedestrian flow, as many factors may influence the properties of pedestrian movement. However, there is much research on specific environments of pedestrian movement, and this is discussed in the following section.

2.2.2.2 Corridors

Corridors are the most basic and common element in pedestrian structural facilities, and many studies have been conducted to understand the characteristics of pedestrian movement within this environment.

Figure 2.3 compares fundamental diagrams from guidelines (represented by lines) and experiments (represented by dots). It shows that there is discrepancy between fundamental diagrams from different

Figure 2.3 Fundamental diagrams for pedestrian movement in planar facilities (such as sidewalks, corridors and halls) [28]. Lines represent results from planning handbooks or guidelines (HD [15], WM [37] and PM [25]) and points represent experimental results (Older [24] and Helbing *et al.* [14])

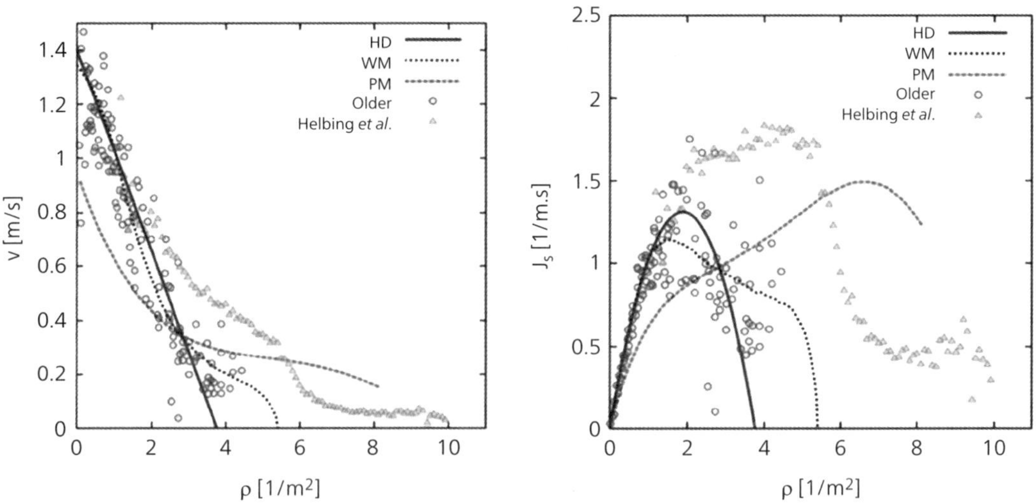

sources. For instance, the maximum J_s varies between 1 $(m.s)^{-1}$ and 2 $(m.s)^{-1}$; similarly, maximum density varies between 3 $(m.s)^{-1}$ and 10 $(m.s)^{-1}$. Various reasons for this discrepancy have been identified, such as cultural and population differences, trip purpose (e.g. commuting, leisure, business), type of participant, psychological factors and incentives for movement. However, Schadschneider *et al.* [28] argued that, considering these discrepancies, all the diagrams agree that by increasing the density, speed decreases.

2.2.2.3 Bottlenecks

Pedestrian movement in bottlenecks has also been thoroughly studied, giving fundamental diagrams which show the relationship between speed, average speed and density. For instance, in one study [8], empirical experiments were conducted to understand pedestrian behaviours and movements in the bottleneck (see Section 2.3.3 for further details on bottlenecks).

During congestion, pedestrians form a funnel-shaped group upstream of the bottleneck (Figure 2.4).

Figure 2.4 Trajectory of pedestrians in bottleneck [8]

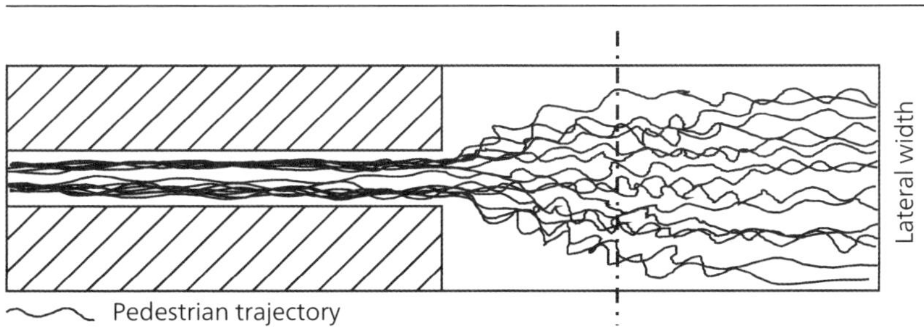

Figure 2.5 Flow–density relations for different widths in bottlenecks [8]

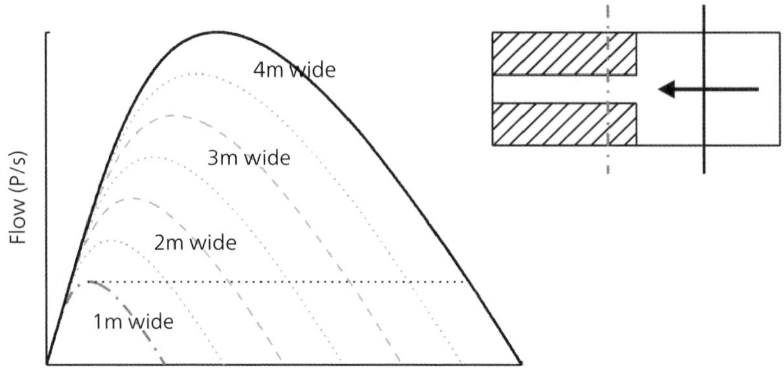

Figure 2.6 Characteristics of pedestrian movements on stairs from guidelines [2] (PM [25], WM [37], NM [23] and FN [10])

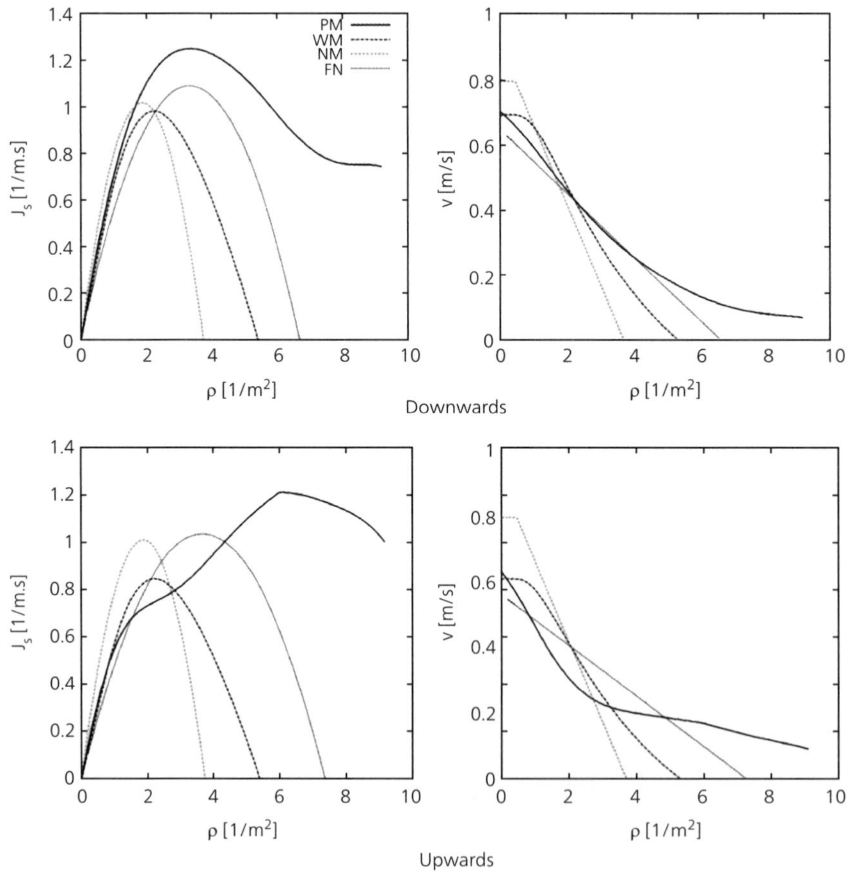

In this experiment, the width of the corridor was $4m$ and the width of the bottleneck was $1m$. The fundamental diagrams for different widths of the bottleneck have been identified, and it has been shown that they have similar shapes but with different capacities (Figure 2.5).

2.2.2.4 Stairs and escalators

Stairs and escalators are very important in pedestrian modelling as they often define egress and ingress points of infrastructure, such as stations or stadia, and are often the main conduits between different floors of activities. Stair and escalator capacity and flow are often critical in design due to their crucial roles in evacuation. Various guidelines exist which provide recommended characteristics of pedestrian flow in stairs and escalators; these are discussed in Section 6.4.3.

Figure 2.6 provides comparisons between the guidelines (PM [25], WM [37], NM [23] and FN [10]) for pedestrian movement characteristics in downwards and upwards movements. Notice that not all the guidelines provide the same diagrams, and in some cases there are significant differences. For instance, this figure shows that PM [25], WM [37] and FN [10] provide different diagrams for downwards and upwards movement, whereas NM [23] gives the same fundamental diagram for both downwards and upwards. Regarding the downwards movement, the comparisons show that for density less than $2m^{-2}$, all four guidelines provide similar flow, and for higher density the changes between the diagrams (i.e. flow–density diagrams) become more significant. The range for maximum density capacity for both downwards and upwards is $3.8m^{-2} - 9.2m^{-2}$, and it should be noted that PM [25] does not provide the maximum density for the case where pedestrians are at a standstill (i.e. the flow becomes zero).

Due to the importance of pedestrian movement on stairs and escalators, a considerable number of experiments have been conducted by researchers to analyse this behaviour. A comparison of these studies is provided in Figure 2.7. For downwards movements, it can be seen that for a capacity of less than $1.5m^{-2}$, all the experiments produced similar results – linear increase of flow. The maximum flow for downwards is above $0.8m.s^{-1}$, and the maximum capacity in the majority of the cases is around $3m^{-2}$. For upwards movements, however, different patterns in the results from [30] and other experiments can be observed. As explained in [2], this is due to the different measurement used in the [30] experiments.

Figure 2.7 Characteristics of pedestrian movements in stairs from the literature [2]

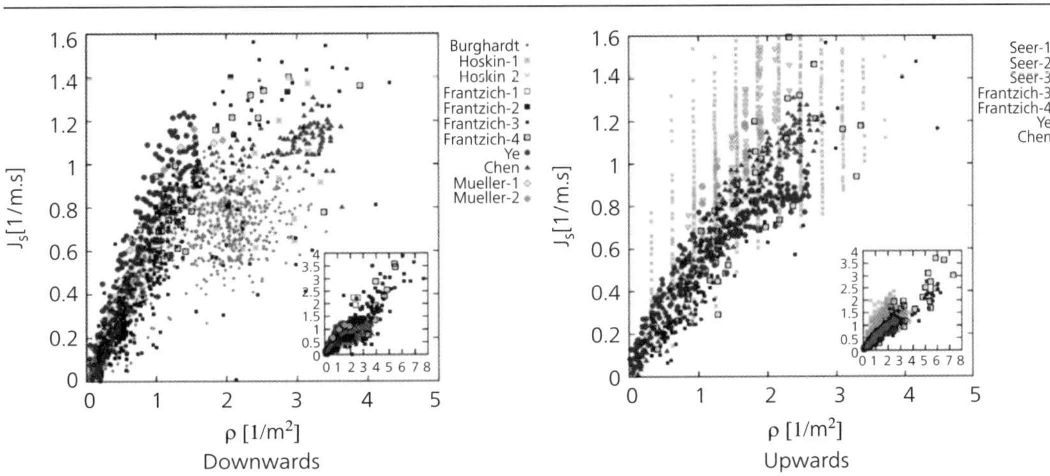

The pedestrian flow characteristics in congested and uncongested conditions can be explained by fundamental diagrams that show the relationship between speed, flow and density. However, despite extensive research in this area, the literature shows wide variations in the results in these fundamental diagrams due to other parameters influencing the speed, flow and density relationship. These influences are discussed further in Section 2.2.3.

2.2.3 Parameters influencing pedestrian walking speeds

Pedestrian walking speed is one of the essential inputs for calculating the capacity of pedestrian facilities. It is simply the speed at which a pedestrian walks in planar facilities under normal conditions, and usually follows a normal distribution $N(1.34, 0.26^2)$ [36]. Fundamental diagrams, as discussed in Section 2.2.2, show that normal walking speed is impeded at higher pedestrian densities. Along with measuring pedestrian speed at higher density, Fruin [9, 10] also introduced a measurement for pedestrian comfort level, which he defined as Level of Service (LoS). LoS is mainly used as a criterion for pedestrian safety in public places. Table 2.2 shows the average speeds at various pedestrian densities according to Fruin's LoS, and it is evident that walking speed reduces as density increases.

Many other parameters impact walking speeds – for example, culture and race, environmental conditions, trip purpose and groups. Figure 2.8 depicts comparisons, from different studies, of walking speeds in different pedestrian facilities with respect to gender and age.

It has been shown that men's walking speed is over 10% faster than the average walking speed of women [37]. This could be due to differences in physical characteristics between men and women, such as larger step lengths and higher step frequencies for men [5]. And the walking speed of the 18–60 age group has been shown to be significantly higher than that of other age groups.

Cultural and racial differences can also influence walking speeds, with studies showing that cultural differences in sensitivity to increases in density lead to different walking speeds between different people [4].

Another parameter is the temperature of the environment. It was found that walking speed is usually increased by decreasing the temperature. For instance, walking speed at 20 °C is almost 92% of the average, whereas this increases to more than 109% at a temperature of 15 °C [37]. As well as atmospheric

Table 2.2 Speeds at various densities, based on Fruin's LoS for pedestrians [9, 10]

Level of service	Space (m^2/ped)	Average speed (m/s)	Flow (ped/min/m)
A = Free flowing	≥ 12.077	≥ 1.321	≤ 6.562
B = Minor conflicts	≥ 3.716	≥ 1.270	≤ 22.966
C = Some restrictions to speed	≥ 2.230	≥ 1.219	≤ 32.808
D = Restricted movement for most	≥ 1.394	≥ 1.143	≤ 49.213
E = Restricted movement for all	≥ 0.557	≥ 0.762	≤ 82.021
F = Shuffling movements for all	≥ 0.557	≥ 0.762	variable

Figure 2.8 Pedestrian walking speeds in various situations with respect to gender and age [34]

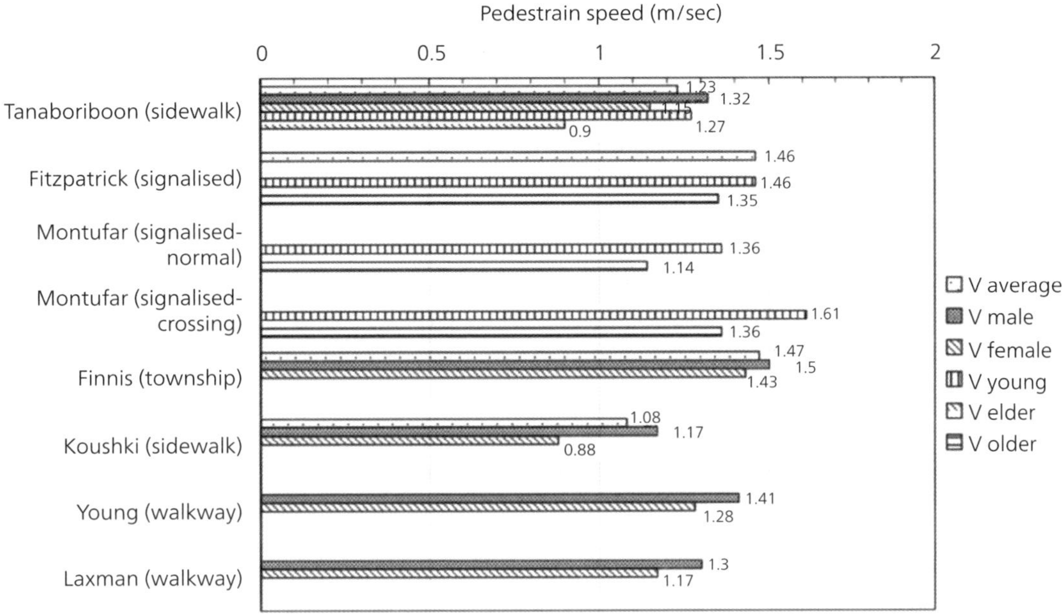

conditions, the physical nature of surroundings can impact on walking speeds. Figure 2.8 shows that signalised crossings generally generate faster walking speeds than those undertaken along sidewalks or walkways.

Trip purpose is important too, as it affects speed. Business and commuting speeds are notably faster than those for shopping and leisure [37]. Trip purposes can change at different stages of a pedestrian's journey, leading to variable walking speeds over the course of a journey. For example, a commuter may walk through a retail area on their way to or from a station, thus becoming a shopper with a slower walking speed for this stage of their journey. Furthermore, large public gatherings, such as religious pilgrimages like the annual Hajj Islamic pilgrimage to Mecca, which attracts thousands, even millions, of people, exhibit certain characteristics with regards to pedestrian speeds, which could be replicated in pedestrian modelling of such events.

The formation of groups can also impact on walking speeds. Studies have concluded that the desire of shared-identity groups to be in close proximity to one another regulates their walking speed, resulting in a reduced speed of the group [33]. Individuals walking in pairs, however, may have higher average walking speeds as they tend to form single lines, allowing the second pedestrian to follow the path opened by the first pedestrian [35].

Pedestrian flows are influenced by the same parameters: gender, age, culture and race, environmental conditions, trip purpose and groups. Figure 2.9 shows the impacts of different factors on fundamental flow diagrams from research by [5].

The effects of the parameters on pedestrian flow are similar to those on pedestrian speed. For instance, higher temperatures and steeper gradients lead to lower pedestrian flows as well as speeds. *Shy away distance* is the minimum distance that pedestrians keep from different elements of the surrounding environment (walls, obstacles, other pedestrians); this is often known as the edge effect in modelling. *Type of infrastructure (grade)* is the different types of walking infrastructure. For example, in a public transport facility, the types of walking infrastructure can be walkways, concourses, hallways, platforms, stairs and escalators; but in an urban environment, intersections and crossings also appear.

2.3. Pedestrian flow behaviour

A classification for pedestrian movement behaviours is provided in [31]. In this classification, there are three categories for pedestrian flow: unidirectional flows, multidirectional flows and bottlenecks. For each category, the relevant behaviour types are discussed. It should be noted that in terms of flow dimension, pedestrian movement can be two-dimensional, such as movements in corridors and along the ground, or

Figure 2.9 Individual pedestrian characteristics and external factors influencing fundamental diagrams (flow–density relation) [5]

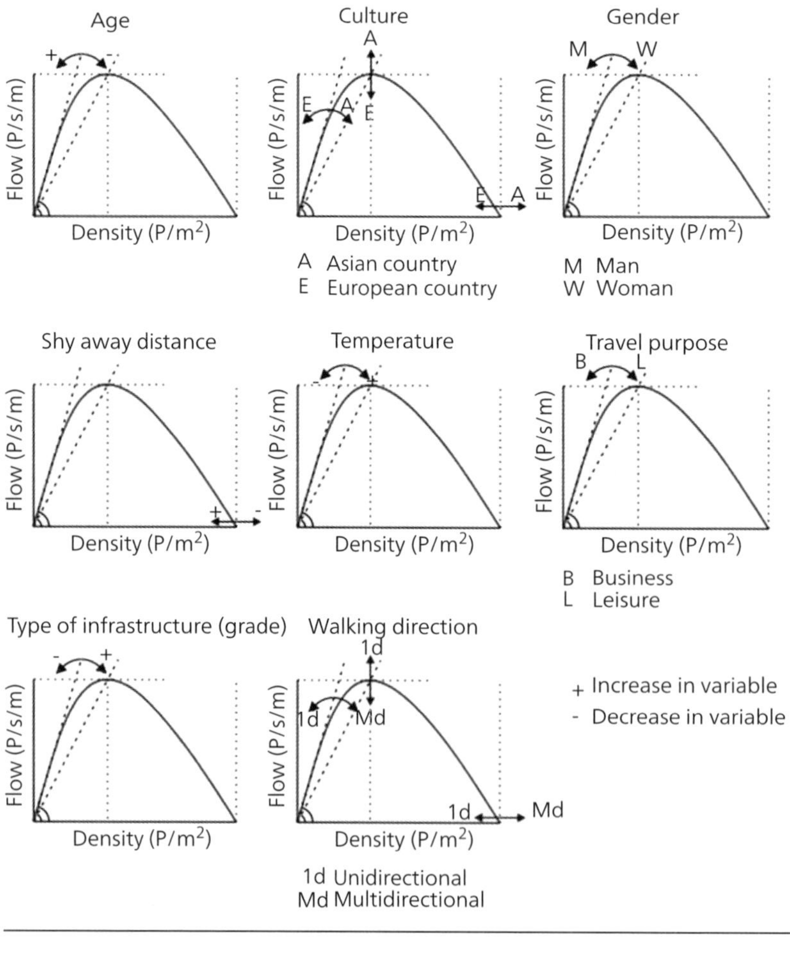

three-dimensional, such as motions in stairways. Studying pedestrian movement in three-dimensions is more complicated to analyse and therefore pedestrian behaviour has mainly been studied in two dimensions, with movement in the third dimension commonly represented by a reduction in two-dimensional speed.

Figure 2.10 depicts the categories of pedestrian movement behaviours discussed in this section.

2.3.1 Unidirectional flows

Unidirectional flow exists when all pedestrians move in the same direction. Overtaking and turning behaviours occur in unidirectional movements and are discussed in this section. It should be noted that overtaking and turning can also occur in multidirectional movements, where the pedestrian behaviours would be similar.

2.3.1.1 Overtaking

Pedestrians with faster desired walking speeds tend to overtake those with slower speeds to maintain their constant walking speed (Figure 2.11). Pedestrian overtaking occurs in different places and is opportunistic. A typical example of overtaking can be seen at railway stations during peak hours, where pedestrians with faster walking speeds overtake slower pedestrians to catch a train, while at the same time avoiding collisions with other pedestrians. Most of the research on overtaking is into evacuations (e.g. [13]). Pedestrian overtaking behaviour in normal situations, however, is different. For instance, [42] shows that competitive behaviours, such as squeezing and pushing, rarely happen in normal situations, as opposed to evacuation situations. It also shows that pedestrian choice of side in overtaking is not influenced by gender or small groups.

Figure 2.10 Classification of pedestrian movement behaviours [31]

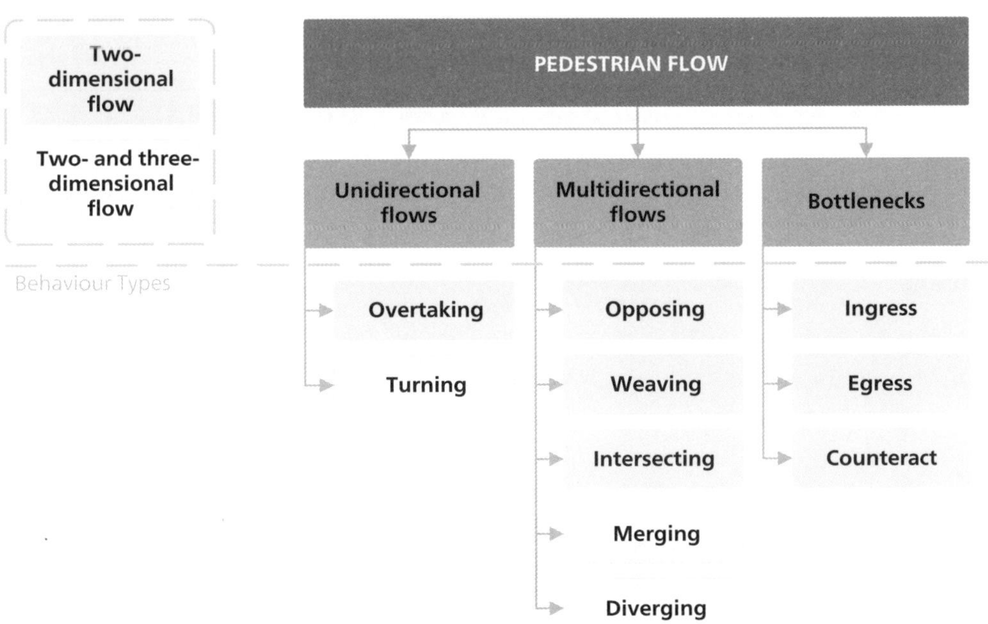

Figure 2.11 Example of overtaking showing that a pedestrian with faster speed needs to change direction to be able to overtake

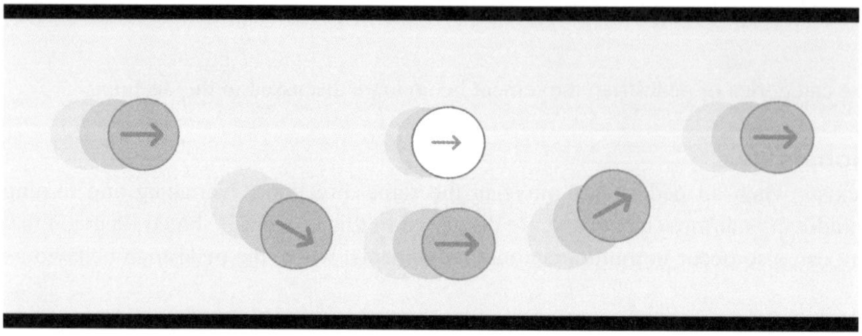

2.3.1.2 Turning

Turning behaviours occur when a change in the direction of motion is needed (Figure 2.12). Empirical and simulation studies have been conducted to quantify and understand pedestrian turning behaviours. For instance, the relationship between turning angle and walking speed was investigated in a cellular automata model in [42]. It concluded that since pedestrians have inertia, walking speed during turning decreases in a similar way to vehicular turning behaviour, but at a smaller scale.

2.3.2 Multidirectional flows

Pedestrian behaviours in multidirectional flows are more difficult to study due to the added complexity of interaction and the large number of possible scenarios. The direction of flows in the streamlines

Figure 2.12 Example of turning behaviour

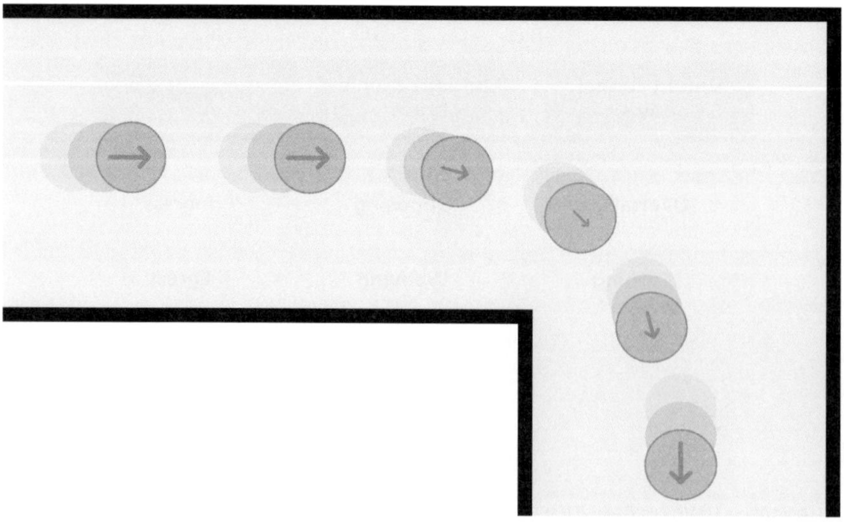

(i.e. pedestrian flow line) are also varied, which further increases complexity. Different behaviours in multidirectional flows, such as opposing, weaving and intersecting, merging and diverging, are discussed in this section.

2.3.2.1 Opposing

Opposing is a common flow in bidirectional pedestrian movement and can be observed in daily situations, such as corridors. When there is an opposing situation in bidirectional flow, pedestrians adjust their speed and change their direction to avoid collision with pedestrians moving in the opposite direction. Opposing flow can also be considered as an extreme case of intersecting, where the angle between the two streamlines is 180°. There are various factors influencing pedestrian dynamics in opposing situations, such as the flow composition, free walking speed, desired walking speed and travel purpose [31].

2.3.2.2 Weaving and intersecting

Pedestrian weaving and intersecting occurs when pedestrian streamlines with different directions come together. In this situation, conflict-avoidance behaviours are observed by pedestrians to prevent collision. The difference between weaving and intersecting is the angle between pedestrian streamlines. When the angle is within the 0–90° range, it is weaving, while movements in streamlines with differences ranging from 90–180° are called intersecting [31]. Figure 2.13 depicts an example of weaving. Weaving and intersecting happen in pedestrian facilities with multiple streamlines, such as in railway stations, shopping malls or passageways.

Pedestrian weaving and intersecting behaviours consist of overtaking and dodging behaviours. In some cases, weaving can be avoided by providing facilities such as placing railings, but this is not always possible [39]. Research has shown that weaving and intersecting behaviours can impact on the quality of pedestrian walking facilities in terms of speed, capacity, comfort and safety. For instance, weaving behaviours can impose a 20–30% reduction in pedestrian walking speed [27]. The capacity of walking facilities can be reduced by 20–40% due to weaving [40]. Due to the reduction in pedestrian speeds and collision-avoidance behaviours by pedestrians, comfort in the facility can also deteriorate. Weaving and intersecting can also reduce safety due to an increase in the possibility of trips, falls and bumping [39].

Figure 2.13 Example of pedestrian weaving

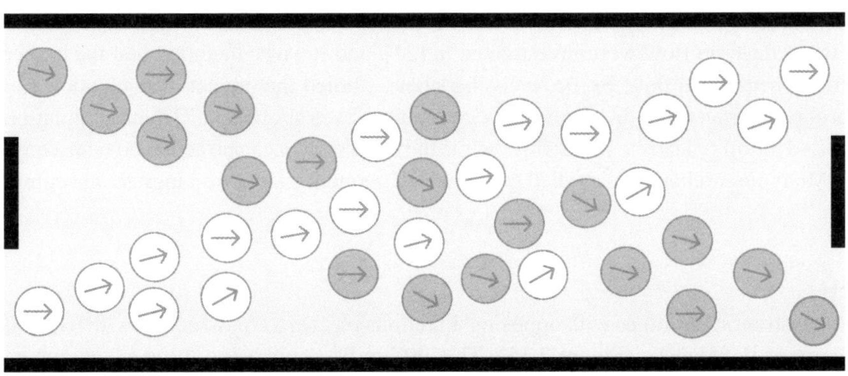

Figure 2.14 Example of merging behaviour. In this figure, it is assumed that the direction of movement is from left to right

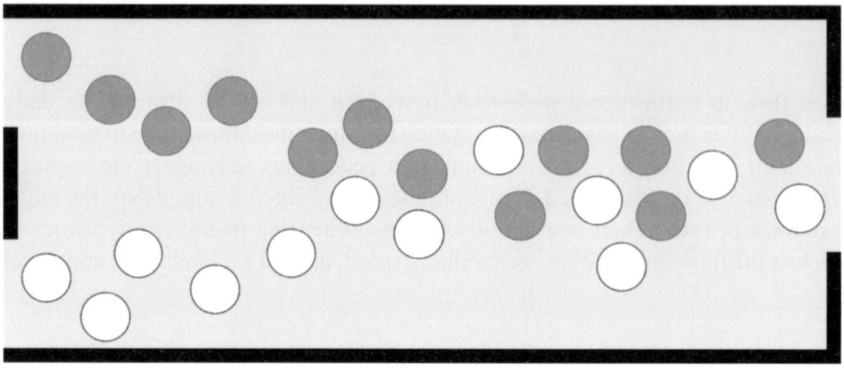

2.3.2.3 Merging and diverging

Merging behaviours occur when multiple pedestrian streamlines join to form one streamline (Figure 2.14). In other words, merging is the combination of weaving and turning [31]. Diverging is the opposite of merging, where one pedestrian streamline divides into multiple pedestrian streamlines with different directions.

2.3.3 Bottlenecks

Bottlenecks are locations where there is a significant decrease in capacity. Bottlenecks are very important in pedestrian modelling and should be analysed carefully. If bottlenecks are not taken into account properly, they can result in jamming. More importantly, in emergency situations, bottlenecks can have a critical impact on evacuation. Bottlenecks can happen in both uni and bidirectional pedestrian flow. Various studies have been conducted to analyse pedestrian movements in bottlenecks.

2.3.3.1 Ingress and egress

Pedestrian ingress and egress at bottlenecks have attracted a lot of attention from researchers, both because capacity decreases at bottlenecks and because of the significance of bottlenecks in evacuation. Extensive research has been done to understand pedestrian behaviour at bottlenecks of different widths and lengths to produce or enhance fundamental diagrams. For instance, the impacts of initial density (ranging from $0.4m^{-2}$ to $0.5m^{-2}$) on the pedestrian flow were investigated in [21], and it was concluded that the higher initial flow would increase pedestrian flow. Some researchers have studied the impact of population and different combinations of pedestrian types on ingress and egress. It has been shown by [7] that a population with 5% persons of reduced mobility leads to lower flow, while the highest flow can be achieved with a high proportion of children. Many researchers also studied the impact of geometry factors on ingress and egress flow [19].

2.3.3.2 Counteract

In counteractive flow, pedestrian streamlines with opposing directions meet at a cross-section with reduced capacity, such as at doors or bottlenecks (Figure 2.15). This differs from opposing movements, where

Figure 2.15 Example of counteract behaviour

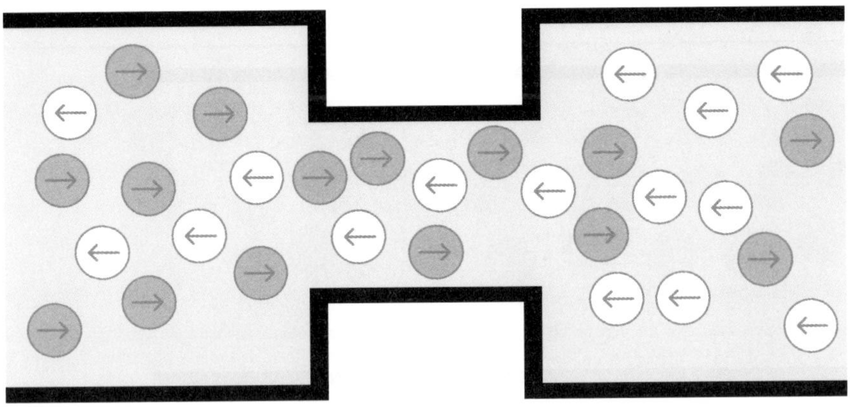

pedestrian streamlines have opposite directions without any reduction in capacity. Pedestrian behaviour in counteract has not been widely studied and requires further research in order to analyse pedestrian behaviours in greater detail [31]. An example of research on counteract is [12], in which behaviour was investigated using empirical experiments. It concluded that the length of bottlenecks should be short, as people in long bottlenecks try to turn the direction in their favour, which will result in reduced flow in the bottleneck and eventually inhibit flow completely. Reducing the flow in the bottleneck can also result in compression in the crowd as pedestrians reduce their distance to the person in front and people become more and more impatient and pushy.

2.4. Movement in crowds – collective effects

Pedestrian dynamics produce interesting collective effects and self-organising phenomena. The collective effects of people in a crowd are more than an aggregation of a group of individuals, and many studies have been conducted to understand these effects. Collective effects occur because pedestrians in crowds have different behaviours. For instance, pedestrians in a crowd change their walking speed and trajectory to avoid collision with other pedestrians by keeping their distance from them. The speed of an individual is generally subject to the speed of the crowd. It is very important for collective effects to be understood properly and taken into account in pedestrian modelling.

Collective effects can be defined in three classes [32]: collective phenomena, collective behaviour and collective action. Collective phenomena are exhibited by groups of people, such as lane formation (Figure 2.16). Collective behaviour is the conscious or unconscious human behaviour that forces people to adapt to change in their environment, such as changing direction to avoid collision with an obstacle. The final class, collective action, is the most complex of the three classes and accounts for people's intentional behaviours, chosen from a series of possible alternatives, in order to meet either goals or social or cultural norms. The classes of collective effects are not strictly defined, and there can be overlap between classes. For instance, lane formation can be interpreted both as a collective behaviour that people make in order to avoid collision with oncoming pedestrians and as the observable phenomenon that results from other collective behaviours or actions. Generally, a combination of collective behaviours or collective actions results in the emergence of collective phenomena.

Figure 2.16 (a) Lane formation in bidirectional movement in a corridor; (b) a frozen, high-density situation. Darker circles represent pedestrians travelling from left to right, and lighter circles represent pedestrians travelling in the opposite direction [11]

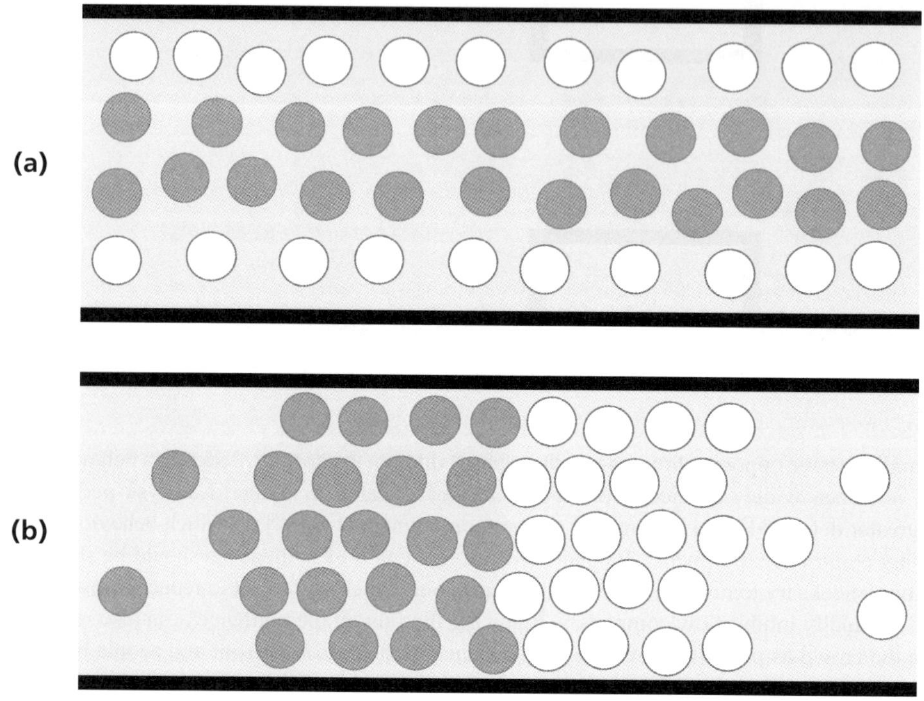

Collective effects include the following [28]:

2.4.1 Lane formation

Lane formation: Lane formation is a phenomenon created in bidirectional movement in environments like corridors and sidewalks. In such cases, pedestrians that are moving in the same direction will form one (or multiple) lane(s), even though no physical signage or barriers to specify defined directions is provided (Figure 2.16(a)). Lane formation can result in capacity loss of 4–14.5% in bidirectional environments [6]. The number of lanes depends on the widths of the corridor, and the boundary of the lanes can also vary dynamically [11]. If the density increases significantly and goes well beyond the capacity, a *frozen* situation could occur, where there would be a cessation of movement (Figure 2.16(b)).

2.4.2 Jamming

Jamming and clogging occur when the inflow exceeds the capacity. An example of where jamming happens in bottlenecks is shown in Figure 2.17. Jamming should be carefully analysed in pedestrian modelling as it has important impacts on pedestrian flow and evacuation time.

2.4.3 Density waves

Dense crowded situations can introduce fluctuations to individual motions, producing stop-and-go effects and turbulent flow. This collective effect is known as density waves and can potentially result in crowd

Figure 2.17 Example of jamming at the bottleneck

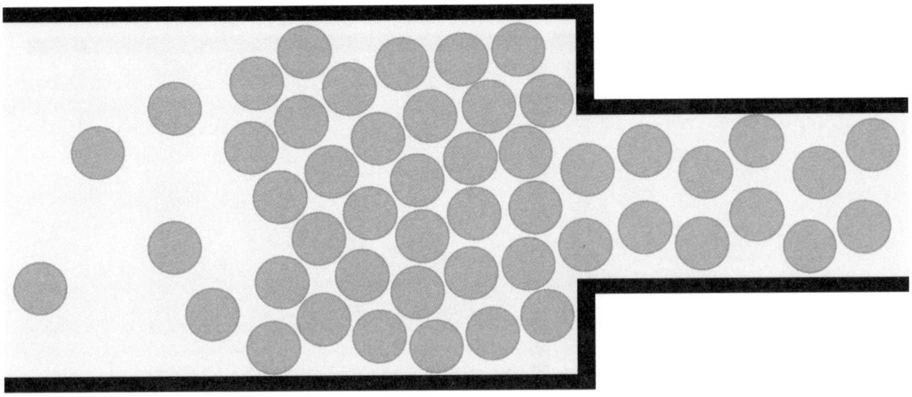

disasters if not taken into proper consideration when considering operation and design [17]. This phenomenon has similarities with density waves experienced in congested road traffic. Density waves are most common in confined pedestrian movement. Figure 2.18 depicts a schematic of density waves.

2.4.4 Oscillations

Oscillations happen in bottlenecks with bidirectional movements. In such cases, only pedestrians moving in one direction can pass the bottleneck, while pedestrians moving in the opposite direction must wait. This will continue until a few pedestrians from the opposite direction manage to pass the bottleneck. Then the number of pedestrians moving in the opposite direction will increase until only pedestrians of that direction are able to pass the bottleneck. Figure 2.19 depicts an example of oscillations where there is only left-to-right movement by pedestrians.

2.4.5 Emergency situations

Panic can occur in emergency situations, where many people compete for limited space, leading to selfish, asocial or irrational behaviour or actions. Collective actions undertaken in emergency situations can also be positive, such as people stopping to help each other during a crowd disaster evacuation, not because they need to in order to escape but because they want, or feel obliged, to help. This desire or obligation may be motivated by a common social identity shared by a group. Analysis of crowd disasters suggests that they occur not as a result of panic behaviour but a lack of capacity of the pedestrian infrastructure involved [28].

Figure 2.18 Example of density waves where there is a stop-and-go situation for pedestrians. Circles with an arrow indicate moving entities, while circles with a line have stopped

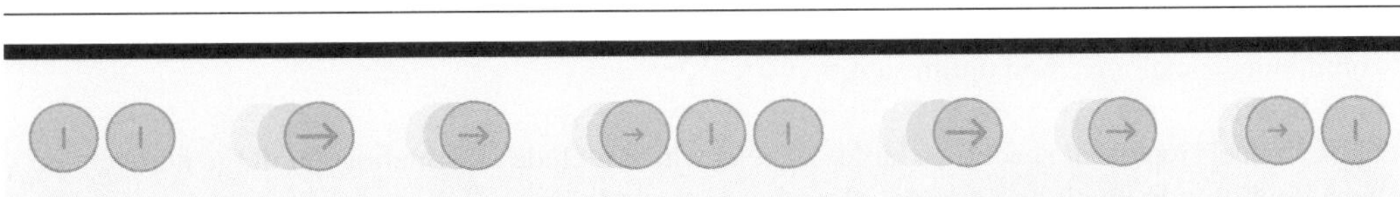

Figure 2.19 Example of oscillations where there is only left-to-right pedestrian movement at the bottleneck [11]

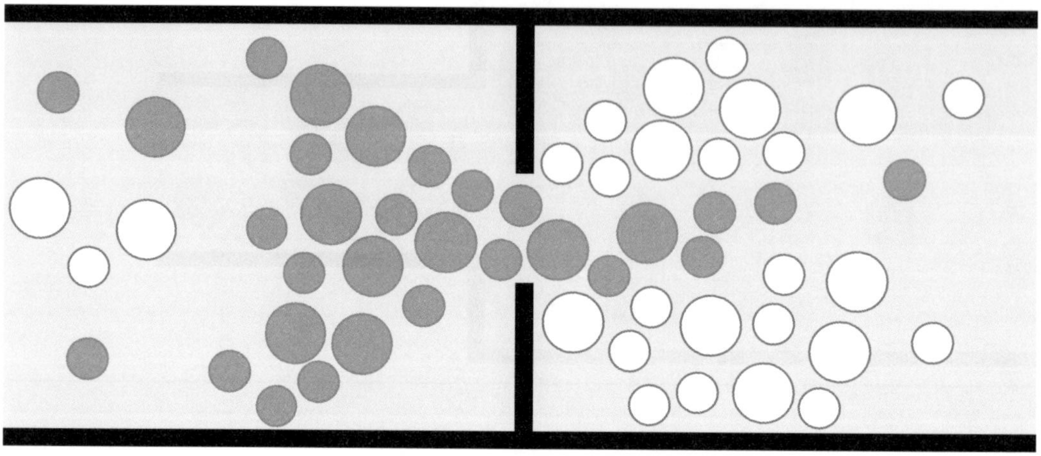

2.5. Approaches to pedestrian modelling

Modelling pedestrian dynamics is complex because of the various parameters that affect pedestrian speeds and walking directions. As previously discussed in this chapter, pedestrian infrastructure is highly diverse (sidewalks, crossings, stairs, shopping malls), and each facility can significantly impact on pedestrian movement. Unlike vehicle movement, pedestrian movement is not limited to a lane and, depending on the underlying facilities, different aspects should be taken into consideration in the modelling approach. In addition, pedestrian group behaviour is different from individual behaviour. Pedestrian walking speed is also subject to the presence of others in the environment as well as various psychological and cultural aspects. To address these challenges in pedestrian dynamics, many approaches to modelling crowd behaviour have been developed over the past 25 years. The inspiration for these approaches has come from a wide range of sources, including fluid mechanics, probability functions and discrete choice models to effectively capture pedestrian dynamics. This section discusses these methods to provide an understanding of the mathematical approach, merits and weaknesses of each.

2.5.1 Types of model

Simulation of individual and crowd behaviour can be broken down into four main categories:

(*a*) Force-based
This includes approaches that use predominantly classical mechanical models as a basis for modelling pedestrian movement as a result of a number of forces acting on an individual. The models may use Newtonian or fluid mechanics as a basis of the approach; however, some are more complex, optimising pedestrian journeys by incorporating a cost or effort function. The force-based model types are continuum, social force and optimisation.

(*b*) Grid-based
These models represent space as a lattice of cells which include information about the number and behaviour of individuals and the spatial environment. Individuals define movements using transition probabilities to neighbouring cells.

(*c*) Discrete choice
This category includes models which use stochastic processes and random decisions as the basis of their approach to modelling pedestrian behaviour. An individual is faced with a set of choices on a journey with different weightings given to each alternative, the highest being most attractive.

(*d*) Agent-based
These models are based on the multi-agent systems (MAS) computational paradigm [38]. This approach focuses directly on individuals, their behaviour and interaction with objects and others. Agents make autonomous decisions in relation to their environment to proactively meet certain goals.

These four main categories and the models that have been developed as a result of the overarching approach are shown below.

2.5.2 Continuum models

In continuum models, individuals are self-driven particles that interact through physical rules. Each individual within the model has a preferred position and velocity and adapts this to avoid collision with others and any environmental boundaries. The current velocity and direction of a single individual in the crowd is a result of the circumstances and interactions. In simple terms, pedestrians are treated as individuals moving according to Newton's law:

$$m_p \frac{dv}{dt} = f, \ \frac{dr}{dt} = v \tag{2.3}$$

Where m_p, v and r denote mass, velocity and the position of the pedestrian, and f the sum of all forces exerted by it or acting on it (Figure 2.20).

Figure 2.20 Schematic diagram for continuum models

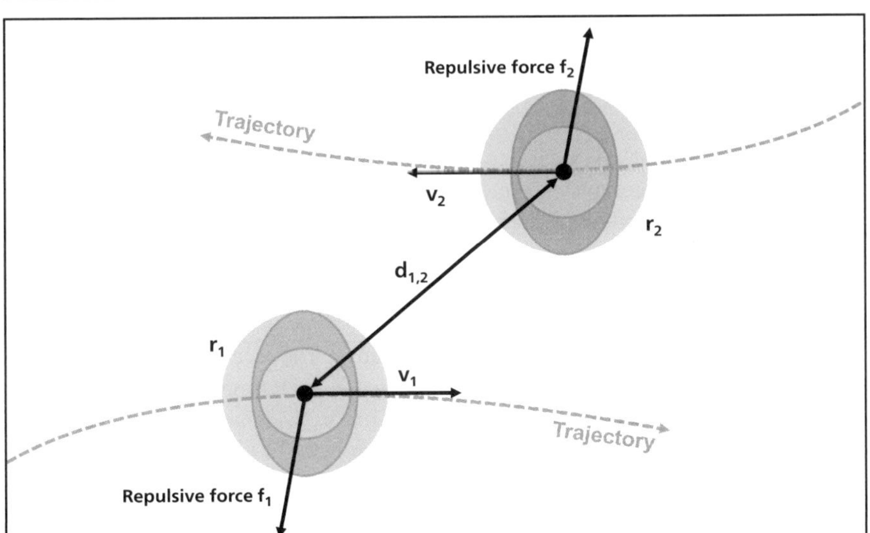

Figure 2.21 Schematic diagram for social force models

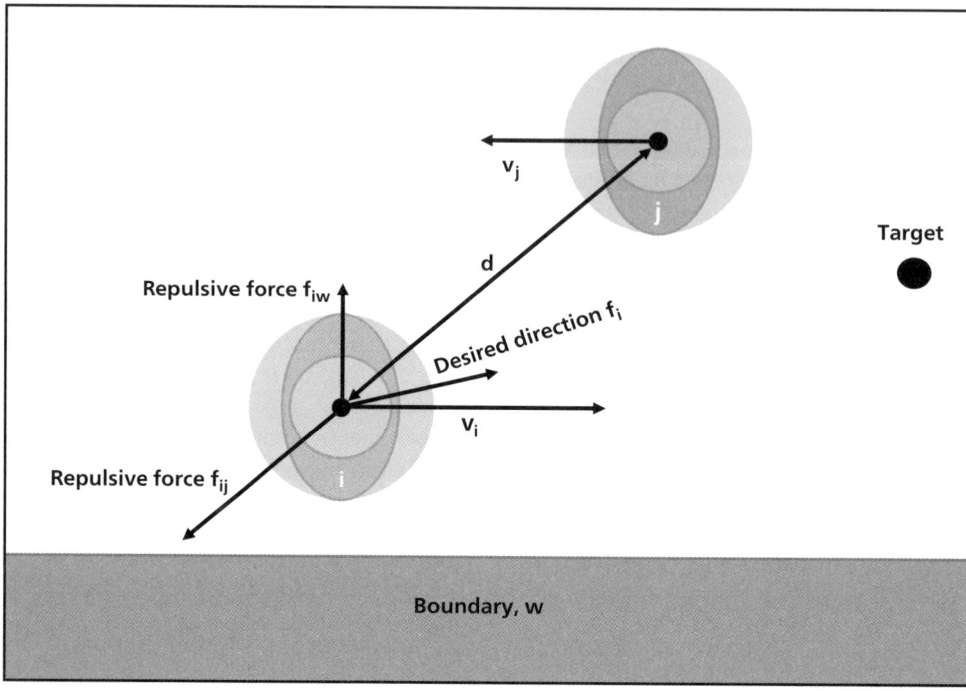

2.5.3 Social force models

Social force models are an enhancement of the continuum models described above. Individuals are driven by three forces: the desired force, the interaction force between pedestrians and the interaction force between a pedestrian and boundaries (walls or obstacles).

$$m_i \frac{d\vec{v}_i(t)}{dt} = \vec{f}_i^{\,0} + \sum_{j(\neq i)} \vec{f}_{ij} + \sum_{w} \vec{f}_{iw} \tag{2.4}$$

Where m_i represents the mass of pedestrian i; v_i is the actual walking velocity of pedestrian i; f_i represents the diagram of the desired force; f_{ij} represents the interaction force between pedestrians i and j; f_{iw} represents the interaction force between pedestrian i and boundary (or obstacle) w. Figure 2.21 shows a schematic diagram for social force models.

2.5.4 Optimisation models

Optimisation models are an enhancement of the concepts outlined above. The acceleration element of the equation incorporates a utility which is evaluated by each individual in each time-step of the model to optimise the path taken. The utility can comprise a number of different considerations but is most commonly a form of cost optimisation. The following components are common:

- Observable environment
- Walking cost

Figure 2.22 Schematic diagram for optimisation models

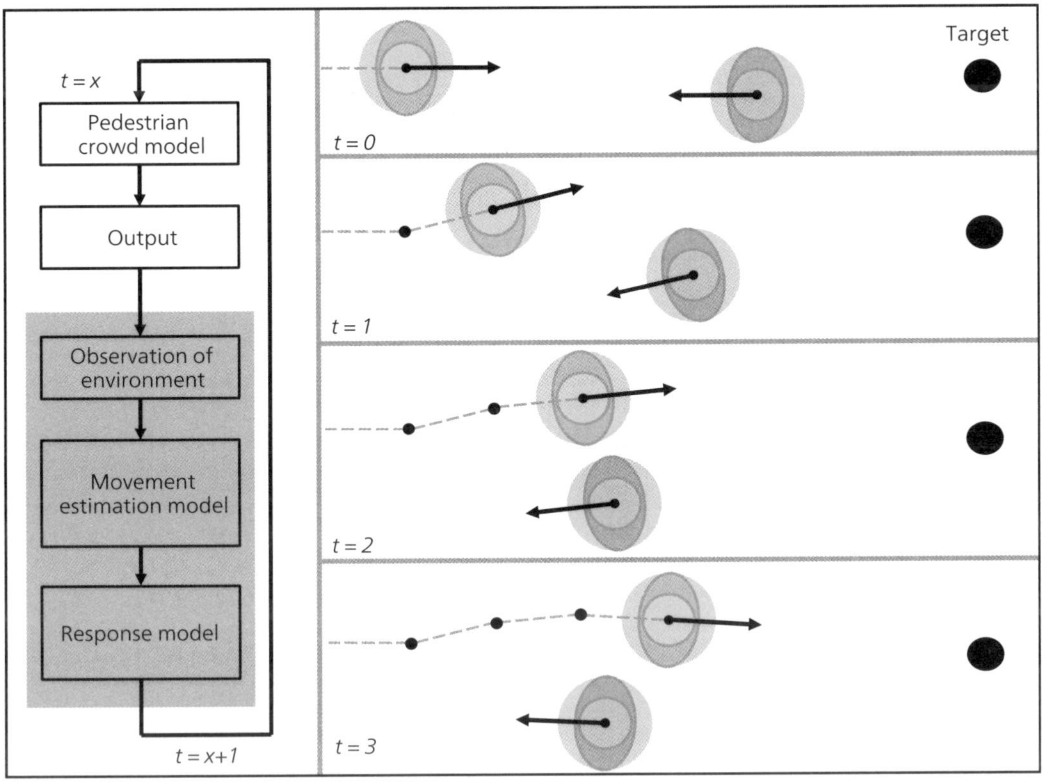

- Anticipation of others' behaviour
- Deviation from route
- Cost of interaction with others.

The initial approach in this field was developed by [16]. Figure 2.22 depicts a schematic diagram for optimisation models.

2.5.5 Grid-based models

In grid-based models, a cell can either be occupied or empty, and pedestrians move synchronously and by only one cell per time-step. Movements are orthogonal, and transitions between cells are probabilistic, with values calculated using different criteria, such as the distance to the target or travel time (Figure 2.23).

The simplest approach is to use one or more static floor fields in which the cells contain the same information along the entire simulation; this is, however, unrealistic for dynamic crowd behaviour. Schadschneider and Seyfried [29] proposed that probabilities of movement are modified by both a dynamic and static floor field. The static field represents the influence of the environment, whereas the dynamic field logs other pedestrians and causes the broadening and weakening of these traces with time.

Figure 2.23 Example of grid-based models, with orthogonal movement shown in (a), (b) and (c)

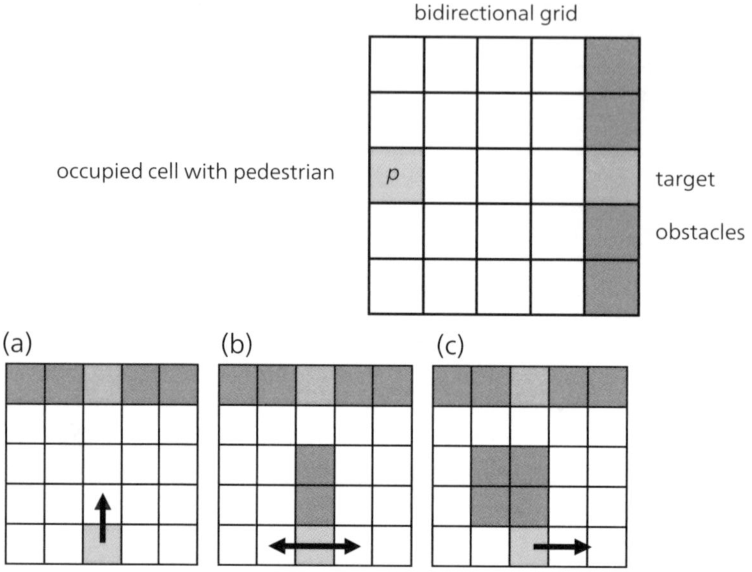

2.5.6 Discrete choice models

Discrete choice models describe, explain and predict choices between two or more discrete alternatives. For example, the choice set for a person deciding which mode of transport to take to work includes driving alone, walking or taking the bus. The choice set is complicated by the fact that a person can use multiple modes for a given trip, such as driving a car to a train station and then taking the train to work. In this case, the choice set can include each possible combination of modes. Alternatively, the choice can be defined as the choice of 'primary' mode, with the set consisting of car, bus, rail and other (e.g. walking, bicycle).

2.6. Summary points

- Understanding pedestrian behaviour is important for designing and planning a safe, comfortable, efficient and enjoyable experience for pedestrians in public places, such as railway stations, shopping malls and stadia.
- Fundamental diagrams provide a graphical representation of macroscopic relationships between observables, and pedestrian movement is often defined by terms such as 'speed', 'flow' and 'density'.
- The primary objective of fundamental diagrams is to provide characteristics and patterns of pedestrian movements in different structural facilities by showing the relationship between speed, flow and density of pedestrians.
- Uni and bidirectional pedestrian movements exist in most pedestrian structural facilities, and there is much research on specific environments of pedestrian movement, such as corridors, stairs, transport terminals and shopping malls.
- Pedestrian walking speed is one of the essential inputs for calculating the capacity of pedestrian facilities, but speeds are impacted by many parameters, such as gender, age, culture and race, environmental conditions, trip purpose (e.g. commuting, leisure, business) and groups.

- The collective effects of people in a crowd are more than an aggregation of a group of individuals, and many studies have been conducted by various researchers to understand these effects.
- Pedestrians in a crowd change their walking speed and trajectory to avoid collision with other pedestrians by keeping their distance from other pedestrians.
- Collective effects can be defined in three classes: collective phenomena, collective behaviour and collective action.
- Modelling pedestrian dynamics is complex because there are various parameters that affect pedestrian speeds and walking directions.
- Many approaches to modelling crowd behaviour have been developed over the past 25 years; however, simulation of individual and crowd behaviour can be broken down into four main categories: force-based, grid-based, discrete choice and agent-based.

REFERENCES

[1] Almejmaj M, Meacham B and Skorinko J (2015) The effects of cultural differences between the west and Saudi Arabia on emergency evacuation – clothing effects on walking speed. *Fire and Materials* **39(4)**: 353–370.

[2] Burghardt S, Seyfried A and Klingsch W (2013) Performance of stairs – fundamental diagram and topographical measurements. *Transportation Research Part C: Emerging Technologies* **37**: 268–278.

[3] Caramuta C, Collodel G, Giacomini C *et al.* (2017) Survey of detection techniques, mathematical models and simulation software in pedestrian dynamics. *Transportation Research Procedia* **25**: 551–567.

[4] Chattaraj U, Seyfried A and Chakroborty P (2009) Comparison of pedestrian fundamental diagram across cultures. *Advances in Complex Systems* **12(3)**: 393–405.

[5] Daamen W. (2004) *Modelling Passenger Flows in Public Transport Facilities*. PhD thesis, Delft University, Delft, the Netherlands.

[6] Daamen W and Hoogendoorn SP (2003) Experimental research of pedestrian walking behavior. *Transportation Research Record: Journal of the Transportation Research Board* **1828**: 20–30.

[7] Daamen W and Hoogendoorn SP (2010) Capacity of doors during evacuation conditions. *Procedia Engineering* **3**: 53–66.

[8] Daamen W, Hoogendoorn SP and Bovy PHL (2005) First-order pedestrian traffic flow theory. *Transportation Research Record: Journal of the Transportation Research Board* **1934**: 43–52.

[9] Fruin JJ (1971) Designing for pedestrians: A level-of-service concept. *Highway Research Record: 50th Annual Meeting of the Highway Research Board*. Washington, DC, USA.

[10] Fruin, JJ (1971) *Pedestrian Planning and Design*. Metropolitan Association of Urban Designers and Environmental Planners, New York, NY, USA.

[11] Helbing D (2004) Collective phenomena and states in traffic and self-driven many-particle systems. *Computational Materials Science* **30(1)**: 180–187.

[12] Helbing D, Buzna L, Johansson A and Werner T (2005) Self-organized pedestrian crowd dynamics: Experiments, simulations, and design solutions. *Transportation Science* **39(1)**: 1–24.

[13] Helbing D, Farkas I and Vicsek T (2000) Simulating dynamical features of escape panic. *Nature* **407(6803)**: 487.

[14] Helbing D, Johansson A and Al-Abideen HZ (2007) Dynamics of crowd disasters: An empirical study. *Physical Review E* **75(4)**: 046109.

[15] Heskestad G (2002) Fire plumes, flame height, and air entrainment. In *SFPE Handbook of Fire Protection Engineering* (DiNenno PJ, Drysdale D, Beyler CL *et al.* (eds)), 3rd edn. National Fire Protection Association, Quincy, MA, USA, pp. 2-1–2-17.

[16] Hoogendoorn SP and Bovy PHL. Simulation of pedestrian flows by optimal control and differential games (2003) *Optimal Control Applications and Methods* **24(3)**: 153–172.
[17] Kuang H, Cai M-J, Li X-L and Song T (2015) Asymmetric effect on single-file dense pedestrian flow. *International Journal of Modern Physics C* **26(6)**: 1550064.
[18] Lam WHK, Lee JYS and Cheung CY (2002) A study of the bi-directional pedestrian flow characteristics at Hong Kong signalized crosswalk facilities. *Transportation* **29(2)**:169–192.
[19] Liao W, Seyfried A, Zhang J *et al.* (2014) Experimental study on pedestrian flow through wide bottleneck. *Transportation Research Procedia* **2**: 26–33.
[20] Martinez-Gil F, Lozano M, García-Fernández I and Fernández F (2017) Modeling, evaluation, and scale on artificial pedestrians: A literature review. *ACM Computing Surveys (CSUR)* **50(5)**: 72.
[21] Nagai R, Fukamachi M and Nagatani T (2006) Evacuation of crawlers and walkers from corridor through an exit. *Physica A: Statistical Mechanics and Its Applications* **367**: 449–460.
[22] Navin FP and Wheeler RJ (1969) Pedestrian flow characteristics. *Traffic Engineering, Institute Traffic Engineering* **39(9)**: 30–36.
[23] Nelson HE and Mowrer FW (2002) Emergency movement. In *SFPE Handbook of Fire Protection Engineering* (DiNenno PJ, Drysdale D, Beyler CL *et al.* (eds)), 3rd edn. National Fire Protection Association, Quincy, MA, USA, pp. 3-367–3-380.
[24] Older SJ (1968) Movement of pedestrians on footways in shopping streets. *Traffic Engineering & Control* **10(4)**: 160–163.
[25] Predtechenskii VM and Milinskii AI (1978) *Planning for Foot Traffic Flow in Buildings*. National Bureau of Standards, US Department of Commerce, and the National Science Foundation, Washington, DC, USA.
[26] Pushkarev B, Zupan JM, Pushkarev B and Zupan JM (1975) Capacity of walkways. *Transportation Research Record: Journal of the Transportation Research Board* **538**: 1–15.
[27] Sano T, Shida K and Tatebe K (2001) An experimental study on pedestrian crossing conflicts by physical index. Study on characteristics of pedestrian crossing flow Part 1. *Journal of Architecture, Planning & Environmental Engineering* **66(546)**:127–132.
[28] Schadschneider A, Klingsch W, Klüpfel H *et al.* (2011) Evacuation dynamics: Empirical results, modelling and applications. In *Extreme Environmental Events: Complexity in Forecasting and Early Warning, Volume 1* (Meyers RA (ed.)). Springer, New York, NY, USA, pp. 517–550.
[29] Schadschneider A and Seyfried A (2009) Empirical results for pedestrian dynamics and their implications for cellular automata models. In *Pedestrian Behaviour: Models, Data Collection and Applications* (Timmermans H (ed.)). Emerald Group Publishing, Bingley, UK, pp. 27–43.
[30] Seer S, Bauer D, Brandle N and Ray M (2008) Estimating pedestrian movement characteristics for crowd control at public transport facilities. *2008 11th International IEEE Conference on Intelligent Transportation Systems: IEEE, Beijing, China*, pp. 742–747.
[31] Shi X, Ye Z, Shiwakoti N and Li Z (2015) A review of experimental studies on complex pedestrian movement behaviours. *Proceedings of the 15th COTA International Conference of Transportation Professionals (CICTP 2015), Beijing, China*, pp. 1081–1096.
[32] Sieben A, Schumann J and Seyfried A (2017) Collective phenomena in crowds – where pedestrian dynamics need social psychology. *PLoS ONE* **12(6)**: e0177328.
[33] Templeton AM (2017) *Physical Crowds and Psychological Crowds: Applying Self-categorization Theory to Computer Simulation of Collective Behaviour*. PhD thesis, University of Sussex, Brighton, UK.
[34] Vanumu LD, Ramachandra Rao K and Tiwari G (2017) Fundamental diagrams of pedestrian flow characteristics: A review. *European Transport Research Review* **9(4)**: 49.

[35] Vizzari G, Manenti L, Ohtsuka K and Shimura K (2015) An agent-based pedestrian and group dynamics model applied to experimental and real-world scenarios. *Journal of Intelligent Transportation Systems* **19(1)**: 32–45.
[36] Weidmann U (1993) *Transporttechnik der Fussgänger: Transporttechnische Eigenschaften des Fussgängerverkehrs*. Institut für Verkehrsplanung, Transporttechnik, Strassen- und Eisenbahnbau, Zürich, Schweiz.
[37] Weidmann U (1992) *Transporttechnik der Fussgänger: Transporttechnische Eigenschaften des Fussgängerverkehrs*. Institut für Verkehrsplanung, Transporttechnik, Strassen und Eisenbahnbau, Zürich, Schweiz.
[38] Wooldridge M (2013) Intelligent agents. In *Multiagent Systems* (Weiss G (ed.)), 2nd edn. The MIT Press, Cambridge, MA, USA, pp. 3–50.
[39] Wu J and Lu S (2013) Feature analysis and operation evaluation of pedestrian weaving zone. *Transportation Research Record: Journal of the Transportation Research Board* **2393**: 66–74.
[40] Xiao-xiong W and Qi-da F (2011) Impact analysis on pedestrian traffic flow in weave area of building. *MACE 2011 Second International Conference on Mechanic Automation and Control Engineering, IEEE, Inner Mongolia, China*, pp. 2215–2218.
[41] Yanagisawa D, Kimura A, Tomoeda A *et al.* (2009) Introduction of frictional and turning function for pedestrian outflow with an obstacle. *Physical Review E* **80(3)**: 036110.
[42] Yuen JKK and Lee EWM (2012) The effect of overtaking behaviour on unidirectional pedestrian flow. *Safety Science* **50(8)**: 1704–1714.
[43] Zhang J and Seyfried A (2013) Empirical characteristics of different types of pedestrian streams. *Procedia Engineering* **62**: 655–662.

ICE Publishing

Urban Space – AECOM
ISBN: 978-0-7277-6381-5
https://doi.org/10.1680/mpm.63815.035

Chapter 3
Data collection

3.1. Overview of the chapter

This chapter describes various sources of data and discusses their practical application when modelling. It explores the differences between data collected during experiments and live observations. Different methods of data collection are outlined and their limitations discussed.

The first topic covered in this chapter is principles of data collection, in Section 3.2. It covers the data-collection process from start to end and what must be considered prior to undertaking a data-collection exercise. The types of data that can be collected are discussed along with the methods for their collection, where the data should be collected and how long for. Section 3.3 discusses sources of data and how data is collected, including manual counting and automated methods, such as video, microwave, infrared, pressure, laser, ultrasonic and mobile-device data. Section 3.4 discusses the limitations of data collection that modellers should be aware of and try to minimise through appropriate planning. Finally, Section 3.5 provides a number of reference texts for further reading on the subject of data collection.

3.2. Principles of data collection

Data forms one of the primary inputs for any pedestrian movement model; therefore, data collection is a critical part of the modelling process. Unlike vehicular-data collection, relatively little research has been undertaken into pedestrian-data collection. The various types of data that can be collected, as well as different methods for obtaining this data, are outlined in this section.

3.2.1 Data-collection process

To effectively undertake data collection, it is necessary to first understand the objective of the pedestrian model. This varies from project to project but usually relates to interventions aimed at improving the safety, security, convenience, continuity, comfort, system coherence or attractiveness of the pedestrian environment [3]. Understating what needs to be modelled and why, as well as the level of detail required [12], defines the purpose of data collection, which allows the appropriate count location(s), duration and other considerations to be determined, and the data-collection method to be selected [9]. In some cases, data may already exist, and the planning of the data-collection process will determine whether it is sufficient to meet the objectives of the project [12].

Figure 3.1 outlines the steps in the data-collection planning process. The outcomes of this planning process form the methodology that must be defined before any surveys are undertaken to obtain relevant data [12].

3.2.2 Types of data

The requirements of the model dictate the type of data that needs to be collected. Most pedestrian models require a combination of several different data inputs. Typical demand data include flow counts, cordon counts and turning movements. Data for specific facility types, such as railway stations, may include such information as entry or exit and boarding or alighting counts [12]. Flow counts record the number of

Figure 3.1 Data-collection planning process [13]

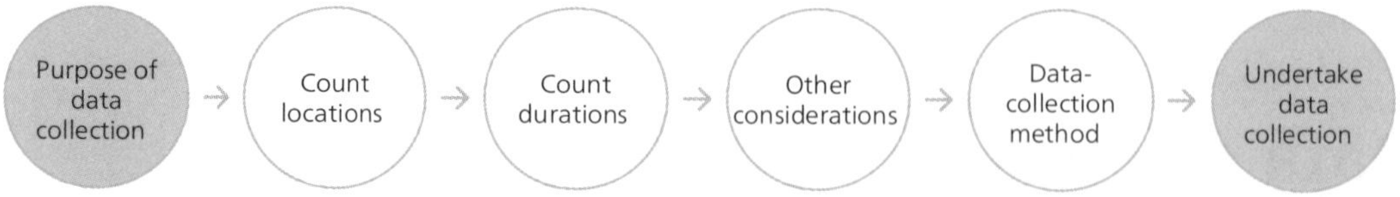

pedestrians that pass a screenline [9], which is an imaginary line where data is collected that can be used to validate model behaviour – for example, the number of pedestrians entering or exiting a building. Cordon counts, like flow counts, observe the number of people entering or exiting a cordon area [3] and can include a snapshot of the number of pedestrians within a cordon area at a given point in time – for example, the number of passengers in a station concourse during peak periods. Turning counts refer to the number of pedestrians that turn right, left or continue straight for each approach leg of an intersection [9] and, as such, include both numerical (counts) and routing information (direction). The routing of pedestrians through the model is another type of data that can be input. This type of data includes *origin* and *destination* data – that is, the point of entry (origin) and exit (destination) of pedestrians entering and exiting a study area, with no regard for routing in between – which is used to create the origin–destination demand matrix [12]. More detailed *routing* data would observe actual routes taken by pedestrians walking through a study area.

Another data type that can be collected relates to the reason why a trip is being made, also known as trip purpose. Trip purposes include commuting, business, shopping, cultural and social, among others [3]. Trip purpose is significant as it affects the behaviour and speed of the pedestrian – for example, a commuter who makes the same trip every day has good knowledge of the route is in a hurry and, as a result, walks faster than a tourist travelling for cultural reasons. Similarly, a business traveller values time more highly, and therefore walks faster, than a shopper. The frequency with which a pedestrian makes a certain journey is also useful data [3] as it can support the trip purpose. For example, a daily trip is most likely associated with a commuter. Trip frequency can also have an impact on pedestrian behaviour. A pedestrian making a one-off trip is likely to be slower as they need to familiarise themselves with their surroundings and safely find their way.

Pedestrian behaviour is the final data type that can be observed and collected for input into a pedestrian model. Behaviours are the specific characteristics exhibited by pedestrians, such as speed, density and requirement for personal space, tendency for group formation and waiting and queuing traits. The particular pedestrian behaviours observed in a study area should be replicated in a pedestrian model to ensure an accurate representation of pedestrian movements in the area.

Given the varied data types that can be collected on pedestrian movement, determining the type of data required is critical in order to select the most appropriate data sources, which are explained further in Section 3.3.

3.2.3 Data-collection methods

There are two main characteristics related to data collection. The first is the nature of the data collection, which can be either manual or automated [9]. Automated counts use detection or recognition technology to automatically obtain data. Manual data collection, on the other hand, requires an observer to physically collect data. Manual data collection can occur both physically on-site or later, from video footage of the project area identified for modelling. Both manual and automated data-collection methods can obtain pedestrian data, although one method may be more effective or appropriate for a specific task than another. Some considerations to be made when deciding between the two include the nature of the data to be collected, the location and duration of counts and the resources available for data collection [13]. In some cases, a combination of manual and automated data collection can be undertaken, or manual processing of automatic data collection may be required, particularly with emerging collection technologies such as GPS (global positioning system) and Wi-Fi, for example. More detail on manual and automated counting methods is provided in Section 3.3.

The second characteristic of data-collection methods is the survey duration. Data collection can be either continuous or short duration. Continuous counts are those that automatically collect data continuously – twenty-four hours a day, seven days a week for an ongoing period of time [9]. Short-duration surveys can be done either manually or automatically and for a certain period of time, as required – from a few hours up to several months [9]. Finding the balance between the amount and type of information required and the available resources is critical when planning survey durations. While continuous counts provide a vast volume of reliable data, the cost associated with setting up permanent count locations may be prohibitive when compared to short-duration counts. More detail on count duration is provided in Section 3.2.5.

3.2.4 Count locations

Pedestrian activity is dynamic and highly variable and can therefore be unpredictable – pedestrians cross roads away from crossings, do not naturally walk in single file and do not always keep to footways. It would therefore be impossible to collect all data types and movement combinations for all possible locations without using significant resources. Given the emerging nature of pedestrian data collection when compared to motorised traffic, large-scale non-motorised data-collection programmes generally have a much smaller number of survey locations, leading to smaller samples [4]. It is therefore important to identify appropriate locations for collection in order to obtain the most accurate data. Count locations are very much dependent on the input requirements of the pedestrian model, but Ryus *et al.* [13] introduce four general approaches to count locations: targeted, representative, control and random locations. Targeted locations are chosen in order to obtain data for a particular purpose and are often related to certain projects or facility types. If data collection is undertaken before and after implementation, they can be used in order to quantify the impact of a specific intervention. Examples of targeted locations are counts at key locations of a railway station, such as entries and exits, gatelines, stairs, escalators and junctions [12]. Pedestrian crossings are also considered important targeted locations in urban areas as they facilitate the interactions of pedestrians and road traffic. In addition to the number of pedestrian crossings, the frequency of crossings is important due to its impacts on the traffic flow. To effectively reflect reality, all types of crossing should be considered – formal crossings (e.g. pelican crossings, zebra crossings and puffin crossings), informal crossings (e.g. dropped kerbs and pedestrian islands) and even jaywalking.

As with targeted surveys, the location of control surveys is carefully considered in order to satisfy a specific need. Control locations are chosen in places where no intervention has been implemented in order to assess the impact of a particular project elsewhere and its effect on changes in demand. Figures 3.2 and 3.3 show four targeted and four control count locations related to a planned intervention in an urban grid network of streets.

Representative locations are selected in order to provide an overall representation of pedestrian traffic across a community and, therefore, instead of focusing on locations that are expected to have the greatest flows, they should be scattered throughout the community, surrounded by different land uses and located on different facilities [13]. An example of four representative count locations in an urban street grid network is

Figure 3.2 Targeted count locations in an urban area

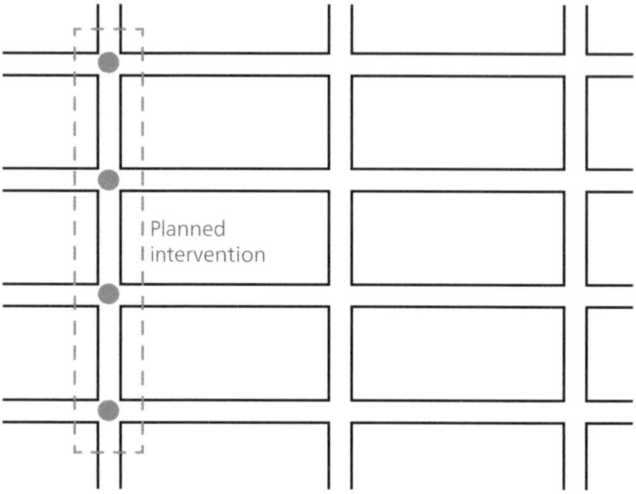

provided in Figure 3.4. This type of data-collection location choice is useful in identifying overall trends and understanding the impact of specific land-use types on pedestrian flows across a wider area. In this case, targeted counts would be unfeasible and less related to a specific intervention or project.

Count locations can also be chosen randomly; however, this is not an effective method as it may not capture strategic locations and may provide data in locations with low volumes of pedestrian traffic, which inherently have higher variability over time. This can lead to reduced accuracy, especially from short duration counts [13]. An example of four random count locations in a street grid network is provided in Figure 3.5.

Figure 3.3 Control count locations in an urban area

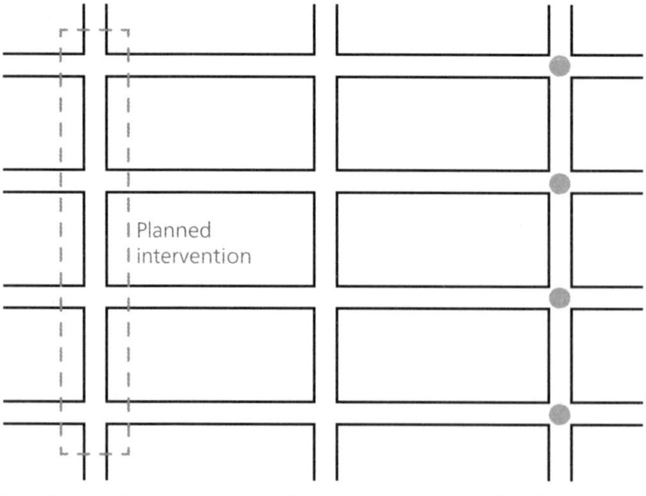

Figure 3.4 Representative count locations in an urban area

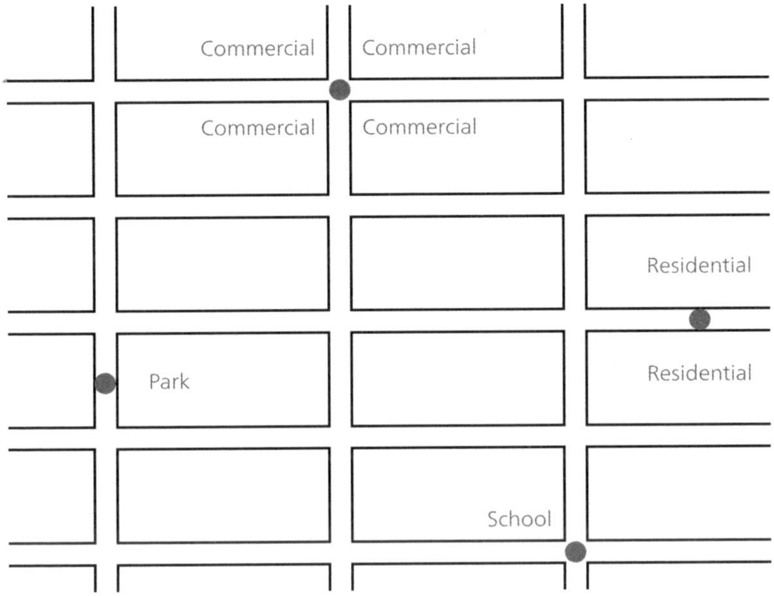

Figure 3.5 Random count locations in an urban area

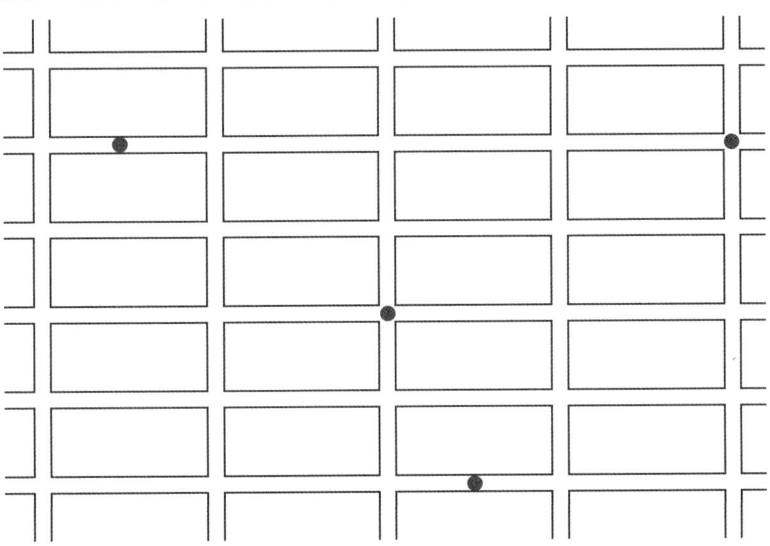

3.2.5 Count durations

It is important that sufficient data is collected in order to meet the requirements of the model. Understanding the purpose of the data collection as well as the resources available for collecting the data, including the budget, staff and equipment, is therefore important when defining count durations [13]. As introduced in Section 3.2.3, data collection can be undertaken over either short or continuous periods of time.

Continuous counts use automated counting technology to obtain long-term pedestrian data. Since data is collected at all times, it provides a high level of accuracy and is less susceptible to variability in flows, which can negatively affect short duration counts. Installing permanent counting technology, however, is expensive, and this volume and accuracy of data may not be required for the model. Continuous count sites are typically set up by public or government agencies to monitor long-term pedestrian activity and trends, and such data may be made available to planners and modellers.

Short-duration counts can be performed manually by enumerators on site, manually using video or manually using portable automated counting devices, and they can take from as little as a few hours to several months [9]. When undertaking short-duration counts it is important to capture the data related to the pedestrian model – typically, the peak flows. Short-term counts should be undertaken on a 'typical' day, avoiding overly hot, cold, wet or windy weather, particularly if the count location is outdoors, as well as any events such as works, special events or holidays that would reduce or increase the typical flow of pedestrians in the study area.

Accidents or unexpected events occurring during the survey period can also render the data collected unreliable; this may include disruption on the wider network which could materially impact normal behaviour at the survey site [3]. Understanding the potential data errors caused by variability during short-duration flows is important in order to accurately adjust or extrapolate the data. It can be useful to undertake a combination of short-duration and continuous counts, using data obtained from the continuous collection to adjust the short-duration counts as required [13].

3.2.6 Considerations

There are many other considerations that must be taken into account when preparing to undertake data collection for a pedestrian model, and a site visit is recommended early in the project to identify and resolve most of these. A site visit is useful to inform or confirm any modelling assumptions as well as to provide an overall understanding of the study area, including the key pedestrian flows and behaviours [12].

The type and location of infrastructure where data is to be collected is important. Different data-collection methods may be appropriate for enclosed spaces, such as corridors, and for open areas or rooms [9], as the movement behaviours of pedestrians walking in these spaces may differ, thereby influencing potential detection techniques. The surface of the ground and width of the facility can further limit the detection technologies that can be used. Different considerations must be made when surveying outdoor locations, compared with those indoors. The weather will have a greater impact on both the effectiveness of collection techniques and the quality of data collected in outdoor locations.

Outdoor locations are also more likely to have safety implications, such as those related to vehicular traffic, as well as the potential for equipment theft, although this may also be present indoors. Another critical consideration when planning data collection is the land use and characteristics surrounding the count locations. Certain pedestrian flow characteristics may be evident in residential areas compared to commercial areas, or near schools, while nearby open spaces, such as parks or squares, may increase the route options for pedestrians, thereby making their movements harder to survey. Pedestrians may, for example, take short cuts that avoid the data-collection point [13]. Pedestrian flow characteristics and volumes are also affected by travel patterns, which may influence the data-collection method, time or duration [9]. A station may be predominantly utilised by commuters during the week, and shoppers and leisure travellers on the weekend, which may determine the peak day and time period for counts to be undertaken.

It should be noted that in pedestrian modelling the collected data is typically used in three phases: model building, calibration and validation. The modeller should carefully identify required data for each of these phases and consider it in the data-collection objectives. Chapter 6 discusses model building, calibration and validation phases, and readers are referred to this chapter for further details.

3.3. Sources of data

This section outlines different data-collection techniques relevant to pedestrian modelling. Various detection techniques ranging from manual counting to advanced technologies are explained first. Details of other sources of data, such as guidelines and experimental research data available in the public domain, are then provided.

3.3.1 Collection techniques

There are various techniques for pedestrian data collection, which can be broken into three main categories: manual, semi-automated and fully automated. Each technique has different merits and disadvantages. A single technique is not suitable for all applications and, depending on the purpose of the project, scale of study area, data accuracy, installation and maintenance cost and reliability, the most suitable technique or a combination of them should be selected [7]. For instance, manual counting is very straightforward and only requires basic equipment, but is labour-intensive. Conversely, infrared technology allows automatic extraction of pedestrian counts, but the accuracy of outputs is dependant on external factors, such as weather conditions.

Another important criterion to identify a suitable collection technique is whether the technology is intrusive or non-intrusive. Intrusive techniques impose additional disruption on pedestrian flow during installation and maintenance. A good example of intrusive techniques is pressure sensors that use pedestrian weights to count pedestrians. These sensors need to be installed under the surface where pedestrians walk, and hence any maintenance of the sensors could potentially cause disruption to pedestrian flow. In addition, some technologies are only capable of detecting pedestrian volume counts, whereas other technologies can also track pedestrian trajectories.

These are only a few examples of different aspects to be taken into consideration in the selection of collection techniques. The most common techniques, including their strengths and limitations, are explained in the following section. A better understanding of these data-collection techniques will help readers to select the most suitable one(s).

3.3.1.1 Manual counting

Manual counting is the simplest collection technique. It requires the presence of an enumerator, or data collector, at the site with only simple equipment such as papers, pens or pencils, clipboards, hand-counters or tally sheets [2]. To conduct manual counting, the enumerator should consider a screenline, which is an imaginary line, and count when a pedestrian passes the screenline. Other useful information, such as direction of movement, gender and age group, can also be collected in the manual counting process. The main limitation of manual counting is that it is labour-intensive, subject to human error and limited to low flow rates.

If collecting further information is required, a questionnaire or postcard survey can be conducted. More detailed information can be collected with these surveys, such as pedestrian route choice, mode choice and trip purpose.

Postcard surveys are less intrusive as a pre-paid postage card is handed to pedestrians which can be completed in their own time and posted back for analysis. More recently, cards handed out have contained only a web link or QR[1] code to an online survey. Response rates, however tend to be low – typically, in the 10–20% range.

To conduct a questionnaire survey, the enumerator should have the questionnaire ready either as a hard copy or on a tablet computer (or PDA[2]) and approach pedestrians individually. If pedestrians are willing to participate, they answer the questions set out in the questionnaire.

The main advantage of conducting a postcard or questionnaire survey is that detailed information from pedestrians can be collected. However, in questionnaire surveys people may adapt their answers to what is more socially acceptable, hence they are subject to inaccuracy and bias.

The following survey, which involves the surveyor visually or physically following a target pedestrian, is another manual approach to monitoring pedestrian routing. It too is labour-intensive, but it is limited to the visible activities of pedestrians – that is, the intentions of pedestrians cannot properly be understood [8].

3.3.1.2 Video imaging

Video-image processing is one of the most popular data-collection techniques. Using cameras, a large area can be covered, and it is used for both counting and tracking pedestrians. This method can be very cost-effective if the surveillance cameras are already installed in the area and therefore no additional cameras need to be installed.

Processing video footage can be conducted either manually or automatically. Manual counting is straightforward and similar to the process explained above (3.3.1.1) but using video footage. Automated video processing has recently attracted a lot of interest as it can reduce human involvement and hence bypass human errors and reduce cost. Automated video processing is based on sophisticated pattern-recognition algorithms to count and track pedestrians from the video footage. This is an active area of research and many pattern-recognition algorithms have recently been developed to increase the level of accuracy of automated pedestrian detection.

The main advantage of video-imaging techniques is that the large area that can be covered can capture reality to a great degree of detail even in very crowded or poor weather conditions. In addition, the video footage can be filed and be used for future reference. The main disadvantage is the accuracy of the detection algorithms, which can be impacted by image resolution and environmental factors, such as rain or low light. Another disadvantage is that it is not possible to understand pedestrian intentions (e.g. trip purpose and frequency) and personal characteristics as this detection technique is limited to the visible activities of pedestrians. This disadvantage is, however, applicable to many collection techniques discussed in this section.

3.3.1.3 Microwave radar sensors

Microwave radar was one of the earliest technologies for detecting objects and was widely used before and during World War II. A radar detector consists of an antenna through which it transmits electromagnetic radiation towards the field of view. When an object passes through the projected beam, a proportion of the radiation is reflected back to the antenna (Figure 3.6). In this way, the presence of the object can be detected and, depending on the size of reflected radiation, the speed of the object can be identified. Generally, there are two types of radars.

The first type transmits continuous waves of constant frequency. Based on the changes in the frequency of reflected radiation, the speed of the object can be identified using the Doppler effect.[3] This type of radar is not able to detect motionless objects.

The second type of radar addresses this limitation by transmitting frequency-modulated continuous waves (i.e. saw-tooth waveform). The frequency of reflected radiation of this saw-tooth waveform always changes in constant rates and is used to detect motionless objects. The main disadvantage of radar sensors is the limited detection distance and that off-the-shelf radars for pedestrian counting are not commonly available [4]. In addition, radar sensors can only be used for counting pedestrians and are not able to track them.

3.3.1.4 Infrared

There are two types of infrared detectors available: passive and active. Passive infrared detectors use the energy emitted from bodies – the heat differential in the field of view – to detect the presence of objects [4]. They are also able to detect the volume, direction of movement and speed of objects in the field of view (Figure 3.7) [2].

Active infrared systems consist of three components: a transmitter, a receiver and a data-collection unit [4]. The detection zone is illuminated by low-power infrared energy and the presence of an object is detected when the beam is broken (Figure 3.8).

Figure 3.6 Microwave radar sensor

Figure 3.7 Passive infrared detector

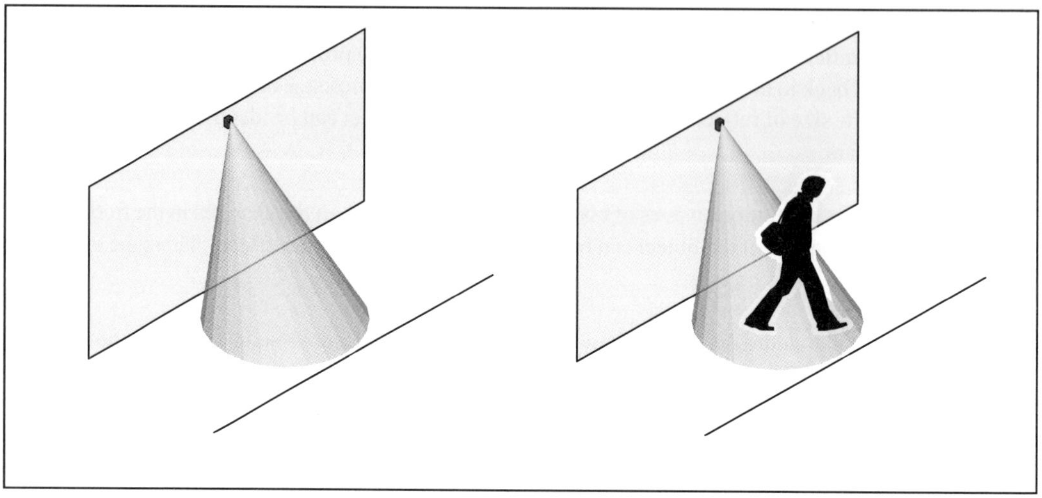

The advantage of infrared systems is that they are low-cost and portable with an unobstructed appearance. The disadvantage is that they can only count pedestrians – they may not be able to accurately distinguish between an individual and a group of pedestrians, and hence they are not suitable for very crowded areas. In addition, the detection area is usually limited and the accuracy of passive sensors is highly dependent on weather conditions.

3.3.1.5 Pressure and seismic sensors

Pressure sensors or mats operate by detecting changes in the force (i.e. weight) of pedestrians. Seismic sensors are similar to pressure sensors, but they identify pedestrians by the passage of energy waves

Figure 3.8 Active infrared detector

generated by feet (Figures 3.9 and 3.10). Both of these sensors need to be installed under the floor of the pedestrian infrastructure to be able to function. However, they are not limited to use in buildings and urban areas. For instance, some train operators use load-weight sensors to capture passenger counts and journey data, and to provide information to passengers on train-carriage occupation (see [10] for an example of the use of these sensors in passenger-data collection).

Even though the maintenance cost for the pressure sensors is low, they have high installation costs, and some pressure and seismic sensors are not able to accurately distinguish between an individual and a group of pedestrians. They are also considered to be intrusive detection techniques as the installation and maintenance process may disturb pedestrian flow.

3.3.1.6 Laser scanners

In laser technologies, laser beams are transmitted to the detection area and the distance to the object is calculated using the gap between the sending and receiving impulses. Laser technologies are also known as LIDAR (LIght Detection And Ranging) or LADAR (LAser Detection And Ranging). The advantages of lasers are that they are able to detect not only objects but their size, speed and trajectory. Using laser technologies, a large amount of data can be obtained in a short period of time. The disadvantages of lasers are their high energy consumption and short detection radius.

3.3.1.7 Ultrasonic detectors

Ultrasonic detectors were first used in the mid 1950s to detect vehicles, and the first users of this technology were in some states in the USA (e.g. Michigan, Illinois and New York). Ultrasonic detectors follow the same principles as microwave radar detectors, but they transmit pressure waves of sound energy instead of microwaves. The frequency of the sound energy is between 25 kHz and 50 kHz, which is above the human audible range [4]. Ultrasonic detectors are capable of detecting volume, presence, classification and speed [7]. The main disadvantage of these detectors is that their accuracy is subject to weather conditions and the sensor must be kept clean to allow smooth emission of sonic waves [2].

Figure 3.9 Underfloor pressure sensor

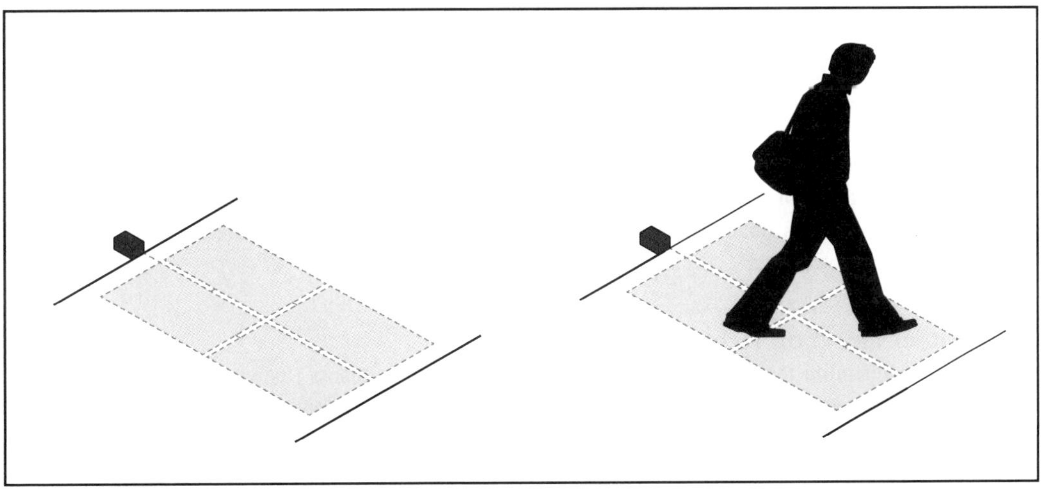

Figure 3.10 Pressure mat

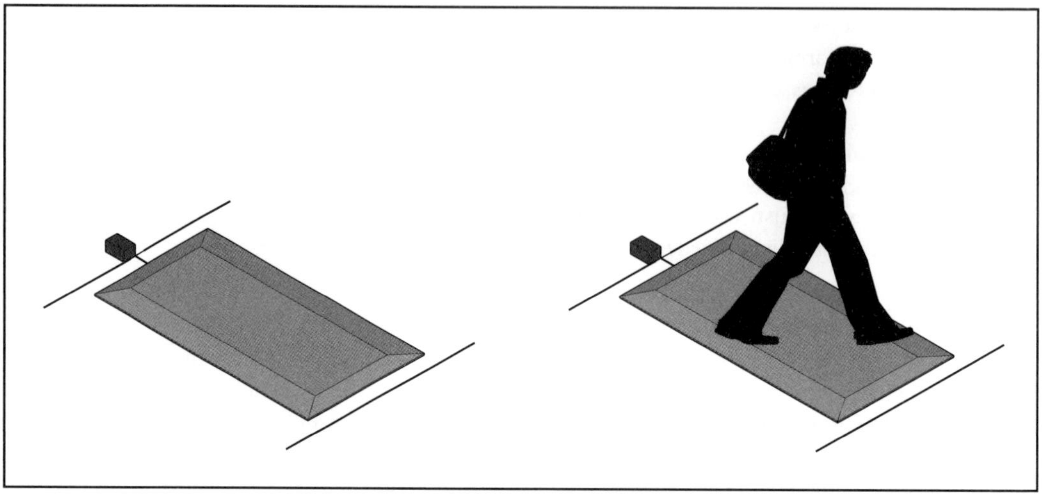

3.3.1.8 Mobile-device data

Data from mobile devices is a relatively new source of data for transportation and pedestrian modelling problems, which recently has attracted a lot of interest from the transport planning and modelling community. This data includes:

1. GPS trace data from mobile phone devices and GPS-enabled devices in vehicles and bicycles;
2. anonymised Call Detail Records (CDR) provided by mobile-phone-network operators; and
3. data from Bluetooth and Wi-Fi detectors [6].

Due to the ubiquity of these devices, this data represents a considerable proportion of the population and therefore can be effectively used in pedestrian modelling problems. Data from mobile devices includes a wide range of pedestrian-related data, such as pedestrian volume counts and trajectories as well as population mobility, activity and lifestyles [1].

The main concern in using mobile-device data is that it represents only a proportion of the population, and hence some adjustment and preprocessing of data is required to scale the data up so that it represents the whole population more accurately. The main reason for this limitation is that not all mobile devices have GPS enabled or CDR data is usually provided by one (or limited) operator(s) and does not include all the network operators. In addition, extracting pedestrian trajectories from mobile data requires additional computation tasks, such as noise filtering, trajectory compression, stay point detection, trajectory segmentation and map matching.[4]

3.3.1.9 Other detection technologies

There are other detection technologies available, such as thermal sensors (using thermal-imaging technology) and stereo camera sensors (using stereo cameras to understand depth in the field of view). These technologies are mainly used in the retail industry to count and monitor pedestrians or customers, and they have shown a good level accuracy in detecting pedestrian counts and direction.

3.4. Limitations of data collection

The previous sections have outlined the numerous types of data-collection techniques. The method used to meet the study objectives is dependent on the resources available and the data required. However, all data collection is limited in some manner, and it is important to understand these limitations in order to minimise the risks during the planning stage of any data-collection exercise. These weaknesses are discussed in brief in the following subsections.

3.4.1 Specification

Detailed planning of a data-collection exercise is critical to ensuring that the data collected meets the study requirements and does not contain gaps which could impact later stages of the study, such as modelling. It is important to carefully review the study methodology to ensure that it is robust and will deliver the required outputs.

3.4.2 Sampling method

Sampling methods for collecting pedestrian data have been discussed in earlier sections, and the limitations of each have been defined in Section 3.2 and 3.3.

3.4.3 Spatial

When considering the spatial scope of a survey, it is important to understand what is required from the study. For example, tracking people's movements over large distances is difficult and can be both time- and resource-intensive. The survey method or methods chosen should be selected with this in mind.

3.4.4 Temporal

When developing a survey scope, it is important to understand the existing behaviour which is being captured. For example, at a busy railway station it is important to cover not only the peak periods but also the shoulder peaks, so that any models are suitably populated to represent peak behaviour. Alternatively, infrastructure close to sports stadia may have different patterns of behaviour when events occur – for example, at a football match you would expect spectators to arrive from two hours prior to a match, but the stadium will clear within 15 minutes of the end of the game.

3.4.5 Span of data

Pedestrian behaviour is subject to variation on a daily, weekly and seasonal basis. It is important to consider the length of data collection in order to capture these variations. Dialogue with the relevant public body or infrastructure operator can provide valuable insights into these variations. Demand development should consider these variations, and adjustments should be considered to correct data that is deemed unrepresentative of typical conditions.

3.4.6 Sample size

Sampling error is the difference between an estimate derived from the sample and one that would have been obtained using the same data-collection method for the entire population. The size of the error can be estimated from the survey results. For pedestrian modelling, this typically relates to using postcard or questionnaire surveys to discover travel behaviour, such as route, mode choice and journey frequency.

3.4.7 Unplanned disruption

There are a number of potential risks associated with undertaking surveys that unfortunately cannot be controlled, but they can be minimised. The major risk with transport-related surveys is network-wide disruption. This can cause a significant change in user behaviour at the survey location and essentially invalidate any ongoing data collection. It can be minimised by undertaking surveys on a number of consecutive days. Similar risks include extreme weather events, fire evacuations and security issues.

3.4.8 Data evaluation and validation

Once the data collection is complete, the quality of data should be evaluated using appropriate statistical approaches. It should be noted that the collected data is always subject to inaccuracy due to either human error or a malfunction in the data-collection technology. Thus, it is essential to conduct data evaluation on the collected data and then properly handle missing data and inaccuracy if needed.

There are a series of statistical approaches to evaluate data, which range from simple methods, like histograms and scatter plots, to more sophisticated methods, like statistical tests and data-correlations analysis. Depending on the type of data and application, suitable data-evaluation approaches should be selected and be applied. Relevant data-evaluation approaches are explained in Chapter 5.

3.5. Further reading

In addition to the primary data collected using the techniques explained in the previous section, secondary data can be used as a complementary source. Secondary data for pedestrian modelling problems is collected from various sources available in the public domain, such as guidance books, periodical data published by public and government agencies and academic and experimental research data. This section briefly explains the types of data available in the public domain which are useful for pedestrian modelling. Guidance books usually provide common methodologies, including details of what needs to be taken into consideration in conducting pedestrian modelling.

The best known and most referenced guidance books in pedestrian modelling are:

- Weidmann [14],
- Predtechenskii and Milinskii [11],
- Fruin [3] and
- Heskestad [5].

Even though these guidelines are relatively old, they are still widely used as reference books in the literature. In addition, public agencies such as government or local authorities and railway station and airport operators usually publish guidelines setting out the steps necessary to conduct pedestrian modelling:

- Network Rail, UK publishes a guideline titled *Station Capacity Assessment Guidance* [12], which includes requirements for collecting specific series of data, such as station entry and exit flows, peak-minute flow at key locations, passenger counts at important locations (e.g. gatelines, stairs, escalators and junctions), passenger journey times and boarding and alighting figures;
- Transport for London (TfL), UK[5] provides online open data for buses, underground services, traffic flow, traffic congestion and cycling counts;
- Department for Transport, UK[6] releases a set of data on a yearly basis entitled WebTAG DataBook (with some additional minor versions throughout the year), which includes aggregated data, such as annual growth rates, number of trips for different trip purposes and modes.

These are only some examples of types of data that can be collected from government and public agencies, and they can vary by country.

3.6. Summary points

- To effectively undertake data collection, it is important to first understand the objective of the study, such as testing interventions related to improving the safety, security, convenience, continuity, comfort, system coherence or attractiveness of the pedestrian environment.
- There are many considerations that must be taken into account when preparing to undertake data collection for a pedestrian model, and a site visit is recommended early in the project to identify and resolve most of these.
- Most pedestrian models require a combination of several different data inputs, such as flow counts, cordon counts and turning movements.
- There are various techniques for pedestrian data collection, which can be broken into three main categories: manual, semi-automated and fully automated.
- Video-image processing is one of the most popular data-collection techniques as a large area can be covered and it can be used for both counting and tracking pedestrians.
- Continuous count data can be a valuable source of data and makes use of automated counting technology to obtain long-term pedestrian data, which is less susceptible to the variability in flows that can negatively affect short-duration counts.
- There are a number of other detection techniques, such as data from mobile devices, pressures sensors and laser scanners, that have applications in different fields.
- There are a number of potential risks associated with undertaking surveys that must be considered, such as network-wide disruption.
- The data collected will be subject to inaccuracy and therefore should be evaluated using appropriate statistical approaches.
- All data collection is limited in some manner, and it is important to understand these limitations in order to minimise the risks during the planning stage of any data-collection exercise.

NOTES

1. Quick response code.
2. Personal digital assistant.
3. The Doppler effect is the changes in the carrier frequency of the reflected radiation from a moving object with a constant velocity.
4. See [15] for details of the required computational tasks for extracting pedestrian trajectories.
5. Transport for London (TfL), UK is a local-government authority responsible for the transport system in Greater London, UK.
6. This is the government department responsible for the transport network in England and transport matters in Scotland, Wales and Northern Ireland that have not been devolved.

REFERENCES

[1] Andrienko G, Andrienko N, Chen W, Maciejewski R and Zhao Y (2017) Visual analytics of mobility and transportation: State of the art and further research directions. *IEEE Transactions on Intelligent Transportation Systems* **18(8)**: 2232–2249.

[2] Caramuta C, Collodel G, Giacomini C, Gruden C, Longo L *et al.* (2017) Survey of detection techniques, mathematical models and simulation software in pedestrian dynamics. *Transportation Research Procedia* **25**: 551–556.

[3] Fruin, JJ (1971) *Pedestrian Planning and Design*. Metropolitan Association of Urban Designers and Environmental Planners, New York, NY, USA.

[4] Federal Highway Administration (2016) *Traffic Monitoring Guide*. US Department of Transportation, Federal Highway Administration, Washington, DC, USA.

[5] Heskestad G (2002) Fire plumes, flame height, and air entrainment. In *SFPE Handbook of Fire Protection Engineering* (DiNenno PJ, Drysdale D, Beyler CL *et al.* (eds)), 3rd edn. National Fire Protection Association, Quincy, MA, USA, pp. 2-1–2-17.
[6] Marković N, Sekuła P, Laan ZV, Andrienko G and Andrienko N (2017) Applications of trajectory data in transportation: Literature review and Maryland case study. *IEEE Transactions on Intelligent Transportation Systems*. arXiv:1708.07193.
[7] Martin PT, Feng Y, Wang X *et al.* (2003) *Detector Technology Evaluation*. Department of Civil and Environmental Engineering, University of Utah Traffic Lab, Salt Lake City, UT, USA.
[8] Millonig A and Gartner G (2007) Monitoring pedestrian spatio-temporal behaviour. *BMI* **296**: 29–42.
[9] Minge E, Falero C, Lindsey G, Petesch M and Vorvick T (2017) *Bicycle and Pedestrian Data Collection Manual*. Minnesota Department of Transportation, St Paul, MN, USA.
[10] Nair N (2015) *Passenger Rail Usage: 2014–15: Quarter 4 Statistical Release*. Office of Rail and Road, London, UK.
[11] Predtechenskii VM and Milinskii AI (1978) *Planning for Foot Traffic Flow in Buildings*. National Bureau of Standards, US Department of Commerce, and the National Science Foundation, Washington, DC, USA.
[12] Network Rail (2011) *Station Capacity Assessment Guidance*. Network Rail, London, UK.
[13] Ryus P, Ferguson E, Laustsen KM, Proulx FR, Schneider RJ *et al.* (2015) NCHRP web document 205: Methods and technologies for pedestrian and bicycle volume data collection. Transportation Research Board, Washington, DC, USA.
[14] Weidmann U (1992) *Transporttechnik der Fussgänger: Transporttechnische Eigenschaften des Fussgängerverkehrs (Literaturauswertung)*. Institut für Verkehrsplanung, Transporttechnik, Strassen- und Eisenbahnbau, Zürich, Schweiz.
[15] Zheng Y (2015) Trajectory data mining: An overview. *ACM Transactions on Intelligent Systems and Technology (TIST)* **6(3)**: 29.

ice Publishing

Urban Space – AECOM
ISBN: 978-0-7277-6381-5
https://doi.org/10.1680/mpm.63815.051

Chapter 4
Macroscopic movement analysis

4.1. Overview of the chapter

The analysis of space has been grappled with for centuries, and much research has been undertaken to rationalise and simplify this field of study. Many theories and methodologies have been devised to make spatial analysis more approachable and useful, which has increased the scope of its use in the planning profession. Spatial analysis is a wide and varied discipline, which is discussed in Section 4.2.

The use of macroscopic modelling for high-level analysis and the value of static analysis as a macroscopic modelling technique are then discussed. Emphasis is placed on the practical benefits of undertaking static analysis early in a project design phase, highlighting the positive outcomes of such analysis prior to detailed modelling (e.g. micro-simulation models). The use of macroscopic simulation and static analysis in pedestrian modelling is discussed in Section 4.3.

Pedestrian routing is the last topic discussed in this chapter, in Section 4.4. Pedestrian routing has applications in both macroscopic and microscopic pedestrian modelling. In the context of pedestrian behaviour theory, pedestrian choice is categorised at three levels: strategic, tactical and operational. Pedestrians make routing decisions based on different criteria, such as the shortest or fastest route, or the path of minimum perceived impedance. These criteria can be used to model pedestrian routing. Approaches based on statistics, such as utility and logit models, to select the most probable routes for pedestrians among all available routes are briefly explained.

In practice, a combination of macroscopic and microscopic pedestrian modelling is applied for studies of pedestrian movement. Macroscopic modelling allows larger study areas to be modelled as they do not require a great level of detail. In contrast, microscopic models require a considerable level of detail, such as interactions between people and the environment, and obstacles which makes them computationally complex and time-consuming to build. Thus, a combination of macroscopic and microscopic modelling allows the modeller to first gain a better overview of the problem using macroscopic models and then focus on areas where more detailed analyses are required using microscopic models.

These chapters aim to provide a general overview and practical approaches for microscopic and macroscopic pedestrian modelling respectively, with the detailed theory behind those approaches provided in the relevant sections.

4.2. Spatial analysis

Much can be gained from the analysis of space in the context of pedestrian modelling, and applying such analysis evaluates the way in which space can, and is expected to, be used. The outcome of spatial analysis can identify, for example, the most appropriate location for a railway station, or the potential routes within an urban area most likely to be used for specific journeys.

In architecture and urban planning, space is a central theoretical discipline, and the challenge is to find a way to study it. The problem of space itself is not, however, confined to architecture. In anthropology, for example, it exists as an empirical problem. The first-hand study of a large number of societies has left the anthropologist with a substantial body of evidence about architectural forms and spatial patterns, which ought to be of considerable relevance to the development of a theory of space. But the matter is far from simple. The body of evidence displays a very puzzling distribution of similarities and differences.

One effective approach to tackle this complexity is to use the mathematics of spatial network analysis to provide in-depth knowledge about the relationship between objects and different spaces. Spatial network analysis is, in fact, an old discipline, dating back at least to Euler, the Swiss mathematician, who in 1736 solved the problem of the Seven Bridges of Königsberg [7]. Euler rationalised the city of Königsberg's (today Kaliningrad, Russia) seven bridges connecting four land masses by representing them as a series of nodes and links. A single node was assigned to each land mass, with connecting links representing the 'bridges' between each node. This representation of the connections between points became known as a graph, which ultimately led to the modern branch of mathematics known as graph theory, as well as the development of both network codification and network generalisation [6].

Figure 4.1 shows a historical view of the city of Königsberg overlaid with four circular nodes, representing the city's land masses, connected by seven links, representing the bridges connecting the nodes. On the right a simplified graph of the same network of nodes and links is provided. In spatial analysis, relevant techniques in graph theory are used to evaluate the study area based on various metrics. In this section, some of the most common measurements or metrics proposed for spatial analysis which can be used for urban studies and consequently for pedestrian modelling have been explained.

4.2.1 Centrality

Centrality is a measure that defines the significance of a point or link within a wider network. The concept of significance varies based on context, the type of network and what is being analysed. There are many indices of centrality, which can be used to define what characterises this significance based on what is being

Figure 4.1 The Seven Bridges of Königsberg – graph theory

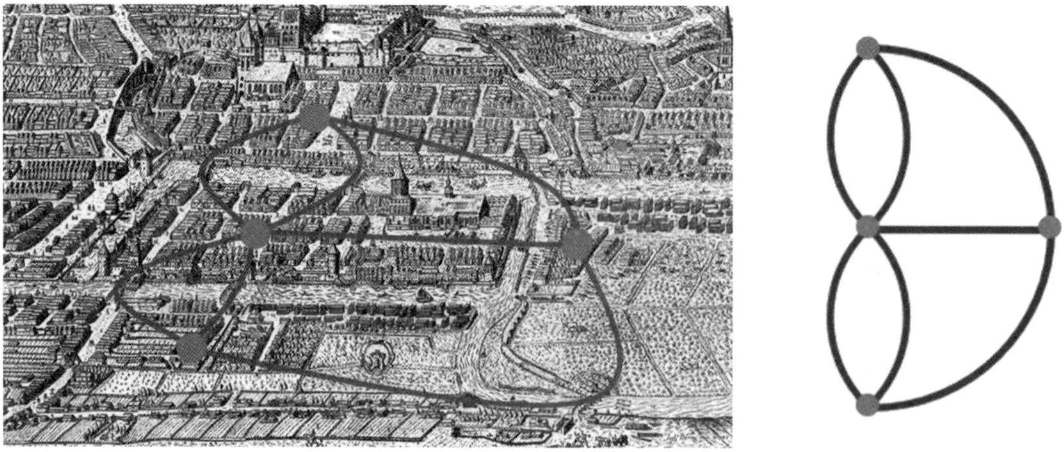

Table 4.1 Summary of centrality indices

Index	Explanation
Connectivity	A count of physical connections between a point and other points in the network
Closeness	A measure of distance between a point and other points in the network
Betweenness	A count of the number of times a point falls on the shortest path between other points in the network
Straightness	A measure of deviation from the imaginary straight line between two points in the network
Accessibility	A measure of the potential for connections to or from a point in the network considering measures of reach and remoteness of that point

analysed [5]. Table 4.1 summarises the most common indices of centrality: connectivity, closeness, betweenness, straightness and accessibility.

The measures of centrality are explained in greater detail in subsequent sections, using an example of a section of the street network of downtown New Orleans, USA, to illustrate the differences between some of the metrics in measuring the centrality of the network.

4.2.2 Connectivity

The most basic measure of centrality is connectivity, which counts the number of connections that a point in a network has to all other points in the network (Figure 4.2). In a network made up of axial (straight) lines, the connectivity index is based on the number of lines that directly intersect each line in the network [15]. The more intersections that exist, the greater the connectivity of the network, since a greater number of points are directly connected to other points. Connectivity can be defined for any kind of network.

For instance, in the network of an underground metro system, an interchange station located on two lines has a greater connectivity than an equivalent station located on a single line only, since the interchange station is directly linked to a greater number of other stations. Conversely, a terminus station at the end of a single line has lower connectivity, since it is connected only to one other station in the network. Such information may be useful in demand forecasts for pedestrian modelling in identifying stations that are likely to be busier than others due to their higher connectivity.

4.2.3 Closeness

Closeness as a measure of centrality considers the inverse of distance between a point (or link) and all other points (or links) in the network along the shortest paths [18]. The lower the total distance from a point to all others in the network, the higher its closeness. Deriving the closeness of a point establishes how close it is or how quickly and easily it can be reached from the rest of the network. A simple linear distance is used in the measure, resulting in an average closeness between all destination points. The closeness measure is defined as:

$$C_i = \frac{N - 1}{\sum_{j = 1; j \neq i}^{N} d_{ij}} \tag{4.1}$$

Where:

- N is the number of points in the network, and
- d_{ij} is the geodesic shortest path distance between points i and j [17].

Figure 4.2 Connectivity of a pedestrian network

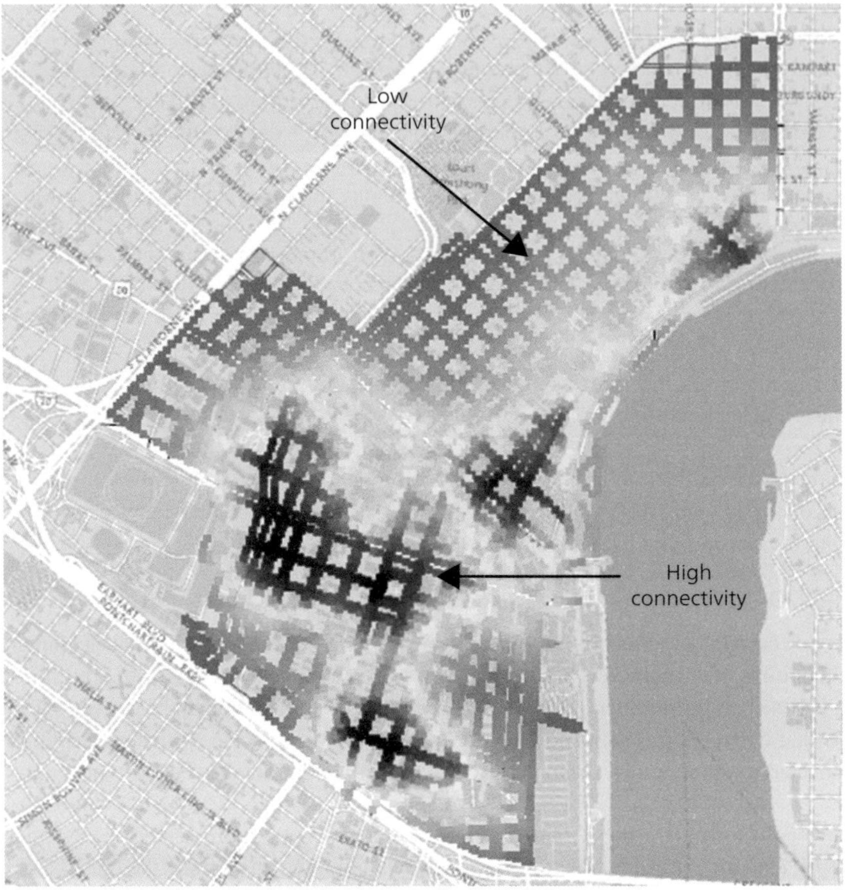

In urban design and planning, the closeness metric can be used for different purposes. For instance, in a study area the closeness can be used to identify the best location for a railway station. Using the closeness metric, a location which has better centrality – that is, a location which is closer to all the points in the network – can be selected as the location of the station.

Figure 4.3 shows an example in which closeness was used to measure centrality in order to analyse the pedestrian experience in downtown New Orleans, USA. The figure identifies and visualises areas that have the shortest distances to all the places in the study area. The gradient shades represent the results of the centrality metric calculation. The areas coloured darkest grey indicate shorter distances, while the areas in lighter grey and outside the white highlight indicate longer distances. Areas of high centrality, which are expected to experience high levels of footfall within the network, are circled.

4.2.4 Betweenness

Betweenness recognises the role of a point (or link) as a connector between other groups of points (or links) by considering the number of times that the point falls on the shortest path between other pairs of points. Quantifying the number of times a point acts as a connector along the shortest path

Figure 4.3 Example of centrality measured by closeness

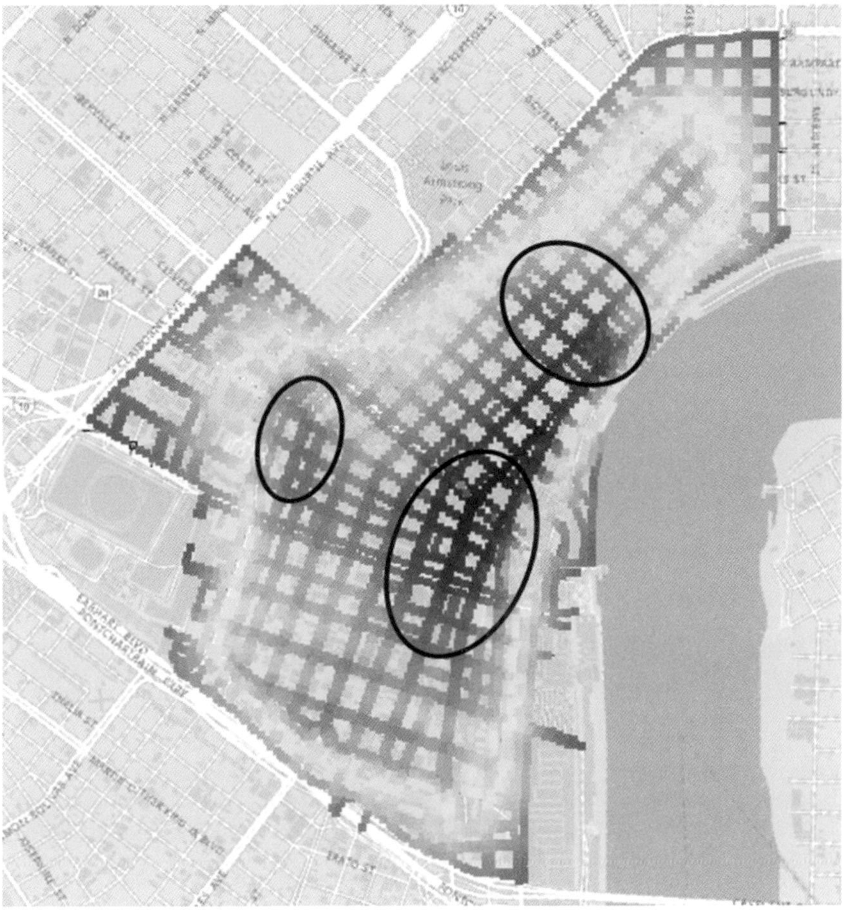

between other pairs of points allows its level of betweenness to be captured [5]. Although numerous earlier studies proposed a similar concept of centrality based on betweenness [2, 4 and 20], it was Freeman [9] who defined a general procedure for measuring betweenness as the fraction of shortest paths between pairs of other points in the network that pass by a particular location. In case more than one shortest path is found, each is given equal weight so that the weights sum to unity. The betweenness measure is defined as:

$$B_i = \frac{1}{(N-1)(N-2)} \sum_{j=1;k=1;j\neq k\neq i}^{N} \frac{n_{jk}(i)}{n_{jk}} \tag{4.2}$$

Where:

- N is the number of points in the network,
- n_{jk} is the number of shortest paths between points j and k and
- $n_{jk}(i)$ is the number of these shortest paths that pass i [17].

Figure 4.4 Example of centrality measured by betweenness

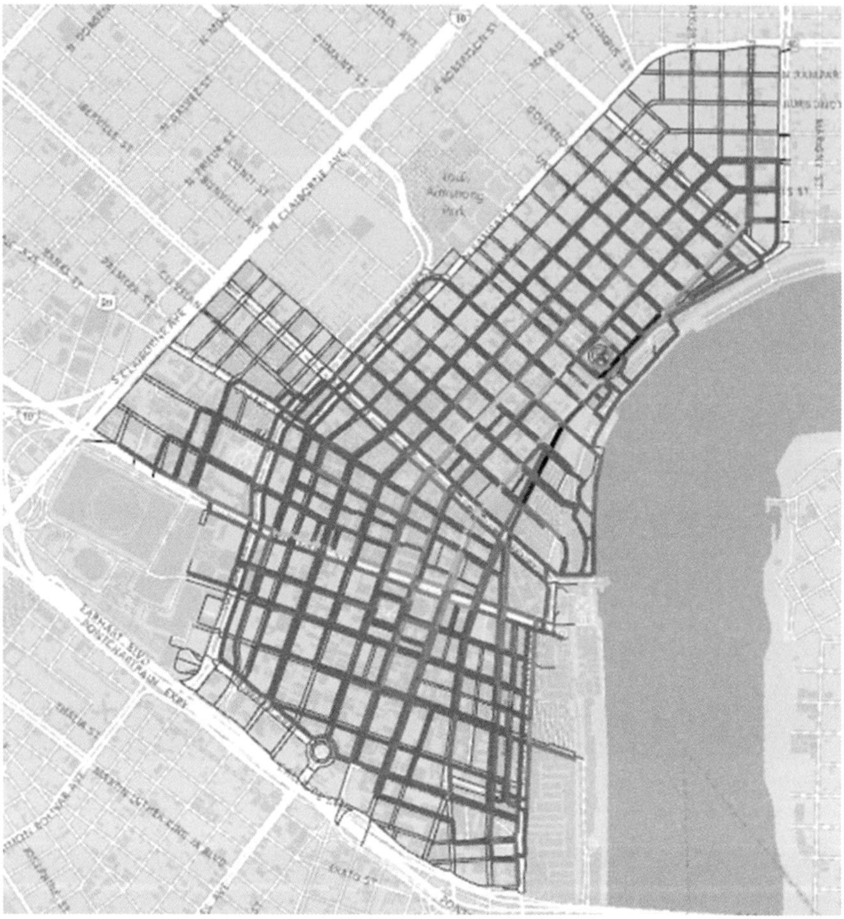

Similar to closeness, betweenness can be used to identify the best location for a railway station. However, the betweenness measure does not find a location for the railway station which is closer to all the locations in the network but one that happens to be on the routes of many pedestrians (e.g. commuters, tourists).

Figure 4.4 depicts an example of betweenness measured between defined attractions in the downtown New Orleans street network. In this spatial distribution, the city's main tourist and nightlife area is to the north, and the convention centre is to the south. Routes shown in black fall on the shortest path between the attractions and therefore have a greater betweenness. These streets are expected to be the busiest with people moving between the attractions, while the other streets are expected to be less busy as they do not fall on the shortest path between the attractions.

4.2.5 Straightness

The straightness metric is based on the theory that the efficiency of movement between two points (or links) increases with decreased deviation of the shortest paths from the Euclidian straight line between those points (or links). The smaller the deviation from this virtual line connecting them, the greater the straightness of the route. The straightness measure takes into account that the geometric constraints of the network prevent straight-line connections between points. The straightness measure is defined as:

$$S_i = \frac{1}{N-1} \sum_{j=1;j \neq i}^{N} \frac{d_{ij}^E}{d_{ij}} \tag{4.3}$$

Where:

- N is the number of points in the network,
- d_{ij}^E is the straight-line Euclidian distance between i and j and
- d_{ij} is the shortest possible route between the same nodes i and j [17].

Figure 4.5 is a measure of straightness in downtown New Orleans. Areas in dark grey indicate areas of the network on the greatest number of straightest routes between all locations on the network, while the areas of the network falling on the least number of straight routes between all locations are outside the white highlight. Since the metric is based on least direction change rather than distance between all locations on the network, the areas of higher connectivity spread further west along the straight parallel streets than in the closeness measure of connectivity. The area of highest straightness connectivity is outlined with a black dashed line.

4.2.6 Accessibility

The concept of accessibility considers the potential for connections from one point to other points in a network. Several measures of accessibility have been introduced in the literature. Below, two of the most common accessibility metrics – *reach* and *gravity index* – are outlined.

The reach measure captures the number of surrounding points that can be reached from a given location within a defined search radius on a network [19]. Reach centrality describes the number of points in a network that are reachable along a shortest-path distance and is defined as follows:

$$Reach_{i,r} = \sum_{j=1}^{n} O_r \tag{4.4}$$

Where:

- $Reach_{i,r}$ is the reach at location i within a distance threshold r and
- O_r is an opportunity that can be reached within the distance threshold [19].

The gravity index [12] builds upon the reach metric, which counts the number of points located within a search area, by further considering the travel costs associated with getting to those points. The gravity measure therefore measures both the weight of the destination point as well as the spatial impedance of travel to that point [19] and is defined as follows:

$$A_i = \sum_{j=1}^{n} \left(\frac{W_j}{d_{ij}^{\beta}} \right) \tag{4.5}$$

Where:

- A_i is accessibility at point i,
- W_j is the attractiveness of location j,

Figure 4.5 Example of centrality measured by straightness

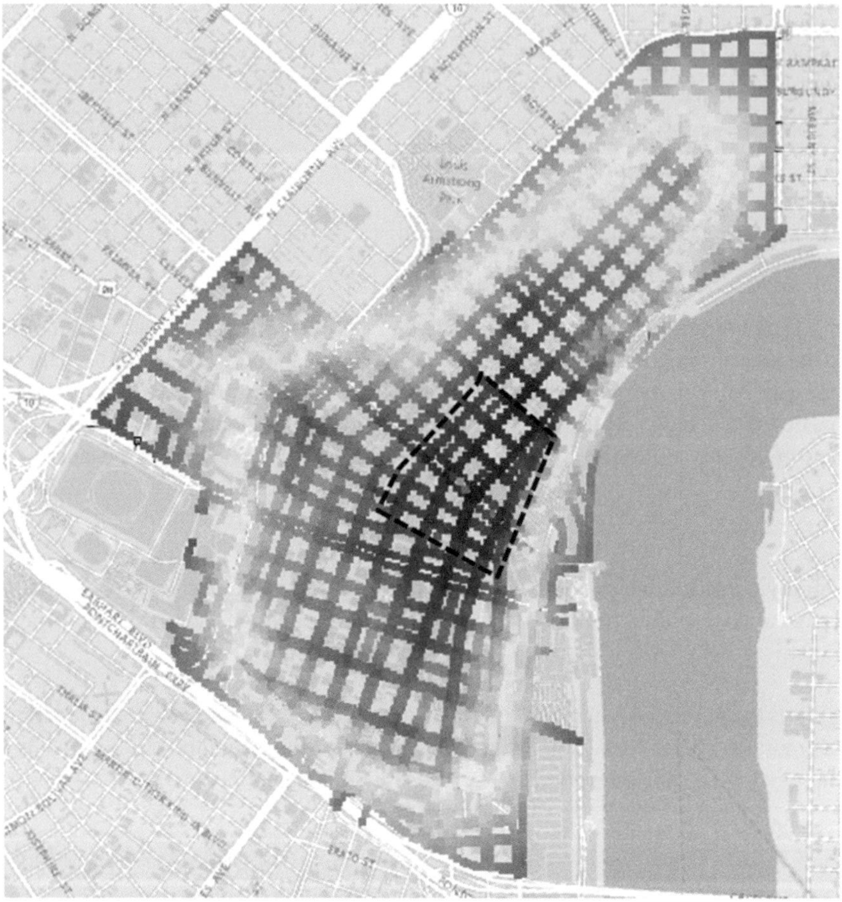

- d_{ij} is the distance between location i and j,
- β is the exponent for distance decay and
- n is the number of points in the study area [19].

Figures 4.6 and 4.7 show the measure of accessibility used to measure the centrality of the street network in downtown New Orleans. A search radius of 500m is defined in both cases. Overall network accessibility is evaluated in Figure 4.6, with areas of greater accessibility shown in black, indicating streets with shorter distances to all other locations in the network. Areas that are less accessible due to greater distances to the rest of the network are outside the white-highlighted area. Figure 4.7 depicts accessibility to public transport stops (bus and streetcar).

4.2.7 Spatial analysis theories and methodologies

There is a significant body of theoretical research on the analysis of spatial networks, which has led to the development of various methodologies and tools for evaluating the structure of cities and other networks. One of these, Space Syntax, is a set of techniques used to simulate changes and analyse their effects in urban environments.

Figure 4.6 Example of centrality measured by accessibility – street network accessibility

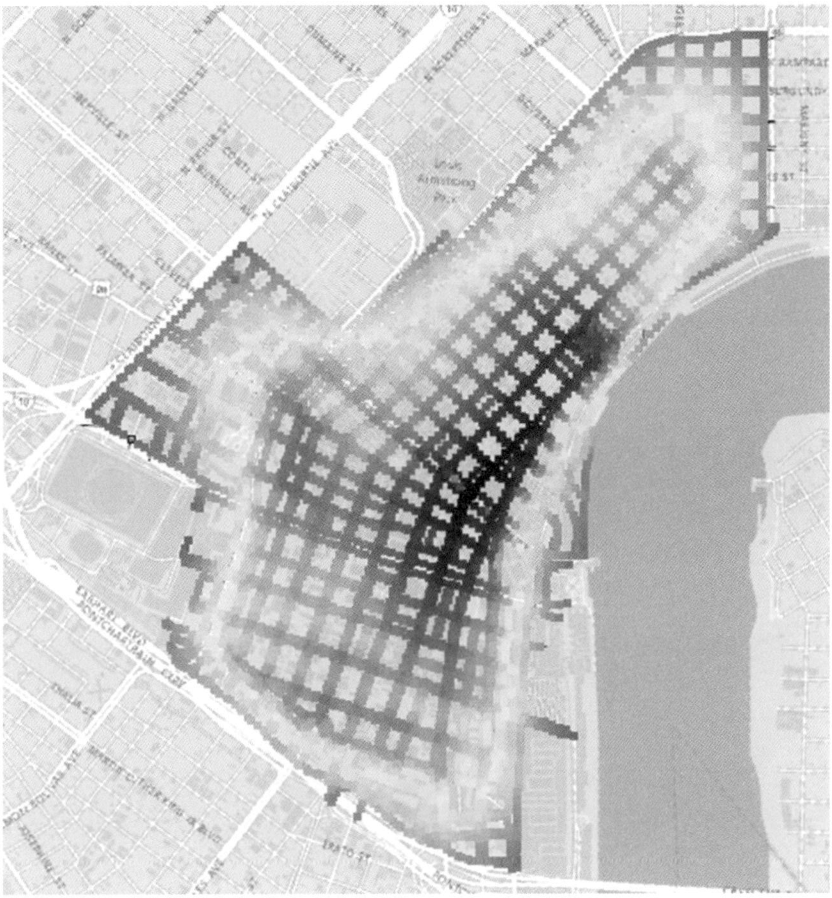

Radius: 500m

Space Syntax analysis considers the relative accessibility of places in a city by measuring various features of the street network on an axial map. This shows the degree of visual accessibility from the perspective of someone moving through the urban environment [15]. Space Syntax also evaluates the impact of spatial configuration on behaviour, communication and interaction. It highlights the importance of human perception on the development of movement patterns in the city, the influence of the geometric and topological structure, the spatial layout of the city and the functional land-use of the urban structure on the development of local movement patterns [1].

Space Syntax theory defines the concept of urban centrality, with the spatial agglomeration of activities, services and employment forming the urban centre. It identifies two types of network, foreground and background, within a city. The foreground contains the longer lines connecting the various parts of the city, and the background network of shorter lines connects the residential areas to each other and the foreground network [1].

Figure 4.7 Example of centrality measured by accessibility – public transport stop accessibility

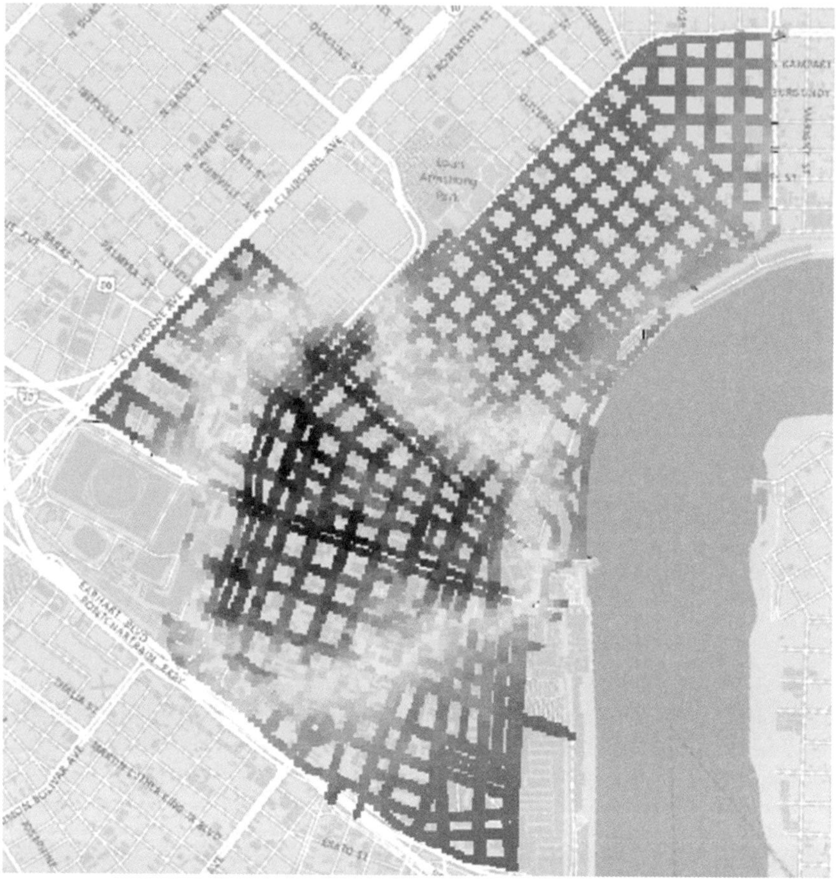

Radius: 500m

Theoretical research on the analysis of spatial networks has led to the development of other tools that allow for easy visualisation of analytical results. sDNA (spatial design network analysis) is a GIS (geographic information system) and CAD (computer-aided design) tool and methodology for analysis of spatial networks that adapts the techniques of Space Syntax, such as closeness and betweenness, to current cartography standards.

One tool for the spatial analysis of networks, the Place Syntax Tool, is a plug-in application that addresses the problems of analysing and predicting urban pedestrian movement that face conventional accessibility and Space Syntax research. Place Syntax aims to combine the Space Syntax description of the cognitive environment – that is, topological axial line distance – with conventional descriptions of attraction in a combined accessibility analysis model [21].

In order to better understand walking issues in urban environments, AECOM has developed the Urban Space Smart City Toolkit. It uses the visual-programming environment Grasshopper 3D, a plug-in for CAD application Rhino, to assess distance and accessibility between points in spatial networks and allows users to merge spatial network analysis with land-use data and origin–destination matrices to produce predictions of pedestrian flows between locations. Outputs from this tool were used in Figures 4.2 to 4.7.

4.3. Macroscopic simulation

Macroscopic approaches deal with pedestrian dynamics by assuming the crowd is the smallest representable entity: in fact, they use the aggregate representation of the crowd through speed, flow and density relationships to describe pedestrian movement [24]. This kind of modelling depicts pedestrian behaviour as a continuous fluid characterised by average quantities, like flow, average speed and area module, without considering the behaviour of the individuals [13 and 24].

In macroscopic models, individuals have no autonomy to change their kinematic state or to control their interactions. Macroscopic models aim to control the mean velocity of the group [16]. Whereas microscopic flow theory considers characteristics such as age, culture, gender, temperature, travel purpose and type of infrastructure, macroscopic models are concerned only with speed, flow and density [8].

Macroscopic models focus on the aggregate representation of pedestrian movements in a crowd through speed, flow and density relationships [10, 22 and 8]. The macroscopic behaviour reveals that the pedestrian average speed is reduced as the density increases. The flow equation, derived from analogy to fluid flow, is given by:

$$Flow\ (Q) = Average\ Speed\ (V) \times Average\ Density\ (K)$$

For instance, using the formula above, the flow of people through a passageway at a railway station can be determined if their average speed and density is known. More commonly, it is the average density of people in the space that is sought, which conversely can be obtained if the average speed and flow of the crowd is known.

4.3.1 Static analysis

In practice, undertaking a static analysis of space as a form of macroscopic modelling can add great value to a project without great expense of time or cost. Static analysis is often undertaken early in a design or master planning process to give a high-level indication of capacity requirements or any issues likely to be encountered. This allows the design team to develop the design process in an informed way and within a reasonable margin of error, long before any detailed modelling is undertaken to evaluate the spatial adequacy of the design.

It is possible to undertake static analysis early in the design process because it requires fewer data inputs than micro-simulation modelling,[1] which would not be possible without a comprehensive data set and operational assumptions. The process of static analysis also provides an opportunity to identify, request and compile all the data that will subsequently be required for micro-simulation modelling. In a larger multidisciplinary project, this may involve a vast data set from many different sources. In addition to collating data, the static analysis allows gaps in the data sets to be identified and assumptions to be defined and agreed with relevant parties as required. At this point, it is also possible to identify and bring to light any disparities between available data, assumptions and any industry codes or standards, allowing these and other issues to be resolved earlier in the project than would otherwise be possible.

An example of undertaking static analysis in a practical context is in the planning and design of a railway station. Many aspects of capacity can be calculated statically using the forecast demand. The widths of entrances, walkways, platforms and stairs can be calculated as well as the number of lifts, ticket machines, ticket barriers and escalators. For example, the required number of ticker barriers at a railway station is given by:

$$N_b = \left(round\ up \left\{ \frac{f_{entry}}{c_b \times 5} \right\} + \left\{ \frac{f_{exit}}{c_b \times 5} \right\} + X \right) \tag{4.6}$$

Where:

- N_b is the number of ticket barriers required,
- f_{entry} is the peak 5-minute entry flow defined as the maximum number of passengers entering the station to board train service(s) within the peak 5-minute period,
- f_{exit} is the flow of exiting passengers defined as the maximum number of passengers exiting the station from train service(s) in a 5-minute period,
- c_b is the throughput capacity of one ticket barrier per minute and
- X is 1 (if the total number of barriers required without X is equal to or less than 10) or 2 (if the total number of barriers required without X is greater than 10).

The outcome of the static analysis can subsequently be tested through a more detailed micro-simulation model to optimise the design.

4.4. Routing

Pedestrian routing is one of the most important elements of pedestrian modelling in both micro and macro modelling. Almost all practical pedestrian modelling problems include pedestrian routing. If routing is not properly taken into consideration, the accuracy and validity of models can be significantly affected.

The dynamic nature of pedestrian routing and high variations of pedestrian route choices make the accurate modelling of pedestrian routing a very challenging task. The variability in pedestrian route choices is a result of pedestrians choosing or changing their route based on a large range of internal (i.e. personal) and external factors, which vary from one person to another. For instance, a pedestrian might suddenly change their route to buy a newspaper (internal factor) or to use staircases instead of lifts due to the large number of people waiting for the lifts (external factor). These simple examples show the variation of pedestrian routing and the complexity required in some models to accurately represent this behaviour.

4.4.1 Pedestrian choice levels

In pedestrian behaviour theory, pedestrian choices are categorised into *strategic*, *tactical* and *operational* levels [13]. At the strategic level are long-term decisions, such as the destination and departure time of journeys. The strategic level is mainly used in the context of pedestrian demand modelling and the estimation of origin–destination matrices. The tactical level comprises medium-term decision-making for the journey, including pedestrian route choices. The last level is the operational level, in which pedestrians make short-term decisions throughout their journey based on the current situation of their surrounding environment. Examples include overtaking slower pedestrians, changing direction due to crowding, waiting and the interaction of pedestrians with other pedestrians and obstacles [23]. Pedestrian behaviour

Figure 4.8 Pedestrian choice levels

during the operational level was discussed in Chapter 2. Figure 4.8 depicts the hierarchy of pedestrian choice levels.

The building blocks of pedestrian route choices are a series of activities that are performed in a defined schedule. Some of these activities are compulsory while others are discretionary [8 and 14]. Based on journey time and desires, the activity schedule can be updated throughout the journey, new activities can be added and others can be updated or even removed. Some activities can be performed in multiple locations, such as getting coffee. Pedestrians can decide when or where to perform the activities, either in advance or spontaneously during the journey. To reach the location of each activity, pedestrians may have a number of alternative routes and make a routing decision based on different criteria, such as the shortest or least congested route. Figure 4.9 depicts different components of pedestrian tactical-choice levels.

The following example is provided to help readers better comprehend activity scheduling and route choice at the tactical-choice level. It is from a pedestrian modelling project to study the gateline capacity of a railway station and describes the activity schedule on a typical commuter journey for Sophia, the subject of our example.

Sophia travels to work through a combination of walking and using public transport. The activity schedule at the start of her journey is as follows:

- walk to the railway station,
- buy a cup of coffee on the way to the train,
- buy a ticket at the station,
- walk to the platform and
- board the train at the designated platform.

Figure 4.9 Components of the pedestrian tactical-choice level

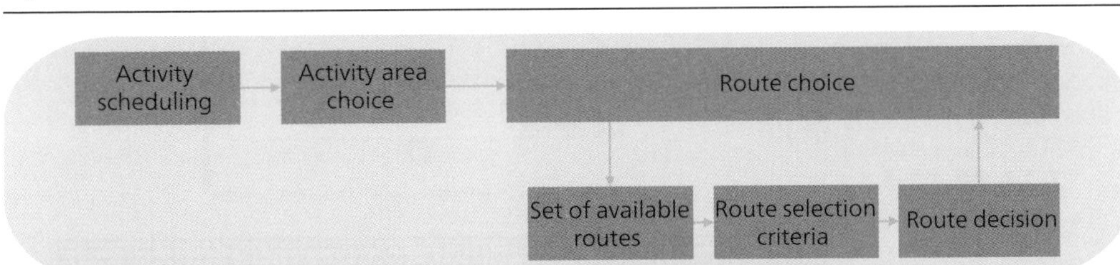

Figure 4.10 Example of tactical pedestrian choices

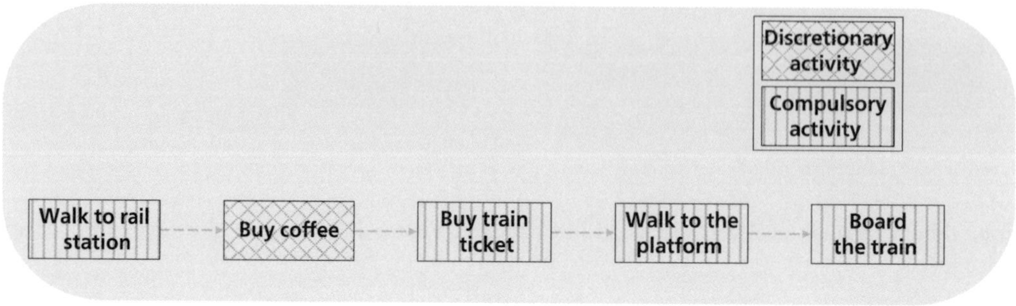

In this scenario, buying coffee is a discretionary activity because if Sophia is in a rush then buying coffee can be skipped, whereas all the other activities are compulsory (Figure 4.10). Furthermore, buying coffee can be performed in different locations (inside or outside the station) and, similarly, tickets can be bought either at the counter or self-service ticket machines. However, some of the activities can be only performed in one place – for instance, boarding the train can only be performed at the designated platform.

The next choice for Sophia is the route choice to reach each activity. Figure 4.11 shows a simple example of routing in a railway station. This shows that she can enter the station through either of the two entrances (shown on the left-hand side) by following a different route (route 1 or route 2) to reach the destination (platform 1). This example shows only two routes; however, other possible routes may exist.

Figure 4.11 Example of pedestrian routing in a railway station

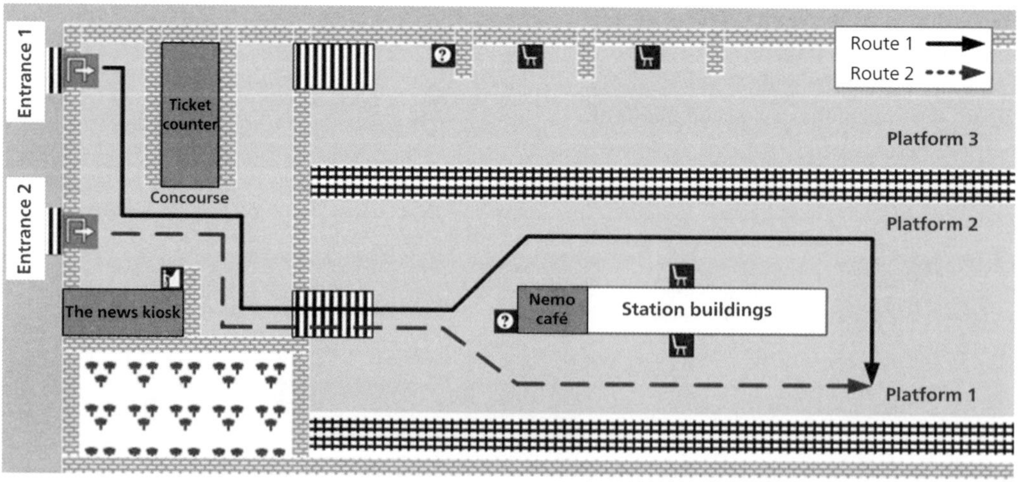

4.4.2 Factors affecting route choice

In order to understand pedestrian route-choice decisions it is important to first have a definition of pedestrian routes. A route can be defined as a series of links connecting a number of nodes or locations along a pedestrian's journey from origin to destination to perform activities as set out in the activity schedule [3]. This definition can easily be represented in a discrete network with nodes and links. The same definition is also relevant to continuous spaces, as the space can be limited to a number of pedestrian routes between different locations. These routes, however, might overlap by sharing some links with each other, and pedestrians can potentially impede each other on their routes. As explained in the previous section, pedestrian route choice in pedestrian modelling is considered at the tactical level, and pedestrian decisions at this level consist of medium-term decisions depending on the activity schedule and prevailing conditions. A number of criteria will influence pedestrian route choice, such as the number of turns, distance existence of obstacles and time limitations.

Pedestrian route choice is very important in pedestrian modelling because different routing decisions will often result in different outcomes. Pedestrian route choice in models should therefore be considered carefully due to the large number of route choices and existing criteria influencing them. Route choices should be identified as part of the data-collection process to ensure the data gathered is fit for purpose. Once all relevant routes are identified and selected, the pedestrian route choice can be modelled based on criteria influencing pedestrian route choice.

Table 4.2 shows the list of common pedestrian route-choice criteria. Taking into account all of these criteria within a pedestrian model is not feasible, and therefore in practice the most common or relevant criteria are used by pedestrians to choose routes.

Route-choice factors can be categorised into external and internal (or personal) factors. Examples of external factors are the built environment, level of congestion, weather conditions; and internal factors can be physical capabilities, personal preferences and familiarity with the environment. For instance, the route choice for a commuter is often different from that of a tourist. A commuter is familiar with routes and may use shortcuts, whereas a tourist tends to follow the map or signage for routing. In addition, the objective of a commuter is to choose the fastest or most convenient routes, whereas the tourist may prefer routes that include more attractions or have better scenery.

Table 4.2 Pedestrian route-choice criteria [11]

Longest leg first
Shortest leg first
Fewest turns
Fewest pedestrian lights
Fewest obstacles or obstructions
Variety-seeking behaviour
Minimising negative externalities (e.g. exposure to pollution)
Avoiding crowded areas
Avoiding detours
Maximising aesthetics, better scenery, tourist-attraction locations nearby
Fastest route

4.4.3 Wayfinding, signage and information

Another factor that affects route choice is wayfinding. Wayfinding becomes more important in confined places, such as buildings, airport terminals and interchange stations. Familiar users, such as commuters, will not use signage, as their route is based on knowledge of the surroundings, and they will follow a learnt or preferred route compared to those unfamiliar with their surroundings [23].

In some infrastructure environments, where management of pedestrians is key to maintaining stable operations, wayfinding is designed to reduce congestion by providing specific routes intended to alleviate cross-flows or pinch-points, even if these routes are not the shortest option. The proportion of pedestrians who use wayfinding will depend on a number of factors. For instance, experiments have shown that wayfinding is more heavily used in airports than in railway stations due to complexity and familiarity with environment. This should be taken into account in the pedestrian modelling process. In Part II of the book, there are examples of how wayfinding has been used in pedestrian modelling for different applications.

4.4.4 Modelling route choice

In the literature, various techniques are proposed to model pedestrian route choice. These range from more sophisticated approaches, such as route-choice logit-based models and utility-based models, to more straightforward approaches, such as routing based on only the shortest route or the fastest route. The former has attracted more interest in academic research by focusing on the theory behind the model, whereas the latter approaches are used more in practice due to simplicity and effectiveness.

The utility-based models define formulas to calculate the utility or disutility of each route, and pedestrians tend to choose routes with higher utility. The disutility, U_r, for route r is defined as follows:

$$U_r = \alpha L_r + \varepsilon_r \tag{4.7}$$

Where:

- L_r is the attribute impacting pedestrian routing (decision criteria), which is usually the length of the route,
- α is the weight of the attribute and
- ε_r is a random number indicating the error term.

Using this formula, the route with higher utility, or lower disutility, has a higher probability of being selected by pedestrians.

Different approaches using statistical methods have also been proposed to estimate the probability of choosing one route against one or multiple routes. These statistical models are mainly based on a logit model, such as a nested logit, multinomial logit, C-logits or binomial logit [8]. This section does not provide details of these methods but recommends that readers refer to the following references: [3 and 24].

In Part II of the book, the modelling of pedestrian routes in different environments or applications is explained in more detail by providing practical examples for each application.

4.5. Summary points

- Macroscopic modelling allows larger study areas to be modelled as they do not require a great level of detail.
- A significant body of theoretical research has been conducted on the analysis of spatial networks, leading to the development of various methodologies and tools for evaluating the structure of cities and other networks.
- Centrality is a measure that defines the significance of a point or link within a wider network; the most common indices of centrality are connectivity, closeness, betweenness, straightness and accessibility.
- In macroscopic models, individuals have no autonomy to change their kinematic state or to control their interactions, as they deal with pedestrian dynamics by assuming the crowd is the smallest representable entity using an aggregate representation of the crowd through speed, flow and density relationships.
- Static analysis is often undertaken early in a design or master planning process to give a high-level indication of capacity requirements or any issues likely to be encountered, which allows the design team to further develop the design process in an informed way and within a reasonable margin of error, long before any detailed modelling is undertaken to evaluate the spatial adequacy of the design.
- Pedestrian routing is one of the most important elements of pedestrian modelling in both micro and macro modelling, but the dynamic nature of pedestrian routing and high variations of pedestrian route choices make the accurate modelling of pedestrian routing a very challenging task.
- In pedestrian behaviour theory, pedestrian route choices are categorised into *strategic*, *tactical* and *operational* levels.
- Various techniques have been proposed to model pedestrian route choice, which range from more sophisticated approaches, such as logit-based and utility-based models, to more straightforward approaches, such as routing based on only the shortest route or the fastest route.
- Wayfinding has had a significant influence on the routing of pedestrians and becomes more important in confined environments, such as buildings and interchange stations, as unlike familiar users, such as commuters, those unfamiliar with their surroundings are unable to use prior knowledge to determine their route and follow a learnt or preferred route.

NOTE

1. Micro-simulation modelling is explained in Chapter 5.

REFERENCES

[1] Al-Shaheen QE (2012) *The Role of Integrated and Diverse Retailing in the Production of Social Vitality: Reshaping Retailing in Kuwait City.* PhD thesis, Heriot-Watt University, Edinburgh, UK.

[2] Bavelas A (1948) A mathematical model for group structures. *Applied Anthropology* **7(3)**: 16–30.

[3] Bovy PH and Stern E (1990) *Route Choice: Wayfinding in Transport Networks*. Kluwer Academic Publishers, Dordrecht, the Netherlands.

[4] Cohn BS and Marriott M (1958) Networks and centres of integration in Indian civilization. *Journal of Social Research* **1**: 1–9.

[5] Colchester J (2016) *Network Theory: An Overview.* Publisher unknown.

[6] Cooper C and Chiaradia A (2015) sDNA: How and why we reinvented spatial network analysis for health, economics and active modes of transport. *GIS Research UK (GISRUK) 2015 Proceedings, Leeds* (Malleson N *et al.* (eds)). University of Leeds, Leeds, UK.

[7] Coupy E (1851) Solution d'un problème appartenant à la géométrie de situation, par Euler. *Nouvelles annales de mathématiques, Serie 1* **10**: 106–119.

[8] Daamen W (2004) *Modelling Passenger Flows in Public Transport Facilities*. PhD thesis, Delft University, Delft, the Netherlands.

[9] Freeman LC (1977) A set of measures of centrality based on betweenness. *Sociometry* **40**: 35–41.

[10] Fruin JJ (1971) Designing for pedestrians: A level-of-service concept. *Highway Research Record: 50th Annual Meeting of the Highway Research Board*. Washington, DC, USA

[11] Golledge RG, Jacobson RD, Kitchin R and Blades M (2000) Cognitive maps, spatial abilities, and human wayfinding. *Geographical Review of Japan, Series B*. **73(2)**: 93–104.

[12] Hansen WG (1959) How accessibility shapes land use. *Journal of the American Planning Association* **25(2)**: 73–76.

[13] Hoogendoorn SP and Bovy PHL (2002) Normative pedestrian behaviour theory and modelling. In *Transportation and Traffic Theory in the 21st Century: Proceedings of the 15th International Symposium on Transportation and Traffic Theory, Adelaide, Australia*. Elsevier, Amsterdam, the Netherlands, pp. 219–245.

[14] Hoogendoorn SP and Bovy PHL (2004) Pedestrian route-choice and activity scheduling theory and models. *Transportation Research Part B: Methodological* **38(2)**: 169–190.

[15] Lerman Y, Rofè Y and Omer I (2014) Using Space Syntax to model pedestrian movement in urban transportation planning. *Geographical Analysis* **46(4)**: 392–410.

[16] Martinez-Gil F, Lozano M, García-Fernández I and Fernández F (2017) Modeling, evaluation and scale on artificial pedestrians: A literature review. *ACM Computing Surveys* **50(5)**: 72.

[17] Porta S, Strano E, Iacoviello V *et al.* (2009) Street centrality and densities of retail and services in Bologna, Italy. *Environment and Planning B: Planning and Design* **36(3)**: 450–465.

[18] Sabidussi G (1966) The centrality index of a graph. *Psychmetrika* **31(4)**: 581–603.

[19] Sevtsuk A (2010) *Path and Place: A Study of Urban Geometry and Retail Activity in Cambridge and Somerville, MA*. PhD thesis, MIT Department of Urban Studies & Planning, Cambridge, MA, USA.

[20] Shimbel A (1953) Structural parameters of communication networks. *Bulletin of Mathematical Biophysics* **15(4)**: 501–507.

[21] Ståhle A, Marcus L and Karlström A (2007) Place Syntax Tool – GIS software for analysing geographic accessibility with axial lines. In *New Developments in Space Syntax Software* (Turner A (ed.)), ITU Faculty of Architecture, Istanbul, Turkey, pp. 35–42.

[22] Still GK (2000) *Crowd Dynamics*. PhD thesis, Dept of Mathematics, University of Warwick, UK.

[23] Stubenschrott M, Matyus T, Schrom-Feiertag H, Kogler C and Seer S (2017) Route-choice modelling for pedestrian evacuation based on infrastructure knowledge and personal preferences. *Transportation Research Record: Journal of the Transportation Research Board* **2623**: 82–89.

[24] Timmermans H (2009) *Pedestrian Behaviour: Models, Data Collection and Applications*. Emerald Group Publishing Limited, Bingley, UK.

ice Publishing

Urban Space – AECOM
ISBN: 978-0-7277-6381-5
https://doi.org/10.1680/mpm.63815.069

Chapter 5
Microscopic movement

5.1. Overview of the chapter

In the previous chapter, pedestrian modelling at the aggregated and macro level was discussed. In this chapter, dynamic pedestrian modelling at the micro level is explained. The chapter starts with an overview of the modelling methods explained in Chapter 2. This is followed by details of micro-simulation model types and an overview of existing commercial and open-source software packages. Finally, the modelling process and uncertainty in modelling is discussed.

5.2. Principles of pedestrian models

In Chapter 2, the different mathematical methods available in the literature to model pedestrian movements were explained: force-based, grid-based and discrete choice. The force-based models use different forces to represent obstacles and other pedestrians, which affect the route choices and speeds of pedestrians. In grid-based models, the simulation area is divided into cells, and each cell can be occupied with only one pedestrian. The movement of pedestrians is permitted only to adjacent cells, with choices depending on the distance and travel time to their targets. Discrete choice models are stochastic-based models in which the choices of pedestrians along their journeys are given greater weights, resulting in a higher chance of selection. These methods are the foundation of the majority of pedestrian micro-simulation modelling platforms.

Pedestrian micro-simulation software packages are tools to model and replicate reality. They encompass intelligent entities, called agents, who act as pedestrians and can dynamically react to the current situations of the modelled environment. Various characteristics of pedestrians and crowds are usually incorporated in the software to make micro-simulation models more realistic. These include physical forces between pedestrians, such as pushing and shoving crowds, which in reality can affect pedestrian route choice and speed. In addition, crowd phenomena such as lane formation, jamming and density waves are incorporated in micro-simulation software, allowing more accurate replication of crowd behaviours.

Micro-simulation software packages have undergone rigorous validation against various sources of data, especially empirical pedestrian data captured from observations or through experiments. The validation process verifies collective behaviours of crowds, such as lane formation and queuing, and flow rates for a range of densities are accurately replicated in the software. The validation process can even be extended to qualitative aspects, such as the shape of queues at bottlenecks, through visual validation [2].

Pedestrian micro-simulation models can be developed for various applications, such as assessing alternative design options, evaluating operational scenarios, enhancing event management and assessing infrastructure upgrades. In other words, micro-simulation models are used to assess design layouts or current situations by considering the interaction of individuals (pedestrians) in models. Various useful outputs can be extracted from these models, such as capacity constraints, efficiency of pedestrian movements in areas with cross-flows, effectiveness of wayfinding and decision points, measurement of the available space, congestion level expected at special events, pedestrian journey times, queue times and areas of high footfall [13].

5.3. Defining a crowd

In almost all pedestrian micro-simulation modelling projects, crowds play an important role; and, in fact, crowds are the main element of any pedestrian study or modelling. Thus, it is important to know the definition of a crowd and what attributes and behaviours related to crowds are considered in pedestrian modelling. In the literature, various definitions for a crowd can be found, and some of them are not precise. For instance, in [11] a crowd is defined as a number of people walking through a railway station, but in [8] it is defined as movement of individuals in front of a podium. More precise are [15], in which a crowd is defined as more than two people at the same place, and [7], as follows: a group of more than 100 people who are present in an area for more than 60 seconds with a density of more than 1 person per m^2.

A more comprehensive and generic definition for a crowd is provided in [4]: 'A sizeable gathering of people in a given location, with a sufficient density distribution, who have come together for a specific purpose over a measurable period of time and who, despite being predominantly strangers or in an unfamiliar situation, feel united by a common identity and are, therefore, able to act in a socially coherent manner'. This definition could be modified slightly for different types of crowds, such as a crowd at a sporting event, an airport or a protest. For instance, for the crowd at the sporting event, the definition might read 'feel unified by the common identity of being sports fan'.

In dynamic and micro-simulation modelling, the definition and characteristics of crowds become more important, as a great level of detail regarding the crowd and environment need to be provided. Lack of proper understanding of characteristics of crowds can result in unreliable or inaccurate models.

In [4], the classification for the attributes of crowd is based on a number of key attributes. Figure 5.1 depicts the categories of crowd characteristics and their subcategories. Understanding these characteristics is important both for building models and the calibration and validation process. In pedestrian modelling projects, the modeller should carefully study the characteristics of the crowd and take those characteristics into account in the model when necessary. Obviously, not all of a crowd's characteristics can be included in the pedestrian modelling due to the limitations of micro-simulation software and available data; the modeller should, however, ensure that important and relevant characteristics are included. For instance, carrying luggage is an important characteristic for airport modelling due to the impacts of carrying luggage on the speed of pedestrians and the space that they occupy; but, of course, this behaviour is not as relevant for modelling pedestrians in stadia. The relevant crowd behaviours for different applications of pedestrian modelling are covered in Part II, and case studies are provided for each application.

In the next section, details of pedestrian micro-simulation modelling are explained, including the steps required in the model-building process. A brief overview of commercially available micro-simulation software packages is also provided to give readers a better understanding of those used in industry and their capabilities.

5.4. Micro-simulation modelling

The pedestrian micro-simulation modelling process can be divided into three parts: model information, dynamic modelling and analyses [9]. The model information stage consists of a combination of data-collection and data-processing tasks. This information includes details of the layout, demand and

Figure 5.1 Various attributes of crowds [4]

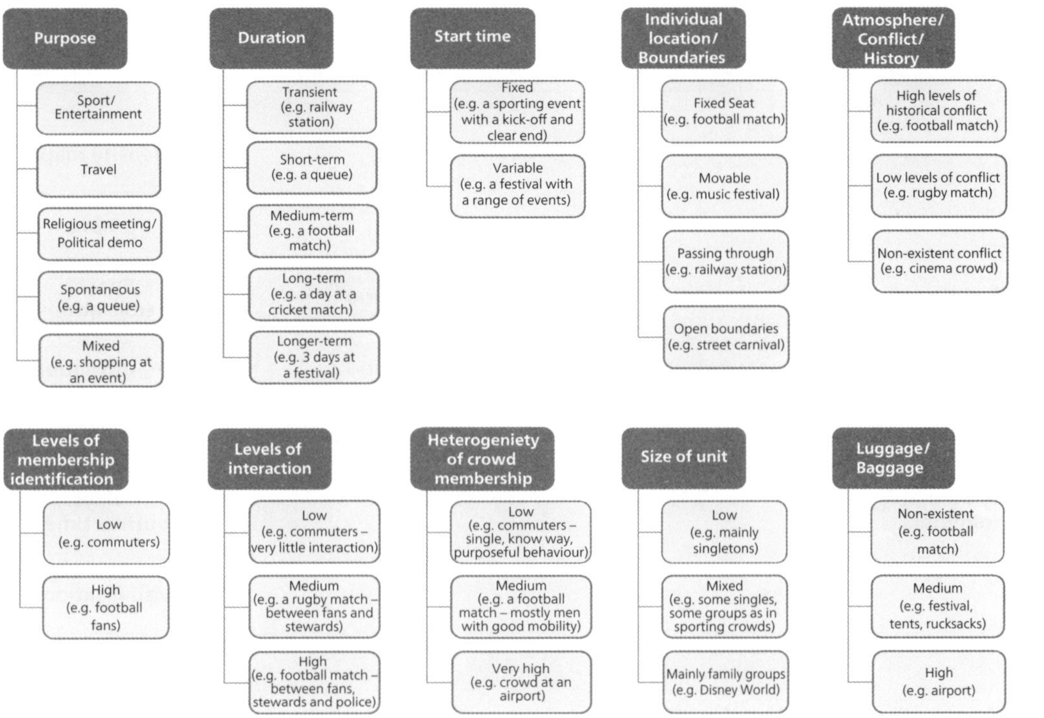

origin–destination matrices, pedestrian behaviour and type, event schedules (e.g. public transport schedules) and operative components (e.g. ticket gates). This information is used to build a model within the pedestrian micro-simulation software.

Typically, a modeller will first develop a base model representing the current situation of the study area. To ensure the base model is able to produce reliable results, the base model has to pass calibration and validation criteria. These criteria are provided in available guidance books and best practices. Once this is achieved, a number of scenarios are built upon the base model to meet the study objectives. The what-if analysis is then conducted, which includes comparative assessment of scenarios based on defined key performance indicators. The summary of this process is captured in Figure 5.2.

5.4.1 Pedestrian micro-simulation software packages

The first step in conducting pedestrian micro-simulation modelling is to select micro-simulation software to build the model. Pedestrian micro-simulation software packages are used to mimic the movement or flow of pedestrians in a defined environment as closely as possible [5]. Micro-simulation models are, in fact, the results of transition from macroscopic fluid-flow models to discrete cell-based and microscopic agent-based models in continuous spaces. With the advance of technology and computation power, there has been an increase in the use of pedestrian micro-simulation software by engineers, transport planners and modellers, asset owners and transport operators to design more efficient and safer places, such as airports, stations, stadia and plazas. As a result, a wide range of pedestrian micro-simulation software is available to

Figure 5.2 Pedestrian micro-simulation modelling process [9]

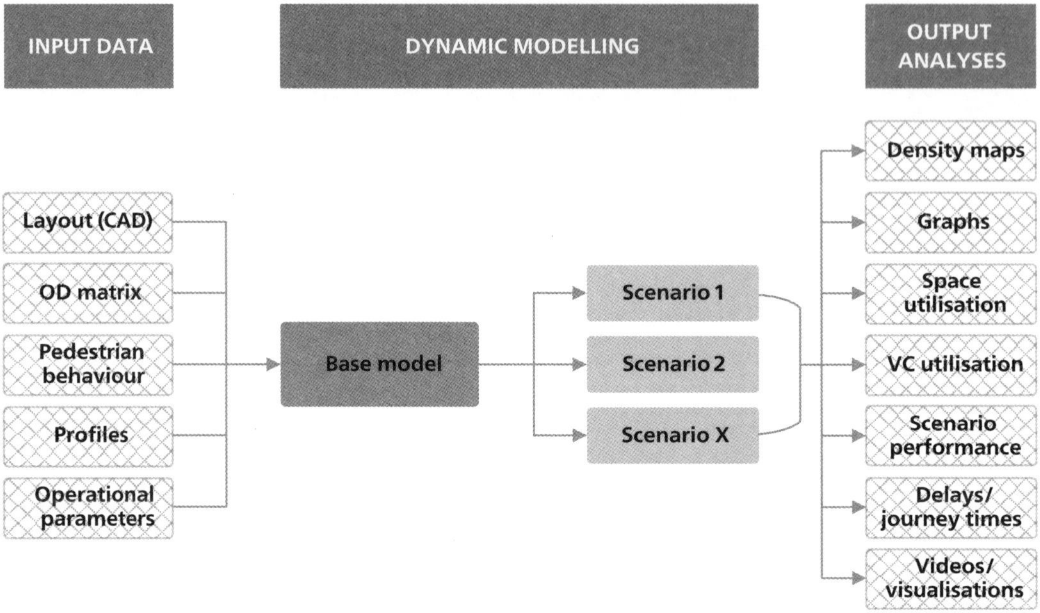

fulfil these requirements. Examples of key functionalities of micro-simulation software which are usually shared between existing packages include the following [2]:

- simulation of cross-flow and counter-flow of pedestrians in various situations (e.g. corridors, concourses and staircases)
- simulation of ingress and egress through bottlenecks for various scenarios, such as evacuation modelling, as well as simulation of realistic queue size and shapes at bottlenecks
- definition of different pedestrian types, speeds and size profiles
- interaction of pedestrians with each other and other simulation objects (obstacles and stairs)
- interaction of pedestrians with public transport and traffic
- 2D and 3D visualisation of the simulation (in recent years, virtual-reality rendering has also been offered by many software packages).

In the remainder of this section, the most commonly used commercial pedestrian micro-simulation software packages are summarised. A brief overview of open-source pedestrian micro-simulation software is then provided.

5.4.2 Micro-simulation software – commercial

5.4.2.1 LEGION

LEGION is a pedestrian micro-simulation software package developed at LEGION International Limited in 1997,[1] and its agent-based pedestrian model was developed at the Maia Institute.[2] In LEGION, each pedestrian (entity) is represented by a circular body moving in 2D space. Pedestrians interact with each other, obstacles and traffic to dynamically update their speed and routes within the simulation. Readers should note that to simulate and visualise pedestrian interactions with vehicular traffic, LEGION software must be used in conjunction with Aimsun simulation software. LEGION is capable of modelling different simulation objects for space and buildings, such as multiple levels, lifts, stairs, escalators and ramps. It can be used for modelling pedestrians in various environments, such as railway stations, airports, stadia, urban areas and large buildings.

In LEGION, each entity moves towards their destination by minimising a cost function. The cost function is the weighted sum of the following:

- inconvenience: cost equivalent of additional efforts by the pedestrian to reach the destination as a result of the crowd;
- frustration: cost equivalent of reduction in the speed of the pedestrian from the desired speed as a result of congestion; and
- discomfort: cost equivalent of deviation from desired clearance of neighbours and obstacles.

The cost function is dynamically calculated throughout the journey as pedestrians dynamically adapt their route choices and speeds to be able to minimise this cost function.

LEGION is capable of taking various input parameters to define pedestrian entities. These include different pedestrian types, behavioural characteristics and speed and size profiles [3]. LEGION offers a wide range of outputs from micro-simulation models, such as ingress, egress, journey time, speed and normalised flow. The outputs can be provided in terms of density or space-utilisation heatmaps, charts and tables.

5.4.2.2 Viswalk

Viswalk was developed by PTV group and released in 2010. It is based on the social force model developed by Dirk Helbing [10]. In this model, pedestrian motion is the result of different forces acting on pedestrians. These forces arise from the desires of pedestrians to reach their destinations, and from distances to other pedestrians and obstacles along the journey. All these forces impact on pedestrian behaviours, such as speed and route choices. Viswalk allows the user to define pedestrian walking-behaviour parameters related to the social force model, such as relaxation time, degree of anisotropy of the forces, strength of forces between two pedestrians and side preferences. This way, the user can adapt the pedestrian behaviours to more accurately represent those under study. In Viswalk, it is possible to define pedestrian behaviour for each area individually, such as ramps and stairs, to take into account the changes of pedestrian behaviours in these areas.

Viswalk is capable of modelling different scenarios, such as the movement of pedestrians through multi-storey buildings, pedestrian crossings, public transport boarding and alighting and bidirectional movements through narrow corridors [14]. To be able to simulate the interactions between pedestrians and vehicles, the PTV Vissim[3] module is required in addition to Viswalk. Different output measurements can be extracted from Viswalk, such as journey time, density levels, LoS, queuing times and waiting times. Viswalk provides both 2D and 3D visualisation, allowing the user to visually monitor the simulation and record videos of simulation.

5.4.2.3 Pedestrian Dynamics

Pedestrian Dynamics was developed by INCONTROL Simulation Solution[4] in conjunction with Utrecht University, the Netherlands, in 1990 [3]. Pedestrian Dynamics is based on the least-effort pedestrian modelling approach. Pedestrian Dynamics is one of the few pedestrian simulation software platforms

Figure 5.3 Pedestrian navigation in Pedestrian Dynamics [1]

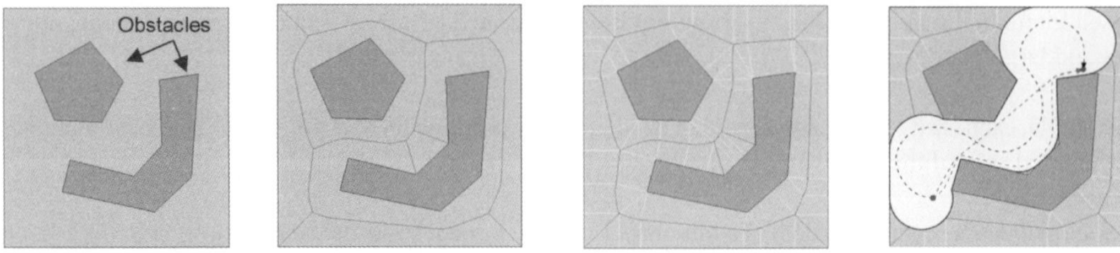

(a) Environment with obstacles (b) Medial axis (c) Explicit Corridor Mapping (d) Path following

that is capable of mesoscopic simulation in addition to micro-simulation modelling. Mesoscopic simulation is a simplified version of microscopic modelling but with greater detail than macroscopic simulation.

Navigation of entities (pedestrians) in Pedestrian Dynamics is based on a specific data structure, called Navigation Mesh. Navigation Mesh is basically a network (or a graph) where the edges form a medial axis (Figure 5.3(b)) – a set of curves representing the walkable space for pedestrians. The edges are annotated with a number of nodes to provide a subdivision of the walkable space into a polygonal area, called an Explicit Corridor Map (Figure 5.3(c)). The polygonal area converts the edges into corridors – free space that pedestrians can use around the route (Figure 5.3(d)). Pedestrians can walk flexibly and efficiently within a corridor and use the free space in the corridor to avoid other pedestrians. Figure 5.3(d) depicts the corridor used for path following of pedestrians.

Pedestrian Dynamics allows the building of models by importing the layout from different formats, such as City GML, CAD, DFX, XML and other industry standards [1]. Both 2D and 3D visualisation and modelling are supported. Pedestrian Dynamics includes domain-specific simulation objects for stadia and transport interchanges, which makes modelling more convenient. To customise the predefined rules, the modeller can use a scripting language, called 4D Script, which gives a great deal of flexibility for customisation. Pedestrian Dynamics is capable of providing various simulation outputs, such as a frequency map showing the number of pedestrians passing a specific area, density graphs, travel-time maps, flow counter maps, travel-evacuation time graphs or histograms, walking distance and delay times.

5.4.2.4 MassMotion

MassMotion is pedestrian micro-simulation software developed in 2005 by Arup.[5] It is based on two pedestrian modelling methods: the social force model and least-effort model (an optimisation-based pedestrian model) [3]. Within the simulation, the social force model determines the speed and routing decision of each agent (individual) influenced by other agents and obstacles in the model. The least-effort model allows the user to define the cost of routes and actions affecting agent's choices to adapt pedestrian behaviours more closely to reality.

MassMotion's models can be built using BIM (Building Information Modelling) or CAD geometry or from scratch using MassMotion's polygon modelling tools. The software allows the programming of individual personalities and behaviours that affect decisions throughout an individual's journeys. This means agents can dynamically respond to the current situation of the modelled environment in the simulation, such as taking stairs instead of a lift because of a long queue for the lift.

MassMotion provides an SDK (Software Development Kit) through which the user can modify the built-in algorithms using programming languages. A wide range of analysis outputs can be pulled out from MassMotion models, such as queuing times, queue length, LoS and footfall. In terms of visualisation, MassMotion offers both 2D and 3D visualisation that can be viewed using the standalone Viewer software [12].

5.4.2.5 Micro-simulation – open-source

There are a number of open-source micro-simulation tools and libraries available within the pedestrian micro-simulation community. These tools are mainly used in academia for research. They are not widely used for commercial projects, mainly because of the lack of a proper graphical user interface, making model building less convenient for users. In fact, some of these tools actually come as pedestrian simulation libraries, which means that building pedestrian micro-simulation models can only be done through coding a programming language. In addition, some of these tools lack high-quality visualisation capabilities – again, making them less convenient to the end user. However, they usually offer advanced built-in algorithms with flexibility for modifications.

As these tools are open-source the user has access to the source code and hence it is possible to modify it to adapt the tool to his or her requirements [6]. This obviously requires time and effort to first understand the code and then apply the modifications. But it has one major advantage: it allows the user to modify the algorithms and functionality of the software more effectively.

The following is a non-exhaustive list of open-source pedestrian software packages and tools:

- PEDSIM is a C++ library for micro-simulation pedestrian modelling. It can be used to study pedestrian density and evacuation time.
- JuPedSim is an open-source framework written in C++ and Python to simulate pedestrian dynamics. It consists of four modules which each serve different purposes, as follows: 1) computing trajectories, 2) analysing trajectories and model validation, 3) visualisation input and outputs and 4) creating and editing geometry files.
- Menge is a modular framework for the simulation of pedestrian movements in a crowd. Menge works by decomposing the crowd simulation into a number of component sub-problems, such as goal selection, plan computation, plan adaptation, motion synthesis and environmental queries [6].

5.5. Uncertainty

Pedestrian modelling projects are inherently prone to uncertainties; these are wide-ranging and relate to all aspects of the modelling process, from data collection to network coding and simulation, to model analysis and outputs. It would be impossible to discuss every possible uncertainty; however, some common uncertainties in modelling projects are discussed in this section.

The availability of reliable and sufficient data for input into the model can lead to uncertainty in the quality of these inputs and therefore its outputs. Poor quality CAD can have either errors or lack essential detail. In some cases, there can be drawings with inconsistencies between levels, areas of the space may be unresolved or details may be missing, such as the length or number of risers in a staircase or escalator. In such cases, when clarity cannot be established, assumptions need to be made regarding the layout of the model.

The same is true for any gaps identified in the data set where obtaining data to fill those gaps is not possible. In projects where a future year is tested in scenario modelling, there is uncertainty related to the future-year demand-forecasting methodology. Incorrect future-year demand inputs will result in unrealistic outputs. In order to verify the reliability of the forecast demand, sensitivity tests are undertaken with various demand scenarios with the intention of covering the entire range of demand options in the model.

Uncertainties can also arise from interaction with the sponsor of the modelling project, who may not define the modelling scope or agree assumptions in a timely manner, and who may request changes to the scope during the project process.

In some software packages, there is a lack of transparency regarding the source and robustness of the underlying research that supports the default model simulation inputs and parameters, such as entity behaviour, and characteristics, such as speed – or there is little scope available to change these inputs. This results in uncertainty in the quality of the research underpinning the pedestrian behaviour in the model, which can reduce confidence in the model outputs.

5.6. Summary points

- Pedestrian micro-simulation software packages are tools to model and replicate reality that encompass intelligent entities, called agents, who act as pedestrians and can dynamically react to the current situations of the modelled environment.
- Micro-simulation models are, in fact, the results of transition from macroscopic fluid-flow models to discrete cell-based and microscopic agent-based models in continuous spaces.
- Micro-simulation software packages have undergone rigorous validation processes against various sources of data, especially against empirical pedestrian data captured from observations or through experiments.
- Pedestrian micro-simulation models can be developed for various applications, such as assessing alternative design options, evaluating operational scenarios, enhancing event management and assessing infrastructure upgrades.
- In almost all pedestrian micro-simulation modelling projects, crowds play an important role, and crowds are the main element of any pedestrian study or modelling.
- In dynamic and micro-simulation modelling, the definition and characteristics of crowds become more important as in this type of modelling a high level of details regarding the crowd and environment need to be provided.
- The pedestrian micro-simulation modelling process can be divided into three parts: model information, dynamic modelling and analyses.
- The first step in conducting pedestrian micro-simulation modelling is to select the micro-simulation software to be used to build the model.
- A wide range of pedestrian micro-simulation software is available with diverse capabilities.
- Pedestrian modelling projects are inherently prone to uncertainties; these are wide-ranging and relate to all aspects of the modelling process, from data collection to network coding and simulation, to model analysis and outputs.

NOTES

1. LEGION is now owned by Bentley Systems.
2. www.maia-institute.org/pedestrian-behaviour/
3. http://vision-traffic.ptvgroup.com/en-us/products/ptv-vissim/
4. www.incontrolsim.com/software/pedestrian-dynamics/
5. www.oasys-software.com/

REFERENCES

[1] www.incontrolsim.com/software/pedestrian-dynamics/ (accessed 08/02/2019).

[2] Berrou JL, Beecham J, Quaglia P, Kagarlis MA and Gerodimos A (2007). Calibration and validation of the Legion simulation model using empirical data. In *Pedestrian and Evacuation Dynamics 2005* (Waldau N, Gattermann P, Knoflacher H and Schreckenberg M (eds)). Springer, Berlin, Heidelberg, Germany, pp. 167–181.

[3] Caramuta C, Collodel G, Giacomini C, Gruden C, Longo G *et al.* (2017) Survey of detection techniques, mathematical models and simulation software in pedestrian dynamics. *Transportation Research Procedia* **25**: 551–567.

[4] Challenger W, Clegg CW and Robinson AM (2009) *Understanding Crowd Behaviours: Volume 1 – Guidance and Lessons Identified*. TSO, London, UK.

[5] Charlton J, Giddings B and Horne M (2008) A survey of computer software for the urban design process. *9th Design Decision Support Systems Conference (DDSS 2008), Eindhoven, the Netherlands*, http://2008.ddss.nl/ (accessed 06/06/2019).

[6] Curtis S, Best A and Manocha D (2016) Menge: A modular framework for simulating crowd movement. *Collective Dynamics* **1**: 1–40.

[7] Duives DC, Daamen W and Hoogendoorn SP (2013) State-of-the-art crowd motion simulation models. *Transportation Research Part C: Emerging Technologies* **37**: 193–209.

[8] Duncan H (2009) Pre-event crowd behaviour prediction system: A study within concert environment. *Journal of Crowd Safety and Security Management* **1(2)**: 37–48.

[9] Federici ML, Manenti L and Manzoni S (2014) *A Checklist for the Evaluation of Pedestrian Simulation Software Functionalities*. arXiv:1404.7717.

[10] Helbing D and Molnar P (1995) Social force model for pedestrian dynamics. *Physical Review E* **51(5)**: 4282–4286.

[11] Hoogendoorn SP and Bovy PHL (2004) Pedestrian route-choice and activity scheduling theory and models. *Transportation Research Part B: Methodological* **38(2)**: 169–190.

[12] Oasys. www.oasys-software.com/products/pedestrian-simulation/massmotion/ (accessed 08/02/2019).

[13] Tanoz-Sargeant T (2019) The assessment of transit oriented developments using microsimulation models. In *Transit Oriented Development and Sustainable Cities* (Knowles RD and Ferbrache F (eds)). Edward Elgar, Cheltenham UK, pp. 172–185.

[14] Viswalk. www.ptvgroup.com/en/solutions/products/ptv-viswalk/ (accessed 08/02/2019).

[15] Wijermans FEH (2011) *Understanding Crowd Behaviour: Simulating Situated Individuals*. PhD thesis, University of Groningen, Groningen, the Netherlands.

ICE Publishing

Urban Space – AECOM
ISBN: 978-0-7277-6381-5
https://doi.org/10.1680/mpm.63815.079

Chapter 6
The model-development process

6.1. Overview of the chapter

In the previous chapter, pedestrian modelling at the micro level was discussed. In this chapter, the approach to pedestrian model development is discussed. While each individual software platform has a slightly different approach, there are common steps in model development.

The pedestrian micro-simulation modelling process can be divided into three parts: model information, model development and analyses [4]. Section 6.2 discusses preparing to develop a model, from establishing its scope to preparing CAD and developing assumptions. The model information stage consists of a combination of data-collection and data-processing tasks. This information includes details of the layout, demand and origin–destination matrices, pedestrian behaviour and type, event schedules (e.g. public transport schedules) and operative components (e.g. ticket gates).

The information is used to build a model within the pedestrian micro-simulation software. Details of micro-simulation model building are discussed in Section 6.3 as well as the typical sorts of network and behaviour coding that are used to control agents within a model.

Finally, various approaches to model calibration and validation processes, as well as details of analysis and dissemination of results modelling, are discussed. The available guidance on model development and common issues is also highlighted.

6.2. Model information

6.2.1 Establishing project scope

A clearly defined project specification is a critical element in any pedestrian modelling project, so parameters, such as model time periods, expected objectives and outputs, and assumptions can be agreed with the project sponsor and any other relevant stakeholders prior to commencing the modelling work. Table 6.1 outlines a list of parameters to be defined; however, there may be other aspects specific to the project that should also be agreed.

6.2.2 Site visits

Prior to building a pedestrian model, it is particularly useful for the modeller to undertake site visits to become familiar with the existing conditions and identify any issues and concerns. The site studies should focus on the area that is to be modelled in order to obtain an overall understanding of the site and its operation and accessibility. Key movement patterns and flows should be observed, along with behavioural patterns in general and particularly at pinch-points. Queuing and any other aspects that may impact crowd movement should also be monitored. Site visits can aid in providing a visual level of validation at key locations within the model.

6.2.3 CAD

The primary step in building a micro-simulation model is defining the layout of the study area. Micro-simulation software packages have built-in tools allowing the user to draw the layout. It is also possible to

Table 6.1 Project scope parameters to be agreed

Model objectives
The extent of the area to be modelled
CAD layouts and any operational nuances
Modelled scenarios
Required outputs

import the layout to the model from existing CAD drawings (the project sponsor often has access to the layout of the area under study in CAD format). If such drawings are available, the modeller can import the layout into the software, rather than drawing the layout from scratch, which can be time-consuming, depending on the size and complexity of the model. The drawings should provide sufficient details of the layout, which could include fire-exit locations, stairs and escalators, entrance points and the placement of all furniture and other obstacles on each floor. CAD drawings can also be used if the area under study is and urban area.

Before using any CAD drawings for pedestrian micro-simulation modelling, CAD preparation process needs to be conducted to ensure that they are appropriate for modelling. The CAD preparation process is undertaken to remove any unnecessary elements from the CAD file and keep only the lines, objects and layers relevant to the pedestrian model. This will enhance the accuracy and processing time of the software's calculations. Many software programs consider additional lines as obstacles for pedestrians and may include them in the calculation of pedestrian route choices and speeds; thus, the higher the number of unnecessary lines, the longer processing time required by the software for calculation [6]. Some software packages offer the ability to incorporate BIM (Building Information Modelling) files; however, some drawing adaptation as described above is still required.

6.2.4 Assumptions

While the process of data collection for pedestrian modelling should focus on providing sufficient data to make the models as accurate as possible, it is not always possible in practice. One of the main reasons for this is that data collection is a very expensive task, especially if it needs to be comprehensive. For instance, collecting pedestrian route-choice data requires either advanced technology to track pedestrians using computer-vision technologies or pedestrian path-following surveys, which are labour-intensive. Therefore, it is often necessary to make reasonable assumptions to fill in gaps in data collection.

For instance, in a pedestrian modelling project to help design a new building (e.g. a railway station or airport), a lot of data is often not available during the design stage. This could be important information, such as the average number of passengers, their arrival profiles and types. In cases with a lack of available data, the modeller should make some assumptions and build the models based on those assumptions.

Assumptions can be made for various missing data, such as facility usage by pedestrians, delays, pedestrian routing and demand growth. The modeller should ensure that the assumptions have been made based on expert knowledge, historical data and data from similar environments or projects, as well as on the specifications and requirements of the specific project.

The modeller should then evaluate the assumptions based on the evidence that was used to make them. It is also good practice to analyse the sensitivity of the model to changes in the assumptions. Conducting such

Figure 6.1 Assumptions analysis

Stability	Sensitivity
A B C D	A B C D
A – Very confident	A – Minor impact
B – Fairly confident	B – Manageable impact
C – Uncomfortable	C – Significant impact
D – Very uncomfortable	D – Critical impact
Will the assumption turn out to be correct?	How much does it matter if the assumption turns out to be true?

analysis on the assumptions can help increase confidence in the accuracy of the model. Figure 6.1 gives an example of categorising the assumptions in terms of the confidence (*stability*) and *sensitivity* of the model.

6.2.5 Demand-matrix estimation

An origin–destination (OD) matrix and pedestrian routings are essential inputs to any pedestrian modelling projects. OD matrices specify the number and type of pedestrians who move between each origin and destination pair within the period of modelling. The origin and destination points within a model constitute all of the entry and exit points of pedestrians entering and leaving the model. All these points must be identified prior to the development of the OD matrix, either through analysis of the study area or information received from a project sponsor. A site visit (see Section 6.2.2) may also help in identifying the model entries and exits as well as observing pedestrian routing. Pedestrian routing defines how pedestrians move between each pair. In cases where multiple routes exist, the intermediate points define the route taken between origin and destination.

In some cases, it may be appropriate to aggregate origins or destinations into larger groups or zones in order to reduce the complexity of a model; however, it is important to ensure that the level of detail required in the model is not compromised in doing so. An example of a set of origins (filled dots) and destinations (open dots) is provided in Figure 6.2. The left plots show five origin points are connected to four destination points, resulting in 20 OD pairs. These are aggregated into two origin and two destination zones on the right, reducing the number of OD pairs significantly – to four pairs.

6.2.6 Pedestrian types

Micro-simulation software allows users to define different pedestrian types. Each pedestrian type can have its own capabilities and characteristics, such as speed profiles, size profiles, routing profiles and facility use (e.g. using lifts or stairs). To build accurate pedestrian models the modeller should ensure that relevant pedestrian types have been identified and the characteristics of each pedestrian type have been accurately modelled.

Pedestrian types usually vary based on gender, age, luggage and mobility. Persons with reduced mobility (PRMs) are categorised into wheelchair users and pedestrians with physical mobility impairment. Pedestrians with luggage are categorised into pedestrians with heavy luggage, pedestrians with large luggage and adults with young children.

Figure 6.2 A set of origins and destinations with pairs linked

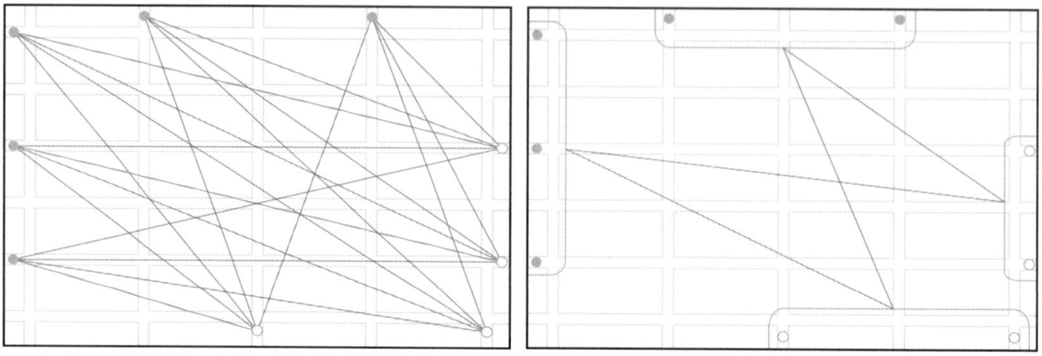

Most of the pedestrian micro-simulation software has predefined speed profiles for different types of pedestrian. The differentiation is based on geography (e.g. Asian, North American, southern European) and the objective of the journey (e.g. commuting, tourism, attending sporting events).

As mentioned earlier, different routing can be defined for different pedestrian types. For instance, pedestrians with wheelchairs should be routed in a way that they do not use stairs or escalators and always use lifts instead. Another example is pedestrians with heavy luggage who have different routing preferences from other pedestrians; they prefer to use lifts, but they might still use stairs if the lifts are too busy.

6.3. Model development

6.3.1 Network coding

This section will provide a general overview of the process and best practice of coding a pedestrian model using micro-simulation software. Since software packages differ in their modelling methods, parameters, and objects, this overview is not intended to replace a coding manual that may be provided by the developer of each software package; rather, it is intended to provide generic guidance that can be followed in the network coding process using any pedestrian modelling software package.

6.3.1.1 Model-naming conventions

It is important to maintain a consistent model-naming convention for model files associated with a single project. This is also the case for the objects within an individual model, and it is relevant to all projects, no matter how complex or simple. Using clear and consistent filenames allows for easier file management, auditing and project handover. Filenames should include identifying information about both the project and the specific file, not least the project name or code, demand year, modelling period and scenario or layout.

A consistent approach to naming model objects should include identifying information about each object, such as the type, location and purpose. Where multiple instances of the same object exits, such as individual ticket gates in a gateline, these should be numbered in a logical order.

6.3.1.2 Layers

In some models, it may be beneficial to utilise the layer functionality provided by modelling software. This is particularly applicable to larger and more complex models but may not be necessary for smaller models.

Layers can be used to separate modelling objects by type, function or location, such as different levels within a multilevel building. Separating objects into different layers makes it easier to locate and edit these objects as required.

6.3.1.3 Pedestrian demand types

Demand types are set up for each model entrance or origin, specifying the number and make-up of pedestrian types that enter the model at each origin bound for each destination. In some cases, it may be necessary to set up multiple demand types for the same origin, such as when different proportions of PRMs are required to go to different destinations in the model.

6.3.1.4 Profiles

Various data profiles can be set up within the modelling software to allow various actions to occur at the appropriate times. Profile inputs can be static – that is, they can be spread evenly and maintain a set distribution over the modelling period – or they can have a variable distribution over time – that is, they can change throughout the modelling period. Table 6.2 summarises the types of data profiles available in pedestrian modelling and their purpose.

Note that not all data profiles are necessary in every pedestrian modelling project. The choice of data profiles is dependent on the requirements of the model.

6.3.1.5 Origins and destinations

Objects that allow entities to enter and exit are modelled and placed at the model's relevant entry and exit points. Entrances and exits should correspond with the origin and destination points specified in the model's OD matrices (see Section 6.2.5), which specify the destinations for each arriving entity. In locations where entities both enter and exit the model, both an entrance and exit object may be required. Each entrance object should be associated with at least one entry profile (see Section 6.3.1.4) that specifies the number, type and distribution of entities arriving into the model.

6.3.1.6 Targets and decision points

Targets and decision points are objects that aid in directing pedestrians along a set route between their origin and destination. Targets are objects that act as intermediate points in the model and can be used sequentially to align pedestrian flows along defined paths where multiple routes exist between two points. Decision points are objects that can be used to change a pedestrian's current target or final destination. These decision points can be associated with an availability profile (see Section 6.3.1.4),

Table 6.2 Model data profile types

Entry profiles	Specify number of entities and their time of entry into model
Availability profiles	Specify time period(s) for when certain actions are possible
Delay profiles	Specify time period(s) and duration of delays
Event profiles	Specify time period(s) of specific event occurrence
Exit profiles	Specify when a destination is available for exiting the model
Size profiles	Specify sizes of entity types and luggage
Speed profiles	Specify walking speeds for each entity type

so that they activate only at specific times. An example of this would be allowing passengers to board a train only once the train has arrived, or operating with a condition that activates only once specific conditions are met, such as directing pedestrians away from a crowded platform at certain density levels. Decision points may also be used to change an entity's type, which may be useful in cases where a pedestrian's behaviour changes part way through a simulation, such as an airport passenger who changes from a PRM with luggage to an unencumbered pedestrian once they have checked in that luggage.

6.3.1.7 Delay points and waiting areas

Delay points and waiting areas may be used in a pedestrian model as intermediate destinations between the model's origin and destination. Both objects accommodate a specified capacity of pedestrians who are dispersed within the object boundaries according to a set of dispersal parameters. Delay points cause visiting entities to be delayed for a specified amount of time, set by a delay profile (see Section 6.3.1.4), and are useful for simulating short stops along a pedestrian route, such as the delay experienced by a pedestrian passing through a ticket turnstile. Conversely, entities visiting a waiting area will be delayed for an indefinite period of time or until they are told to do something else by a decision point linked to an event profile (see Sections 6.3.1.4 and 6.3.1.6). Waiting zones are useful for simulating locations where pedestrians must wait until a specific event occurs, such as railway-station platforms, where pedestrians wait until the arrival of a train service before boarding the train. Waiting areas can also be used as destination objects for entities which, rather than exiting the model, remain in the model until the end of the simulation period.

6.3.1.8 Vertical circulation

Vertical-circulation objects allow pedestrians to move between levels in multi-storey buildings and include both static infrastructure (stairs, ramps) and technical installations (escalators, travellators, lifts). Not all software packages provide all kinds of vertical circulation objects, or their use may not be practical in all situations, such as modelling curved or circular staircases. In these situations, other objects may be used to replicate the level-change behaviour that would occur.

6.3.1.9 Movement-adjustment zones

A movement-adjustment zone allows the movement characteristics of pedestrian behaviour to be adjusted while the pedestrian is moving within the boundaries of the zone. These zones can be used to change pedestrian speed or direction. Direction change can be in either a linear or circular direction, facilitating, for example, a more natural, circular movement to be made around a corner. These zones may also be used to replicate specific walking behaviours, such as walking up or down stairs or ramps, using escalators, and to prioritise certain pedestrian types or activities over others.

6.3.1.10 Conditional analysis

Analysis lines and zones can be used to measure the flow of entities across a point or count the occupancy of an area. Apart from the use of these for analysis purposes, obtaining these metrics can add conditional parameters to other objects – for example, activating them only once a certain condition has been met.

6.3.2 Pedestrian interactions with traffic

There has been a shift in attitudes towards walking as a legitimate transport mode with significant cost, health and environmental benefits. For decades, cities have been built with the intention of improving traffic flow and efficiency, but, as a result of this shift, improving conditions for walking in urban environments is increasingly being prioritised by urban designers and planners. Re-prioritising the often limited space in cities away from cars and towards walking means that there are increasing conflicts between modes, requiring planners to balance the conflicting demands of safety, pedestrian flow efficiency and traffic flow efficiency. Just as pedestrians crossing roads can reduce the traffic flow of a road, traffic adds time and distance to pedestrians. Roads act as barriers to pedestrian activity, requiring pedestrians to detour from their shortest path to improve their safety by crossing the barrier at appropriate crossing facilities [11].

In order to create accurate micro-simulation models of the interaction between pedestrians and traffic a sound understanding of the behaviour of pedestrians around traffic is required. Currently most micro-simulation software is able to replicate pedestrian behaviour at signalised crossings, where the flow of pedestrians and traffic is segregated; however, it is unable to simulate the interaction of pedestrians crossing roads at uncontrolled locations between formal crossing points [11], since the behaviour of pedestrians in these situations is not well understood.

Differences in pedestrian behaviour as a result of driving culture may affect the interaction of pedestrians and traffic, even at controlled crossing locations. In some countries – for example, China – traffic is less disciplined and rarely gives way to pedestrians at zebra crossings, despite pedestrians having right of way at these facilities [11]. This influences the behaviour of pedestrians at these types of crossing facilities, such that they will still wait for a gap in traffic in order to cross the road, whereas in other countries, where the traffic rules are more closely observed, pedestrians can cross with minimal delay, with vehicular traffic yielding to allow safe passage.

Most pedestrian simulation tools focus on pedestrian behaviour at road crossings, rather than combining walking behaviour with the crossing process at uncontrolled locations. Vissim was the first simulation tool and the first multimodal microscopic simulation software to include the interaction between pedestrians and vehicles, using their Viswalk module [11]. Other software packages now facilitate modelling this interaction, such as Aimsun for LEGION. For more information on micro-simulation software packages and their capabilities, see Sections 5.4.1–5.4.3.

6.3.3 Pedestrian interactions with public transport

Pedestrian interaction with public transport is concerned predominantly with interaction in urban environments with street-based forms of public transport, such as buses, trolleybuses and trams. Hence, the treatment of such interaction in pedestrian modelling is the same as for regular vehicular traffic. Special consideration, however, needs to be made for the interaction of pedestrians alighting from and, particularly, boarding various forms of public transport.

The behaviour of passengers waiting at a bus stop can vary based on the number of passengers waiting, weather conditions and cultural context, all of which should be considered to accurately represent this interaction. Passenger behaviour at stations and on platforms can also be influenced by a variety of factors relating to both pedestrian characteristics and those of the infrastructure at the station. The types of passengers using a station will affect behaviour. A commuter, for example, will behave differently to a tourist. The commuter, who, unlike the tourist, is highly aware of their surroundings, will know exactly where to go and the best place to board their train based on factors such as train crowding and exit location at their destination. Whether or not there is a platform choice can influence whether passengers go directly to a platform or wait on a concourse until the platform information becomes available. The location of platform entry and exit points, as well as the presence of platform screen doors, can influence the locations, positioning and behaviour of waiting passengers on the platform. Train frequencies and service patterns can affect how long passengers dwell on a platform. In cases where trains to multiple destinations depart from the same platform, some passengers may choose to board the first arriving train and change services further down the line, while others may choose to wait until their preferred train service arrives. It is important that these factors are understood for the boarding of public transport services by passengers to be accurately represented in the pedestrian model.

6.4. Analysis

6.4.1 Calibration and validation

The main purpose of micro-simulation models is to support engineering or architectural design by analysing new schemes and alternative options. Before models can be used for such purposes, it is important to ensure the models produce reliable outputs. This assurance is achieved by a process called calibration and validation. The calibration process includes inputting verified data into the model to replicate the reality and observed situations. The validation process uses a series of measurements to ensure validity of the simulation models. In the validation process, sources of data that have not been used in the model building and calibration process are used [9].

The validation process ensures that the results of the model are reliable. It can be the case that a model has been accurately calibrated but is not reliable as the model cannot meet the validation criteria. Thus, it is important to ensure the model is validated properly. The criteria for the calibration process is more straightforward than those for validation. This is because in the calibration process models need to be verified against the input already in use in the model-building process, such as pedestrian counts, pedestrian speeds and pedestrian routes. Validation is the process of verifying the model against criteria and data not in use in the model-building process. Therefore, the validation process can be more challenging and more diverse than the calibration process. In the literature and in practice, various approaches for the validation of models have been provided.

Some pedestrian modelling guidelines provide specific validation criteria, such as [7], where the model validation should be conducted against journey time, flow rate, origin–destination counts and visual validation. The validation process might, in fact, vary for different pedestrian modelling projects depending on the purpose and scope of the project. A list of generic approaches and criteria for validation is provided in [8]. These criteria are relevant to any micro-simulation modelling, but they are also easily applicable to pedestrian modelling. In the following section, a summary of the criteria suitable for pedestrian modelling is provided.

In practice, the validation process is conducted by evaluating the model against quantitative measures using real data, such as delay, journey time, flow rate, pedestrian density and capacity. Depending on the objectives of the pedestrian modelling study, different criteria might be used. To compare the output data from the model with real data, different statistical methods have been proposed in the literature [2 and 8]. The most common of these is provided below (there is also a good overview in [3]).

The most common measurement in traffic and transport studies is the GEH statistic, named after its inventor, Geoffrey E. Havers, a transport modeller in the UK. The GEH statistic is derived from Chi-squared statistics that take into account both relative and absolute errors in the calculation. GEH is usually used for validation of pedestrian survey counts (i.e. flow).

$$GEH = \sqrt{\frac{(M - C)^2}{(M + C)^2}} \tag{6.1}$$

Where:

- *GEH* is the GEH statistic,
- *M* is the modelled value and
- *C* is the observed value.

In the GEH statistic measurement, the smaller values show better 'fit' between the modelled and observed values [9].

Another commonly used measurement is the mean absolute per cent error (MAPE), calculated as follows:

$$MAPE = \frac{1}{N}\sum_{n=1}^{N}\left|\frac{M_n - C_n}{C_n}\right| \tag{6.2}$$

Where:

- *N* is the number of observations and
- *M* and *C* are as defined for the *GEH* statistic.

Normalised root mean square error (RMSN) can be used to quantify overall error between modelled and observed values [1]. The formula for the calculation of RMSN is as follows:

$$RMSN = \frac{\sqrt{N \sum_{n=1}^{N} M_n - C_n}}{\sum_{n=1}^{N} C_n} \tag{6.3}$$

6.4.2 Other methods of calibration and validation

- Comparison with other calibrated or validated models. One way to validate a simulation model is to compare it with other existing validated models. Such a comparison is only possible if other models are available and accessible, and it must be undertaken carefully as existing models may have been developed for other objectives, which is when a comparison should not be drawn.
- Visual validity. Most pedestrian micro-simulation software packages benefit from 2D and 3D visualisation. This visualisation can be used to verify whether the simulation can produce behaviours similar to those observed in reality. In the data-collection process, videos of pedestrian movement in the study area are usually recorded and are a good reference for comparison and validation. In addition, experts who are familiar with pedestrian behaviour in the study area can review the model and indicate whether the model replicates reality accurately.
- Historical data validation. If historical data is accessible, it can be used to verify the outputs of the models. Historical data may show some variations over time due to changes in demand and other inputs, and these variations should be taken into consideration when a comparison is undertaken.
- Sensitivity analysis. This is a strong tool for validating micro-simulation models. It evaluates the response of the model to changes in input parameters. In other words, sensitivity analysis checks whether the model can sensibly respond to changes expected in reality. For instance, increasing

demand in one specific route should potentially increase the travel time of pedestrians on that route. If the model shows otherwise, the modeller should suspect that the results of the model are not reliable and that the source of the issue needs to be investigated.

- Internal validity. Running the simulation models using different random seeds and observing the variation of results is a useful validation approach. If the model has a high level of robustness, validating the variation against the randomness should not be significant. Thus, if randomness results in significant difference between the output results, this is a sign of the model's lack of validity.

6.4.2.1 Analysis and reporting

The analysis is the most important part of any pedestrian modelling project as the main purpose of the modelling is to provide results and outcomes. Thus, it is important to present the analysis results in a clear and consistent way. There are different metrics in the industry to disseminate the analysis results, and the most common ones are explained in this section. The modeller should select the metrics which are pertinent to the project and can better reflect its objectives. These metrics are provided below and are adapted from [7] and [10]:

- Cumulative mean density (CMD) plots. A very effective way to communicate the results of pedestrian micro-simulation modelling is to depict the mean levels of pedestrian density of different areas within the study area from the beginning of the simulation. CMD plots are usually used in combination with Fruin's LoS[1] to demonstrate the performance of an area in terms of pedestrian comfort over 15-minute periods.
- Cumulative high density (CHD) plots. CHD plots can be used to calculate and visualise the density of the areas that have reached a specific limit. Such plots are colour-coded to show the amount of time throughout simulation (or in 15-minute periods) that has experienced such high density, based on the defined density limit. The density limit can be defined either by experts or by referring to pedestrian guidance books.
- Space utilisation plots. These diagrams demonstrate how space is used by pedestrians throughout the simulation, including popular and unconventional moves.
- Desire line plots. These diagrams monitor the movement of each pedestrian type. The modeller should consider different colours to clearly differentiate between the movements of pedestrians. These plots can also be used to check the routing of each pedestrian type. For instance, it can be ensured that pedestrians with wheelchairs have used the lifts and not other facilities.
- Journey time. Journey-time analysis can assess the design performance from the user perspective, using three different approaches:
    - Point-to-point journey time: this metric is used to measure the journey time between two points of interest, broken down into walking, waiting and using facilities (lift and escalator).
    - Clearance time: this metric measures the time required for a crowd to clear a specific area and is particularly useful for evacuation-scenario assessment.
    - Generalised journey time: using generalised journey time, more complex analysis can be conducted. It takes into account different factors impacting pedestrian journey in terms of time and comfort. These are as follows: journey time, walking, delay, queuing, up or down stairs and up or down escalators. The factors are weighted and contribute to the generalised journey time based on their weights. The modeller should be careful in choosing the correct weights; consulting guidance or best practice documents to identify the most suitable weightings is recommended.
- Model videos and screenshots. Pedestrian micro-simulation software commonly provides 2D, and often 3D, visualisation for the effective use of videos or screenshots of the layout and simulation at points of interest, such as gates, entrances, exits, stairs and ramps. Using screenshots, various aspects of the models can be presented to engage an audience.

6.4.3 Guidance

Guidance documents are used to specify recommended or mandatory standards and guidelines to be followed in undertaking pedestrian modelling projects. While no comprehensive guidance exists to cover all aspects of pedestrian modelling, various public bodies produce and maintain guidance documents relating to aspects of pedestrian modelling relevant to their industry and projects. Much guidance is produced by transport-related agencies for transport stations and interchanges, and guidance also exists for other industries, such as event venues and sporting stadia.

In the UK, guidance is produced by transport agencies, such as Transport for London (TfL) and Network Rail (NR), as well as major infrastructure development authorities, such as High Speed 2 (HS2) Limited and Crossrail Limited. Guidance for other industries is produced by bodies including the Department for Digital, Culture, Media and Sport (DCMS) and the Events Industry Forum (EIF). It is important to note that no consistency applies in terms of content or frequency of updates for guidance documents, although most are updated as required, such as when software changes occur that affect the content of the guidance. A list of selected guidance documents commonly used in pedestrian modelling is given in Table 6.3.

6.4.4 Most common issues – lessons learnt

Even with best-practice network coding practices, issues tend to arise in most pedestrian micro-simulation models. Despite significant improvements in recent years that allow for the development of highly robust

Table 6.3 Pedestrian modelling guidance documents

Transport for London (TfL)

- *Station Modelling with LEGION SpaceWorks: Best Practice Guide*, version 3.2 (2016)
  A summary of the key steps for modelling TfL Rail & Underground (R&U) stations using LEGION to produce consistent and accurate outputs.
- *G371A Station Planning Standards and Guidelines* (2017)
  A definition of space requirements for passengers and staff accommodation in stations. Includes guidance on calculating peak passenger flow and ticket gating and issuing requirements.

Network Rail (NR)

- *Guide to Modelling Gatelines in LEGION SpaceWorks* (2016)
  An overview of gateline model coding, covering object-naming conventions, delay times and passenger throughput, object sizing, placement, parameter settings and linking and methods for dealing with congestion.
- *Station Capacity Planning Guidance* (2016)
  Guidance for undertaking capacity assessments for all stations on the National Rail network in the UK. The guidance promotes a consistent and best-practice approach to assessing the performance of passenger space within stations.

Department for Culture, Media and Sport

- *Guide to Safety at Sports Grounds* (5th edn) (2008)
  Guidance for ground management and technical specialists (e.g. architects, engineers) and representatives of relevant authorities on the assessment of the number of spectators that can safely be accommodated within a sporting facility.

Events Industry Forum

- *The Purple Guide to Health, Safety and Welfare at Music and Other Events* (2nd edn) (1999)
  Guidance on the safe organisation and running of all kinds of events.

and reliable models, modelling software packages are still commonly affected by issues that cause models to block up and become unresponsive, and cause entities to behave in unexpected or unintended ways. Common model issues can be grouped into two categories, relating either to network coding or the software package.

Poor or inconsistent network coding often results in modelling issues. Inconsistencies in coding occur as a result of multiple modellers working on the same model or different scenario models for a single project. Inconsistencies pertaining to object-naming conventions, filtering and linkages, as well as the overall approach to coding, such as methodology in routing, entity types and connectivity, can create confusion and simple errors in the model that lead to unexpected model outputs, which can sometimes go unnoticed and the cause of which can be difficult to identify.

Stability issues with some software packages can cause model simulations to become unresponsive. This often occurs when crowds of pedestrians become congested in tight spaces and are unable to move, often a result of over-capacity demand or opposing or cross pedestrian flows. As this occurs, more pedestrians continue to attempt the same movement over time and the congestion builds until the model simulation slows or stops completely.

Model simulation times can also be prohibitively slow in some software packages due to a number of reasons, including the complexity or size of the model environments in terms of accessible space, number of entities or the number and complexity of objects in the model. Simplifying the model inputs, such as the CAD, demand and overall model scope, as much as possible is recommended to minimise the chance of excessively slow simulation times. Some types of behaviours and movements, such as queuing, can also increase model simulation times in some software packages. This may result in inconsistent model simulation times between different scenario models in the same project, making it difficult to plan the process of running multiple simulations.

Given the nature of pedestrian modelling as a relatively niche industry, the modeller has limited access to peer support – for example, through online user forums. Support and guidance related to overcoming common issues, best practise in network coding and optimising settings for reducing simulation speeds may be available from the developers of the various software packages. Some more specific pedestrian modelling challenges, as well as strategies for overcoming them, are covered in the case studies in Part II of this book.

6.5. Summary points

- The pedestrian micro-simulation modelling process can be divided into three parts: model information, dynamic modelling and analyses.
- Prior to building a pedestrian model, it is particularly useful for modellers to undertake site visits to familiarise themselves with the existing conditions and identify any issues and concerns.
- Typically, a modeller will first develop a base model that passes calibration and validation criteria to ensure its results are reliable, evaluating the model against quantitative measures using real data such as delay, journey time, flow rate, pedestrian density and capacity.
- A number of scenarios are built upon the base model to meet the study objectives, and what-if analysis is then conducted, including comparative assessment of scenarios based on defined KPIs.
- Guidance documents used to specify recommended or mandatory standards and guidelines to be followed in undertaking pedestrian modelling projects are often maintained by various public bodies relevant to their industry and projects.
- In order to create accurate micro-simulation models of the interaction between pedestrians and traffic a sound understanding of the behaviour of pedestrians around traffic is required.

- Pedestrian modelling projects are inherently prone to uncertainties, and these are wide-ranging and relate to all aspects of the modelling process, from data collection to network coding and simulation, to model analysis and outputs.

NOTE

1. See Chapter 2 for further information regarding Fruin's LoS.

REFERENCES

[1] Balakrishna R, Antoniou C, Ben-Akiva M, Koutsopoulos HN and Wen Y (2007) Calibration of microscopic traffic simulation models: Methods and application. *Transportation Research Record: Journal of the Transportation Research Board* **1999**: 198–207.

[2] Dridi MH (2015) Pedestrian flow simulation validation and verification techniques. *Current Urban Studies* **3(2)**: 119–134.

[3] El Esawey M and Sayed T (2011) Calibration and validation of micro-simulation models of medium-size networks. *Advances in Transportation Studies* **24**: 57–76.

[4] Federici ML, Manenti L and Manzoni S (2014) *A Checklist for the Evaluation of Pedestrian Simulation Software Functionalities*. arXiv:1404.7717.

[5] Hoogendoorn SP and Bovy PHL (2004) Pedestrian route-choice and activity scheduling theory and models. *Transportation Research Part B: Methodological* **38(2)**: 169–190.

[6] LEGION. www.bentley.com/en/products/brands/legion?skid=CT_WRO_LGN_REDIRECT_18.

[7] Network Rail (2011) *Station Capacity Assessment Guidance*. Network Rail, London, UK.

[8] Sargent RG (2009) Verification and validation of simulation models. *Proceedings – Winter Simulation Conference* **7(1)**: 162–176.

[9] Smith J and Blewitt R (eds) (2010) *Traffic Modelling Guidelines: TfL Traffic Manager and Network Performance Best Practice*, 3rd edn. Transport for London, London, UK.

[10] TfL Rail & Underground – Station Modelling (2016) *Station Modelling with LEGION SpaceWorks: Best Practice Guide*, version 3.2. Transport for London, London, UK.

[11] Wang T (2012) *Study of Pedestrian–Vehicle Interaction Behaviour by Microscopic Simulation Modelling*. PhD thesis, Faculty of Engineering and the Environment, Civil and Environmental Engineering, University of Southampton, Southampton, UK.

ICE Publishing

Urban Space – AECOM
ISBN: 978-0-7277-6381-5
https://doi.org/10.1680/mpm.63815.093

Chapter 7
Looking to the future

7.1. Overview of the chapter

In this chapter, the future of pedestrian modelling and some of the factors that may affect the industry in years to come are discussed. There are a number of areas where pedestrian modelling can benefit from technological advances. The availability of data from mobile devices is already starting to change the field, and further advancements in data are discussed in Section 7.2. Technological improvements to both hardware and software are then discussed in Section 7.3. The chapter concludes with a discussion on urban mobility and how autonomous vehicles and an ageing population could affect how pedestrians are modelled in the future.

7.2. Data

As current technologies evolve, data-collection methods will improve and new techniques will be developed. Existing data-collection methods have various limitations, ranging from lack of accuracy to prohibitive costs. These limitations are discussed in general in Section 3.4. The factors that may influence changes to data-collection methods are further discussed in this section.

7.2.1 Changing travel patterns

Currently, pedestrian modelling, particularly transport interchange modelling, is concerned with typical weekday morning and evening peak pedestrian movements. Crowd numbers are highest at these times and place the greatest demand on walking infrastructure. However, working culture in offices and other places of employment is shifting, with employers commonly promoting flexible working environments that allow staff to adapt their working hours and locations to better suit their personal needs and lifestyles. This flexibility has resulted in the spread of travel demand over a longer period of the day as well as a reduction in overall demand by reducing people's need to travel, such as when working from home.

7.2.2 Pattern-recognition software

Much research and redevelopment is currently being undertaken on pattern-recognition algorithms to increase the level of accuracy of automated pedestrian detection from recorded video footage. Pattern-recognition algorithms can detect additional information, such as the direction, gender and age group of passing pedestrians, as well as individual counts within the study area. As this process becomes more automated, accuracy and speed will increase, while improvements in computing power will reduce processing times.

In the coming years, live data could feed predictive models to trigger operational adjustments. For example, train-load-weight data for arriving trains could inform operational decisions regarding crowd management at terminus stations in advance of train arrivals. In fact, such models are becoming more commonplace in the field of traffic modelling and traffic management. Madrid and San Diego are among the cities that have implemented predictive traffic modelling systems.

7.2.3 Sensors

Existing technologies are limited to the use of microwave radar, infrared, pressure, laser, and ultrasonic sensors for collecting pedestrian movement data. These are discussed in greater detail in Section 3.3.1. Advances in sensor technologies may lead to the development of new techniques for counting pedestrians that overcome the limitations of existing methods.

7.2.4 Wi-Fi, Bluetooth and mobile-phone connectivity

As computing becomes ever more mobile, more data relating to pedestrian movement will be obtainable using Wi-Fi, Bluetooth and other forms of mobile connectivity. Currently, captured mobile data represents only the portion of the population that uses mobile phones and other cellular devices – that is, traditionally younger, more technologically conversant people; however, device ownership is increasing. According to the Spring 2018 Global Attitudes Survey [3], 76% of the adult population in the UK and 81% in the USA owned a smartphone in 2018. The same research shows that a wider demographic now owns smartphones, so the data obtained is more representative of the population as a whole. The trends of rising smartphone ownership and smartphones becoming more commonly used by a wider subset of the population are even more prevalent in emerging economies. Furthermore, the number of Wi-Fi hotspots is constantly increasing, so greater accuracy may be obtainable from a denser network of Wi-Fi connection points in the future.

A further limitation of mobile-phone data is that the telecoms infrastructure lacks sufficient granularity to accurately track pedestrian movements. Future developments in mobile-phone technology may allow tracking in much greater detail, increasing its potential for use in pedestrian modelling [1].

7.2.5 Location analytics

Location analytics provide spatial insight from the location information of collected data. When used alongside a geographical information system, location analytics allow data to be analysed and visualised in new ways, which may be useful in analysing the ever increasing data available on pedestrian flows for modelling purposes.

7.2.6 Big data and privacy

The term ‘big data’ has several definitions but generally refers to any data set too complex to be processed and analysed by traditional data-processing applications [1]. Data can be generated in three ways: first, by people – for example, taking photos and writing text; second, by machines – for example, data generated by computers and GPS devices; and, finally, from objects – for example, digital signals acquired by camera [4]. There is significant potential in the use of this kind of data in the tracking of pedestrian movements on a scale and accuracy much greater than is currently possible from traditional forms of data and data-collection methods.

With the rapidly growing availability of easily accessible data, however, people are becoming more concerned about personal data security and privacy. This has resulted in increased privacy regulation, such as that introduced by the European Union in the form of the General Data Protection Regulation (GDPR) in 2018. This kind of regulation will affect the ability to obtain data relevant for pedestrian modelling in the future. Methodologies will need to be developed to ensure that data collected from available sources, such as Wi-Fi and mobile phones preserves the anonymity of individual users.

7.3. Hardware and software

Future technological advancements in both hardware and software systems will be necessary to collect, process and store the data captured by the methods covered in Section 7.2.

7.3.1 Data capture and storage

The availability of vast amounts of data, such as that from the big data sources discussed in Section 7.2.6, will generate demand for equally large data-storage capabilities. Most data is already stored in digital form, on internal or external data hard drives, cards or other forms of storage, such as optical storage media. Advancements in data-storage capability in terms of cost, security, transfer speeds and storage size will allow for the storage and use of the increasing amounts of data being captured.

7.3.2 Cloud computing

Pedestrian models can be very large and complex. The scale of models is likely to further increase in the future as the volume and accuracy of available data increases. Complex models require intense computing power in order to simulate pedestrian movements in a useful period of time. Cloud computing allows for the pooling of shared computer-processing and data-storage resources, providing much greater processing power than could be provided by a single computer.

7.3.3 Software

Advances in computer-processing capability will allow for more complex modelling. In order to realistically model pedestrian behaviour in complex environments, the patterns which control behaviour need to be understood and replicated. Current pedestrian modelling software offers the ability to change the physical behaviour of agents relating to size, speed, luggage and other attributes; and much research is currently ongoing to understand pedestrians with reference to:

- the impact of the environment on planning and route choice;
- the influence of emergent stimuli on planning and behaviour; and
- the familiarity of the environment on reactions to an emergency situation.

The development of more advanced behaviour algorithms will allow more accurate replication of human behaviour and provide a better representation of the choices made on a journey. This will make pedestrian modelling more attractive to a wider range of uses, such as in retail environments.

7.4. Urban mobility

Technological advancements are constantly changing the way people move around cities. Walking, which was once the primary mode of transport for all, declined as a result of the rise of the motor vehicle. However, it is once again gaining prominence as a legitimate means of transport as congestion and health concerns increase.

Other forms of transport are also evolving, and the interaction between pedestrians and both existing and new forms will be relevant for pedestrian modelling as urban mobility changes. Developing transport modes that will need to be considered in pedestrian modelling in the near future are discussed in this section.

7.4.1 Autonomous vehicles

There are many challenges arising from the proliferation of autonomous vehicles, the foremost of which are technological and regulatory. However, the potential benefits of autonomous vehicles in terms of safety, emission reduction, congestion reduction, urban space reallocation, and impact on travel behaviour are considerable. The spread of autonomous vehicles will result in increasing interactions with pedestrians.

The term 'autonomous vehicle' (AV) refers to a number of vehicle types, from self-driving regular passenger vehicles, similar in scale and shape to traditional cars, to self-driving vehicles similar in typology to trams or buses, to small-scale pod-like vehicles. AVs can operate in various contexts, such as regular streets, segregated roadway networks and pedestrian zones, although in the latter they are most likely to be the smaller and slower pod-like passenger vehicles. The interaction between pedestrians and AVs on regular streets or segregated networks is unlikely to be significantly different to the current interaction between pedestrians and regular private and public transport vehicles, as discussed in Section 5.4. Modelling the pedestrian interaction with AVs where vehicles travel among pedestrians in a shared environment will create a greater challenge, one that will require further research as AVs become more widespread.

7.4.2 Demand-responsive transit and travel planning

Mobile-phone-based and online travel-planning applications are changing the way people plan their route choices prior to travel. Applications such as Google Maps and Citymapper better inform pedestrians in real time, providing information about departure times, delays and even congestion levels, so that more informed routing choices can be made at all pedestrian choice levels. This kind of information is particularly useful in informing tactical choices related to medium-term routing decisions, such as which train service to catch on a particular day. As pedestrians increasingly rely on real-time information to inform their route planning, they are able to react to unplanned changes or disruptions by changing their routes to optimise their travel. As a result, pedestrian routes may become less predictable, so it may be necessary to code more spontaneous route choices in pedestrian models to reflect this. Public transport services may also become less static, adapting to changing demand patterns and disruptions in the network. This may need to be considered in modelling the interaction between pedestrians and public transport in microsimulations.

7.4.3 Smart cities

The term 'smart cities' is increasingly used to describe cities that make use of digital technologies to increase the efficiency of urban networks and services. While many aspects of a city, such as its governance, housing, energy grid, industry and healthcare can be described as smart, of most relevance to pedestrian modelling is the growth of smart transport networks. Increasing connectivity between infrastructure, vehicles and users is transforming urban networks from a static combination of roads, tracks and signals to a dynamic interconnected system that responds in real time to changing conditions, such as user demand, weather, incidents and disruptions. Connected infrastructure will not only lead to an optimisation of capacity utilisation and asset performance but also allow much more data to be gathered on pedestrian movements, which in turn will provide more accurate and easily accessible modelling inputs for pedestrian modelling.

7.4.4 Planning for an ageing population

Improvements in nutrition, healthcare and safety coupled with declining birth rates have led to an ageing population in many developed countries. Providing the necessary infrastructure to meet these changing demands will become an increasing priority in order to enable older people to live without extensive care.

As many active older people will be unable, or choose, not to drive, it will be important to recognise that an ageing population demands an efficient, accessible public transport system and the urban and rural infrastructures to accommodate it. Simple infrastructure designs, such as dropped kerbs for wheelchair users, are an obvious example of how the urban landscape can be improved for an ageing population. In a pedestrian modelling context, it is important that this ageing population's characteristics and behaviour is appropriately researched and defined within the software, so that infrastructure design caters for all users.

7.4.5 Application of artificial intelligence and machine learning

Since the 1990s, advances in computing technology have allowed the development of faster computers which can focus on solving more specific problems. Faster computers, advances in deep learning and access to more data have made artificial intelligence (AI) popular throughout the world. Generally, the term 'artificial intelligence' means a machine which mimics human reasoning. An ideal (perfect) intelligent machine is a flexible agent which perceives its environment and takes actions to maximise its chance of success at some goal or objective.

Approaches to simulating pedestrian movement behaviour based on AI are becoming more common. These approaches generally require the collection of a large volume of microscopic pedestrian movement data, which is then fed into an artificial neural network via network training. The trained network is then fed back into a simulation environment to predict pedestrian movement.

Ma, Lee and Yuen [2] describe how two simulation experiments were conducted to evaluate the performance of this type of approach. The experiments were on collision avoidance and road crossing, and the results were compared with the real-life scenario. The comparison revealed that the pedestrian distributions, erratic trajectories and density–speed fundamental diagram in the simulation were reasonably consistent with the real-life scenario.

The development of more intelligent software will allow the user to more accurately simulate complex movements and journeys but also to change how models are developed. Instead of introducing sensors and commands into the model network, such as decision points, a range of entities with objectives will be modelled rather than hard-coded routes. These models once calibrated and validated with real-time data will be able to replicate future scenarios more accurately and thus more accurately predict the impacts of schemes or management measures.

7.5. Summary points

- Pedestrian modelling is commonly focused on typical weekday morning and evening peak pedestrian movements, when demand is greatest on walking infrastructure, but there are a number of factors that will impact how people move and how pedestrians are modelled in the future.
- Greater working flexibility has resulted in the spread of travel demand over a longer period of the day and a reduction in overall demand in people's need to travel, such as when working from home.
- Current mobile data relating to people movement represents typically younger, more technologically conversant people; however, device ownership is increasing.
- Mobile data will be used to track pedestrian movements on a scale and accuracy much greater than is currently possible from traditional forms of data and data-collection methods.
- 'Smart cities' make use of digital technologies to increase the efficiency of urban networks and services, allowing much more data to be gathered on pedestrian movements and, in turn, providing more accurate and easily accessible modelling inputs for pedestrian modelling.
- The development of more advanced behaviour algorithms will allow more accurate replication of human behaviour and provide a better representation of the choices made on a journey.
- Other forms of transport are also evolving, and the interaction between pedestrians and both existing and new modes will be relevant for pedestrian modelling as urban mobility changes.

- Mobile-phone-based and online travel-planning applications are changing the way people plan their route choices prior to travel and better inform pedestrians in real time, providing information about departure times, delays and even congestion levels, so that more informed routing choices can be made at all pedestrian choice levels.
- An ageing population will require changes to how infrastructure is planned in order to ensure it is fully accessible; and appropriate algorithms to represent this behaviour will need to be present within pedestrian modelling software.

REFERENCES

[1] Leeson A, Alvarez P and Ghosh S (2014) Understanding how big data and crowd movements will shape the cities of tomorrow. *Proceedings of the European Transport Conference, Frankfurt, Germany.* https://aetransport.org/past-etc-papers/conference-papers-2014?abstractId=4289&state=b (accessed 04/06/2019).

[2] Ma Y, Lee EWM and Yuen RKK (2016) An artificial intelligence-based approach for simulating pedestrian movement. *IEEE Transactions on Intelligent Transportation Systems Archive* **17(11)**: 3159–3170.

[3] Pew Research Center (2018) *Spring 2018 Global Attitudes Survey, Q45 & Q46.* www.pewresearch.org/global/2019/02/05/smartphone-ownership-is-growing-rapidly-around-the-world-but-not-always-equally/ (accessed 04/06/2019).

[4] Zhang D (2018) Big data security and privacy protection. *Advances in Computer Science Research, volume 77, 8th International Conference on Management and Computer Science (ICMCS 2018), Shenyang, China.* Atlantis Press, Beijing, China.

Part II

Case studies

ICE Publishing

Urban Space – AECOM
ISBN: 978-0-7277-6381-5
https://doi.org/10.1680/mpm.63815.101

Chapter 8
Interchanges

8.1. Overview of the chapter

In most European and densely developed countries the most common mode of transport in urban areas is metro rail services. The demands of a growing, changing population have driven the need for cost-effective design of new metro rail networks and the improvement of older ones. In many urban centres, railway stations have become more than just an access point to a mode of transport; they are now often a hub of multiple economic and leisure activities and thus a destination in their own right in the overall urban fabric.

This chapter discusses the benefits of pedestrian modelling for interchange planning and operation in three different contexts and settings. It discusses how pedestrian modelling has been used in a range of studies for different elements of station design, such as capacity improvements, evacuation, retail, disruptions and events. It describes the analysis and outputs which show how pedestrian modelling science and techniques help in ensuring a safe, efficient and smooth operation of station interchanges amid growing travel demand and people's increasing mobility needs.

8.2. Case study: interchange station, Middle Eastern city

8.2.1 Setting the scene

The station is located on the underground section of the metro rail network, which is a recently planned rapid transit system. Pedestrian flow simulation modelling was undertaken to understand and address potential issues regarding the capacity and pinch-points in the layout of this busy interchange and its ability to safely accommodate the forecast demand in the horizon year of 2026.

The station is a central interchange point between services on two lines of the metro network:

- Metro Yellow Line – which operates on the lower platforms and runs east–west; and
- Metro Blue Line – which operates on the intermediate level platforms and runs north–south.

The station has five levels:

- Street Level;
- Concourse Level;
- Intermediate Level – Blue Line platforms;
- Transfer Level; and
- Yellow Line platforms

There are three entrances to the station from the street, one each to the north-west, north-east and south-east of the road junction at Street Level. Each of the street entrances leads down to a large ticket hall on the underground Concourse Level.

From Concourse Level, two sets of escalators and stairs lead down to the Intermediate Level – Blue Line platforms. A third east–west-oriented set of escalators and stairs leads down directly from the concourse to the Transfer Level, skipping the Intermediate Level. A single set of stairs and escalator also connect the Intermediate Level – Blue Line platforms with the Transfer Level below.

From the Transfer Level, two further sets of stairs and escalators lead down to the lowest level of the station, where the Yellow Line platforms are located.

A total of three lifts connect Street Level (one at each entrance) and Concourse Level. Three more lifts connect the Concourse Level with the Intermediate Level – Blue Line platforms, two of which continue down to the Transfer Level. From the Transfer Level, a further set of two lifts provides access to the Yellow Line platforms below. Figure 8.1 shows the layout of each of the station's five levels.

8.2.2 Defining the problem

A review of the station's forecast demand identified the morning peak hour – 08:00–09:00 – as the busiest time during normal operations. Table 8.1 provides an overview of peak-hour demand at the station. It was assumed that pedestrian numbers to and from the street would be evenly distributed between each of the three street entrances.

The total number of passengers boarding and alighting during the peak hour was anticipated to be 14,800 passengers. This demand puts a significant strain on the vertical circulation elements of the station, particularly the escalators and stairs connecting the platforms with the Concourse and Transfer levels.

Cumulative mean density (CMD) maps, as discussed in Section 6.4.2.1, are commonly used as a method of visualising walking infrastructure performance.

It can be seen in Figure 8.2 that the density levels of boarders waiting adjacent to train carriages are at an acceptable level (black indicating LoS C); however, there are large areas of high density (reaching as high as LoS F) at the base of the escalators and stairs, which are caused by alighting passengers queuing while they wait to exit the platform.

These queues can become large and potentially pose a risk to passenger safety, as there are high densities very close to the platform edges. Additionally, a safety risk is posed to passengers arriving on to the platform via the escalators as their run-off areas are partially (and sometimes completely) blocked by passengers queuing to use the escalators in the opposite direction. The queues do, however, dissipate before the arrival of the next train.

Figure 8.3 shows that there are high densities (up to LoS E (black) and F (segmented grey)) at the base of the escalators to both the Yellow Line and Blue Line platforms; this is mainly due to interchanging passengers.

Figure 8.4 shows a similar situation to Figure 8.2 (for the Yellow Line platforms), namely that while the density for boarders is generally acceptable, the densities experienced by passengers alighting at the base of the stairs and escalators leading from the platforms grow very large (as high as LoS F (black)); in particular, it can be seen that the greatest densities are due to interchanging passengers attempting to go to the Transfer Level.

The Blue Line platforms were found to suffer slightly more than the Yellow Line platforms because the interchanging passengers have only a single staircase and escalator available, as opposed to the two sets of

Figure 8.1 Interchange-station layout

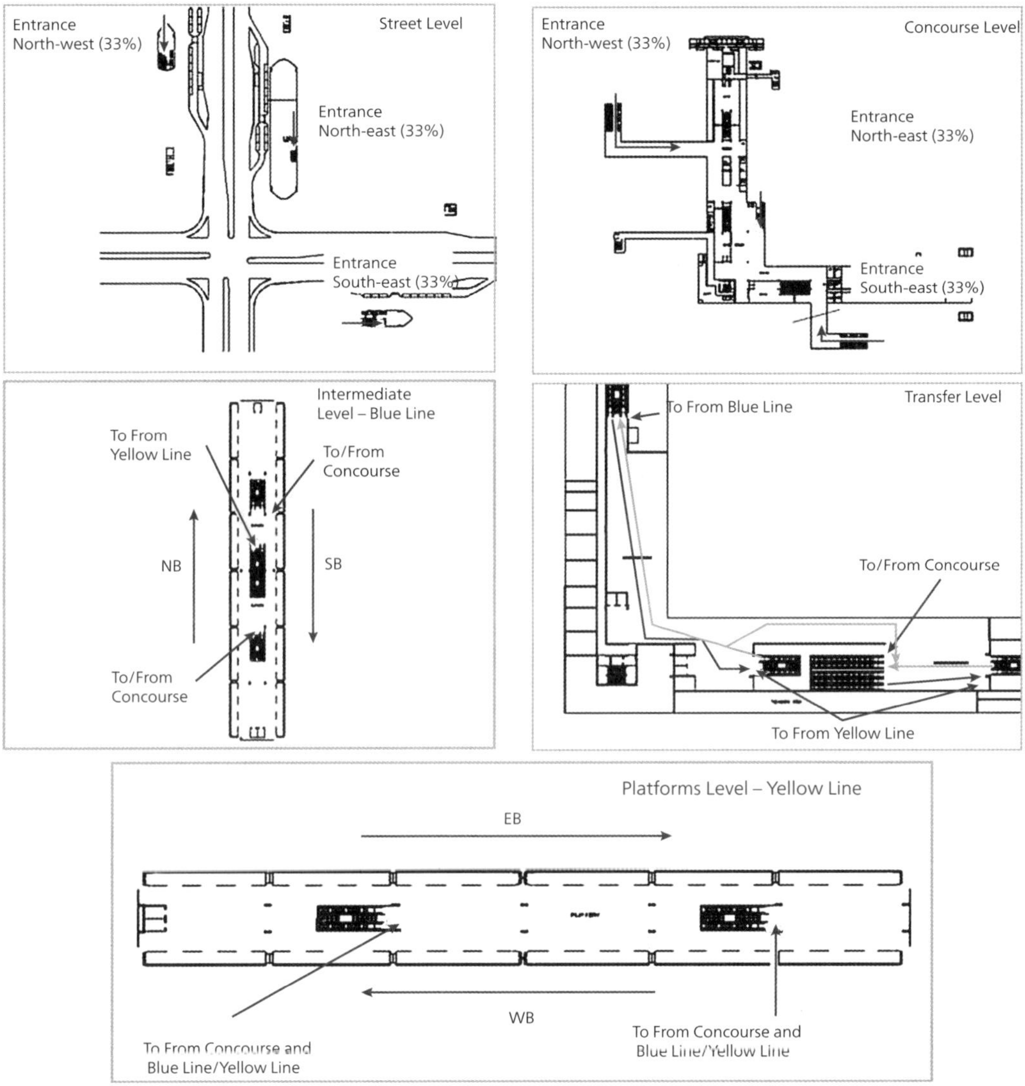

stairs and escalators for passengers alighting on the Yellow Line. This most likely accounts for the fact that on at least one occasion a handful of passengers who had alighted were still queuing at the base of the stairs when the next trains arrived on the platforms. Once again, the high densities experienced very close to the platform edges could pose a risk to passenger safety, as could the blockages in the run-off areas of escalators.

Figure 8.5 shows that the density levels experienced by passengers on the Concourse Level are generally at an acceptable level (LoS B); however, there are some relatively small areas which experience densities as high as LoS E (black). The most significant of these is at the exit gates used by passengers alighting from the Blue Line.

Table 8.1 Origin–destination (OD) matrix

	Street	Yellow Line Eastbound	Yellow Line Westbound	Blue Line Northbound	Blue Line Southbound
Street		600	600	900	500
Yellow Line Eastbound	600			1,550	850
Yellow Line Westbound	600			1,550	850
Blue Line Northbound	500	900	900		
Blue Line Southbound	900	1,500	1,500		

Figure 8.2 CMD map – Yellow Line platforms

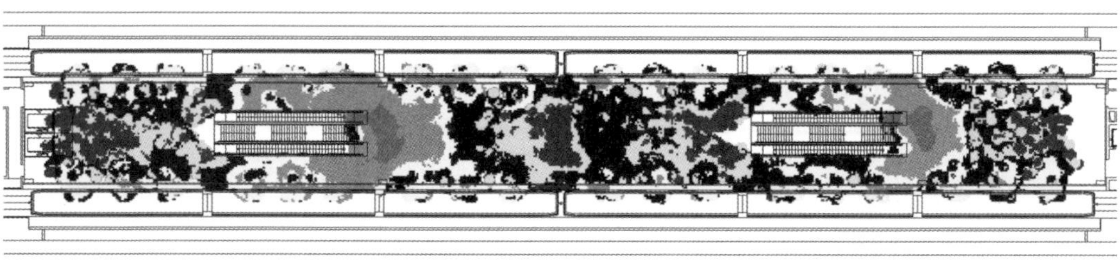

Model screenshots shown in Figure 8.6 portray the worst levels of congestion experienced at the station.

The highest levels of congestion at the station are evident on the platforms and Transfer Level, which can be largely attributed to interchanging passengers from both the Yellow and Blue lines. The large number of interchanging passengers combined with the relatively small capacity available to them results in very high densities (up to LoS F (dark grey dots)) seen in multiple locations. The most concerning areas of congestion are close to platform edges, which could be a risk to passenger safety. There is a potential additional risk to

Figure 8.3 CMD map – Transfer Level

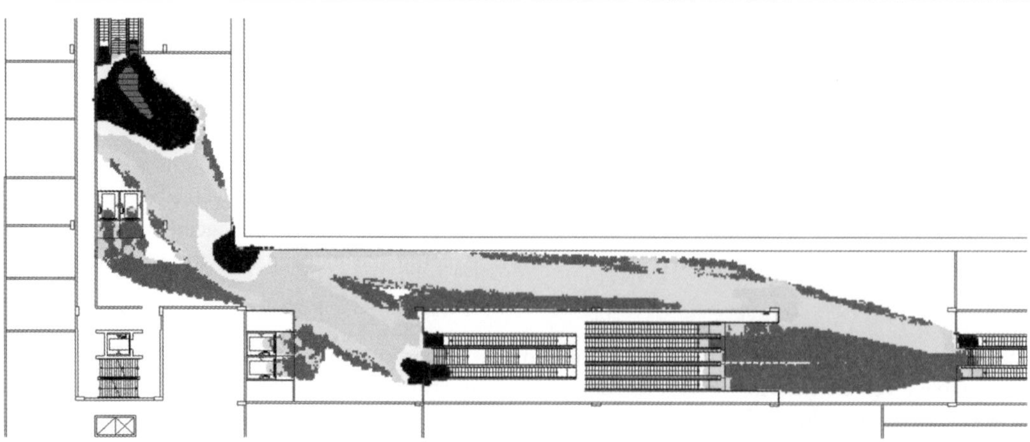

compromise the run-off area of the adjacent escalator delivering people to that level. This poses a serious safety risk, if passengers are unable to disembark a moving escalator due to crowding at its exit.

The concourse and entrance links to Street Level were found to operate comfortably within acceptable levels of service and with sufficient capacity to accommodate future increases in demand. The provision of gateline capacity and ticket-office facilities should be designed in accordance with anticipated demand, and the station's concourse layout has scope for this without compromising general circulation through the concourse.

8.2.4 Proposed mitigation

The performance of the platform levels and the high level of congestion brought into question the ability of the station to accommodate growth in demand beyond the horizon year of 2026. The congestion levels on the platforms would also not meet user expectations of a major new modern transportation hub.

The following mitigation measures were therefore recommended:

- increase platform widths for both the Blue Line and Yellow Line platforms, as currently the platforms narrow at some points to widths below 3m (mostly adjacent to stairs and escalators);
- increase the vertical circulation capacity between the Yellow Line and Blue Line levels;
- introduce crowd-flow barriers at the base and top of the escalators to protect and keep clear the run-off areas of adjacent escalators delivering passengers to those levels;
- identify the locations of ticket-purchasing facilities (TIWs and POMs) so any queues that form do not interfere with the normal flow of passengers within the Concourse Level.

The design layout of the station and the crowd model were amended to incorporate the mitigation measures, and then the model was re-run to assess the adequacy of the mitigation measures and get assurances that they will alleviate the congestion issues highlighted in the previous model run.

8.3. Discussion: degraded operation

While the focus of station-based pedestrian modelling is often normal operation during peak hours, as any experienced commuter will know, railways are frequently susceptible to disruption and delay. This is known as degraded operation, or perturbation. It is defined by guidance documents to cover a range of typical operational situations of varying severity, such as:

- Low severity
    - reduced headway
    - consecutive arrivals or departures
    - loss of vertical circulation or access element (lift/escalator/stair)
- Medium-severity
    - crush-loaded service
    - 15-minute delay to services
- High-severity
    - station used as terminus (planned/unplanned)
    - system shutdown and reopening

Degraded-operations studies are undertaken to help develop or update station crowd-management plans. These plans detail several controls for each level of the station, which may include closing entrances, opening automatic ticket gates, removing retail seating areas, stopping or reversing escalators and calling for police assistance. It is important to engage with the appropriate station management team during this type of study as they will have valuable information about station behaviour.

8.4. Case study: Waterloo, Wessex Capacity Alliance

8.4.1 Setting the scene

London Waterloo is the largest and busiest railway station in the UK. The station is managed by Network Rail, with train services provided by South Western Railway (SWR). The aim of the Wessex Capacity Programme (WCP) is to deliver capacity improvements across the Wessex route into London Waterloo as a programme of infrastructure interventions, which interface with SWR's Rolling Stock programme and align with the Wessex Route Strategic Business Case.

In 2014, AECOM collaborated with Colas Rail Ltd, Skanska, Mott MacDonald and Network Rail to deliver significant improvements at London Waterloo station. The track and infrastructure improvements will enable longer and more frequent trains to run on the Wessex route, which covers the major commuter area of south-west London as well as the south and south-west of England. Usage of Waterloo station is expected to increase because of the improved rail infrastructure, and pedestrian modelling played an integral role in informing the design work. There are two main areas of improvement:

- the reintegration of the existing Waterloo International Terminal (WIT) within the main station; and
- the suburban ten-car scheme on platforms 1–8, which comprises platform extensions and new staircases on Platforms 1–4 – these modifications will enable longer train configurations on platforms 1–8 and platforms 20–24.

A plan of the station is provided in Figure 8.7.

Significant areas of the station had to be modified because of these improvements, and there was an impact on fire safety. This case study outlines the dynamic evacuation modelling undertaken for London Waterloo station to inform the development of the fire strategy and to support decision-making.

The station forms part of the wider Waterloo transport hub, with direct connections to two London Underground ticket halls at this site – via the Colonnade to the Jubilee Line services and via the Bakerloo ticket hall for Bakerloo and Northern Line services – as well as a connection via the Peak Hour Subway (PHS) to Waterloo & City Line services. There is a taxi rank along Station Approach to the porte-cochère and direct step-free connection to the concourse. There are also connections at street level to numerous bus routes.

8.4.2 Defining the problem

As part of the design process for the reintegration of WIT, a range of evacuation scenarios were undertaken for the existing and future design of Waterloo station, including WIT, in order to observe more clearly the potential passenger movements and crowding behaviours within the station complex.

Figure 8.7 Plan of London Waterloo station

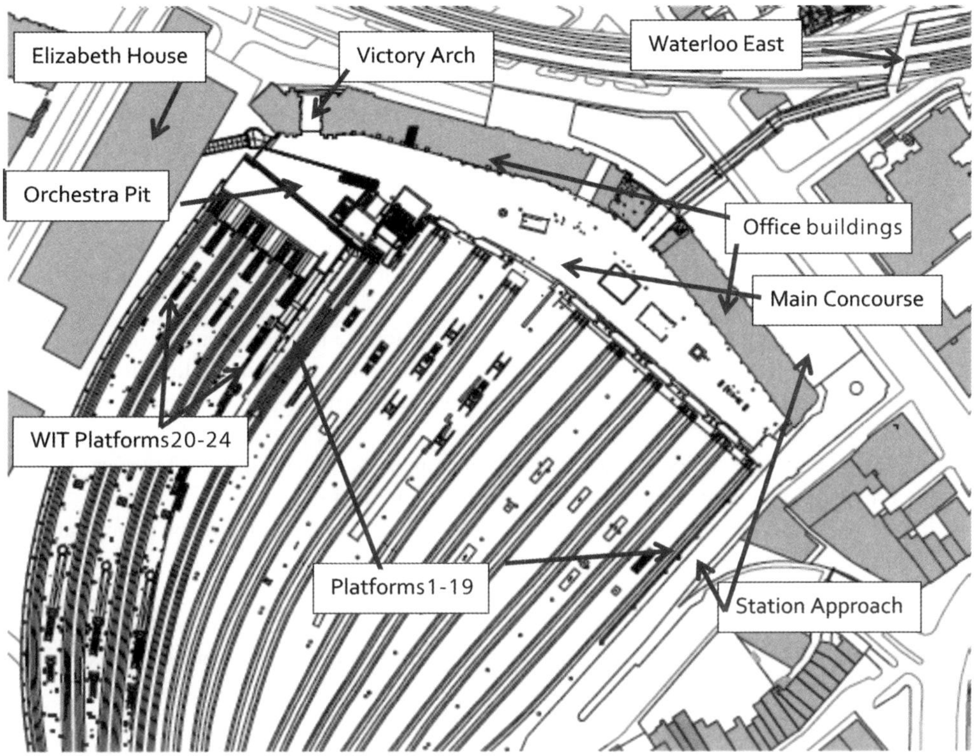

For railway stations, specific guidance documents exist with standards for safe evacuation from station environments:

- Office of Rail and Road *Railway Safety Principles and Guidance* (RSPG), the *Guidance for Fire Precautions on Existing British Rail Surface Stations*; and
- Network Rail *Station Capacity Assessment Guidance* (SCAG).

The recommendations of Network Rail's SCAG are:

- surface station platforms should be evacuated in a maximum of 8 minutes;
- the whole station should be evacuated in no more than 12 minutes.

In accordance with Network Rail's SCAG (section 4.4), there are two main fire scenarios that need to be assessed when developing a fire strategy for a station:

- fire on a train;
- fire in the station.

The fire-on-a-train scenario considers the arrival at the station of a train on fire. The train is assumed to be crush-loaded, requiring evacuation of the passengers on the train and the station. All evacuation routes are available, though the effect of thermal radiation from the fire and its impact on the evacuation of the incident platform can be further considered.

The RSPG and SCAG stipulate that when assessing escape capacity, the fire-on-a-train scenario should assume a train arriving on fire at the busiest platform. The number of passengers to be evacuated is to consist of the number of passengers on the train on fire (a crush-loaded train in this instance) and the number of passengers waiting for that train after a delay in the service of 5 minutes. Other services are to be assumed to be operating a normal service.

The fire-in-the-station scenario considers a single fire anywhere within the station. At a complex station like Waterloo, this scenario is representative of a variety of fires that could occur within a station, such as:

- Fire near to an exit – resulting in this exit being unavailable for use in the evacuation. To assess the adequacy of escape routes, the fire is assumed to discount the most capacious route. This could be caused by a fire within an escalator or lift, in a ticket machine, a temporary concourse exhibit, a stand-alone kiosk or a piece of luggage.
- Fire within a retail unit – the station already has several retail units, with the potential for further retail areas to be provided in the long term (beyond the scope of the works).
- Fire within the back-of-house area – including the plant rooms and LCR (London & Continental Railways Limited) areas. In addition to people's safety, the impact on station and route operation is important.
- Fire in an office – considering the office building and its impact on the operation of the station.
- London Underground – the interface between the two station operators will alert one station to a fire in the other.

On most large infrastructure projects, a pedestrian modeller works with a fire engineer to determine the scenarios which are considered most critical for analysis. This case was no different, and only the regular alighting and boarding passengers – a smaller number than in the fire-on-a-train scenario – were assumed to be within the station.

In order to assess the worst case for the station, it was essential to understand how many people are in the station during peak periods. For evacuation modelling at stations, there are three areas to focus on:

- passengers in the unpaid areas (e.g. concourse, retail);
- passengers on the platforms; and
- passengers on trains, assumed crush-loaded.

Table 8.2 Comparison of static analysis results

	Eurostar		2024 services	
Platforms	20	21–24	20	21–24
Egress capacity	816.8	600.8	949.8	737.6
No of passengers				
750	55 sec	1 min 15 sec	47 sec	1 min 1 sec
1800	2 min 12 sec	3 min	1 min 54 sec	2 min 26 sec
3600	4 min 24 sec	6 min	3 min 47 sec	4 min 53 sec

The initial rough assessment and comparison of the former Eurostar operation and proposed evacuation routes for the WIT platforms is included as Table 8.2.

Table 8.2 shows that the platform-clearance times are all within the required 8-minutes SCAG guidance. However, the results do not robustly reflect platform clearance as some of the routes are at the far end of the platforms, so passengers might not be aware of them or want to walk towards them. Therefore, dynamic pedestrian modelling was undertaken to verify the evacuation time.

8.4.3 Evaluating performance

Models of the normal operations for the existing and future Waterloo station (with WIT) were developed by the Wessex Capacity Alliance (WCA) pedestrian modelling team as part of the early design process. These considered 2014 (baseline) and 2024 (design year) passenger demand and train timetables. The 2024 normal-operations passenger-flow modelling was used in the evacuation-modelling study to estimate passenger distribution within the future station:

- occupancy in the unpaid areas was based on the maximum occupancy recorded in the normal-operation model during both the AM and PM peak 15 minutes;
- platform-waiting volume was based on the number of boarding passengers on the paid side before service disruption, which was estimated using the modelled train load, and platform-announcement assumptions (e.g. Windsor trains have their platform announced 7 minutes prior to departure); and
- the train crush loads were 1,800 for both the new ten-car (suburban and Windsor) and existing 12-car (mainline) trains.

In calculating passenger movements, the software for the study utilised the concept of minimum cost as its cornerstone. Agents attempt to minimise their personal dissatisfaction (i.e. frustration, inconvenience and discomfort) when choosing their next step. Navigation decisions are made at three levels: strategic (choice of target), tactical (route selection) and local (next step).

The five WIT platforms were assessed in all design stages for insight into their evacuation and clearance. Four escape routes are available from each platform island:

- country-end stairs (2 × 1.1m)
- mid-platform stair (1 × 1.7m)
- mid-platform escalators (2 escalators)
- Orchestra Pit route (2 stairs + 3 escalators)

The incident train is assumed to be the first train to arrive. As the proceeding train for platform 20 arrives within 5 minutes of the incident train, it too is assumed to arrive. Both trains are assumed to be crush-loaded. This is a conservative assumption but allows both island and side platforms to be assessed simultaneously. Figure 8.8 shows the WIT platform layout and the available escape routes.

The routing of passengers during the evacuation was optimised to demonstrate the best evacuation times that can be achieved with the proposed infrastructure and the management of crowds – directing passengers to the routes available from their positions at the time of the incident – via station announcements or the presence of station staff.

Figure 8.8 WIT platform layout – escape routes

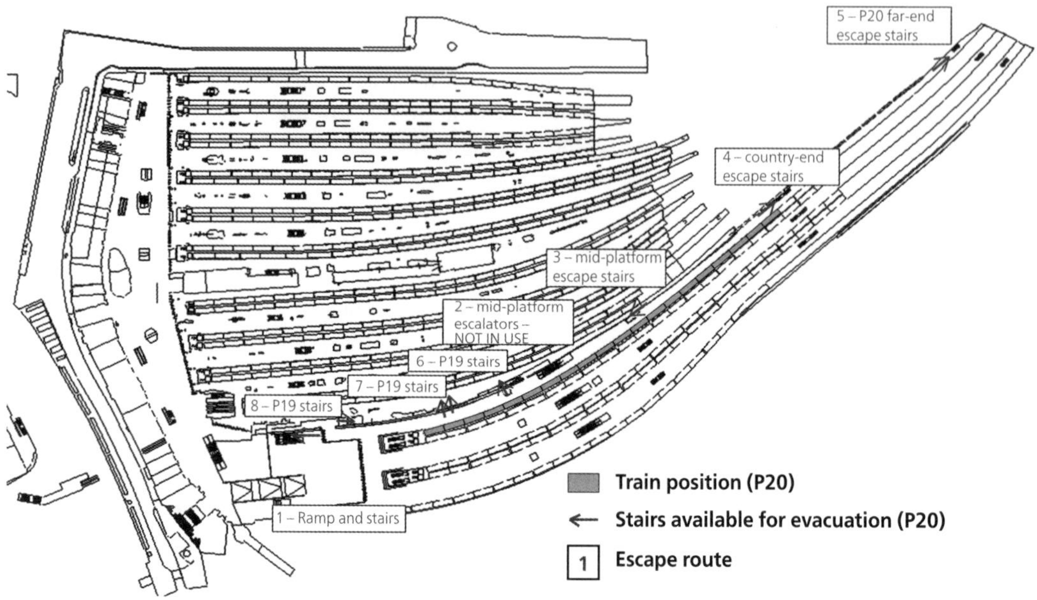

Table 8.3 shows the scenarios modelled and the routes available to the passengers in each scenario.

The total evacuation population is estimated to be 4,520 people on platforms 20–24 and simultaneous crush-loaded train arrivals on platforms 20 and 21/22. A summary of the evacuation population is shown in Table 8.4.

As shown in Table 8.5, in a fire-on-a-train scenario, platform 20 can be cleared within the required 8 minutes regardless where the incident occurs on the train.

These results show the difference in results between static analysis and dynamic analysis. Static analysis shows that a crush-loaded, 1,800-passenger train arriving on platform 20 would expect to clear in 2 minutes 12 seconds. However, depending on the exit routes available, dynamic modelling shows that this can vary from just under 4 minutes to 7 minutes 31 seconds. This demonstrates the value of dynamic micro-simulation modelling over first-principle-based static analysis for a range of safety-related eventualities and for the fire engineer responsible for signing off the design; and it highlights the limitations of static analysis for complex situations.

8.5. Discussion: retail

The commercialisation of concourse and fringe areas has become more common in recent years. The emphasis on retail is evident in the redevelopment of Birmingham New Street and London Bridge, where the stations are an extension of their surroundings rather than a separate operating entity. This section discusses retail in the station environment.

Shopping opportunities provided by large terminus stations can mean a busy transport hub is occupied by many different types of people, from shoppers to commuters and people who live or work nearby.

Table 8.3 Range of tests undertaken to assess evacuation times

Scenario	1		2	3	4	5 only P20	6 only P20	7 only P20	8	Evacuation direction of carriages		
	Ramp (9.8m)	Stair (3*1.5m)	Mid-platform esc	Mid-platform escape stair (1.8m)	Country-end escape stair (2*1.1m)	Escape stair (1.5m)	P19 Stair (1.5m)	P19 Stairs (2*2m)	P19 Stair (2m)	Upstream ⟵	Incident ⟷	Downstream ⟶
Fire on train Country-end												
E1a	✓	✓		✓	✓	✓	✓	✓	✓	–	10	1–9
E1b	✓	✓			✓	✓	✓	✓	✓	–	10	1–9
Fire on train Mid-platform 7												
E2a	✓	✓		✓	✓	✓	✓	✓	✓	8–10	7	1–6
E2b	✓	✓			✓	✓	✓	✓	✓	8–10	7	1–6
E2c	✓	✓			✓	✓		✓	✓	8–10	7	1–6
Fire on train Mid-platform 6												
E3a	✓	✓		✓	✓	✓	✓	✓	✓	7–10	6	1–5
E3b	✓	✓		✓	✓	✓		✓	✓	7–10	6	1–5
E3c	✓	✓		✓	✓	✓		✓	✓	7–10	6	1–5
Fire on train Mid-platform 4												
E4a	✓	✓		✓	✓	✓	✓	✓	✓	5–10	4	1–3
E4b	✓	✓		✓	✓	✓			✓	5–10	4	1–3
NOTE	For the above simulations, P21–24 have all evacuation routes available except for the escalators, which are discounted on these platforms.											
Fire on train City-end												
E5a				✓	✓	✓	✓	✓		2–10	1	–
NOTE	It should be assumed that for P21–24 the main concourse is out of use due to smoke.											

Table 8.4 WIT evacuation population

Platform	Event	Time (hh:mm)	Occupancy (passengers)
20	Train arrival (on fire)	08:33	1,800 (crush load)
21	Train arrival	08:33	1,800 (crush load)
20	Waiting	08:33	143
21/22	Waiting	08:33	456
23/24	Waiting	08:33	321

These different user groups often occupy the space at the same time. Individuals may be part of one or more user groups at the same time or switch their purpose of visit.

When assessing retail and station environments with micro-simulation modelling, the different pedestrian characteristics of visitors, such as size and speed, should be considered. This is arguably even more important when modelling a dual space, as pedestrian characteristics are likely to differ substantially in such spaces.

The aim of the modelling work should be to ensure passengers and retail visitors can safely occupy the space at the same time. Therefore, appropriate station-capacity guidelines should be used. Consumers will exhibit different behaviours and have different characteristics within the shopping environment, and this should be reflected in the modelling process by adjusting the speed and size of pedestrians. The number of visits per shop in the model should be validated against existing data, if possible, or by using comparable locations.

As part of the European Commission's aspirations to create an integrated transport and railway network throughout the EU, the Rail Baltica project aims to construct a trans-European railway in the Baltic region. The line will link major cities and capitals in the region: Helsinki, Tallinn, Riga, Kaunas and Warsaw, with onward connections to Berlin envisaged.

Table 8.5 WIT clearance times

Scenario	Platform 20
E1a	3 min 58 sec
E1b	5 min 26 sec
E2a	5 min 22 sec
E2b	5 min 25 sec
E2c	5 min 24 sec
E3a	4 min 52 sec
E3b	4 min 53 sec
E3c	4 min 43 sec
E4a	7 min 09 sec
E4b	7 min 31 sec
E5a	5 min 12 sec

Figure 8.9 Riga Central Station precinct

The route of the rail line passes through the existing Riga Central Station (Figure 8.9), with plans for a new multimodal hub, including all modes of transportation, as well as urban improvements that open the study area up towards the canal, the market area and the Old City. The existing shopping centre, which is integrated and shares an entrance with the station, is expected to experience a shift in pedestrian flows because of the project's planned improvements, as well as increased numbers of people passing through it on their way to the station.

Figure 8.10 CMD map – ground level, preferred option, 2025

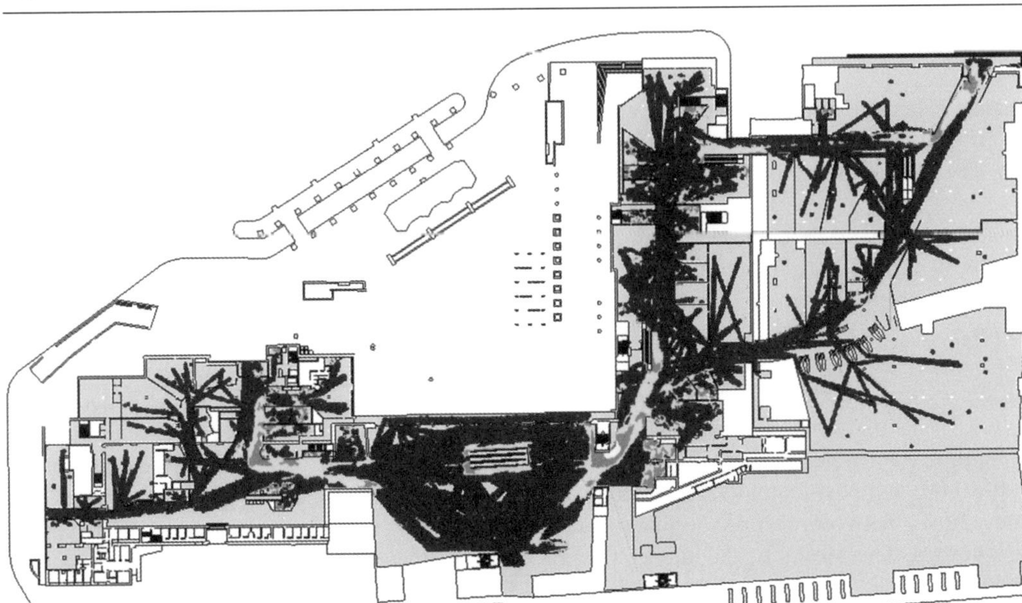

As part of a wider multi-stage assessment involving current-situation analysis, option analysis and implementation planning, a pedestrian flow assessment was undertaken of the urban area containing the station precinct. The multilevel station entrance and shopping centre were modelled for the PM peak-hour period based on 2015 passenger demand, using the existing layout of the infrastructure, as well as separate models testing different layout options.

The CMD modelling outputs (Figure 8.10) showed that there is enough space for free circulation in the common station/shopping-centre entrance as well as on the upper levels of the shopping centre. There were some localised areas of minor congestion at the conjunction of the station and shopping centre; overall, however, there was enough space for comfortable movement throughout. It was recommended to relocate the staircase adjacent to an area of higher congestion – opening the space and reducing crowding – to improve the pedestrian experience between the shopping centre and the station.

The outputs from this modelling informed the development of the final design and gave the project team confidence that the preferred option would efficiently cater to the future demands of pedestrians moving through the shopping centre into the station.

8.6. Case study: Twickenham, Solum regeneration

8.6.1 Setting the scene

The current station at Twickenham has been in operation since 1954. It is in south-west London, in the London Borough of Richmond upon Thames, approximately 18km from London Waterloo. Its proximity to Twickenham Stadium puts the station under the significant stress of high-volume tidal flow on rugby match days. To mitigate this and provide improved facilities for rail passengers, the station is being redeveloped in conjunction with an over-site development.

AECOM was commissioned by Solum to undertake dynamic pedestrian modelling of the existing layout, final design and construction phasing of the station redevelopment. The modelling focused on match-day operations, with the busiest-time scenario modelled for a non-emergency.

To maintain smooth and safe operations at the station on match days, South Western Railway (SWR) deploy a team of crowd-management staff and operate rigid crowd-management plans both pre- and post-match. The operations pre- (a) and post-match (b) are shown in Figure 8.11.

Pre-match, passengers are managed out of the station through either the ticket office or car park via the footbridge or past the end of platform 2. Post-match, London-bound passengers are queued through the car park and fed on to platform 3. Westbound passengers are queued through the ticket office and on to platform 5.

A full survey was undertaken on Saturday 17 March 2012. It recorded station entry and exit, demand to and from each platform, passenger interchanging via the London-end footbridge and boarding and alighting figures per carriage. The match-day surveys were conducted in two periods: pre-match (13:00–17:00) and post-match (18:30–21:00). In these time periods, all main passenger movements, journey times between station entrance and platforms and train arrival and departure times were recorded. The results of the survey for pre- and post-match are shown in Tables 8.6 and 8.7 respectively.

Figure 8.11 Pre- and post-match station management

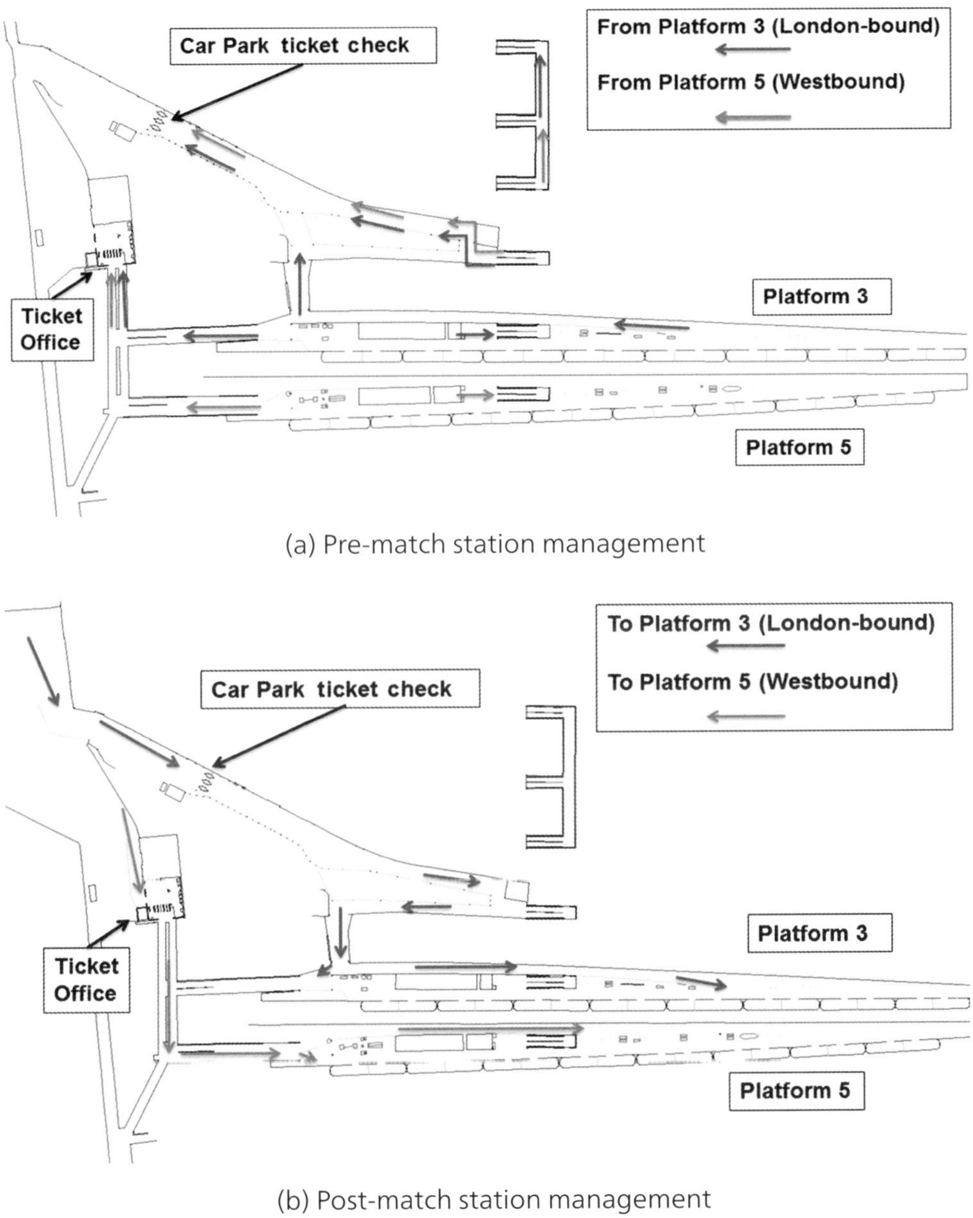

The survey was conducted for England vs Ireland in the Six Nations rugby tournament, which was attended by 80,567 spectators. As can be seen from the survey results, almost half of all spectators get to Twickenham via the station. Station use is lower, yet still considerable, during the shorter post-match survey window as people often stay in the town to enjoy the atmosphere.

Table 8.6 Pre-match OD matrix (13:00–17:00)

	Exit	Platform 3	Platform 5	Total
Entrance	-	1,082	352	1,434
Platform 3	8,772	-	-	8,772
Platform 5	28,400	-	-	28,400
Total	*37,172*	*1,082*	*352*	*38,606*

Table 8.7 Post-match OD matrix (18:30–21:00)

	Exit	Platform 3	Platform 5	Total
Entrance	-	24,437	5,962	30,399
Platform 3	310	-	-	310
Platform 5	960	-	-	960
Total	*1,270*	*24,437*	*5,962*	*31,669*

8.6.2 Defining the problem

The existing situation has been shown to be ill equipped to deal with the large volumes of passengers heading in and out of Twickenham pre- and post-match. The platforms can only accommodate eight-car trains, which do not have the capacity to allow passengers to leave the station quickly enough.

For this assessment, the design year was agreed to be 2026. As Twickenham Stadium was already at capacity during the survey and there are no planned expansions, it was agreed that there would be no match-day demand growth.

Base models were developed of the pre- and post-match situation for the existing Twickenham Station to act as a reference for comparison against the final design and intermediate phases.

During normal operation, the future station was considered using LoS criteria set out in SCAG, supported by London Underground's *Station Planning Standards and Guidelines* (G371A). For match days, poorer levels of service were expected irrespective of the solution, due to very high demand. Therefore, results for

Table 8.8 Post-match overall movement accuracy

Entrance	Destination	Survey	Model	Model accuracy
Car Park	London-bound (P3)	24,256	24,217	–0.2%
Ticket Office	Westbound (P5)	6,189	6,147	–0.7%
Ticket Office*	London-bound (P3)	229	181	–20%
Total		*30,445*	*30,364*	*–0.3%*

* Discouraged route; typically, passengers had queued in the incorrect line – not a significant overall error

Figure 8.12 Post-match model vs observation

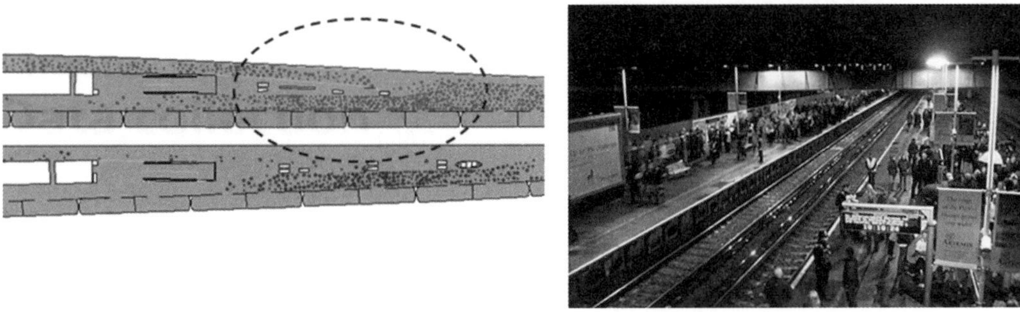

match days focused on the safe and efficient processing of high volumes of passengers, primarily going to and from London, rather than on achieving a specified LoS or flow rate.

Models were developed and validated against passenger movements, boarding and alighting volumes and visual records taken by video camera. The overall validation of the model is shown in Table 8.8, and a comparison of model and actual platform-queuing behaviour is shown in Figure 8.12.

8.6.3 Evaluating performance

The models used to test the designs were developed in consultation with SWR to reflect the way they manage the station for pre- and post-match operations. A critical part of these operations was that

Figure 8.13 Routing options in the proposed station design

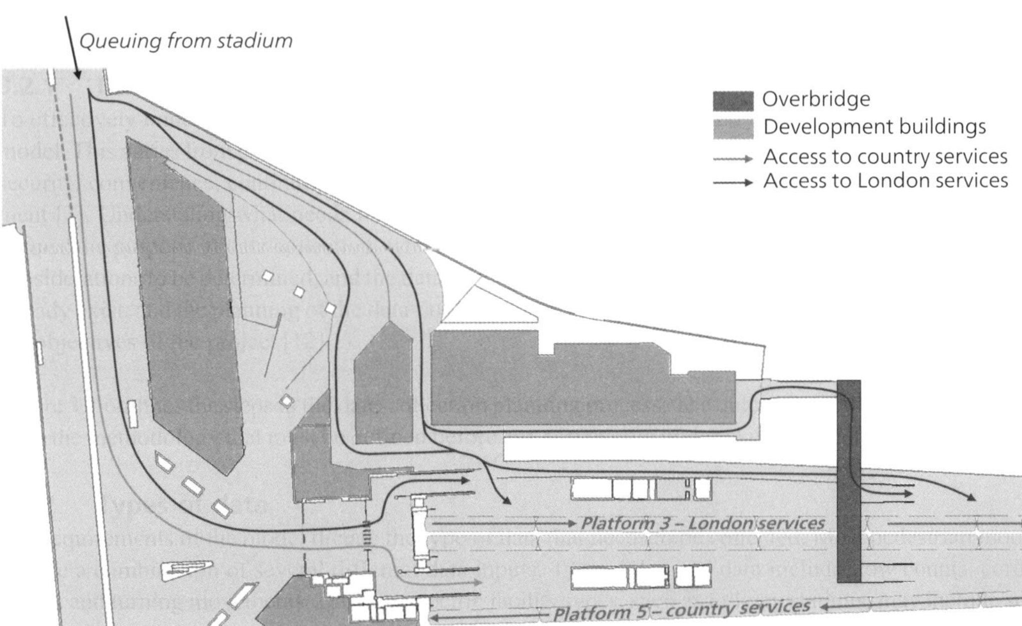

Table 8.9 Comparison of existing and options throughput

Scenario	Number of passengers boarding on platform 3
Current station	14,210
Routing Option 1	18,850
Routing Option 2	19,400
Routing Option 3	18,850
Routing Option 4	17,760
Routing Option 5	17,260

post-match London-bound trains do not leave Twickenham full, as capacity is needed for passengers to board further down the line. The new station development offers several routes to supply platform 3, which are shown in Figure 8.13. Combinations of these routes have been modelled to understand the best routing.

Five routing options were modelled, all of which offered viable solutions to match-day operations and were improvements on the existing station. The options also accommodated smaller passenger movements, such as those of non-match travellers. A comparison of the results for the options modelled is provided in Table 8.9.

In conjunction with the SWR events-management team, it was agreed that the new design could be managed to support one or more means of access to platform 3, allowing flexibility in how access to the platforms is managed. As such, the new design was expected to enable a higher overall post-match throughput than the existing layout, improving sightlines between the multiple access points onto platform 3, as well as substantial operational flexibility and resilience in the event of any incidents.

It must be noted that it was not part of the brief of this project to solve the queuing situation that occurs at Twickenham station on match days. The analysis showed that the station users would be better off on match days with the new design, but that queue management, both inside and outside of the station, would still be required; however, queues would clear more quickly than under current arrangements.

Figure 8.14 Phase 1 – post-match layout

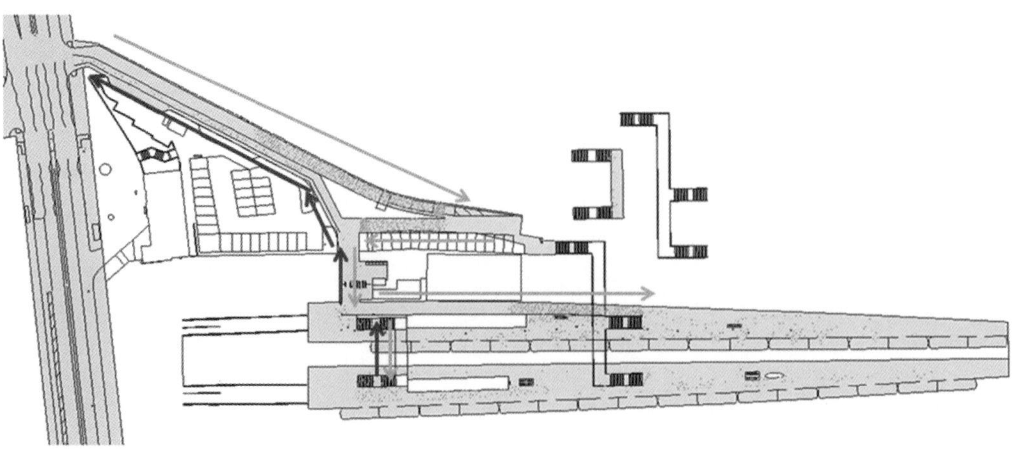

Table 8.10 Construction-phase throughput

Scenario	Number of passengers boarding platform 3
Current station	14,210
Construction phase 1	15,840
Construction phase 2	17,970

Redevelopment of the station was phased so that it would remain operational throughout the two-year construction period. It was important to ensure that the station had capacity to cope with match-day demand during construction, which was split into two major phases.

In Phase 1, when the main station building was under construction, a temporary ticket hall was built in the car park space, and a temporary bridge was provided linking the two platform islands. Passengers entering the station went through the car park, into the temporary ticket hall, using any of the ATGs (automatic ticket gates), and either on to platform 3 or over the temporary bridge to platform 5 (Figure 8.14). Passengers leaving the station used the temporary bridge then exited through the WAG (wide-aisle gate) and through a segregated lane.

Figure 8.15 Phase 1 – New ticket hall

Figure 8.16 Final station design

In Phase 1, there were limited options for entering and leaving the station. The temporary ticket hall and bridge provided adequate capacity pre-match to stop passengers filling up the station, without overly limiting the passengers reaching platform 3 post-match. A comparison of the existing layout and Phase 1 capacities is provided in Table 8.10. Construction of the ticket hall can be seen in Figure 8.15, along with the temporary bridge linking the platform island.

In Phase 2, the main ticket hall is being built, and passengers can more easily reach their destinations, though there are still inevitably queues post-match, which is as much a reflection of limited train-service capacity as of the station itself. Passengers for platform 3 enter through the car park and into a small waiting zone, where passengers can be held if the platform is full. To exit platform 3, passengers, use the ticket hall. To go to and from platform 5, passengers use the ticket hall as normal. A summary of the capacities is shown in Table 8.10, and Figure 8.16 shows the final station design from above the platforms.

8.7. Summary points

- Pedestrian modelling is commonly used to assess movement and operations at stations and interchanges.
- Movement patterns within a station will put pressure on different elements of a station, such as the gateline, vertical circulation and platforms.
- Pedestrian modelling can be used to identify unsafe operating conditions and test design improvements.
- Pedestrian modelling is often used to understand how stations and buildings will be evacuated.
- At large terminus stations the number of evacuation scenarios can be considerable because of the number of passengers, evacuation routes and platforms.
- Stations are becoming extensions of shopping areas and, as such, they can impact the behaviour of passengers.
- It is important to consider the potential impact of introducing retail in terms of comfort, cross-flows and evacuation.
- Some stations are subject to abnormal operating conditions outside of normal weekday operations due to entertainment events.
- Pedestrian modelling is used to inform crowd-management plans that ensure passenger safety and facilitate efficient boarding of and alighting from train services.

ICE Publishing

Urban Space - AECOM
ISBN: 978-0-7277-6381-5
https://doi.org/10.1680/mpm.63815.123

Chapter 9
Events

9.1. Overview of the chapter

One of the primary objectives of any major event, whether sporting or cultural, is to successfully facilitate the gathering of large crowds of people. To safely and efficiently cope with tens of thousands of participants and spectators at such events, organisers are increasingly concerned with satisfying a range of criteria, such as visitor comfort and safety. These factors are considered critically while planning for events in existing venues, during the design of new venues and when refurbishing or expanding existing infrastructure.

Micro-simulation modelling is used to assess events involving large-scale crowd movements, such as the Olympic Games, the Commonwealth Games and the FIFA World Cup. Layouts and operational plans can be optimised to support the design or planning, and the organisers and stakeholders can be confident that the event will operate as planned. The use of facilities, such as toilets and services, as well as retail inside events, can also be assessed to optimise their layout and locations.

This chapter discusses the modelling of large-scale event venues and presents several case studies in which pedestrian micro-simulation has been used to assess the movement of people in crowded event environments. A range of scenarios are discussed, including event ingress, egress, circulation models and how static analysis has been used in assessing the capacity of sporting facilities.

9.2. Discussion: major sport and entertainment venues

The organisers and approving authorities for major events often undertake assessment of crowd movement to achieve the following objectives:

- aid the development of a venue design with adequate capacity and resilience in terms of crowd management and within acceptable performance criteria;
- develop optimum crowd-circulation plans for arrival, departure and during intervals;
- assess the adequacy of the venue layout to meet emergency evacuation standards; and
- develop optimum crowd-management plans and strategies for alternative demand and operational scenarios.

As with the interchanges discussed in Chapter 8, static analysis is often undertaken first to estimate the spatial requirements of a stadium – or cluster of venues – and their precincts. The outputs of these static analyses may be used to:

- provide an initial estimate for the sizing of various design features which can be supplied to the design team; and
- identify which demand and operational scenarios are to be taken forward for dynamic modelling to provide a robust understanding of the safety and capacity of the venue.

Static analysis results are generally used to check and/or inform design characteristics such as:

- the number of personal search areas (PSAs) and turnstiles required to serve arrivals demand within a reasonable time;
- the maximum accumulation of people within the precinct outside the stadium during the arrivals peak;

Figure 9.1 Static analysis outputs

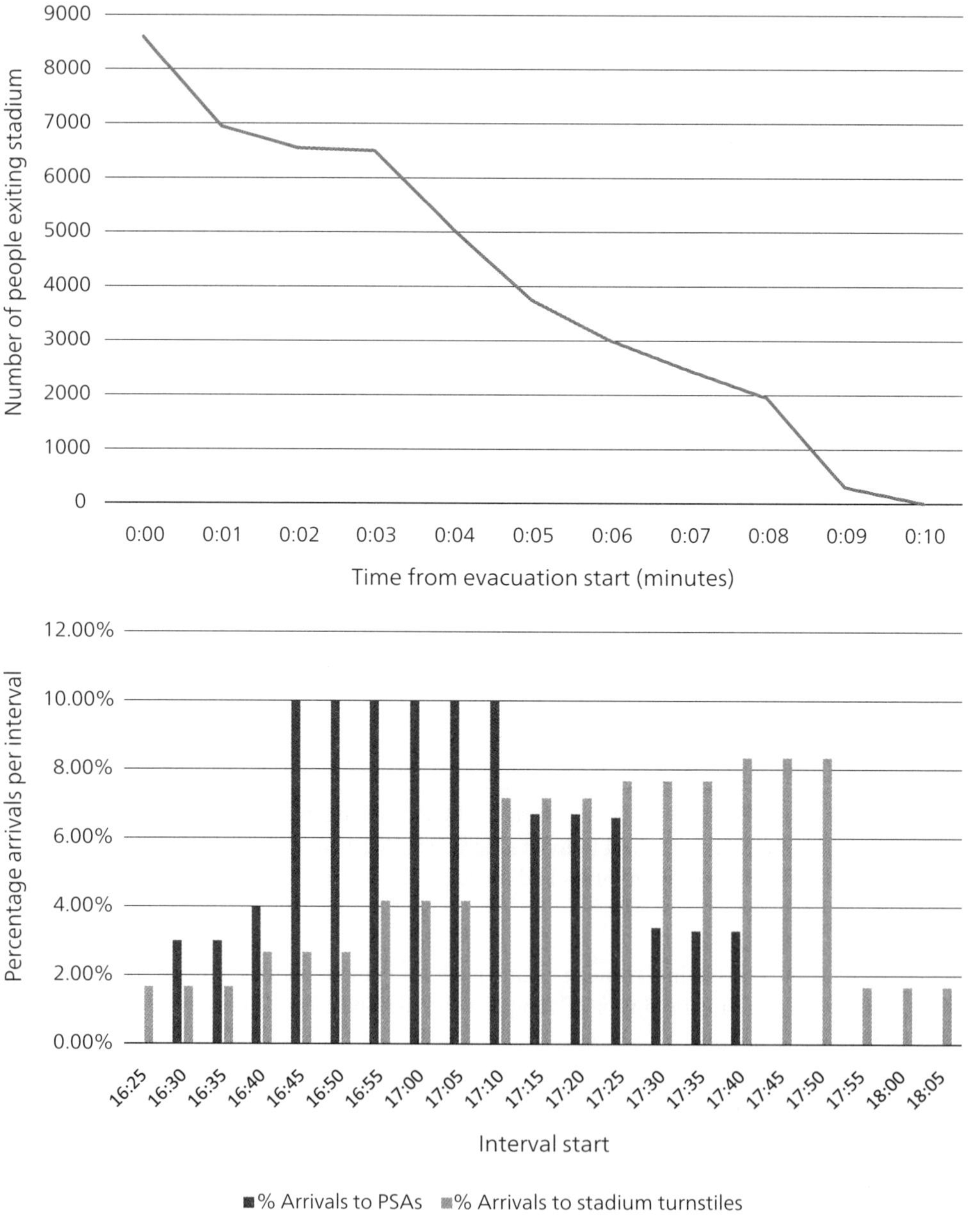

Figure 9.2 CMD output example – arrivals

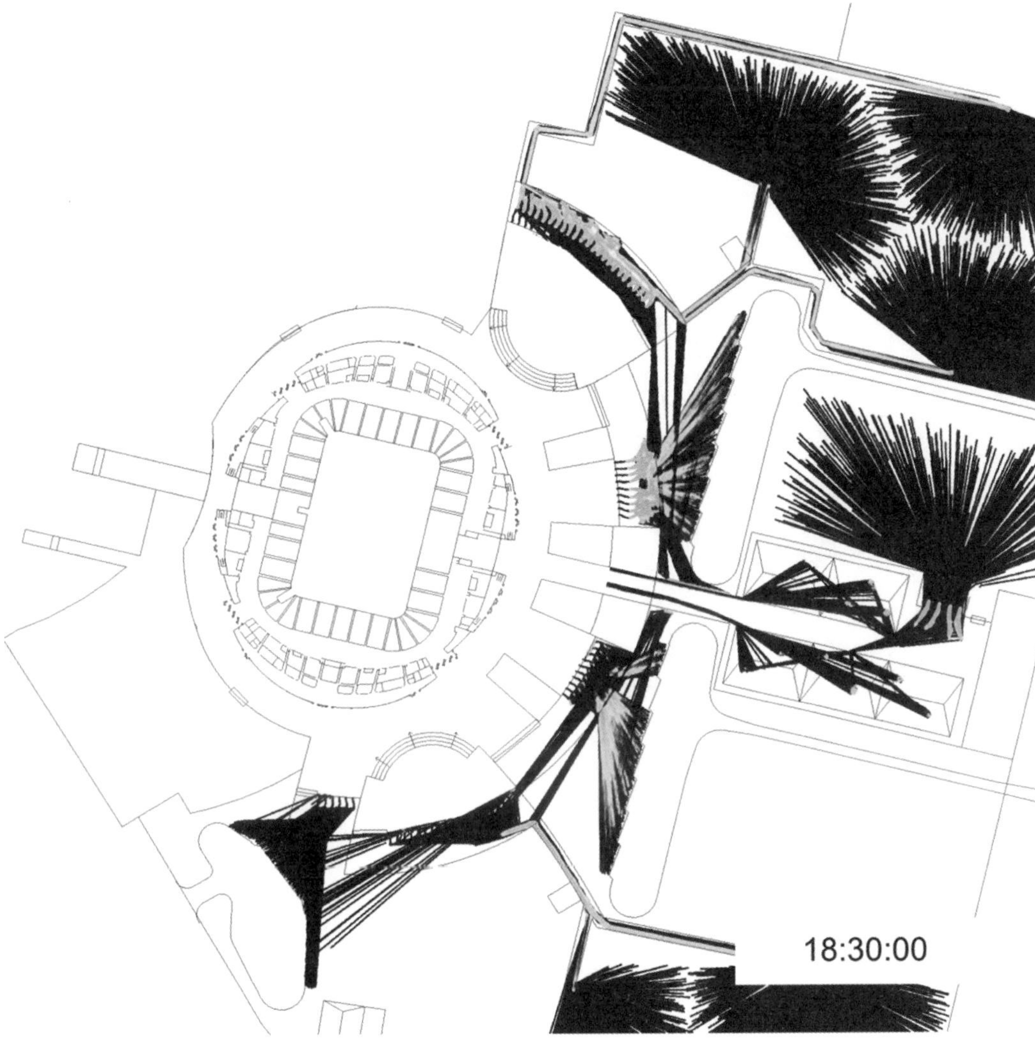

- the number of concessions stands and toilets required to serve demand;
- the stadium exit widths required to serve peak departures and evacuation demand; and
- the estimated time taken by the final person to reach a place of safety in an evacuation.

The above characteristics can be assessed against industry standards and appropriate guidance for maximum queuing or waiting times. The outputs of static analysis include recommended numbers for the assessed metrics, as well as graphs plotted over time, such as for the evacuation of spectators from the venue, as shown in Figure 9.1.

Dynamic pedestrian micro-simulation models of large-scale venues are undertaken to test the most critical scenarios for arrivals, departures and evacuation. Certain aspects of a precinct design, such as

Figure 9.3 CMD output example – departures

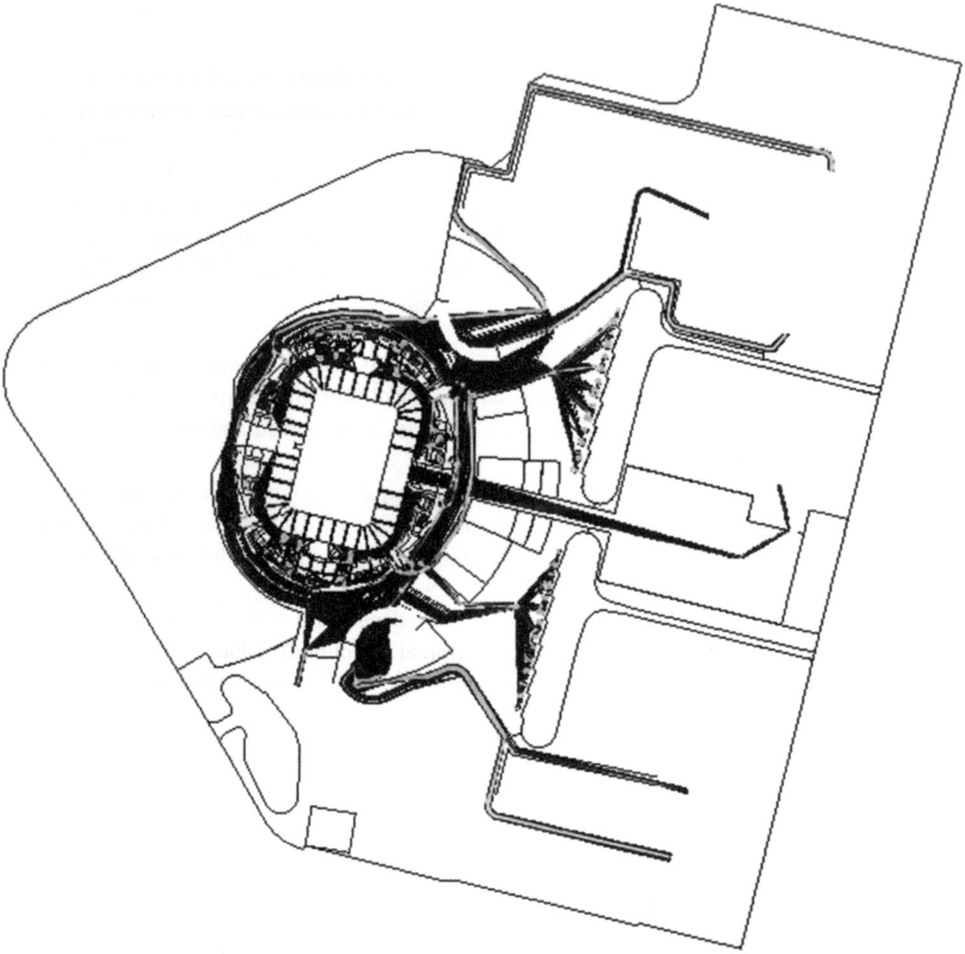

entry-procedure elements, are also assessed for some venues. In some cases, a scenario may need to be split into smaller models, localised to the elements being tested, to manage model complexity, scale and computer-processing time. Examples of CMD outputs for a major stadium are shown in Figures 9.2 and 9.3 for arrivals and departures respectively.

Figures 9.2 and 9.3 show that modelling different scenarios highlights different stress points within the design of a venue precinct. For the arrivals, the critical aspects tested are the gateline areas, where congestion at PSAs is most likely to occur. These areas are opened, allowing the free flow of spectators in a departures model, which then highlights congestion at exit pinch-points and bus-departure bays.

The same is true for modelling scenarios within the stadium-concourse operation, under normal, interval and evacuation scenarios. Typical examples of normal operation and emergency evacuation scenarios are shown in Figures 9.4 and 9.5.

Figure 9.4 CMD output example – concourse circulation

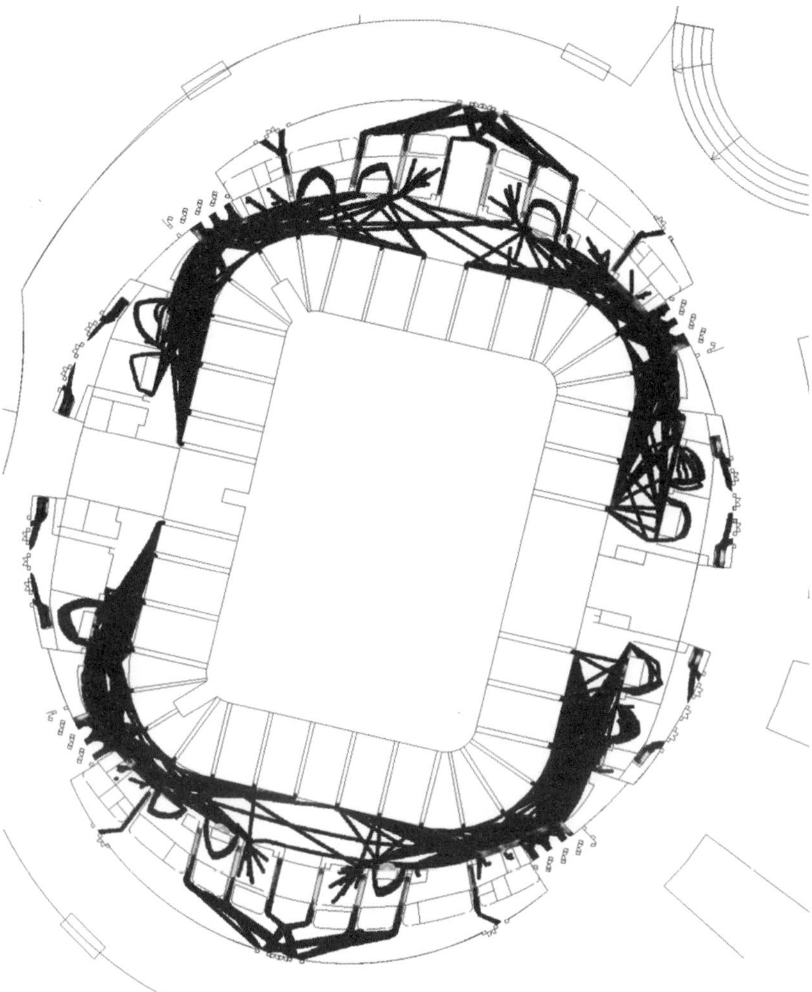

In the evacuation scenario, the vomitories (entrances to the banks of seating), stairs and stair tops and bottoms experience a LoS that indicates crowding, which is to be expected during this type of emergency evacuation scenario. The normal circulation scenario highlights pinch-points in the concourse design. These are typically the narrow corridors and vomitories in stadia, which can be assessed to improve the design. Outputs are also able to identify any obstructions to the flow of pedestrians, such as columns, furniture or retails stands, which can be evaluated and relocated to aid spectator movement.

Dynamic modelling outputs provide a greater level of detail than is possible in static analysis. Figure 9.6 shows the decreasing number of spectator numbers over the course of an evacuation, which provides a greater degree of confidence as it considers the specific geometries of evacuation routes.

Figure 9.5 CMD output example – evacuation

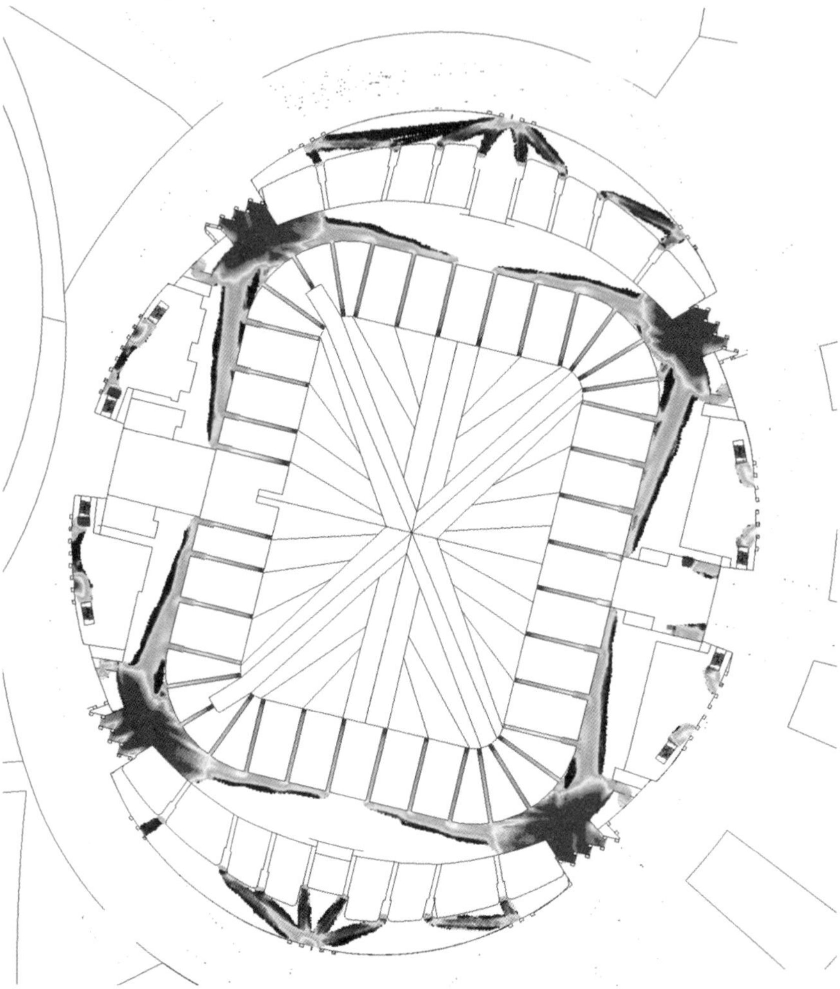

Crowd modelling of major event venues is undertaken in an iterative way to inform and evaluate the stadium and precinct design layout. Often the assessment is undertaken in two stages, first a static analysis, followed by micro-simulation-based dynamic pedestrian modelling. The following sections provide case studies of projects where pedestrian modelling has been used in stadium design and operation.

9.3. Case study: St James' Park, Newcastle United Football Club

9.3.1 Setting the scene

St James' Park is a large football stadium located in Newcastle-upon-Tyne, UK, and the home ground of Newcastle United Football Club (NUFC). The stadium has a capacity of over 52,000 spectators and is one of the largest and oldest in the UK (Figure 9.7). Since football was first played on the site in 1880, the stadium has undergone successive expansions, the most recent of which resulted in additional capacity

Figure 9.6 Example of spectator numbers during evacuation

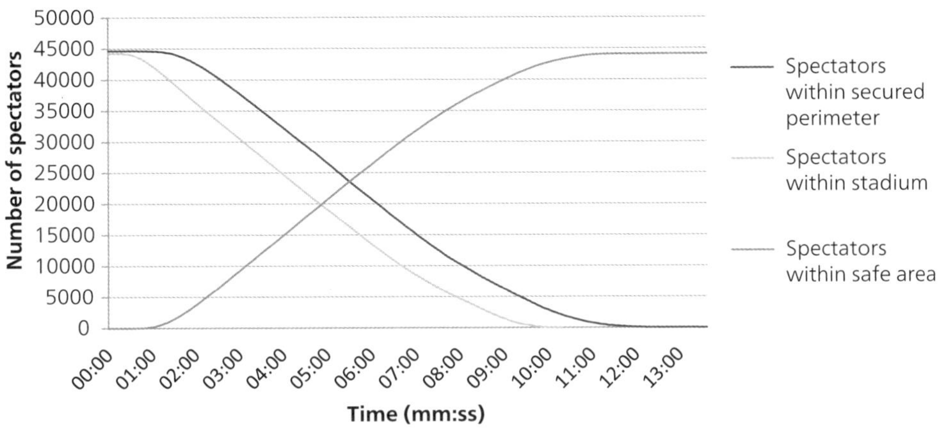

being built above the existing seating bowl on three sides. This approach resulted in a unique architectural layout, with segregated entrances and stairwells serving the new seating sections.

Following pressure from fans' lobby groups and TV broadcasters, a new directive by the Premier League was put in place requiring that, by the beginning of the 2017/18 Premier League season, all clubs with a stadium capacity of over 30,000 spectators would provide pitch-side seating for 3,000 away supporters in the stadium. At St James' Park, the 3,200 away supporter seats were in the North Stand on level 7, as shown in Figure 9.8, which did not satisfy the Premier League's new pitch-side-seating requirement.

9.3.2 Defining the problem

The location of away seating has ensured that safety and security staff at NUFC have been satisfied with the demonstrably safe egress of 3,200 away supporters, alongside more than 49,000 home supporters, at the end of each match. This is helped by the segregated nature of the seated position and the exit patterns of home and away supporters, which are largely towards the south-east and north-west respectively. However, there were concerns over the safety of supporters should a new pitch-side location be enforced by the Premier League, due to the likely introduction of cross-flows in a heavily congested area. Two options for repositioning away supporters were modelled to identify the impact this would have on crowd movement.

The scale and direction of pedestrian flows immediately outside the stadium are unique, as St James' Park is located at the edge of Newcastle city centre. As a result, the predominant mode of transport to and from the stadium is by foot to and from the city centre, where parking and public transport options are located. This is the case for the majority of the 49,000 home supporters, who approach and leave to the south-east. Coach parking for away supporters is located to the north-west of the stadium, in the opposite direction to the city centre. Hence the flows of home and away supporters occur in predominantly opposite directions and work well with the existing seating positions due to the avoidance of cross-flows. Relocating away fans would inevitably lead to creation of conflict in what is a confined physical space in the circulatory corridor underneath the upper-seating tier, particularly in the north-west corner. The rivalry between football teams could further heighten any pedestrian conflict at the end of a match. The potential conflict locations are shown in Figure 9.9.

9.3.3 Evaluating performance

Three scenarios were modelled to evaluate the potential impacts of moving away-supporter seating to a pitch-side location: the existing situation and two relocation options. First, the existing situation was modelled with spectators seated on level 7 of the North Stand. The first relocation option relocated away supporters to the north-east corner of the stadium, the second to the north-west corner of the stadium. A plan of the three modelled away-seating areas can be seen in Figure 9.9.

Modelling large numbers of people in micro-simulation software can lead to slow model run times and added complexity, which in turn are likely to lead to a longer development time and consequently higher project costs. A pragmatic approach is to reduce the model size and to focus only on those areas that are needed, provided this will not impact the outputs and objectives of the study.

Due to the segregated nature of the stadium's seating areas and exits, only the northern part of the building needed to be modelled; the southern exits led immediately to a large public space and so pedestrian flows from these parts of the stadium would not interact with the areas of concern. Hence only the exit gates from the central line of the pitch and above were modelled. These comprised 35 exits, with a current full capacity of 30,119 supporters, representing more than half of the full stadium capacity of approximately 52,000 but allowing the removal of 22,000 people from the model – a significant saving in model-building and simulation time.

The worst-case scenario was modelled in each case. This assumed maximal capacity, movement towards each exit commencing simultaneously and most away supporters being routed to coaches parked in the coach-parking area to the north-west of the stadium.

The complexity and scale of the model required some simplification to be made through a series of assumptions that would not affect the analysis, such as:

- no vehicle movement occurs around the stadium in areas used by spectators during the period of egress, such as the car park and access roads;

Figure 9.7 St James' Park, NUFC – aerial view

Figure 9.8 St James' Park – location of away-supporter seating on level 7 (outlined in white)

- staff and police are on hand but do not actively manage the crowd flow, resulting in people moving to their destinations via the shortest distance, as they would naturally do without intervention from others;
- most away supporters who use wheelchairs exit alongside other away supporters, whereas home supporters who use wheelchairs typically either leave before the main period of egress (exit) or wait until afterwards to proceed to their vehicles; and
- VIP spectators remain in the stadium for a period immediately after the end of the match (and were therefore not part of the egress analysis).

These assumptions were based on typical observed protocols and behaviours and approved by NUFC safety and security staff.

An initial assessment of the stadium had been carried out by the Safety Officer for NUFC, resulting in a report that identified constraints and limitations in certain areas of the ground and rejected these areas on safety or other grounds as possible sites for relocating away fans.

The remaining areas, north-east and north-west pitch-side, were suspected to be worse than the existing location in terms of safety and quality of experience for the spectators. The visual nature of the modelling would confirm whether either of these proposed locations could be a feasible option. The outputs of the modelling identified unexpected additional congestion which could be seen and understood by watching the model video outputs.

The results of modelling egress simulations of each option clearly showed that the proposed relocation of away-supporter seating would significantly increase local crowd densities due to phases of repeated crowd

Figure 9.9 Predominant pedestrian flows, existing and options

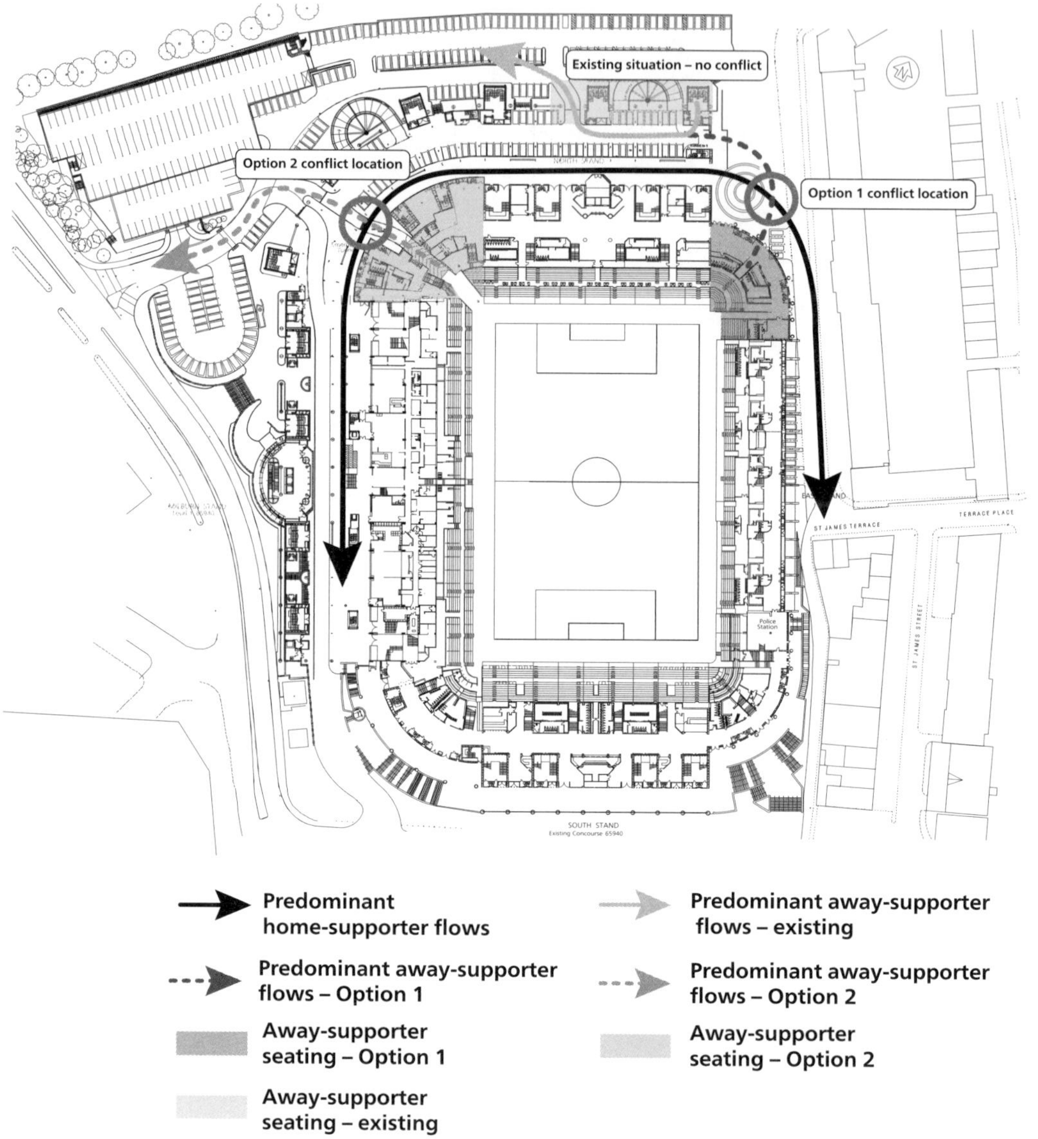

crossover. This would adversely impact egress performance by increasing the time taken to clear the stadium, creating new bottlenecks when moving home supporters to the north stand and reducing the comfort of egress. Furthermore, in areas with high cross-flows it would become harder for safety staff and emergency services to reach certain areas within the stadium. Greater crowd congestion would also increase the number of potential conflicts between home and away supporters.

The increased crowd congestion because of away-supporter relocation is evident in the output CMD maps, which show significantly more congestion and cross-flows in the concourse, particularly at narrow

Figure 9.10 CMD map – existing situation

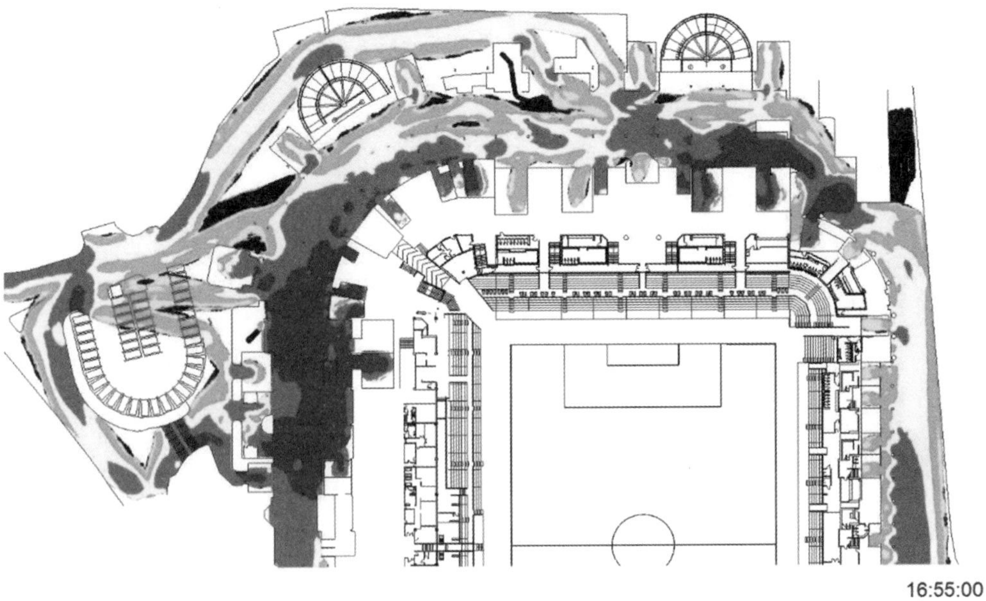

Figure 9.11 CMD map – Option 1

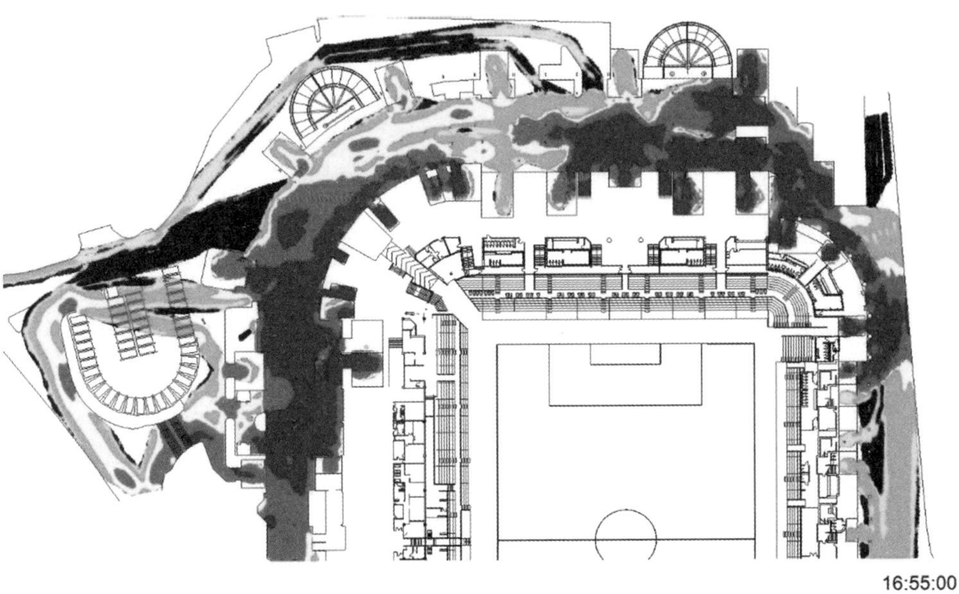

Figure 9.12 CMD map – Option 2

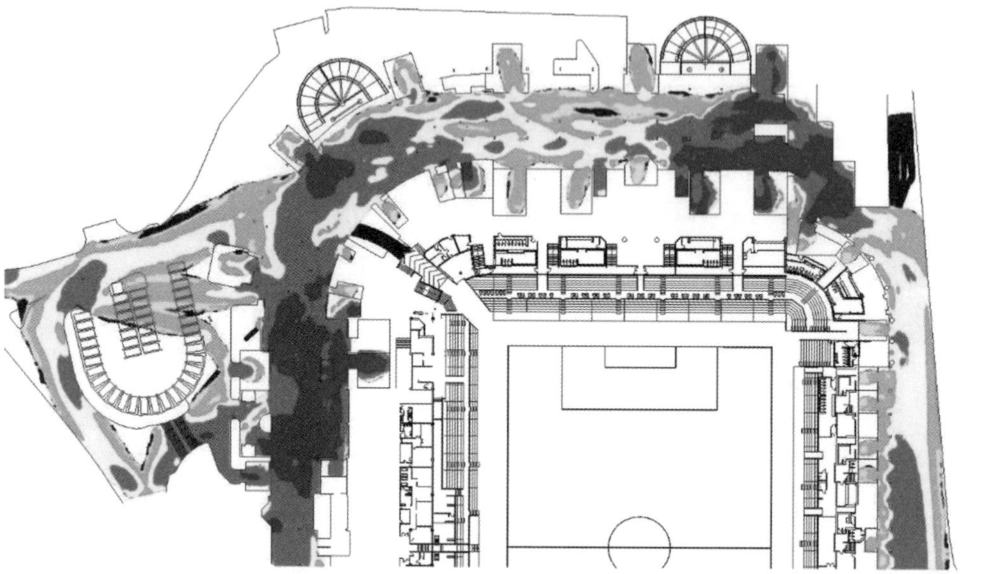

locations. CMD maps of the existing situation and proposed relocation options are shown in Figures 9.10 to 9.12.

The results of the pedestrian modelling showed that the effects of relocating away supporters to a pitch-side location at St James' Park would have significant negative effects on crowd safety and comfort during egress after a match. The modelling provided a solid evidence base that the existing location of away supporters was the optimal location, which enabled NUFC to successfully argue a case against the Premier League requirement. As such, the stadium obtained an exception to the requirement of pitch-side seating for away supporters, and seating for those spectators remained in the existing location.

9.4. Case study: Rio 2016 Olympic Park

9.4.1 Setting the scene

The 2016 Summer Olympics, hosted by Rio de Janeiro in August 2016, were the first Olympic Games ever to be held in South America. A total of 7.5 million spectator tickets were sold to events at the Games, where 10,500 athletes from over 200 countries competed in 42 sports in venues spread across four distinct geographic clusters in Rio de Janeiro. The largest cluster of sporting venues was in the neighbourhood of Barra, with nine venues located in the Barra Olympic Park alone.

Crowd-flow analysis and modelling was used as an integral part of the master planning of the Barra Olympic Park (Figure 9.13), both in terms of providing design inputs and assessing the design layouts at various stages of the masterplan development to inform and assess the adequacy of operations within the park for the duration of the Games.

The design of Olympic parks and large venues often requires crowd modelling to ascertain the safety and comfort levels for spectators using these areas during peak times. The most critical periods that modelling informs are normally the ingress, egress and busy crossover periods under normal operation, as well as the evacuation of venues in an emergency.

Crowd modelling was undertaken to ensure that the spectator experience would be enjoyable through well thought-out connections between venues, sufficient space to enjoy the Park atmosphere and minimised waiting times to enter and leave the Park.

To deliver the vision for the Park, the crowd-modelling team undertook extensive analysis of the 'front of house' area, working closely with the master planners and venue-design teams. A combination of static analysis and dynamic modelling was undertaken to provide an understanding of how the Park might work during the three peak periods expected during the day:

- arrivals, when spectators are entering the Park for events in the morning;
- inter-peak, when morning-session spectators are leaving the Park and afternoon-session spectators arrive; and
- departures, when all spectators leave at the end of the evening session.

9.4.2 Defining the problem

Due to its size and the anticipated daily numbers of people moving in and out of the Park, it was important to determine the required performance of elements such as the PSAs and ensure circulation spaces and exits were appropriately sized to minimise delays and maximise comfort for spectators. There were also significant safety implications in the case of an emergency evacuation if enough capacity was not provided in public areas.

A four-stage modular crowd-modelling approach was adopted for this project, with a view to developing a robust simulation tool that could help in sizing the front-of-house design elements and assess the interface between main circulation domain and the attraction points. It was critical that the outcome could adopt various changes in the designs to act as an iterative tool and that alternate masterplan design scenarios could be assessed with minimum run time. The results of the analyses were fed back to the design team and incorporated as part of the masterplan being developed for the Park.

Figure 9.13 Rio 2016 Olympic Park visualisation

Due to the scale and complexity of the area to be modelled, the project was split into stages. Before any further analysis could be undertaken, the spectator numbers and flow profiles were forecast based on the competition schedule and various operational assumptions – this formed the initial stage of the project. The peak-day Olympics demand profiles are shown in Figure 9.14. This is a very important stage of the assessment as agreement is required between several key stakeholders:

- the Games organisers, who determine the event schedule;
- the International Olympic Committee, which controls VIP movements;
- the media, which determine the coverage of the events; and
- the local transport authorities, which are responsible for running local transport services for spectators.

The next stage involved analysing the design in terms of active circulation, permeable and non-permeable areas and static analysis to inform the initial masterplan layout. Static analysis focused on assessing the sizing of the publicly accessible Olympic Way through the centre of the Park at 12 key sections along its length and on secondary circulation routes. The LoS across the minimum width of each section was calculated based on flow of people per minute. The key sections are shown in Figure 9.15.

After static analysis, dynamic stress testing of Olympic Way was undertaken to determine the sizing and finalise its layout. Detailed micro-simulation modelling was undertaken for defined time periods using Olympics demand with a view to assessing the adequacy of the PSA facilities during the AM peak arrivals, as well as the adequacy of the front-of-house permeable area within the Olympic Park during the inter-peak and PM departures peaks. The extent of the inter-peak and departures models are shown in Figure 9.16.

Finally, micro-simulation crowd modelling was undertaken for the crossover period, using Paralympics demand to ensure enough capacity was available for the higher number of mobility-impaired persons expected during the busiest Paralympics period.

Figure 9.14 Day 7 Olympics demand

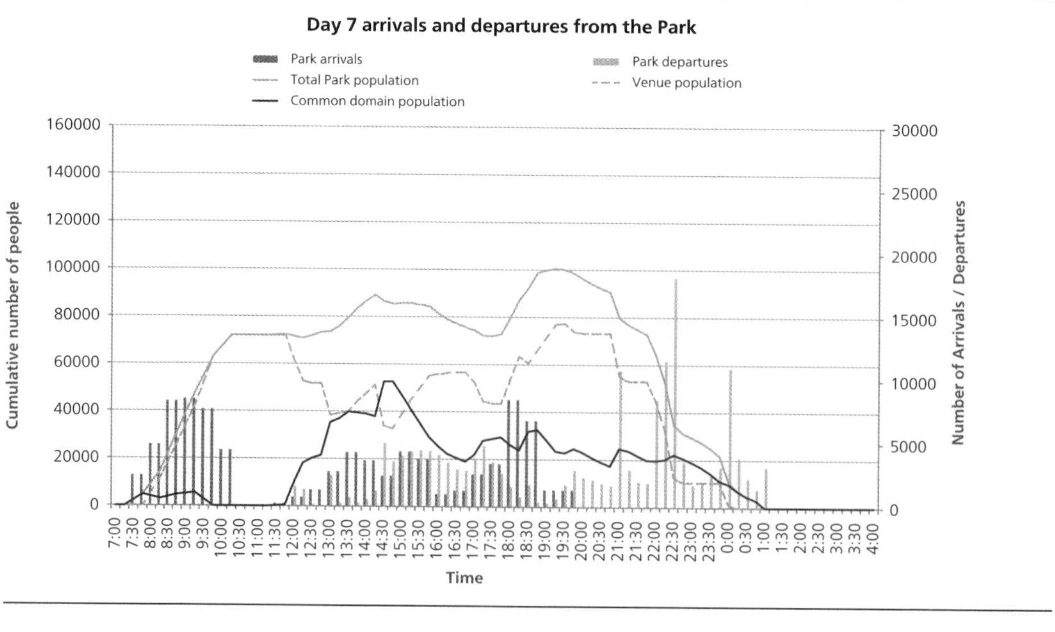

Figure 9.15 Static analysis sections in the common domain

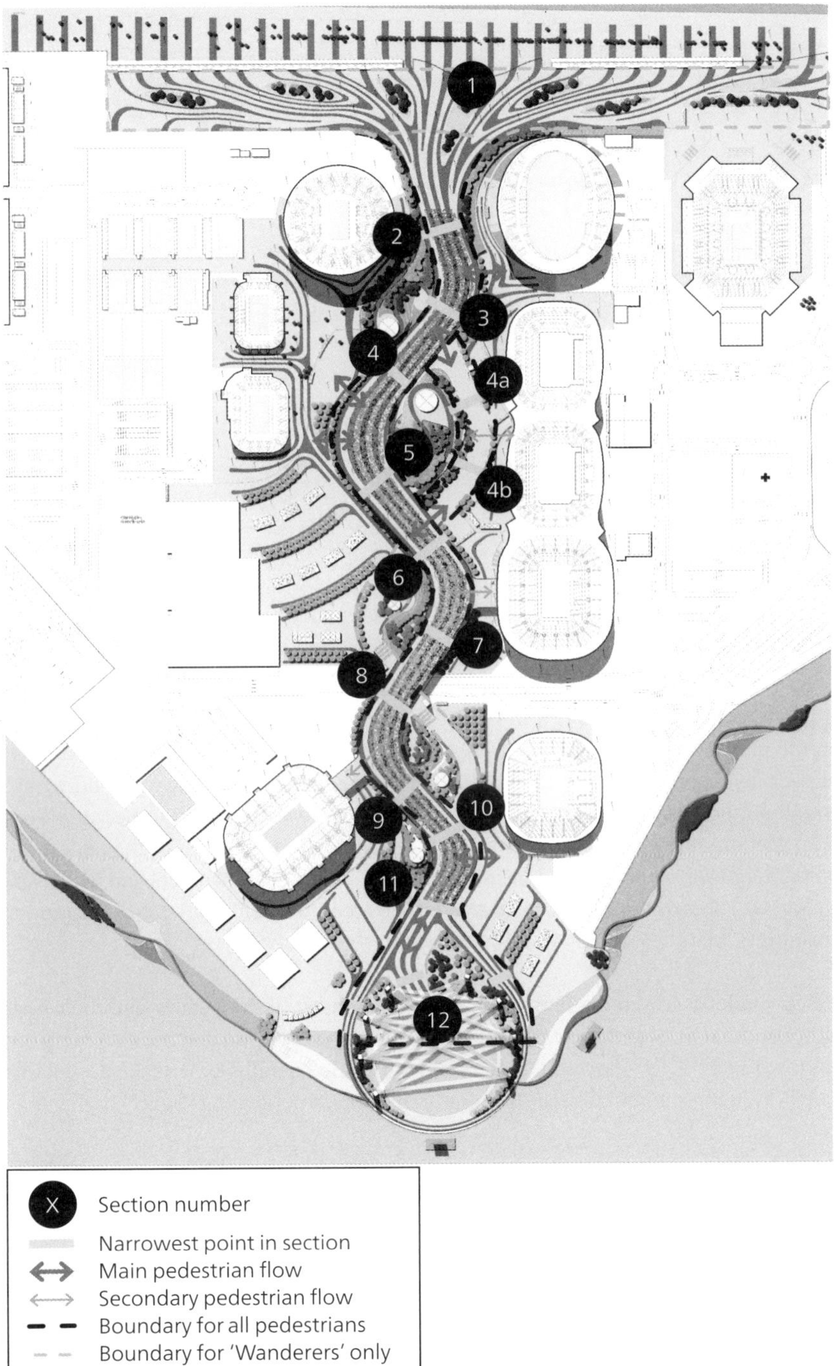

Figure 9.16 Extent of the Rio 2016 Olympic Park crossover model and departures model

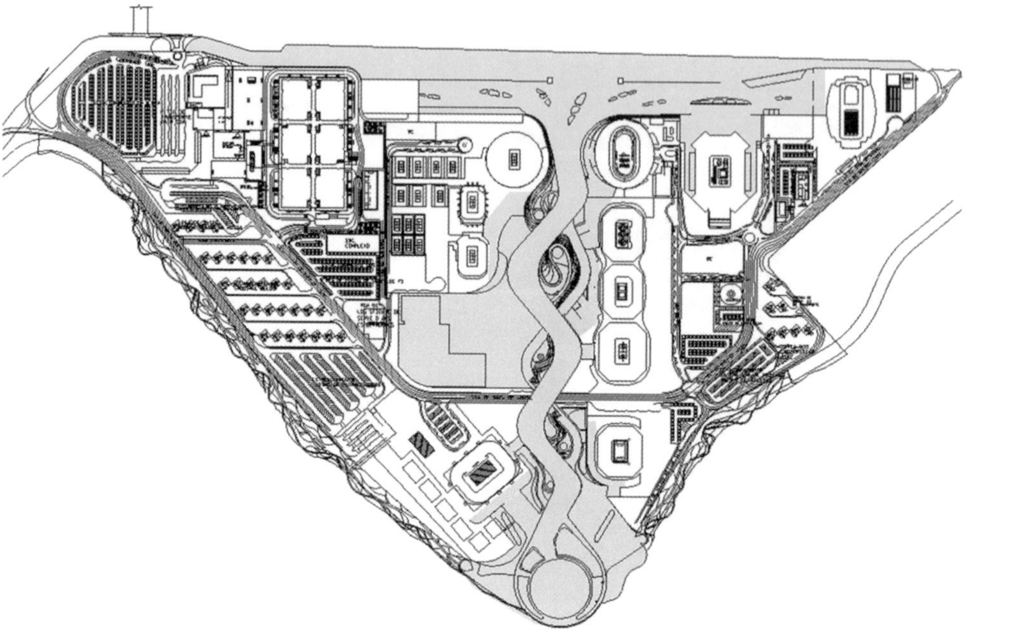

9.4.3 Evaluating performance

Modelling outputs from the inter-peak model, shown in Figure 9.17, highlighted that Olympic Way was expected to operate at lower levels of congestion than indicated by previously tested layouts and within acceptable capacity, though on the peak 15 minutes of the busiest days it was predicted to be very busy, especially on the southern section of Olympic Way. As a result of this finding, the span of the pedestrian bridge along the common domain was widened, and the outputs helped the master planners and landscape designers to finalise the layout, sponsor areas and spectator service areas, as well as crowd-management plans to minimise congestion.

The peak departures-period model assessed the late-night peak, when most of the venues end their final sessions, which is the same time that most of the spectators wandering in the common domain also exit from the Park. Modelling outputs (Figure 9.18) showed that spectators would generally experience acceptable density levels on Olympic Way, with some small areas near the arena exits experiencing slightly higher than desirable densities.

Modelling outputs from the two busiest peaks of the Paralympics suggested that, given the assumptions, the Park would be able to accommodate spectator flow on the busiest day of the Paralympics.

A range of sensitivity tests were also undertaken, and the outputs from these tests were compared with the central case of the inter-peak Olympics demand model. Some of the results showed slightly higher densities on Olympic Way, depending on the variation in demand due to holding capacities or lingering time pre- or post-event within the Olympic Park. However, Olympic Way did not experience extreme densities, suggesting that there was enough capacity to allow for slight variations in demand.

Figure 9.17 Day 7 inter-peak model: CMD map based on Fruin's LoS (walkways) (14:45–15:00)

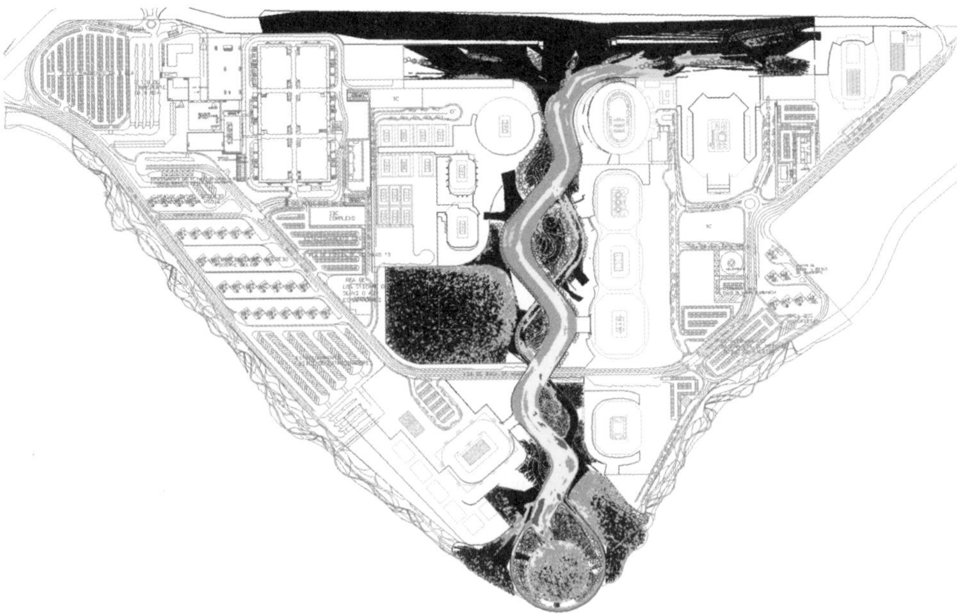

Figure 9.18 Day 7 departures model: CMD map based on Fruin's LoS (walkways) (22:30–22:45)

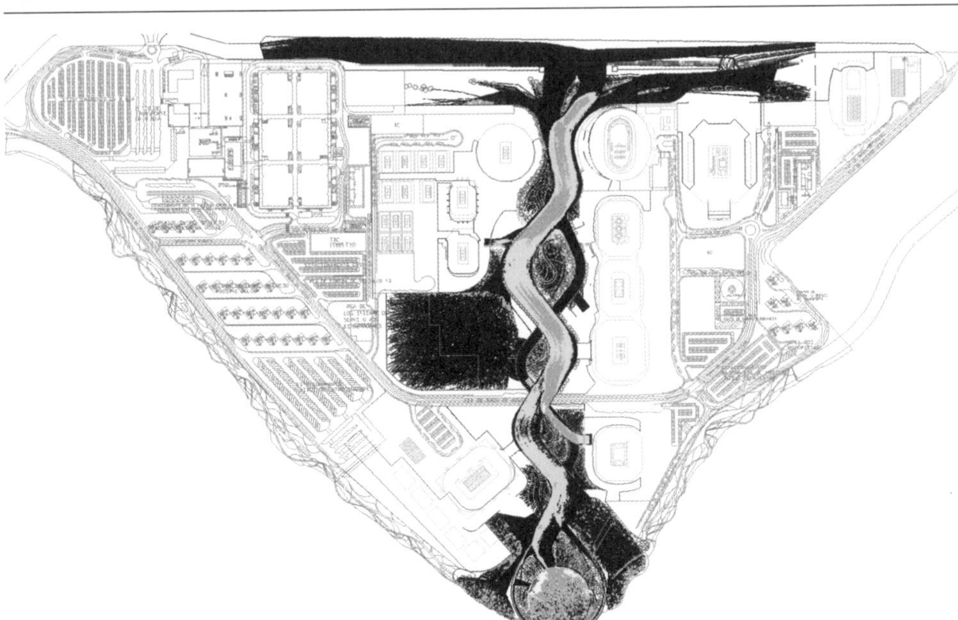

This led to the conclusion that the common domain provided in the masterplan provided the required capacity but, as with all large and busy crowd movements, was sensitive to crowd behaviours and the operational management of spectator service areas. In response to the crowd-flow analysis undertaken over the course of the study, the size of the common domain was slightly increased and Olympic Way widened throughout.

The sensitivity-test outputs showed that more detailed modelling would be critical at the operational planning stage to test various crowd-management measures and overlay designs. The crossover model was then able to form the basis for developing various stages of operational decisions beyond the master planning stage of the project. Thus, the model became a valuable tool for the organisers of the Games during the development of operational details in the run-up to the Olympic and Paralympic Games.

9.5. Summary points

- Pedestrian modelling is commonly used in the design of stadia, entertainment venues and events.
- Several common processes within events, such as security screening, arrivals, departures and crossover, are often assessed.
- At large multi-day, multi-venue events, demand will often change throughout the day and across the site.
- It is key to first understand demand before assessing pedestrian movement.
- Pedestrian modelling can also be used to assess changes to historic venues and ensure spectator safety.
- Outside of event venues, it is important to consider how people access and exit the wider site of large events.
- Spectator comfort levels need to be defined early to help designers provide suitable space for movement.

ice Publishing

Urban Space – AECOM
ISBN: 978-0-7277-6381-5
https://doi.org/10.1680/mpm.63815.141

Chapter 10
Public realm

10.1. Overview of the chapter

As cities develop, they must plan for future growth to ensure they function efficiently. The competing demands of business, transport, the environment and infrastructure must all be considered. As attitudes towards sustainable modes of transport change, pedestrians and cyclists are putting pressure on city planners to account for their needs. The term 'public realm' is often used loosely, sometimes interchangeably with 'public domain', to refer to external urban spaces that are publicly accessible. Whether it is a road between rows of housing, or an urban park, the qualities of public realm are integral to its use, and considering them shapes the urban environment. Public realm plays a key role in the economic and social value of communities. It provides meeting places and contemplative and resting areas, and it creates a sense of place, with connections to local historic and cultural elements, as well as a safe and attractive environment for the public who live and work in and visit the area.

In the previous two chapters, the use of pedestrian modelling in stations and stadia for planning and crowd management was discussed. In an urban context, areas that require crowd modelling can take several forms, such as busy shopping streets, spaces surrounding stations and areas with high interaction with vehicles. This chapter discusses how pedestrian modelling is used to plan, optimise and manage public realm spaces.

10.2. Case study: downtown New Orleans

10.2.1 Setting the scene

Pedestrian modelling for public realm spaces will often be one element of a larger scheme to regenerate or improve an area. To appreciate the rationale for the scheme, it is vital to understand the existing conditions which the scheme should overcome.

The Department of Public Works of the City of New Orleans, USA wanted to complete an inventory and evaluation of downtown transportation, walking infrastructure and conditions in order to analyse potential multimodal improvements. The study area included the French Quarter from Calliope Street to Elysian Fields Avenue and from North Claiborne Avenue to the Mississippi embankment.

A qualitative assessment of the area was undertaken, which included analysis of field and digital data related to pedestrian safety and mobility, such as the condition of sidewalks, public transport availability and other parameters.

A desktop review of the study area was undertaken to supplement the sidewalk-quality data. This included more detailed investigation of 'poor' quality links in Google StreetView, to quantify any reduction in pedestrian link capacity.

Figure 10.1 Downtown New Orleans – main attraction points identified in the study area

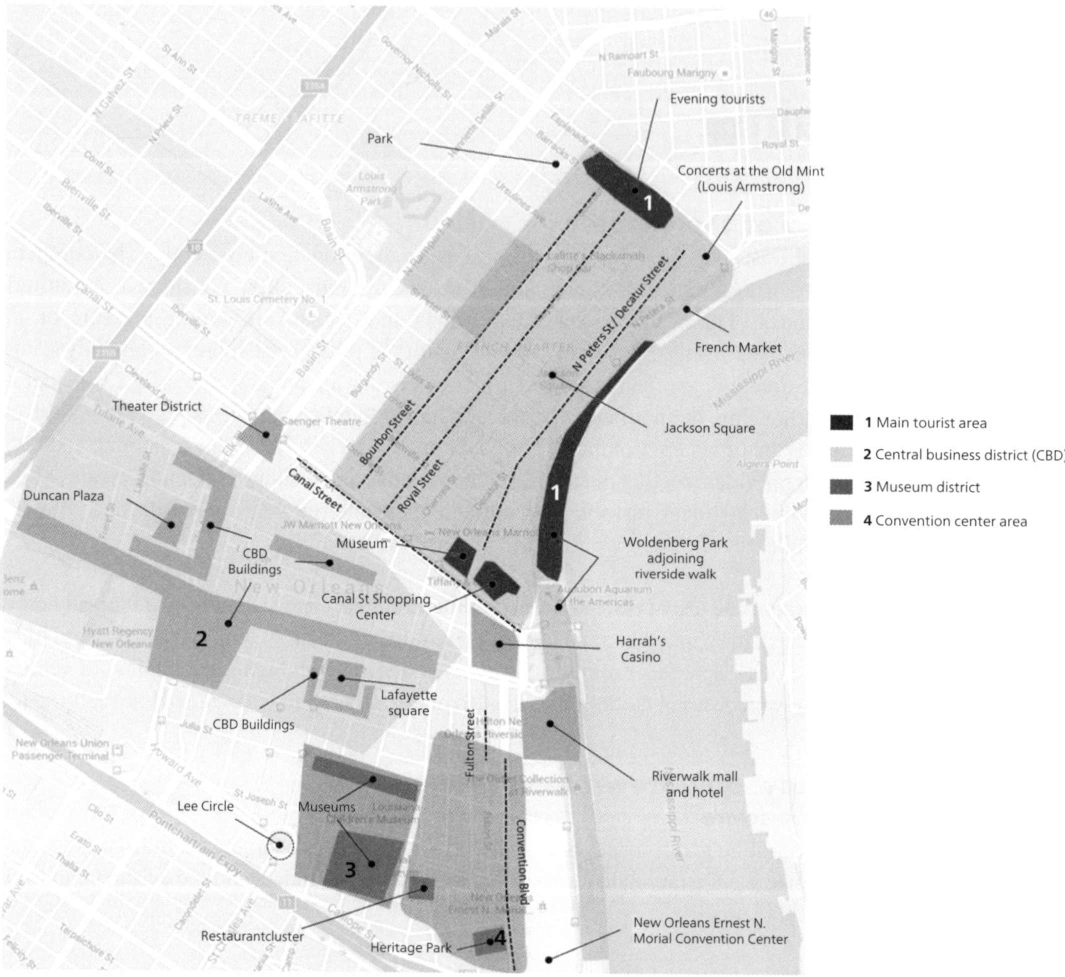

10.2.2 Defining the problem

The distribution and movement of pedestrians in the study area was analysed to assess the connectivity and completeness of the pedestrian network, access to transit stops and major attractions and other important parameters of pedestrian walkability. A combination of Excel-based static analysis and scripting-based network analysis was used to evaluate the pedestrian network in the study area.

Local knowledge informed the major pedestrian attraction points in the study area (such as museums, tourist attractions and shopping arcades) as well as the relative attractiveness of these locations for three different time periods.

Figure 10.1 shows the location of the identified attractions. Some are represented by more than one point to more accurately represent the actual area they occupy.

Figure 10.2 Assigned weights for recreational attractions for early PM in the study area (left) and weights for central business district (CBD) buildings and museum district (right)

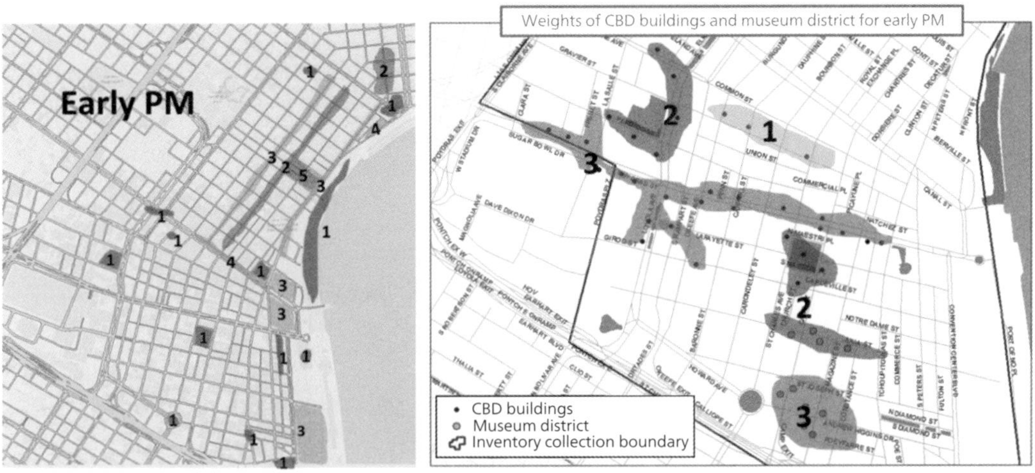

The time periods for which different pedestrian attractions have been identified are:

- AM – the morning commuter peak time, around 9am;
- early PM – the afternoon, leading into commuter peak time, around 5pm; and
- late PM – evening, after Bourbon Street closes to traffic, around 8pm.

Each pedestrian attraction has been allocated an attractiveness weight between 0 and 5, with 0 being the lowest, 3 being the default average, and 5 being the highest grade of attractiveness (i.e. the strongest draw to pedestrians). Figure 10.2 illustrates the assigned weights for all attraction points during different peak periods of the day. The map on the left shows the weights for recreational attractions only, while the map on the right shows weights for the business attractions and museum district.

All other factors being equal, areas of higher connectivity are expected to see higher pedestrian usage than those with lower connectivity. Connectivity analysis of the study-area network was undertaken to try and establish a relationship between actual pedestrian usage of an area and connectivity. Therefore, link connectivity is one of the variables checked when undertaking regression against surveyed pedestrian count data, to try and predict pedestrian flows on the network using known network link characteristics.

Connectivity of the whole network was analysed using the straightness of the route as a metric, as shown in Figure 10.3. Areas of high 'straightness' connectivity denote links which lie on routes with the fewest direction changes, shown in dark grey. Areas of low 'straightness' connectivity denote links which lie on routes with the highest number of direction changes, shown in grey, outside the white-highlighted area. The region with highest straightness connectivity in the study area is outlined in black.

Pedestrian survey data, counted inbound towards each junction, was processed for 44 junctions within the study area. Pedestrian flows outbound from each arm of the junction were also synthesised from this data

Figure 10.3 Downtown New Orleans – whole-network connectivity
Note: The algorithm will underestimate connectivity at the edges of the network.

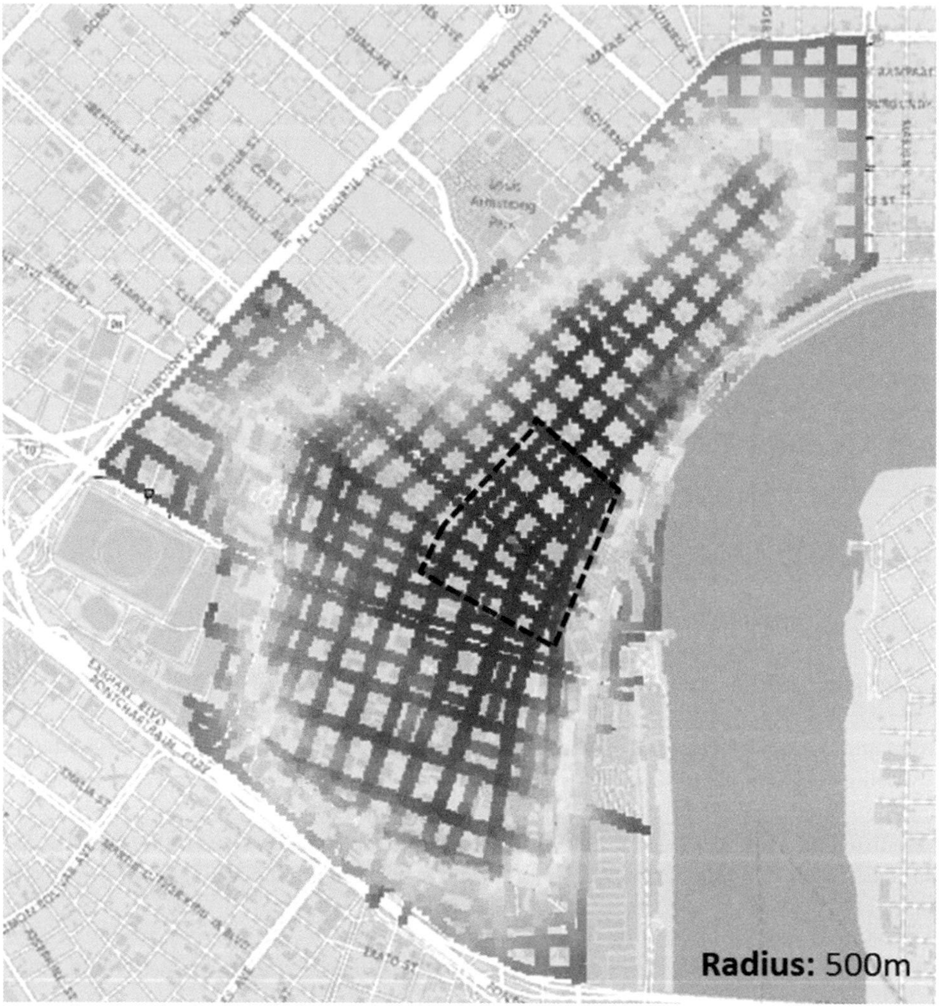

and then plotted for the AM and PM peak 15 minutes. The peak 15 minutes for AM and PM were identified from the survey data as:

- AM – 08:45 to 09:00; and
- PM – 17:15 to 17:30.

Next, regression was undertaken against the following variables to try and extrapolate likely pedestrian flows over the rest of the study area network:

- CBD and museum-district pedestrian attractions (separately for AM and PM);
- other main pedestrian attractions (separately for AM and PM);

- whole-network connectivity (straightness);
- distance from public transport;
- sidewalk width; and
- sidewalk condition.

For the AM, the most significant relationship was a negative correlation between flows and distance from public transit stops. For the PM, the most significant relationships to flows were a negative correlation with distance from the main tourist attractions and, again, a negative correlation with distance from public transit stops. An attempt was made to take sidewalk widths into account for both the AM and PM.

Producing the final flows on the network was a two-step process due to the relative sparsity of data available for the regression, both in terms of counts and input variables. First, the above relationships were used to determine the expected relative usage of each link in the study area by pedestrians. Second, a normalisation curve was fitted to the existing survey data, so that the surveyed links expected to see the lowest usage corresponded to the lowest flows surveyed, and the highest-usage surveyed links corresponded to the highest flow data surveyed. The curve that best fitted the data, for both AM and PM peaks, has the following equation:

$$Normalised\ predicted\ flow = \frac{e^{(predicted\ flow/14)}}{e^{(maximum\ predicted\ flow/14)}}$$

The flows predicted in regression were then also fitted to this normalisation curve to produce the final predicted flows for all links in the study area. Finally, where survey data was available, it overrode the 'predicted' flows with the actual recorded data.

10.2.3 Evaluating performance

Since pedestrian LoS is not solely dependent on densities, a more detailed investigation into sidewalk conditions and widths was undertaken (Figure 10.4) before producing the final LoS results from the flow plots.

Research indicates that for links which provided a fair to very good quality environment for pedestrians, sidewalk condition would have a minimal impact on the routing of pedestrians. However, on some links which provided a poor-quality environment for pedestrians, such as those with obstacles, broken pavement or a sufficiently unpleasant aesthetic to cause security concerns, there could be a notable negative impact on pedestrian flows.

Therefore, for all links recorded as being in poor condition, the link environment was interrogated to try and estimate the scale of impact this would have on the capacity/usage of these links by pedestrians.

Since some of these links appeared to be under construction – and therefore were only in 'poor' condition temporarily – two versions of the plots, the first showing the effect of capacity reduction during construction, and the second assuming no capacity reduction on these links after construction was completed. These capacity reductions were applied, together with the minimum sidewalk widths for each link, to calculate the densities pedestrians were expected to experience on each link during the AM and PM peaks.

As mentioned, the wide links and relatively low pedestrian flows on large parts of the network result in flows that are considered LoS A according to Fruin's assessment, which is based on density. This is true also in the case of the key corridors identified within the study area. Figure 10.5 indicates the key links assessed within the study area, and Tables 10.1 and 10.2 summarise the width, predicted flow and LoS for these corridors within the study area network for AM and PM respectively.

Figure 10.4 Downtown New Orleans – sidewalk conditions and widths

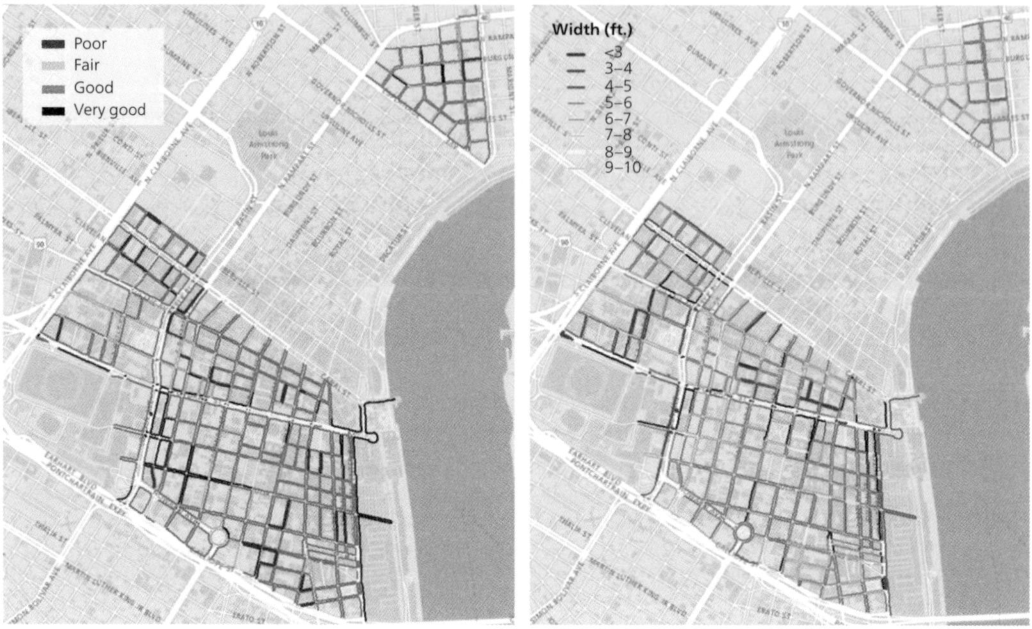

For both the AM and PM, the highest flows are predicted to occur on Canal St and Convention Center Blvd. Additionally, the highest flow rates per unit of sidewalk width occur on Canal St, Tchoupitoulas St, Convention Center Blvd and Common St. In general, flows and flow rates are higher in the AM than the PM except for on Common St.

While all the links demonstrated LoS A during normal peak flows on an average weekday, it is envisaged that if this model is applied with the forecast demand for special event days or for an average weekday 20 years in the future, there will be a wide range of variability observed of densities across the various links of the pedestrian network.

10.3. Case study: Southall Broadway, London Borough of Ealing

10.3.1 Setting the scene

In 2010, the London Borough of Ealing, UK wanted to develop a scheme to significantly improve a local commercial centre. The Southall Broadway Boulevard scheme was based on shared-space principles, which aimed to improve safety, pedestrian connectivity and journey-time reliability along The Broadway in Southall.

Southall Broadway Boulevard was an innovative scheme that created a safer and more comfortable environment for pedestrians while maintaining the capacity of the busy A4020 corridor. The development and assessment of these measures required the application of innovative Vissim modelling techniques to replicate the potential impact of pedestrian movement due to the proposed interface with the traffic flows. Modelling crossing behaviour where pedestrians choose not to use zebra crossings and decide to walk across the road between flowing traffic, through gap acceptance, was particularly challenging. The project also involved the development and testing of bus-priority measures, signal-timing reviews, bus-gating strategies and urban realm measures, including courtesy crossings and Boulevard Zones.

Figure 10.5 Downtown New Orleans – key pedestrian flow corridors in study area (light-grey lines)

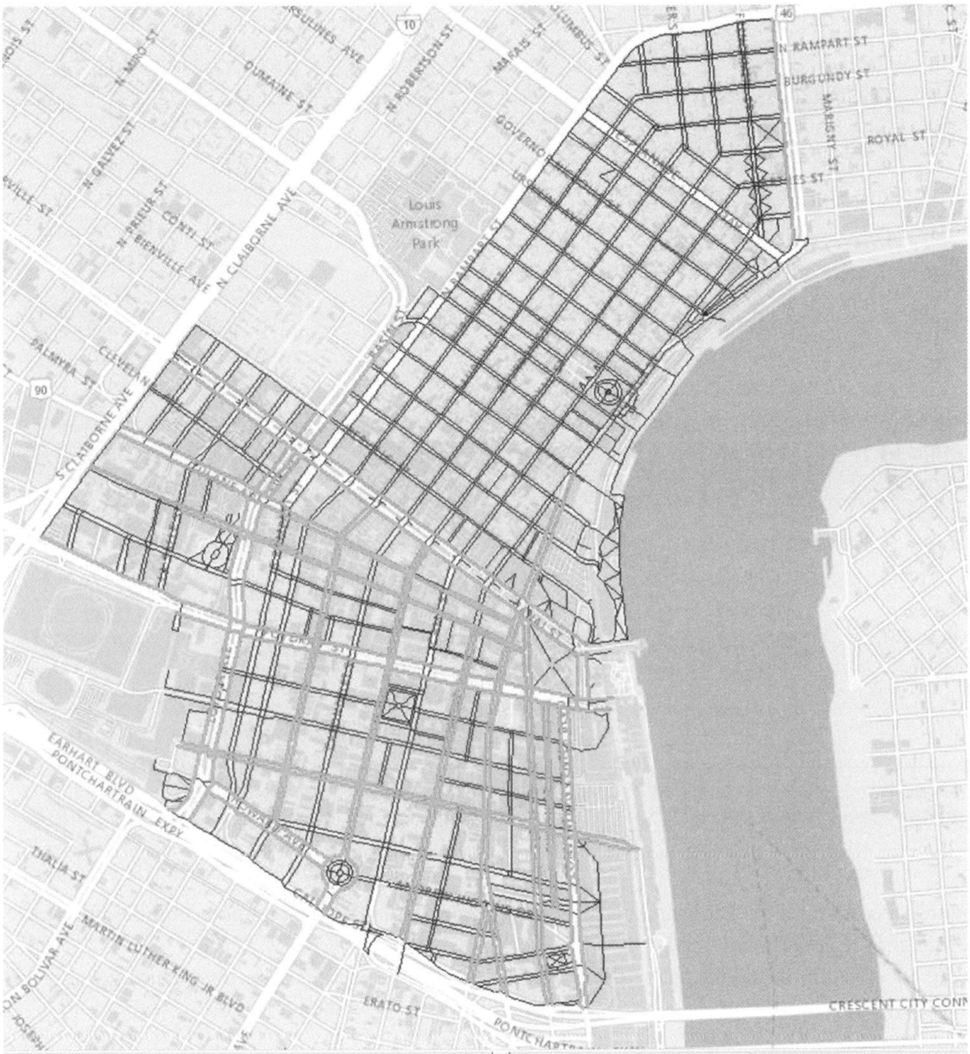

10.3.2 Defining the problem

The adapted section of The Broadway forms the main shopping hub in Southall as well as being a major east–west route in the area. The construction and implementation of the scheme was completed in early 2015. The scheme objectives were to be achieved through the provision of the following measures, which are shown in Figure 10.6:

- amending the cross-section to provide additional space for the safe and commodious movement of pedestrians by widening footways;
- implementing four Boulevard Zones, to include four courtesy crossings along The Broadway;
- removing four existing signalised pedestrian crossings;

Table 10.1 Downtown New Orleans – AM flows and LoS for key corridors within study area network

Corridor name	Condition of corridor overall	Average width (ft)	Average 15-min flows from regression	Flow rate (ppmpm)	Adjusted* LoS (A = best, F = worst)
Elysian Fields Ave	Good	10.9	7	0.014	LoS A
Basin St	Good	8.6	10	0.023	LoS A
Canal St	Good	16.7	24	0.031	LoS A
N Peters St	Good	12.0	8	0.013	LoS A
Tulane Ave	Good	10.7	6	0.020	LoS A
Poydras St	Good	12.1	10	0.026	LoS A
Loyola Ave	Good	10.2	3	0.016	LoS A
Baronne St	Very good	9.5	8	0.030	LoS A
Julia St	Very good	11.1	13	0.024	LoS A
Convention Centre Blvd	Very good	15.6	26	0.048	LoS A
Tchoupitoulas St	Good	12.4	15	0.034	LoS A
Annunciation St	Good	11.2	3	0.007	LoS A
St Charles Ave	Good	11.7	11	0.019	LoS A
Howard Ave	Good	10.8	3	0.010	LoS A
Elk Pl	Good	12.0	13	0.033	LoS A
Common St	Good	10.0	21	0.069	LoS A
Gravier St	Good	8.9	19	0.046	LoS A
S Peters St	Good	13.4	25	0.037	LoS A
Fulton St	Good	14.7	10	0.014	LoS A
Magazine St	Good	12.4	17	0.032	LoS A

ppmpm = pedestrians per minute per metre width.
* Adjusted by sidewalk condition.

- removing an eastbound bus lane and installing a bus gate linked to the operations at the South Road/ Lady Margaret Road junction; and
- reducing the speed limit to 20mph, self-enforced by vertical deflections, narrower carriageways and enhanced urban realm design.

As part of the wider pedestrian modelling of Southall Town Centre, a thorough static assessment of the area was undertaken, including pedestrian area widths as well as the general quality of the urban realm. This assessment aimed to identify any areas which could potentially impede pedestrian circulation and was considered important, given the opportunity to invest in the wider urban realm.

10.3.3 Evaluating performance

The objective of the study called for the development of an integrated local area model able to represent both pedestrian and vehicular activity, and space sharing along and across the main traffic routes.

Monitoring is key to understanding the performance of schemes once they have been delivered. The post-implementation analysis of the Southall Broadway Boulevard scheme relied heavily on the availability of pre-implementation and post-implementation data. Data was collected at various locations along Southall

Table 10.2 Downtown New Orleans – PM flows and LoS for key corridors within study area network

Corridor name	Condition of corridor overall	Average width (ft)	Average 15-min flows from regression	Flow rate (ppmpm)	Adjusted* LoS (A = best, F = worst)
Elysian Fields Ave	Good	10.9	11	0.021	LoS A
Basin St	Good	8.6	11	0.025	LoS A
Canal St	Good	16.7	48	0.057	LoS A
N Peters St	Good	12.0	15	0.025	LoS A
Tulane Ave	Good	10.7	9	0.026	LoS A
Poydras St	Good	12.1	18	0.048	LoS A
Loyola Ave	Good	10.2	8	0.022	LoS A
Baronne St	Very good	9.5	10	0.034	LoS A
Julia St	Very good	11.1	11	0.023	LoS A
Convention Center Blvd	Very good	15.6	29	0.054	LoS A
Tchoupitoulas St	Good	12.4	47	0.082	LoS A
Annunciation St	Good	11.2	5	0.013	LoS A
St Charles Ave	Good	11.7	22	0.038	LoS A
Howard Ave	Good	10.8	4	0.015	LoS A
Elk Pl	Good	12.0	11	0.030	LoS A
Common St	Good	10.0	22	0.065	LoS A
Gravier St	Good	8.9	21	0.047	LoS A
S Peters St	Good	13.4	32	0.047	LoS A
Fulton St	Good	14.7	35	0.044	LoS A
Magazine St	Good	12.4	26	0.043	LoS A

ppmpm = pedestrians per minute per metre width.
* Adjusted by sidewalk condition.

Broadway Boulevard during 2010 and 2012, and in 2015 as a part of this post-implementation study. The performance of the scheme was judged against several criteria:

- pedestrian performance;
- traffic flow;
- travel time;
- speed analysis;

Figure 10.6 Southall Broadway scheme

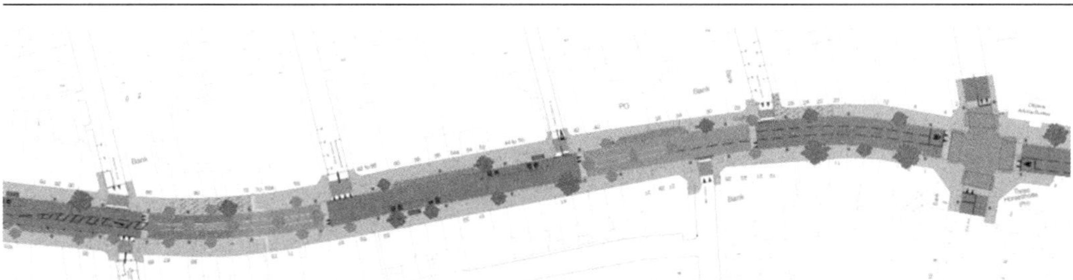

Figure 10.7 Southall Broadway Boulevard – model (a), visualisation (b) and final design (c)

(a) (b) (c)

- illegal parking; and
- loading-zone utilisation.

Pedestrian movements were assisted by increasing the perception of pedestrian priority at junctions along The Broadway by raising pedestrian footways across side roads. Additionally, pedestrian crossing movements were facilitated through the introduction of Boulevard Zones, shown in Figure 10.7, which incorporate refuge islands and shared-space design features. This has resulted in pedestrians assuming right of way in 42% of all crossing movements within Boulevard Zones on The Broadway.

The removal of signalised pedestrian crossings resulted in a spreading of crossing movements along The Broadway during the PM and Saturday pedestrian peak hours, with crossing movements more evenly distributed (Figure 10.8), suggesting that pedestrians are better able to follow 'desire-lines', a measure of increased permeability.

The implementation of the scheme has been shown to have had minimal impact on traffic flows and volumes on The Broadway, although there was a slight decrease in overall traffic volumes relative to 2012.

Figure 10.8 Southall Broadway Boulevard zonal crossing movement

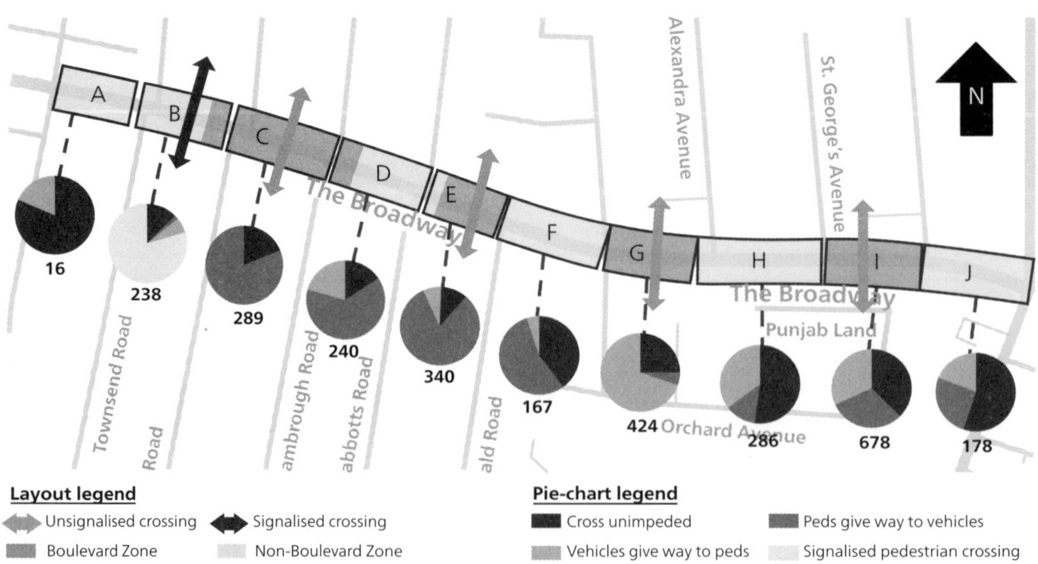

10.4. Discussion: planning for disruption

The built environment is constantly changing as cities evolve and grow. Construction works often impinge on public realm spaces as space is required for access and site works. Analysis of pedestrian behaviour can help planners and designers forecast how roadworks or construction projects will impact pedestrian flows. This section looks at how to approach this from a pedestrian modelling and analysis perspective.

To enable planned construction works in February 2017 along Victoria Embankment in London, UK, the southern footway of Victoria Embankment between Blackfriars Bridge and Temple Avenue was to be closed, including the 'south stairs' leading to the underground footpath on the riverside. Existing pedestrian movements between Blackfriars Bridge and Temple Avenue were to be diverted to the northern footway of Victoria Embankment, as shown in Figure 10.9.

To enable this work, changes to Victoria Embankment were required to provide a safe environment for all users of Victoria Embankment, including pedestrians, cyclists and drivers. The scheme included the following key elements:

- extension of the existing closure of the southern footway of Victoria Embankment to Middle Temple Lane;
- closure of the existing pedestrian crossing at Temple Avenue junction;
- provision of a new Middle Temple Lane pedestrian crossing;
- diversion of the existing cycle lanes between Temple Avenue and Middle Temple Lane so that they would be along the northern side of Victoria Embankment; and
- remodelling of the traffic lanes between Temple Avenue and Temple Place with new road markings and kerbs.

LEGION modelling was undertaken to assess the existing and proposed diversion layouts. Future pedestrian flows (scheme models) were assumed to be the same as the existing flows as the diversion was to be implemented within six months of the assessment being completed and in place for only 12 months.

The modelling considered the following two peak-hour periods:

- AM – 08:00–09:00; and
- PM – 17:30–18:30.

The following two layouts were considered (Figure 10.10):

- existing; and
- proposed (construction scheme).

Figure 10.9 Victoria Embankment – proposed pedestrian diversion/footway closure

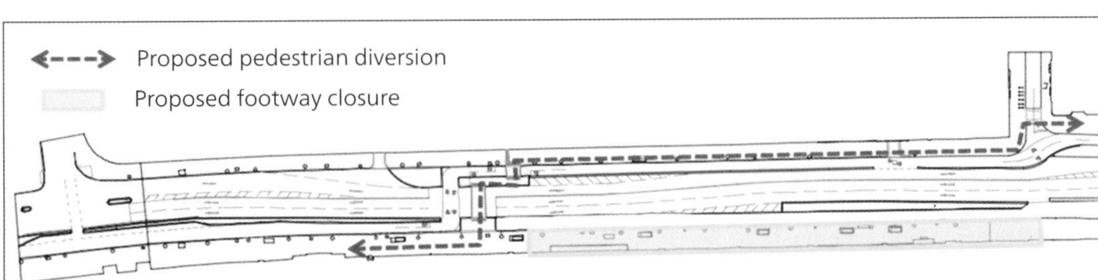

Figure 10.10 CMD maps – existing (a) and proposed (b) layouts

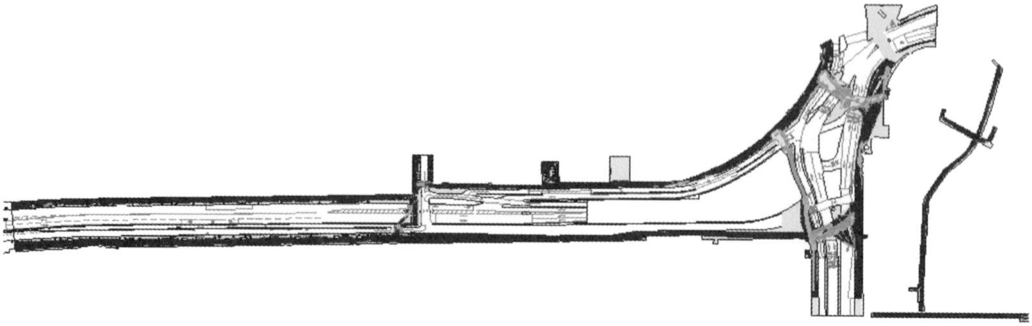

(a)

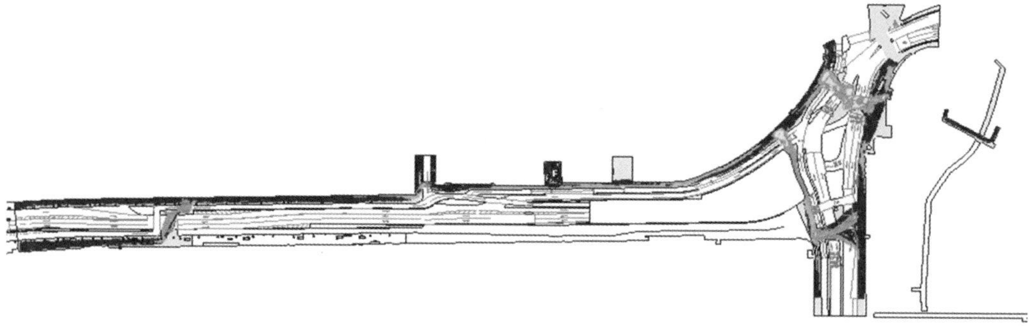

(b)

Signal timings for the relevant junctions were obtained from Transport for London (TfL) for both the existing and proposed crossing points. Pedestrian demand was obtained from TfL's Road Space Management department. The CMD results for the PM peak, which was the busiest, are shown in Figure 10.10(a) and (b) for the existing and proposed layouts respectively.

The proposed closure of the southern footway of Victoria Embankment between Temple Avenue and Middle Temple Lane would only affect approximately 2% and 5% of the current pedestrian demand in the AM and PM peak hours respectively.

The modelling showed that the proposed scheme would have minimal impact on current pedestrian conditions in both the AM and PM peak hours. The footways and crossings in the study area would operate at levels of service comparable to the existing conditions. The new Middle Temple Lane junction would operation within LoS C in both peak hours.

10.5. Summary points

- The study of pedestrian movement on a wider scale is often considered for urban areas as part of an urban regeneration scheme.
- There are several important considerations, such as attractiveness and connectivity, which influence wider movement.

- Macro pedestrian modelling can be used to determine the most attractive routes of movements for pedestrians and identify the optimal locations for transit stops and amenities.
- As shared-space schemes give ever greater priority to pedestrians in urban areas, the interactions between pedestrians and vehicles are expected to play an increasingly significant role in pedestrian modelling and analysis.
- Pedestrian–vehicle interaction is complex but can be modelled realistically to understand and predict behaviour to assess the benefit of urban realm schemes.
- In cities that are constantly developing and changing, it is also important to consider the impact of construction works to ensure enough space remains for people to walk safely.

ice Publishing

Urban Space - AECOM
ISBN: 978-0-7277-6381-5
https://doi.org/10.1680/mpm.63815.155

Chapter 11
Buildings

11.1. Overview of the chapter

Urban areas across the globe are made up of buildings fulfilling a plethora of functions. At their most primitive, buildings provide for the basic physiological human needs of shelter and security, offering space for people to carry out their daily activities. A survey by the US Environmental Protection Agency conducted in the 1990s showed that people spend 87% of their time indoors: at workplaces, schools, shops, hospitals, entertainment complexes and, of course, inside their homes.

As much of our lives are spent inside buildings, ensuring the safety and comfort of building occupants is essential. This is especially relevant in large public buildings, which attract many people and where crowds and congestion are likely to form. Managing people inside buildings can help the optimisation of both space and vertical circulation elements, as well as maintain safe environments for occupants in both normal and emergency situations.

Pedestrian modelling can be used to assess buildings in terms of the safety and comfort of their occupants. The focus of this chapter is to present, through case studies and the guidance with which they were undertaken, a range of building typologies that have been assessed using pedestrian modelling.

11.2. Case study: National Pavilion, Astana Expo 2017

11.2.1 Setting the scene

Hosted in Kazakhstan's capital city, Astana Expo was held from June to September 2017. Visitor numbers exceeded expectations, with almost 4 million people visiting the expo site during this period, an average daily attendance of over 22,000. Over 100 countries participated, with exhibits across many pavilions. The central attraction was the National Pavilion of Kazakhstan, an eight-storey spherical building, hosting a range of thematic exhibits based on the concept of Future Energy. Given the anticipated high demand of visitors, a balance needed to be struck between aesthetics, the interactivity of the exhibits and the functional operation of the building, so that the experience was engaging and yet had enough capacity to meet the designers' aspirations. The National Pavilion is shown in Figure 11.1.

A micro-simulation study was undertaken to assess the operation of pedestrians entering, moving through and leaving the National Pavilion, particularly relating to spaces and aspects of the building expected to be problematic. These included lift operation, potential queuing hotspots and locations where crowd management was required for efficient operation.

The National Pavilion building is unique, each of its eight floor plates differing in both scale and layout. The floorplans for each level are depicted in Figure 11.2.

The layouts and anticipated dwell time for each exhibit were incorporated into a model that allowed each visitor's movement to be followed from entry, through each layer via the exhibits and shows, through to exiting the National Pavilion 150–180 minutes later.

Figure 11.1 The spherical National Pavilion of Kazakhstan

The scenarios assessed were based on two levels of peak demand and tour-group entry size:

- groups of 60 visitors departing on tours every 2.5 minutes; and
- groups of 60 visitors departing on tours every 5 minutes.

The analysis focused on density levels, journey times and queue times to ensure conditions were not too crowded and delays did not adversely impact visitor experience.

Figure 11.2 National Pavilion floorplans (top row: levels 1–4, left to right; bottom row: levels 5–8, left to right)

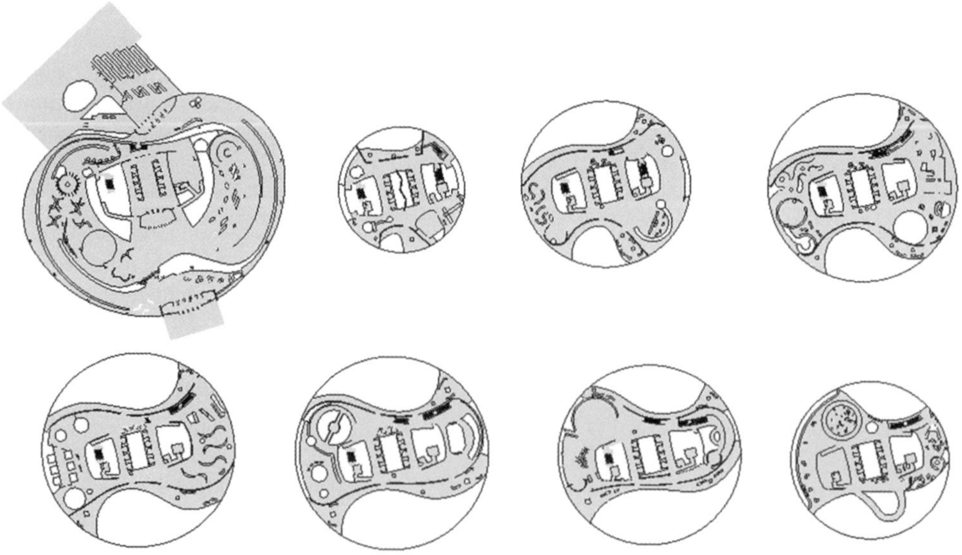

Figure 11.3 Visitor types

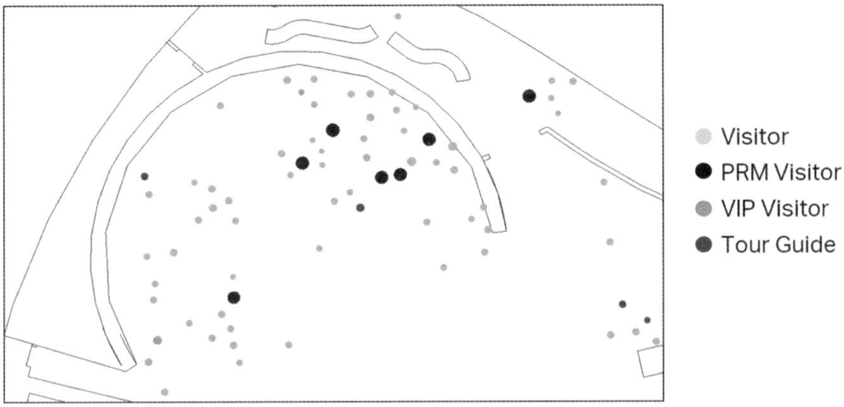

11.2.2 Defining the problem

Modelling the National Pavilion presented several unique challenges related to the complexity of the building design, its function as a tour-led exhibit space and the split of expected visitor types. Journeys, entries and exits, tour groups, viewing exhibits and general circulation all had to be accurately modelled. The four key visitor types were standard Visitors, PRM Visitors, VIP Visitors and Tour Guides. These are depicted in a model screenshot in Figure 11.3.

Visitors to the National Pavilion were led through the exhibits in tour groups. Each 60-person group comprised three tour groups made up of one PRM, four children and 15 adults. The group nature of tours through the building brought challenges resulting from the lack of any grouping-behaviour capability within the modelling software. Groups of people were kept together using additional modelling objects, as well as time profiles, that directed them through each exhibition area as groups at the correct timings based on their tour schedule.

Routing through the building was also unusually complex. Visitors took lifts to level 8 and then typically used stairs to travel down the building to level 3. Not all tour groups visited the exhibits in the same order, and not all tour groups were made up of the same composition of visitor types. Where there were exceptional differences, such as a group of schoolchildren, or a group of PRMs, additional tour support was required, and lift priority was adjusted as necessary.

From the main entrance, two paths were defined for Visitors, with an additional, more direct path for VIP Visitors. The tour entrance paths are shown in Figure 11.4.

With eight publicly accessible levels, ensuring that there was sufficient vertical circulation within the National Pavilion was of critical importance. A cross-section of the building is shown in Figure 11.5. A total of eight lifts provided the key vertical circulation; however, their use was divided between floors and visitor types to facilitate efficient circulation. The eight lifts were allocated as follows:

- four initial shuttle lifts to level 8;
- one VIP Visitor lift;
- one PRM Visitor lift; and
- two shuttle lifts down from levels 3 and 2 to 1.

Figure 11.4 Level 1 routes

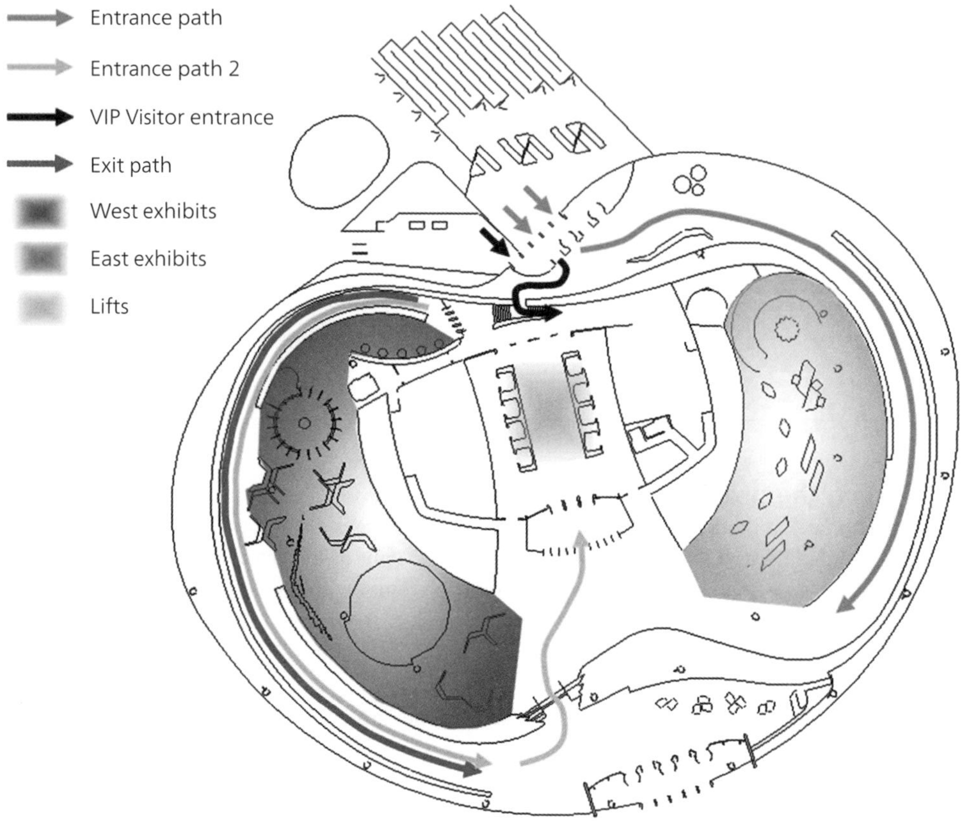

The allocation of lifts is shown in Figure 11.6. The initial shuttle lifts were available for visitors from levels 1–8. The VIP Visitor lift was available for VIP Visitors only, also from levels 1–8. The PRM Visitor lift took PRM Visitors in single-level shuttles from level 8 to 7, level 7 to 6, level 6 to 5, level 5 to 4 and level 4 to 3. The downward shuttle lifts took visitors from level 3 to 2 and from level 2 to 1. In addition, a series of stairs enabled downward movement between levels 8 and 3, with additional capacity available via the emergency stairs between levels 3 and 2. Given the different pedestrian types and lift allocations, lift use in the building was complex to model, requiring accurate filtering of objects and availability time profiles for each lift based on travel speeds and distances between floors.

11.2.3 Evaluating performance

The outputs from this modelling played a vital part in determining the operational strategy as well as the layout of exhibition content within the National Pavilion. Feedback provided to the designers and architects enabled changes to be incorporated into the design which optimised visitor movement. Some recommendations were made as a result of the modelling study:

- maintaining LoS C in circulation areas will promote safe crowd circulation and requires active crowd management by tour guides;

Figure 11.5 Cross-section of the National Pavilion showing levels 1–8

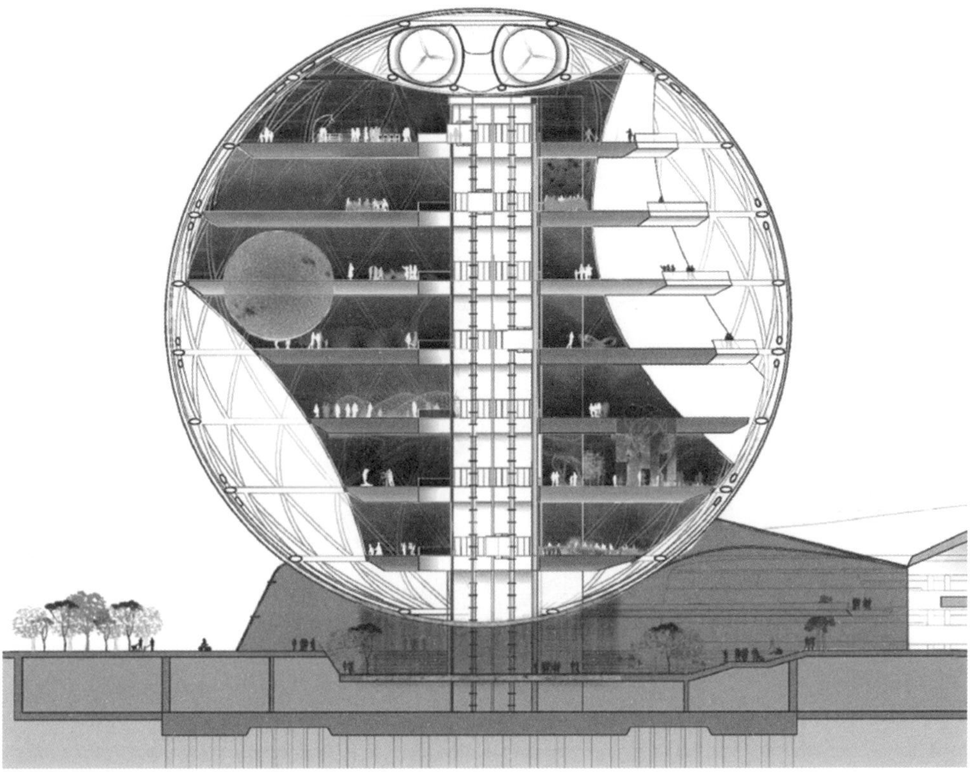

- significant queuing levels in the lift vestibules indicate that crowd management is particularly important in these areas;
- low levels of service indicate that tours need to be carefully organised to avoid localised congestion at exhibits with greater dwell times;

Figure 11.6 Lift allocation

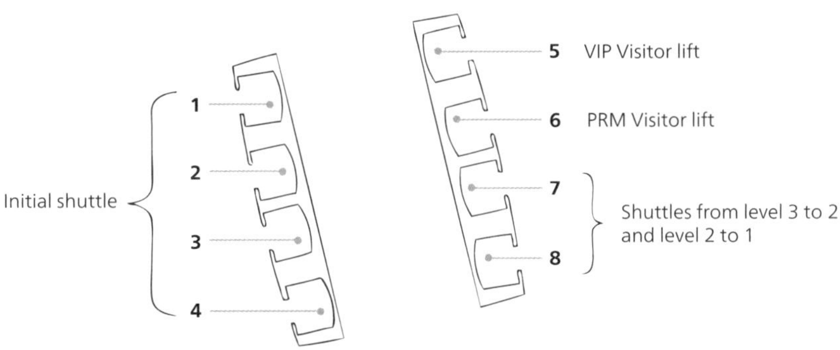

Figure 11.7 National Pavilion CMD map – level 4

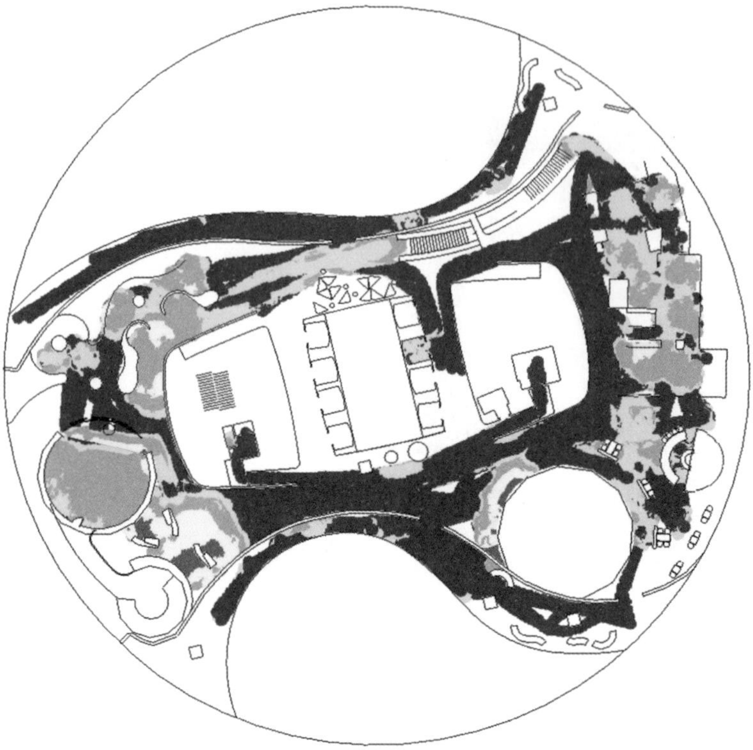

- iterative modelling indicates that a unidirectional crowd-management strategy for visitors and staff entering and departing through the lift area on level 1 is required to ensure effective movement in this critical area;
- if stairs are provided between levels 1 and 3, more lifts can be used for the shuttles, which will increase the capacity of the overall vertical circulation system; and
- there is enough capacity for 60 visitors every 2.5 minutes when the lifts are allocated optimally; this requires careful management, particularly in anticipating cases where visitors miss their allocated time (i.e. an 'overflow' lift shuttle at regular intervals).

The pedestrian flow modelling predicted that under normal operation the crowd-density levels would be within acceptable parameters, provided there was enough capacity for lift management and that stairs were used throughout the structure (including levels 3, 2 and 1). (Figure 11.7 shows acceptable density levels in circulatory spaces, which operate at predominantly LoS A/B. Some congestion of predominantly LoS C/D, down to LoS E in isolated areas, is expected in the exhibition spaces.) Journey times indicated that, given the assumptions around time spent at each exhibition, most visitors would spend 110–150 minutes in the National Pavilion.

Considering optimum entry group size, modelling indicated that either four groups of 15 visitors or three groups of 20 visitors could be comfortably accommodated for entry (alongside up to six tour guides) given the lift constraints.

Based on the scenarios modelled, optimal entry flow rates were established:

- with the use of stairs between levels 2 and 1, 60 visitors per 5 minutes can be accommodated: and
- with the reallocation of lifts to optimise usage, 60 visitors per 2.5 minutes can be accommodated.

This indicates that 1.36m visitors could be accommodated during the entire Expo if the stairs were used as the primary means of descent between levels 3 and 1.

11.3. Case study: School of Science, Health and Wellbeing

Vertical circulation in buildings is often assessed using a static analysis method prior to micro-simulation modelling. In some cases, undertaking static analysis provides sufficient understanding of and confidence in a proposed design, or certain features of it, for which no further assessment is required. Faced with the challenge of accommodating 1,260 students on a highly restricted site, the designers of the new School of Science, Health and Wellbeing (SHaW) in Bromley, south London, UK, proposed a vertical school set over ten levels (Figure 11.8). The challenge of moving over 1,200 students between classes on different levels in a multi-storey building warranted a static assessment of vertical circulation provision in the building's design as well as its proposed operational connectivity strategy.

The aim of the assessment was to evaluate the efficiency of both the vertical transportation elements and the circulation spaces on each level. A review of demand, floor circulation and vertical circulation was undertaken for the peak-time scenarios of morning arrivals, evening departures, lunchtime and movements between classes. A set of demand assumptions was agreed with the project team, with a total of 1,260 students split between three year groups, called KS (Key Stage) 3 (360 students), KS4 (540 students) and

Figure 11.8 SHaW Building, Bromley

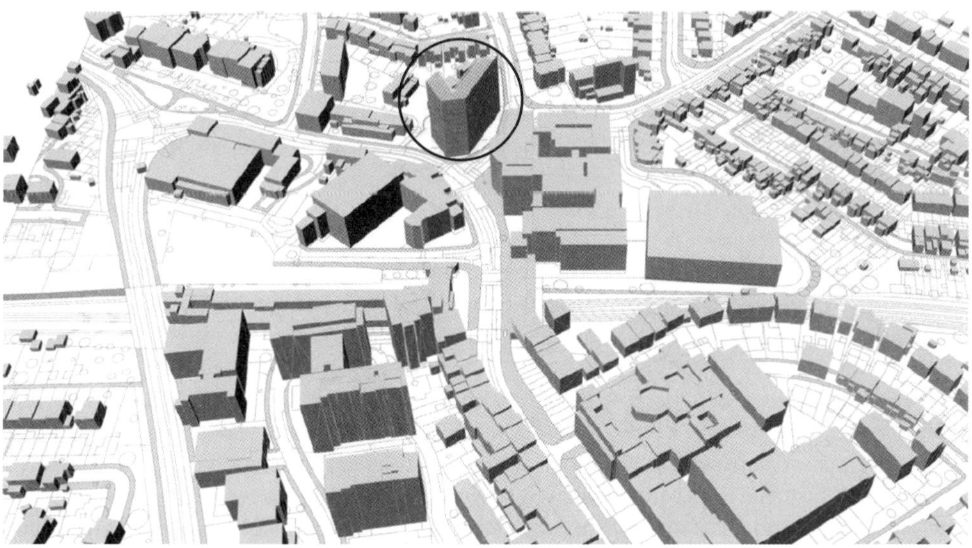

Figure 11.9 Arrivals routing plan

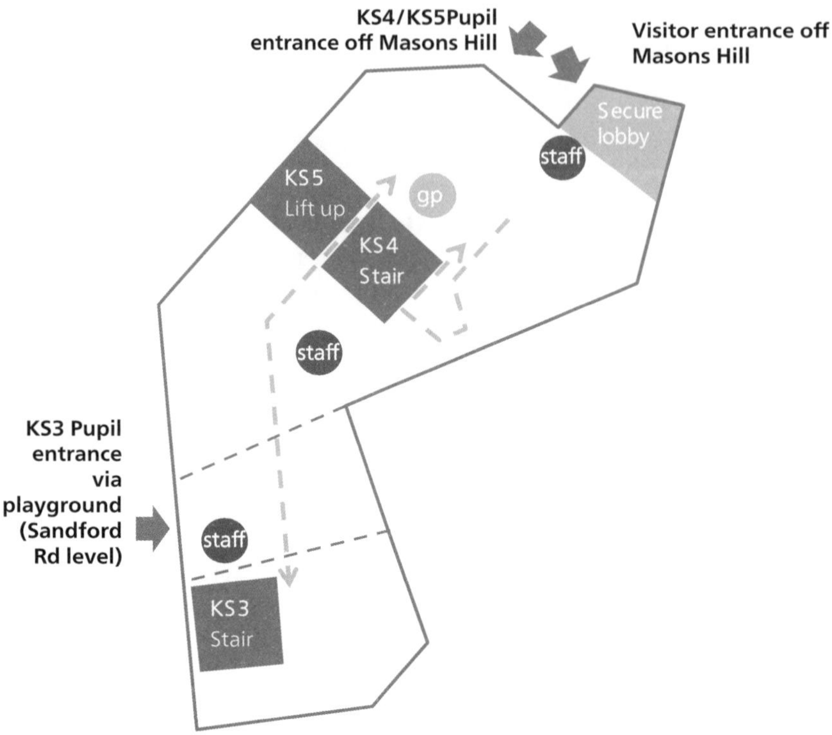

KS5 (360 students), each arriving at and departing from designated learning 'hubs' at staggered intervals based on class start and end times.

The location of the learning hubs for each year group are segregated by floor level, with KS3 located on the lower ground, KS4 on level 3 and KS5 on level 8. KS3 and KS4 students enter the building via entrances on opposing sides of the building and use the stairs to reach their respective hubs. KS5 students use the same entrance as KS4 students but are directed to lifts to reach their hub on level 8. This distribution of hubs and co-located learning spaces facilitates an arrival and circulation strategy that minimises conflicting movements between year groups. These are shown in Figures 11.9 and 11.10.

Circulation and movement patterns between floors were determined using class timetables and associated room schedules for all KS3, KS4 and KS5 students. It was assumed that lifts were only used during lunchtime and peak crossover movements if students were moving up or down by four or more floors.

Static models were developed of each floor, stairwell and lift bank based on the proposed design to identify capacity constraints. Circulatory capacities were calculated for all scenarios using the widths of the building entrances/exits and corridors, as well as the stair and lift capacities between each floor. The stair and lift capacities were assumed to be constant between all levels, while the narrowest sections of corridors were used in calculating corridor capacities. The circulatory capacities of each building element were then compared against student demand and routing patterns to determine locations where capacity constraints exist.

Figure 11.10 Connectivity strategy

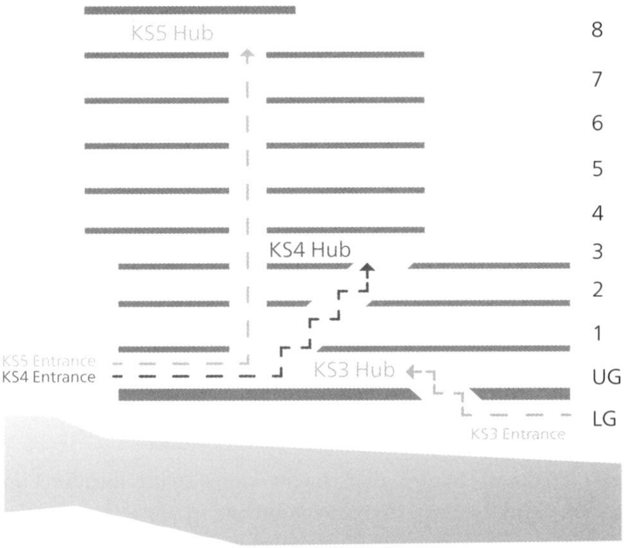

In general, the capacity of the facilities, including stairs, lifts and doors, were found to be sufficient for the assumed arrival profiles, which were staggered to differentiate arrival profiles for each student group. It was found that the maximum number of students entering the building through the lower and upper ground during the morning arrival period is 180 and 450 people respectively in 15 minutes. The doors at both entrances can accommodate this number of students without any crowding issues. In the afternoon departure period, the maximum number of students exiting the building through the lower and upper ground is 180 and 315 people respectively in 10 minutes. Again, the doors at both entrances can accommodate this number without any crowding issues.

The lunchtime and peak crossover routing was used to create spreadsheet analyses, calculating peak circulation movements on each floor and vertical movements on staircases and lifts. The peak crossover analysis demonstrated occasions when many classes circulated on the same floor or used the same set of stairs between floors. The two busiest lunchtime and two busiest crossover scenarios were selected for more detailed analysis.

In all the scenarios assessed, stair capacity is exceeded in situations when many KS3 and KS4 classes are changing levels. While lifts were assumed to be unused for movements of four floors or fewer, they could potentially be put into use under a school circulation strategy to alleviate the situation by providing additional capacity on the cores, enhancing the speed of student movement during the 'peak' times. Furthermore, while some stairs were found to be over capacity, others had enough capacity to cater to overflow from other stairs and could be used to balance demands with a suitable circulation strategy that distributes demand equally to all stairs.

Ultimately, the study determined that there was enough capacity on the staircases, lifts and doors for arriving and departing students, based on the staggered times of classes and registration in the existing timetable. The demand for movement during school hours, however, especially before and after breaks, was found to be higher than the available capacity in specific locations. Because of these capacity issues, a recommendation was made to implement a school circulation-management strategy that incorporated lift usage during the four busiest times and distributed students using the stairs equally between all cores to spread demand.

11.4. Case study: Galeries Lafayette, Champs-Élysées

Pedestrian behaviour in retail locations, such as shopping centres and malls, differs significantly from that in other building types. In a typical building, after arriving, people move from one specific place to another in a transient way based on a set of routing assumptions or knowledge; but shoppers behave in a less predictable manner. People moving through retail locations tend to stop randomly to browse, shop and move through spaces based on their shopping preferences or needs, rather than necessarily taking the shortest path to their destination. Simulating this behaviour allows retail operators to optimise the layout of their stores to provide visitors with a comfortable shopping experience, one that allows free movement and easy access and circulation between floors. It can also identify areas of highest footfall from passing shoppers, which can help determine locations for various shopping facilities, as well as set rental prices as appropriate for the level of footfall expected.

Galeries Lafayette is a French international chain of luxury department stores with locations in Paris and other cities in Europe, Asia and the Middle East. A redevelopment of the store on the Champs Élysées in central Paris, France, prompted the need for pedestrian analysis of the redevelopment's impact on pedestrian capacity at entrances and escalator run-off areas. The open centre is shown in Figure 11.11.

Figure 11.11 Galeries Lafayette, Paris

Figure 11.12 CMD maps – escalator run-off at basement level

Figure 11.13 CMD maps – ground floor main entrance

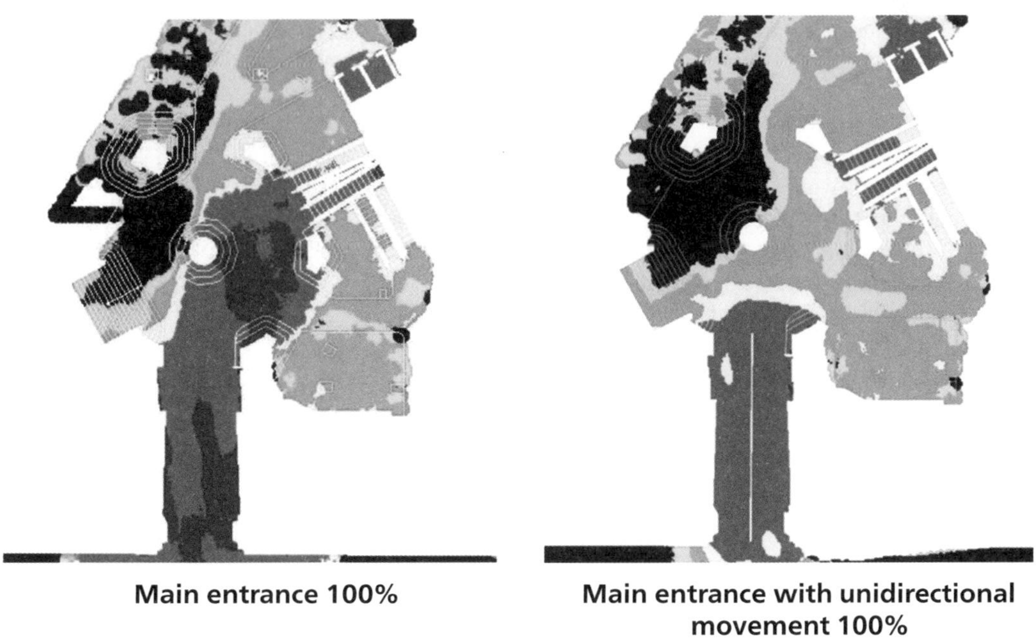

Figure 11.14 Galeries Lafayette evacuation times – all levels

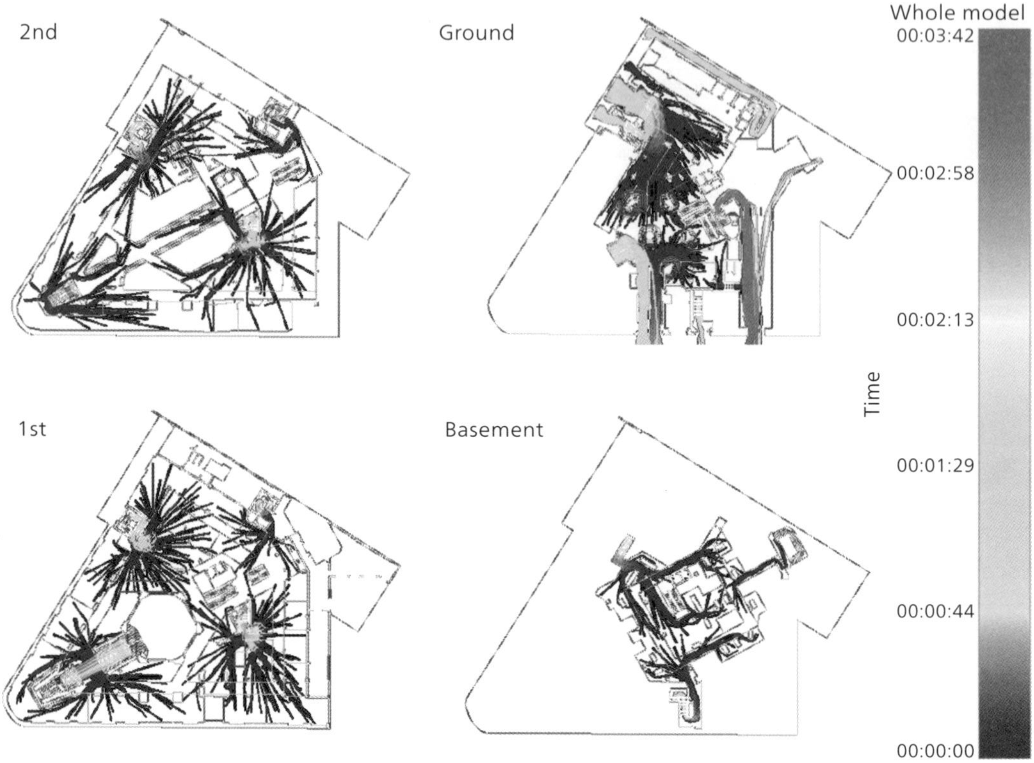

Each of the centre's four retail levels (basement, ground, first and second) were modelled using a set of assumptions agreed with the architect. On all levels it was assumed that shoppers would arrive and depart using the same vertical circulation component. From ground level it was assumed that:

- 25% of shoppers would circulate on the ground level before exiting the centre;
- 42% of shoppers would go up to the first floor via the escalators;
- 10% of shoppers would go up to the first floor via the grand staircase;
- 18% of shoppers would go down to the basement; and
- 5% of shoppers would go to the lifts.

From the first floor, it was assumed that 50% of shoppers would go up to the second floor, and 50% would return to the ground floor.

Three modelling scenarios were assessed using gradually increasing proportions of demand:

- 50%;
- 75%; and
- 100%.

CMD map outputs of the shopping centre showed that, although greater densities of people would be expected in the higher-demand scenario, no serious congestion issues were present in any demand scenario at escalator run-off and other circulatory areas. These are shown for the basement level in Figure 11.12. Run-off and other circulatory areas on other floors were also found to operate at acceptable levels of congestion.

An analysis of the main entrances to the centre on the ground floor showed that at 50% and 75% of full demand the entrances operate well, with some minor friction along the right-hand wall of the main entrance passageway due to shoppers being attracted to the escalators. Significant congestion was evident, however, at the main entrance in the full-demand, 100% scenario model. As a result, it was proposed to erect a barrier along the centre line of the entrance passageway and implement a one-way entrance and exit system to reduce pedestrian conflict. This was subsequently modelled and found to be an effective way of significantly reducing congestion in the main entrance area, as well as the central gallery of the shopping centre (Figure 11.13). The throughput capacity of the entrance passageway was also found to increase by around 24%, from a two-way capacity of 10,500 people per hour with no barrier, to a two-way capacity of 13,000 people per hour with a barrier installed.

Evacuation modelling showed that the shopping centre can be evacuated in under four minutes with an even distribution of emergency-exit use. Figure 11.14 shows the time taken to evacuate all areas of the shopping centre. At just under four minutes, the exit nearest the escalators on the ground floor took the longest to clear, but overall evacuation times were found to be at acceptably safe durations.

The pedestrian modelling of Galeries Lafayette showed that congestion in the proposed layout of the centre would be at acceptable levels and that the building could be safely evacuated in an emergency. It also allowed for the entrance passageway to be redesigned to reduce conflicting movements and minimise congestion in both this area and further inside the ground floor of the centre.

11.5. Summary points

- Various buildings typologies, such as museums, offices, schools and shopping centres, can benefit from crowd-modelling studies.
- For museums and similar buildings, it is important to consider grouping behaviour as this can have a significant impact on space and vertical circulation.
- Pedestrian modelling can aid in understanding peak visitor capacity and areas which require management to ensure smooth operation.
- For schools and university buildings, it is critical to understand demand to and from the building and movements to, from and between classes.
- Pedestrian modelling can also help in developing evacuation plans and strategies for buildings in case of emergency.
- For shopping centres, it is important to ensure that busy-period demand, such as the run-up to Christmas, is appropriately considered to stress-test the design.

ice Publishing

Urban Space – AECOM
ISBN: 978-0-7277-6381-5
https://doi.org/10.1680/mpm.63815.169

Chapter 12
Airports

12.1. Overview of the chapter

Air travel is becoming more popular each year. The aviation industry has experienced dramatic growth over the past 20 years, with passenger numbers rising from 1.47 billion in 1998 to 3.97 billion in 2017. The International Air Transport Association (IATA) forecasts the number of passengers transported by airlines will reach 8.2 billion in 2037, with compound annual growth forecast to be 3.5%. Airlines are growing to serve more passengers by improving their fleets and adding new aircraft that are more fuel-efficient. This growth in demand has put significant pressure on airports, hence the way people arrive, use and experience airports has become more significant.

Pedestrian modelling often plays a role in the design of airports to meet guidelines while considering operational requirements. This chapter discusses how pedestrian modelling can help airport operators manage how people arrive at airports and how they pass through check-in, security and emigration.

12.2. Case study: airport forecourt, Middle East

There are two main modal groups of travel for reaching the landside of an airport: rail and road. As stations were discussed in some detail in Chapter 8, this section will focus on accessing airports by road.

12.2.1 Setting the scene

In the UK, as in many developed economies, private motorised modes dominate the airport surface-access split. At some UK regional airports, as many as 95% of surface-access journeys are made by private vehicles. Surface access is a vitally important element of any airport's infrastructure: it directly impacts on operational efficiency and commercial performance as airside activities rely on a continuous, unimpeded and timely flow of passengers, staff and goods accessing and egressing the site. One of the key constraints on airport capacity is the ability of passengers to travel to an airport terminal in a timely and predictable manner.

The operating characteristics of an airport terminal pick-up and drop-off area – the airport forecourt – are unusual from both the perspective of pedestrians and drivers. The discussion will focus on the drop-off area as this has a greater impact on flight departures and therefore airport operation.

The design of forecourt areas is relatively common across modern airport terminals, and an example is provided in Figure 12.1. A cursory review of large European airports shows this type of arrangement is used at London Heathrow Terminal 5, UK, London Gatwick, UK, Amsterdam Schipol, the Netherlands and Frankfurt, Germany.

These areas typically comprise:

- two one-way road links, which split at the entrance;
- three or four traffic lanes;

Figure 12.1 Typical airport-forecourt drop-off area design

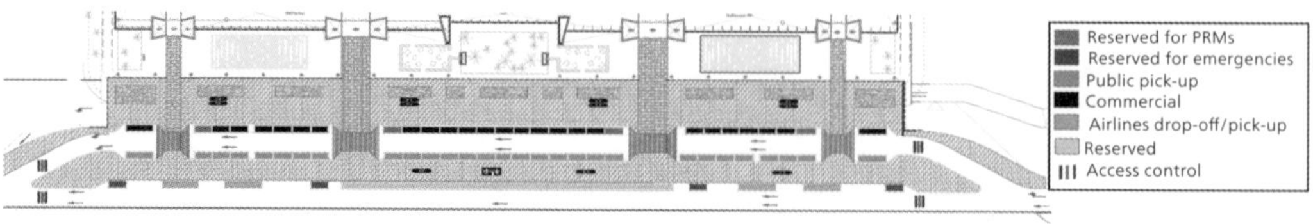

- three or four kerbs for vehicles to park and passengers to exit;
- defined parking for buses, taxis, private vehicles and emergency vehicles; and
- zebra crossings to reach the terminal from the outer kerbs.

The capacity of a forecourt is defined both by the number of vehicles that can be accommodated while stopping to pick up or drop off passengers and the number that can be accommodated while travelling past the kerbside in the through-lanes. Depending on the allocation of gates to respective airlines, certain gates can become more popular and attract more passengers and cars at certain points of the day, so demand is rarely uniformly distributed.

To undertake a micro-simulation study of such a complex area it is important to review available data and collect any necessary additional data. Vehicular demand, occupation, kerbside dwell time, delays at barriers and parking behaviour will affect capacity and operation. Passenger demand, age, gender, group size, walking speed and trolley usage will also impact operation. This is arguably even more important when modelling interactions with vehicles as pedestrian characteristics are likely to differ substantially.

12.2.2 Defining the problem

A model developed to optimise the forecourt operations of an airport in the Middle East will be discussed. Of particular importance to this operation was trolley usage, as 73% of passengers used trolleys after alighting from vehicles. To create a realistic representation of passenger processes, both on the kerbside and while crossing roads, the model considered various characteristics of passengers, such as:

- presence of luggage – in the model, this influenced the speed of movement and journey time of passenger services; and
- use of trolleys – this defined the size of the area around a passenger and the speed of movement.

As there are no off-the-shelf modelling products that can model vehicles, pedestrian journeys where characteristics change (e.g. bringing the luggage to a trolley, loading a trolley, continuing as a person with a trolley) and vehicle–pedestrian interaction, it was necessary to pick a platform that was adaptable and develop a working solution.

Pedestrians are categorised in different pedestrian types and pedestrian classes. Within a pedestrian type, pedestrians have similar technical characteristics and physical walking behaviour. For each pedestrian type, the maximal acceleration, the appearance of the pedestrian and a behaviour parameter file are defined. These attributes are adaptable to meet user requirements, so a pedestrian type was developed for people with trolleys with the corresponding size and speed characteristics. The next step was to develop a script to model trolley usage for passengers with luggage collecting trolleys after alighting from vehicles, to create a realistic representation of passenger behaviour. A summary of the pedestrian type changes is shown in Figure 12.2.

Figure 12.2 An example of model inputs of pedestrian parameters

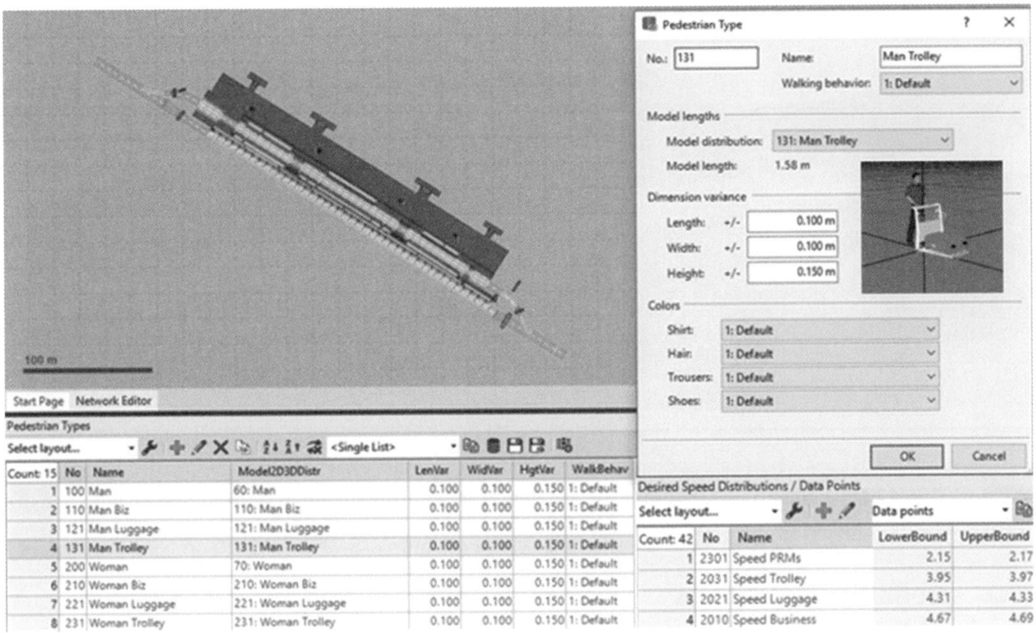

Vehicles accessing the drop-off departures area stop for passengers to alight. It is important to model realistic dwell times to accurately assess congestion within the forecourt and at the barriers. Detailed surveys were carried out to capture these dwell times in the forecourt area, which showed an average dwell time of 120 seconds and a delay at the barriers of 6 seconds on entry and 11 seconds on exit. The entry barrier is an automatic ticketing machine, while the exit barriers are manned operations, hence the dwell time at each location is different.

12.2.3 Evaluating performance

Three parking-allocation scenarios were evaluated. Each scenario had different vehicle groups located in different areas, such as for Scenario 3, in which:

- commercial and public parking facilities are provided at the inner kerb, closest to the terminal; and
- heavy vehicles, like public transport buses and shuttle buses, are provided at the outer kerb.

The modelling identified that Scenario 3 produced the best performance in terms of both traffic and average pedestrian journey time from vehicles to the terminal. Further recommendations were made on vehicular operations, such as the location and number of barriers and the clarity of signage in advance of forecourt lane choice, which would aid the smooth operation of the forecourt. A visualisation of the model is shown in Figure 12.3.

12.3. Case study: hub airport in Africa

12.3.1 Setting the scene

The airport is a major hub, with connections to domestic, regional and international destinations. As one of the largest airports in Africa, the airport's growing passenger numbers have justified a significant expansion

Figure 12.3 Visualisation of the final model

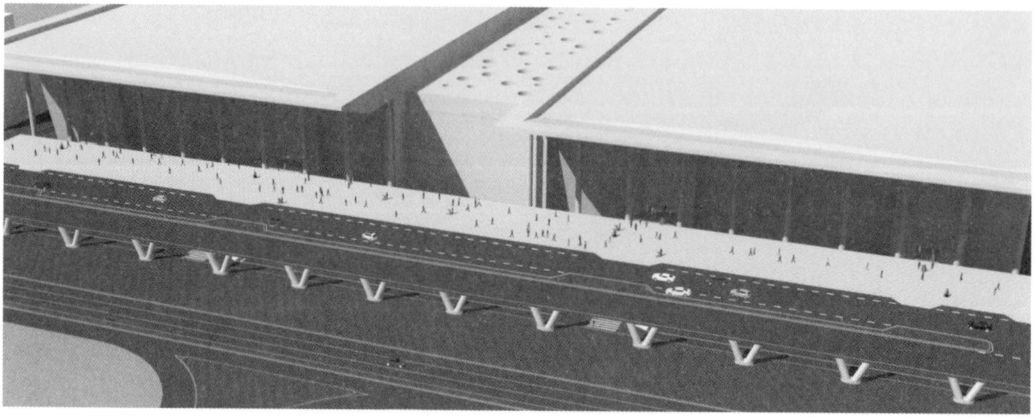

to the departures hall. Pedestrian modelling of the departures hall was undertaken to assess whether the proposed design could be optimised to aid flow conditions and handle forecast passenger numbers.

12.3.2 Defining the problem

12.3.2.1 Service requirements

IATA is the governing association for many of the world's airlines. It provides guidelines to ensure global standards for airline safety, security, efficiency and sustainability. IATA's LoS concept is the benchmark for optimal passenger-terminal facilities. As with the measurement of facility types against other commonly used concepts of LoS, such as Fruin's, it is important that the modelled results are compared with industry-wide standards to accurately comment on whether the new design will be suitable for the predicted increase in capacity.

The LoS concept consists of four conditions: under-provided, sub-optimum, optimum and over-design. A summary of these categorisations and the responses required is shown in Figure 12.4.

To assess whether the new terminal design was suitable for the predicted increase in passenger numbers, the modelled results needed to be compared against the standards defined by the LoS. A summary of the quantitative standards is shown in Figure 12.5.

The project required an analysis of passenger flows and processing times in the departures terminal, specifically the check-in and security areas, for the future capacity scenario. The major outcomes included:

- the development of a simulation model to understand the LoS and capacity of the airport's proposed layout;
- an assessment of the capacity and service levels, in line with IATA's recommended LoS for each element in the terminal; and
- recommendations to improve service levels and the passenger experience.

The departures terminal was assessed against a series of key criteria. These included:

- check-in desk queuing and processing time;
- self-service check-in desk queuing and processing time;

Figure 12.4 IATA's LoS concept from the 10th edn of their *Airport Development Reference Manual*

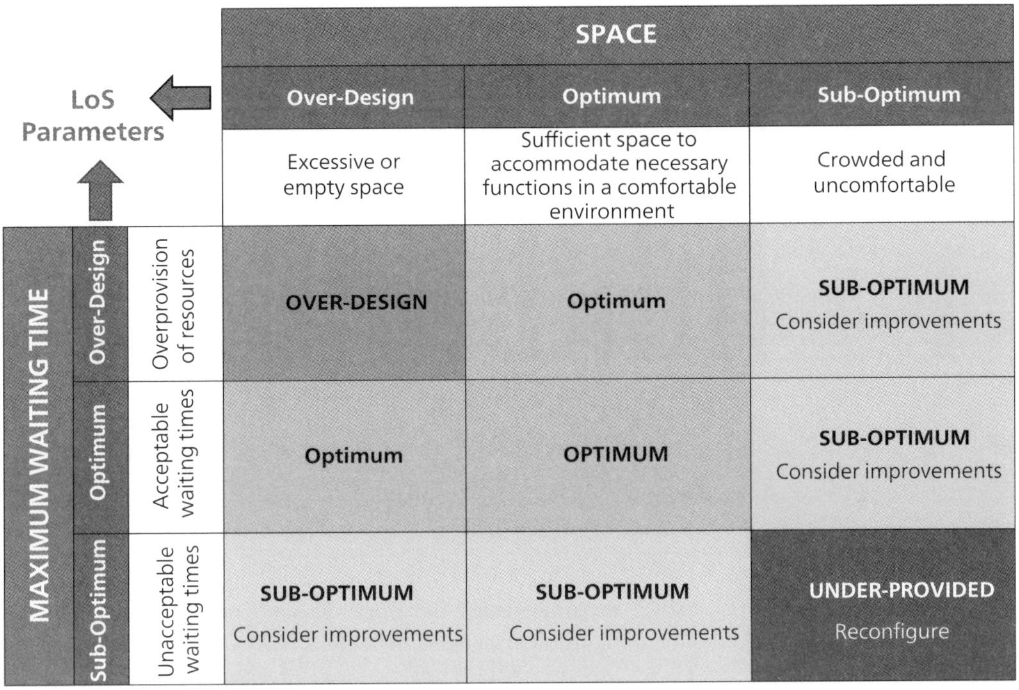

- security queuing and processing time; and
- passport-control queuing and processing time.

12.3.2.2 Demand

A 24-hour flight schedule for the year 2025 was provided by the airport operator. This was a combination of the 2016 timetable and additional flights planned for 2020 and 2025. The schedule included information

Figure 12.5 LoS benchmark values for key airport processes

	SPACE STANDARDS FOR WAITING AREAS (m^2/pax)			WAITING TIME STANDARDS FOR PROCESSING FACILITIES (minutes)			WAITING TIME STANDARDS FOR PROCESSING FACILITIES (minutes)			PROPORTION OF SEATED OCCUPANTS %		
Passenger Terminal Processor				Economy Class			Business Class / First Class					
ADRM 9th Edn	A B	C	D E	A B	C	D E	A B	C	D E	A B	C	D E
***ADRM* 10th Edn**	Overdesign	Optimum	Suboptimum	Overdesign	Optimum	Suboptimum	Overdesign	Optimum	Suboptimum	Overdesign	Optimum	Suboptimum
Public Departure Hall	>2.3	2.3	<2.3									
Check-in: Self-service Boarding Pass/Tagging	>1.8	1.3 - 1.8	<1.3	0	0-2	>2	0	0-2	>3			
Check-in: Bag-drop Desk (queue width 1.4-1.6 m)	>1.8	1.3 - 1.8	<1.3	0	0-5	>5	0	0-3	>3			
							Business Class Check-in desk					
Check-in: Check-in Desk (queue width 1.4-1.6 m)	>1.8	1.3 - 1.8	<1.3	<10	10-20	>20	<3	3-5	>5			
							First Class Check-in desk					
							0	0-3	>3			
							Fast Track					
Security Checkpoint (queue width: 1.2 m)	>1.2	1.0 - 1.2	<1	<5	5-10	>10	0	0-3	>3			
							Fast Track					
Emigration (Passport control) (queue width:1.2 m)	>1.2	1.0 - 1.2	<1	<5	5-10	>10	0	0-3	>3			
Boarding-gate Lounge: Seating	>1.7	1.5 - 1.7	<1.5									
Boarding-gate Lounge: Standing	>1.2	1.0 - 1.2	<1							>70%	50% - 70%	<50%

Figure 12.6 Flight capacity by hour in the 2028 schedule, including added flights

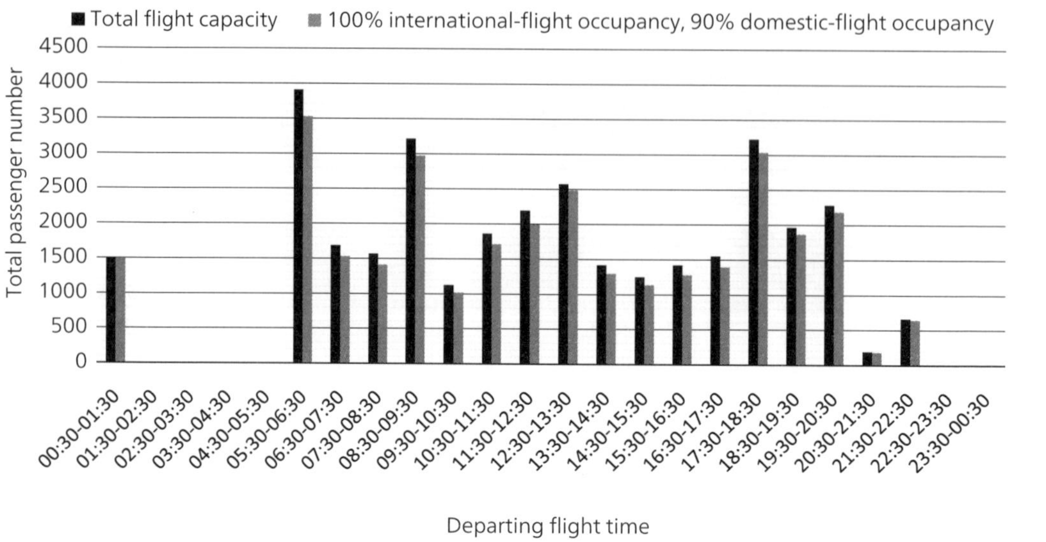

such as departure times, flight operators, departure gates and flight destinations. To account for the increased service in 2028, three additional flights were included in the 2025 AM schedule and one in the 2025 PM schedule.

The peak hours for departures were identified as 05:30–06:30 (AM peak) and 17:30–18:30 (PM peak) by combining the capacity of flights departing each hour, assuming 90% occupancy for domestic flights and 100% occupancy for international flights. The combined flight capacities for each hour are provided in Figure 12.6.

The times for peak occupancy of the terminal do not, however, match the peak flight capacity. This is because passengers arrive at varying times before their departures, based on preferences, travel distances and individual needs. The arrival profiles applied to departing passengers are provided in Table 12.1 and Figure 12.7.

To account for the varying arrival times of passengers, the peak modelled hours within the terminal are defined as the 2.5 hours preceding the peak hour:

- 03:00–05:30 (AM modelled); and
- 15:00–17:30 (PM modelled).

Table 12.1 Proportion of passengers arriving before flight departure time

Time before departure	Over 3 hours	2–3 hours	1.5–2 hours	1.25–1.5 hours	1–1.25 hours	0.75–1 hours	0.5–0.75 hours	Less than 0.5 hours	Total
Domestic	0%	5%	17%	10%	16%	26%	26%	0%	**100%**
International	5%	25%	22%	14%	16%	18%	0%	0%	**100%**

Figure 12.7 Arrival profiles of international and domestic passengers

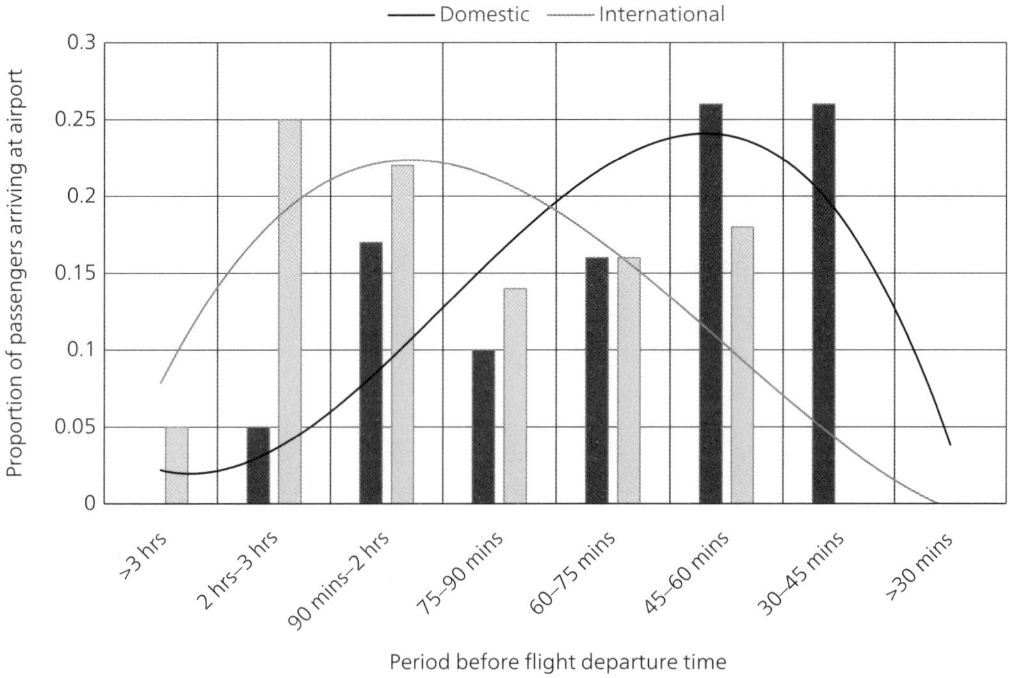

The passenger numbers arriving at the terminal could be defined by applying the arrival profiles to departing flights, and these are shown in Figure 12.8.

Aside from departing passengers, there are people arriving at the airport to escort the departing passengers. As these escorts are not actually flying, they are not represented in the flight-capacities or check-in queuing data. It was therefore necessary to calculate the number of escorts based on the number of departing passengers. The number of escorts depends on whether they are escorting an international/ regional or domestic departing passenger, and their arrival time at the airport is based on the arrival time

Figure 12.8 Passenger arrival profiles for modelling periods

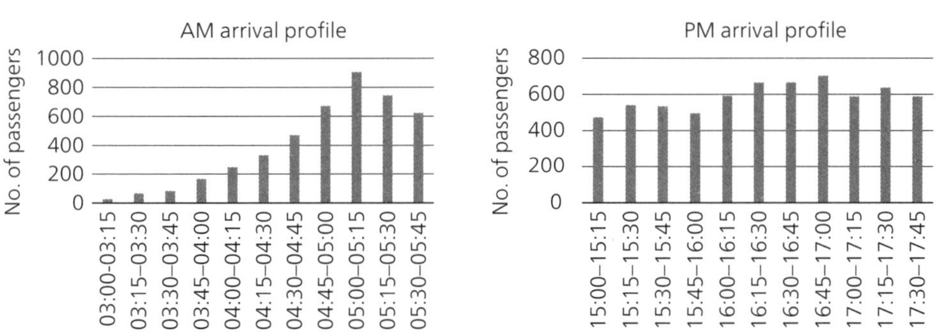

Table 12.2 Assumptions made of passenger types, speeds and luggage requirements

	INT PRM	INT with luggage	INT with trolley	INT hand luggage	DOM PRM	DOM with 1 luggage	DOM with trolley	DOM with hand luggage
Speed	0.6 m/s	1.2 m/s	1.1 m/s	1.4 m/s	0.6 m/s	1.2 m/s	1.1 m/s	1.4 m/s
Proportion	1%	50%	34%	15%	1%	27.5%	34%	37.5%
Luggage	Large	Large	Large	Small	Large	Medium	Large	Small

of the departing passenger. The assumed proportion of passengers with accompanying escorts was defined as:

- domestic – 52%; and
- international/regional – 65%.

Since airline crews also go through security and access the departure gates in the same way as passengers, it was necessary to include the arrival of crew members for departing flights in the model. Arriving airline crew demand and movement patterns were developed based on site-visit observations and information from the airport operator. Crew numbers were calculated based on whether the flight is domestic, international or regional, and are defined as:

- international – 20;
- regional – 10; and
- domestic – 6.

It was assumed that airline crews arrive at the airport 1:30 hours prior to flight departure time.

12.3.2.3 Routing

To prepare passenger demand (as described in Section 12.3.2.2) for simulation, a number of assumptions were made about the profiles of the international or domestic passengers (here a regional passenger is

Figure 12.9 Modelled passenger journey routes through departure terminal

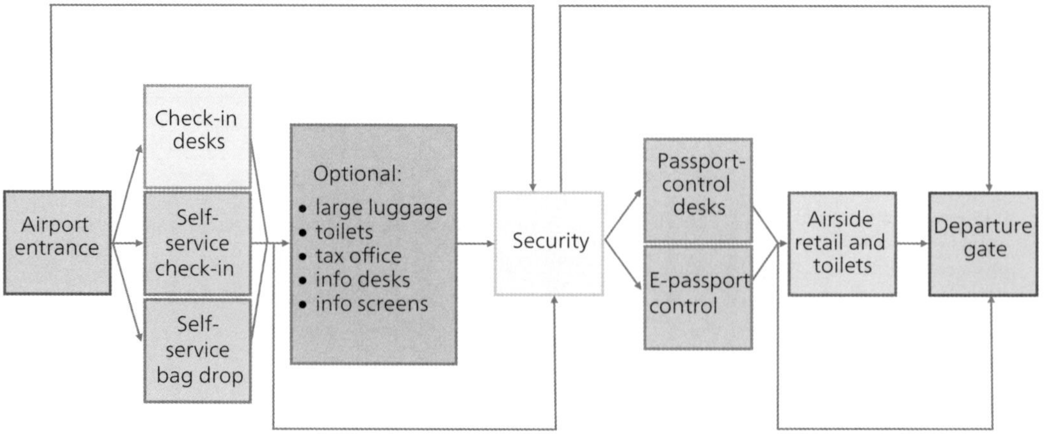

Table 12.3 Summary of check-in for international and domestic passengers

Check-in method	International	Domestic
Check-in desks	50%	37.5%
Self-service check-in kiosks	25%	31.25%
Mobile/online check-in	25%	31.25%
	100%	100%

classed as international). A summary of passenger speed profiles, proportions and luggage requirements is provided in Table 12.2, where PRM is a person with reduced mobility, INT is an international departing passenger and DOM is a domestic departing passenger. Airline crews and well-wishers were assumed to have no luggage and a speed of 1.4 m/s.

Passengers can choose between various routes through the airport. Figure 12.9 shows the potential routes modelled passengers can take from arrival at the airport to boarding their flights.

To represent airport behaviour accurately, the assumed proportions of passengers using various elements in the airport were agreed with the airport operator prior to modelling. Passengers can choose from several check-in options: traditional check-in desk and bag drop, self-service check-in and bag-drop and online check-in prior to airport arrival. The proportion of passengers using these three methods of check-in varies between international and domestic passengers and is summarised in Table 12.3.

Figure 12.10 Modelling results

Terminal subsystems	Overdesign Rating at **both** Space AND Maximum Waiting Time	Optimum Rating at **either** or **both** Space AND Maximum Waiting Time	Sub-Optimum Rating at **either** Space AND Maximum Waiting Time	Sub-Optimum Rating at **both** Space AND Maximum Waiting Time
	Over-Design	**Optimum**	**Sub-Optimum**	**Under-Provided**
Conventional check-in (Economy)	✓			
Conventional check-in (Business)	✓			
Self-service check-in Kiosk		✓		
Security control (International)	✓			
Security control (Domestic)	✓			
Emigration counters	✓	✓	✓ (Peak)	
Emigration e-gates	✓			

Table 12.4 Modelling results commentary

Location	Comments
Conventional check-in (Economy)	■ Over-designed for the majority of the time ■ Busy points become optimal for both waiting space and time
Conventional check-in (Business)	■ Consistently over-designed for both waiting space and time
Self-service check-in kiosks	■ Optimal for waiting time; enough kiosks to minimise waiting time ■ Over-design for waiting space; very few queues forming, which means over provision of kiosks
Security control (International)	■ Consistently over-designed for both waiting space and time ■ Forecast passenger demand does not cause queuing with 100% of checkpoints available
Security control (Domestic)	■ Consistently over-designed for both waiting space and time ■ Forecast passenger demand does not cause queuing with 100% of checkpoints available
Emigration counters	■ Mainly over-designed for both waiting space and time ■ Waiting space becomes sub-optimum at busy points, but waiting time remains over-designed, assuming 100% of desks are manned
Emigration e-gates	■ Consistently over-designed for both waiting space and time ■ Forecast passenger demand does not cause queuing with 100% of checkpoints available

Other key routing assumptions included:

- 15% of international, regional and domestic passengers with a luggage trolley who use a traditional check-in desk use the oversized luggage room;
- 50% of all passengers are subject to a search during the security process; and
- 50% of international passengers use the e-gates at passport control, as opposed to the traditional desk (all regional passengers use manned desks).

12.3.3 Evaluating performance

Outputs from the model simulation were compared against IATA requirements. The results of this analysis are summarised in Figure 12.10.

It should be noted that while most facilities were found to be predominantly over-designed, including check-in, security and emigration e-gates, results were averaged over the modelling period. Furthermore, the modelling assumed that all lanes in these facilities were operational during the entire model run, as agreed with the airport operator. There is scope therefore for a reduction in the number of manned security and emigration lanes – for example, under an operational strategy employed by the airport during less busy periods. Further notes on the results are summarised in Table 12.4.

12.3.4 Proposed mitigation

Pedestrian modelling at the airport showed that the planned facilities have enough capacity to cope with forecast passenger demand at the airport when operating at full operational capacity. These results relate to IATA guidance on density and time requirements for optimal passenger-terminal facilities. The modelling was also able to highlight areas in pedestrian routing and flow that could be improved in the airport design.

Figure 12.11 General routing for domestic (grey) and international (light grey) passengers

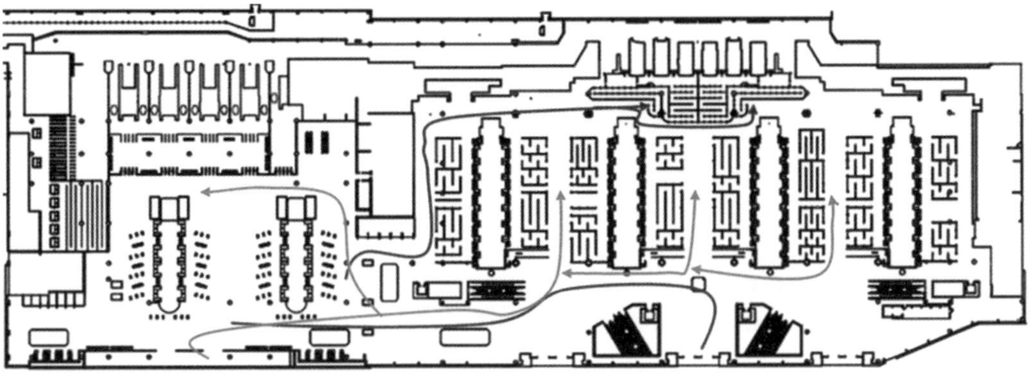

Passenger movement in an airport is complex. The multitude of passenger entrances, stop-off points and decisions made during journeys create cross-flows within the terminal. An outline of the key cross-flows highlighted in the model is shown in Figure 12.11, where the grey and light grey highlight some of the longest domestic and international routes respectively.

Most of the cross-flows evident in the terminal arise from passengers entering the airport at the opposite end from their check-in or security areas. This was especially evident for international passengers using traditional check-in then returning to international security, and domestic passengers using self-service kiosks then going to domestic security.

Pinch-points at various locations were also found to hinder the free flow of passengers. Figure 12.12 highlights these pinch-points (grey) and the walkways with highest passenger flows (light grey). The pinch-points combined with the cross-flow of passengers moving in opposing directions resulted in

Figure 12.12 Areas of restricted movements and areas of highest passenger flows

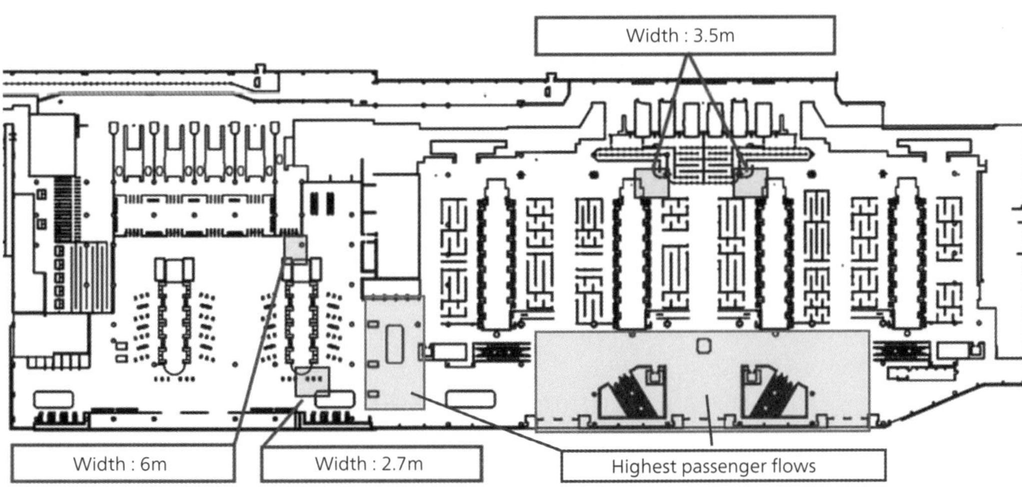

restricted movement for passengers and areas of significant congestion within the circulation space inside the terminal.

These findings led to the recommendation that passenger flows could be improved through the introduction of self-service check-in and fast bag-drop provisions towards the right side of the terminal, targeting domestic passengers. This would allow shorter routes through the airport and lead to a reduction in cross-flows, which cause congestion. Where pinch-points cannot be resolved with design changes, it was proposed that crowd management or signage be used to optimise passenger movements. Adequate signage will ensure that passenger movement in circulation areas is improved.

12.4. Summary points

- Pedestrian modelling plays a vital role in the design of airports to meet guidelines while balancing operational requirements with value engineering.
- Vehicular demand, occupation, kerbside dwell time, delays at barriers and parking behaviour will affect capacity as well as passenger demand, age, gender, group size, walking speed and trolley usage.
- The IATA's LoS concept is the benchmark for optimal passenger terminal facilities, consisting of four conditions: under-provided, sub-optimum, optimum and over-design.
- To assess whether a terminal design is suitable for the predicted increase in passenger numbers, the modelled results need to be compared against the standards as defined by the LoS.
- To represent airport behaviour accurately, the assumed proportions of passengers using various elements in the airport should be agreed with the airport operator prior to modelling.

Urban Space – AECOM
ISBN: 978-0-7277-6381-5
https://doi.org/10.1680/mpm.63815.181

Index

Page numbers in italics refer to illustrations and graphs. Page numbers followed by t refer to tables.

Commercial Uses of Space and Space Tourism

Legal and Policy Aspects

Edited by

Jan Wouters

Full Professor of International Law and International Organizations, Jean Monnet Chair ad personam *EU and Global Governance and Director, Leuven Centre for Global Governance Studies and Institute for International Law, KU Leuven, Belgium*

Philip De Man

Postdoctoral Fellow, Research Foundation – Flanders and Lecturer in International and European Space Law and Senior Researcher, Leuven Centre for Global Governance Studies, KU Leuven, Belgium

Rik Hansen

Financial Services and Markets Authority, Belgium

LEUVEN GLOBAL GOVERNANCE

Edward Elgar PUBLISHING

Cheltenham, UK • Northampton, MA, USA

Published by
Edward Elgar Publishing Limited
The Lypiatts
15 Lansdown Road
Cheltenham
Glos GL50 2JA
UK

Edward Elgar Publishing, Inc.
William Pratt House
9 Dewey Court
Northampton
Massachusetts 01060
USA

A catalogue record for this book
is available from the British Library

Library of Congress Control Number: 2017947081

This book is available electronically in the **Elgar**online
Law subject collection
DOI 10.4337/9781785361074

MIX
Paper from responsible sources
FSC® C013056
www.fsc.org

ISBN 978 1 78536 106 7 (cased)
ISBN 978 1 78536 107 4 (eBook)

Typeset by Columns Design XML Ltd, Reading
Printed and bound in Great Britain by TJ International Ltd, Padstow

Contents

Contributors

Philip De Man is a postdoctoral fellow at the Leuven Centre for Global Governance Studies and Institute for International Law at KU Leuven, Belgium. His research is funded by the Research Foundation – Flanders and covers multi-level interactions between national, European and international space law. De Man is a lecturer in international and European space law at the Master of Science in Space Studies of the Universities of Leuven and Ghent, Belgium. His recent publications include the monograph *Exclusive Use in an Inclusive Environment: The Meaning of the Non-Appropriation Principle for Space Resource Exploitation*, as part of the Space Regulations Library of Springer. De Man is core expert of the International Space Law Group at the MILAMOS project and a member of the Belgian delegation at the Legal Subcommittee of the United Nations Committee on the Peaceful Uses of Outer Space.

Mike N. Gold is the Vice President of Washington Operations and Business Development at Space Systems Loral (SSL). Previously, he was Bigelow Aerospace's Director of DC Operations and Business Growth. He was also appointed by the US Secretary of Transportation to serve on and chair the Commercial Space Transportation Advisory Committee (COMSTAC). Additionally, he was appointed by the National Research Council to serve on the Space Technology Industry–Government–University roundtable which provides direction and advice to NASA's Space Technology Mission Directorate. He dedicates his chapter to Meredith Francis, Darron Lincoln Francis-Gold, Richard White, Michelle Kley, Al Tadros, Ralph Bittelari, and David Ortiz, all of whom have made his time dealing with export control issues bearable.

Rik Hansen is policy officer at the Policy and International Relations Department of the Belgian Financial Services and Markets Authority (FSMA).

Andrea J. Harrington is Air and Space Law Instructor at the University of Mississippi School of Law and Erin J.C. Arsenault Fellow in Space Governance at the Institute of Air and Space Law, McGill University,

Canada. She is also a licensed attorney in the Commonwealth of Massachusetts. She has served as an Assistant Editor for the Annals of Air and Space Law, and as project manager and researcher on numerous projects, including for the FAA Center of Excellence for Commercial Space Transportation and the Secure World Foundation.

Christopher M. Hearsey is Director of Legislative Affairs for Bigelow Aerospace, LLC in Washington DC. He previously served as Corporate Counsel for Bigelow Aerospace from 2013 to 2016. Prior to working at Bigelow Aerospace, he served as Deputy Chief Executive Officer for Programs and Special Projects for the Astrosociology Research Institute (ARI) from 2009 to 2013. He continues to serve as Chairman of the Board of Directors of ARI and Editor-in-Chief of *The Journal of Astrosociology*.

Thierry Herman is Legal Officer at the European Space Agency (ESA) where he provides legal advice for the execution of several ESA programmes, including the navigation and earth observation programmes. He previously held the position of expert in international relations at skyguide, the Swiss air navigation service provider. Prior to that he held the position of Strategy Officer at ESA, in which capacity he was seconded to the National Aeronautics and Space Development Agency of Japan (NASDA, now JAXA), and that of Executive Secretary of the European Centre for Space Law (ECSL). He studied law in Brussels, Belgium, and earned an LLM from the Institute of Air and Space Law, McGill University, Canada.

Armel Kerrest is Emeritus Professor of Public Law. He is Vice Chairman of the European Centre for Space Law of the European Space Agency. He has taught international law with particular focus on the law of the sea and space law at the Universities of Western Brittany and Paris South.

Jean-Bruno Marciacq is an aerospace safety expert at Swiss Space Systems (S3), a private company based in Payerne, Switzerland. From 2007 to 2014 he worked at the European Aviation Safety Agency (EASA) in Cologne, Germany. In parallel to being Project Certification Manager and Rulemaking Officer, he acted as the Sub-orbital and Orbital Aircraft (SOA) Coordinator for EASA. From 2001 to 2007 he worked as ESA Astronaut Safety Manager at the European Astronaut Centre (EAC) in Cologne, Germany.

Jean-François Mayence holds a Law degree from University of Brussels (ULB) and a special Master in Air and Sea Law. He has worked as legal advisor for the Belgian Federal Office for Science Policy (Belspo) since 1 January 1999 and has covered the legal aspects of a variety of scientific and technological areas in which international cooperation plays a key role, including outer space, Antarctica, oceans, biology and biodiversity. He has chaired several working groups and committees in the European Space Agency and in UNCOPUOS. He has been member of the Belgian Delegation in numerous intergovernmental bodies and conferences dedicated to international cooperation in outer space.

Ward Munters (Master of Laws, MSc in Space Studies) is a doctoral researcher at the Leuven Centre for Global Governance Studies and the Institute for International Law at KU Leuven, Belgium. His research is aimed at identifying crucial legal contours of legal terms of art and definitions for 'space debris' and 'active debris removal'. In an effort to integrate the findings into international law proper, and as a means to ascertain the boundaries such legal terms of art should respect, his current focus is on the relationship between international environmental law and international space law.

Daniel P. Murray has 20 years of experience in the space industry, including the past 13 years with the FAA's Office of Commercial Space Transportation, where he currently manages the Space Transportation Development Division. He serves as the co-lead of an FAA Joint Space Operations Group that addresses issues associated with integrating commercial space launch and re-entry operations into the National Airspace System. Prior to joining AST, he was a flight controller and software engineer supporting the Space Shuttle program at United Space Alliance in Houston, Texas.

Katrin Nyman-Metcalf is Head of the Chair of Law and Technology at Tallinn University of Technology and Head of Research at the Estonian e-Governance Academy. She is also active as an international consultant, working globally primarily in the area of communications law, including ICT regulation, privacy and data protection, media law, e-governance and various cyber issues. She has published extensively on issues related to law and technology as well as on regional integration. She represents Estonia in the International Relations Committee of the European Space Agency.

Lesley Jane Smith is Visiting Professor of Space Law at the University of Strathclyde, United Kingdom. She is a qualified solicitor, admitted to practise in the United Kingdom and Germany, and is a partner in the law firm Weber-Steinhaus & Smith in Bremen. She has an extensive command of the field of national and international space law, including developments within Europe at agency and Union level. She is an expert on commercial space activities, particularly contracts, licensing of intellectual property, earth observation, as well as European competition and space procurement. In 2014 she was appointed General Counsel to the International Astronautical Federation in Paris.

Alexander Soucek is legal officer for public international law and space law at the European Space Agency (ESA). He previously held the position of programme coordinator in ESA's Directorate of Earth Observation Programmes. He advises ESA's International Relations Committee and represents ESA at the Legal Subcommittee of the UN Committee on the Peaceful Uses of Outer Space. He studied law in Salzburg, Austria, where he also worked as a scientific assistant in international law, and earned a master's degree in space studies from the International Space University in Strasbourg, France. He teaches space law at various universities in Europe and has published books, commentaries and articles on the subject, including *Outer Space in Society, Politics and Law* (2011, co-edited with Professor C. Brünner) and *Space Law Essentials* (2015), an introduction to space law for law students and space practitioners.

Jeremy Stubbs is Programme Lead for Commercial Spaceflight Regulation at the Civil Aviation Authority. Previous positions at the Civil Aviation Authority include those of Flight Standards Officer and Manager of Flight Operations Policy (Commercial Aeroplanes). He has extensive practical experience as a captain at British Airways and CityFlyer Express.

Steven Wood provides technology commercialization services to more than 20 campuses of the State University of New York (SUNY) as the Assistant Director of Innovation Services with The Research Foundation for SUNY. He pursued an Advanced LLM program at the International Institute of Air and Space Law at Leiden University, the Netherlands, where he also worked in technology commercialization at the Leiden University Office of Research and Innovation Services. Prior to his current position at SUNY, Steven held positions at the US Department of Energy Brookhaven National Laboratory, the US Patent and Trademark Office, and in several start-up companies. He is a licensed S. patent attorney.

Jan Wouters is Full Professor of International Law and International Organizations, Jean Monnet Chair *ad personam* EU and Global Governance, and founding Director of the Institute for International Law and of the Leuven Centre for Global Governance Studies, an interdisciplinary research centre with the status of Jean Monnet Centre of Excellence, at KU Leuven, Belgium. As Visiting Professor at Sciences Po (Paris), Luiss University (Rome) and the College of Europe (Bruges), he teaches EU external relations law. As Adjunct Professor at Columbia University he teaches comparative EU–US perspectives on international human rights law. He is a Member of the Royal Academy of Belgium for Sciences and Arts, President of the United Nations Association Flanders Belgium and practises law as *Of Counsel* at Linklaters, Brussels. He has published widely on international and EU law, international organizations, global governance and financial law. Apart from his coordination of large research projects and participation in international scientific networks, he advises various international organizations and governments, trains international officials and is often asked to comment on international events in the media. In the academic year 2016–2017 he was Visiting Professor at the Hebrew University of Jerusalem, and Herbert Smith Freehills Visiting Professor at the Lauterpacht Centre for International Law, Cambridge University.

Commercial uses of space and space tourism: setting the scene

Jan Wouters, Philip De Man and Rik Hansen

Space applications have become indispensable in our modern society, as they account for an ever-expanding number of facilities of a very diverse nature all over the world. Whereas space services used to be, for a long time, primarily scientific or military in nature, commercial applications of spacefaring are rapidly accounting for the most exciting developments in the fields of space sciences, technology and law. Satellite applications facilitate the provision of such diverse services as global communication, remote sensing, climate observation, data collection, resource exploitation, transportation and space tourism.

The astonishing scientific and technological progress in this respect has been made possible in large part by the enabling and, some would argue, ambiguous legal framework that has been put in place at the international level to govern space activities. The fundamental principles underpinning this legal regime are the free exploration and use of outer space, including the Moon and other celestial bodies, by all States, including their nationals, and the corollary prohibition of national appropriation of outer space, by any means whatsoever, including sovereignty.[1] These principles stand in stark contrast to those of the legal regime governing the airspace above the territory of States, who have complete and exclusive sovereignty over the areas directly superjacent to their land and territorial sea.[2]

---

[1] See Articles I and II of the Treaty on Principles Governing the Activities of States in the Exploration and Use of Outer Space, including the Moon and Other Celestial Bodies, 27 January 1967, 610 *U.N.T.S.* 205 (hereinafter: 'Outer Space Treaty' or 'OST'); Article 11(2) of the Agreement Governing the Activities of States on the Moon and Other Celestial Bodies, 18 December 1979, 1363 *U.N.T.S.* 21 (hereinafter: 'Moon Agreement' or 'MA').

[2] See Article I of the Convention Relating to the Regulation of Aerial Navigation, 13 October 1919, 11 *L.N.T.S.* 174; Article I of the Convention on

It is clear that the application of the rules of two legal regimes, outer space and airspace, which are guided by fundamental concepts and principles standing in such apparent opposition to each other, cannot be reconciled without determining the scope of their respective fields of application. The benefits of drawing a clear boundary between airspace and outer space seem self-evident, for this appears to limit the amount of international legal disputes between competing users of outer space while assuring a continued, secure and predictable development of space activities. While legal certainty should hence be served by a clear demarcation, the question of a dividing line between airspace and outer space itself has resulted in significant legal turmoil that is yet to abate, even after more than half a century of heated discussions.

Intergovernmental discussions on the lower limit of outer space started as soon as the first colloquium on the law of outer space was held in 1958, barely a year after the launch of the first satellite. The related matter of the upper limit of state sovereignty had already been a topic of some contention in the years prior to this landmark event, without a clear outcome. In 1967 the question of the definition and delimitation of outer space was formally placed on the agenda of the main organ of the United Nations tasked with the development of space law, the Committee on the Peaceful Uses of Outer Space (UNCOPUOS). Over the course of the following decades, both the Legal Subcommittee (LSC) and its counterpart, the Scientific and Technical Subcommittee (STSC), of the UNCOPUOS have kept the item on the agenda.

At these forums, many different views, opinions and proposals have been put forward regarding the need, timeliness and criteria for determining a dividing line between airspace and outer space that would be, at the same time, accurate, universal and easy to control. However, no consensus on any of these issues has been reached to this point.[3] The most recent discussions in the LSC have been moving toward a boiling point where the very methodology of the Working Group on the Definition and Delimitation of Outer Space has been questioned as sub-optimal and perhaps even as a factor actively contributing to the persisting deadlock in the discussions on the demarcation issue at intergovernmental level.

International Civil Aviation, 7 December 1944, 15 *U.N.T.S.* 294 (hereinafter: 'Chicago Convention').

[3] For an overview of the different opinions expressed by delegates at the UNCOPUOS, see in general Historical summary on the consideration of the question on the definition and delimitation of outer space, 18 January 2002, UN Doc. A/AC.105/769.

Indeed, views that have traditionally centred on the need for a definition of outer space as the means of clarifying the application of the air and outer space legal regimes have equally traditionally been countered by declarations that the current framework should be retained 'until such time as there was a demonstrated need and a practical basis for developing a definition or delimitation of outer space'.[4] The view is therefore increasingly being expressed that, rather than trying to find a consensus on where airspace ends and outer space begins, analysis should start from the legal questions that are raised by the recent developments in various fields of spacefaring, and to then move forward by looking at possible convergences in the replies to these varied questions.

This book, one of the scientific outputs of an interdisciplinary research project on the governance of space and the air–space distinction, led by Professor Jan Wouters, Professor Dirk Vandepitte and Professor Christoffel Waelkens at KU Leuven, aligns itself firmly with this practical view, and argues for an inductive approach to the demarcation question as a practical means of breaking down the highly contentious issues underlying the overarching controversies surrounding the applicability of the legal regimes of airspace and outer space. For if there is one thing that the ongoing discussions at UNCOPUOS have made clear, it is that the delimitation issue is inextricably linked with other long-standing points of discussion that are of major practical importance. Any analysis of the vexed problem of delimitating airspace and outer space should therefore pay considerable attention to related issues compounding the regulation of specific space activities.

Areas in which legal, political and technological developments have taken flight in recent years are mainly related to the (in)direct militarization of outer space, the industrial exploitation of natural resources of outer space, including the Moon and other celestial bodies, and, most importantly, the commercial uses of outer space and the development of hybrid craft for these purposes, with particular attention devoted to liability for space objects. Our research in each of these areas has proceeded along parallel lines, successively covering the relevance of the relevant area with regard to the demarcation issue; the implications of the adoption *viz.* absence of a demarcation line for the case at hand; and

[4] UNCOPUOS, Report of the Chair of the Working Group on the Definition and Delimitation of Outer Space, Annex II to the Report of the Legal Subcommittee on its fifty-fifth session, held in Vienna from 4 to 15 April 2016, 27 April 2016, UN Doc. A/AC.105/1113, para. 7.

the inferences that can be drawn from the specific requirements of the case for the establishment of a general criterion of demarcation.[5]

Although our research at the Leuven Centre for Global Governance Studies covers each of the above areas and related space activities, as well as other key issues of international and European space law, the present book focuses solely on the implications of the legal and policy challenges posed by ongoing commercial spacefaring developments to the demarcation question.

The greatest contemporary challenge in space law is adapting the ageing treaty framework to today's very different realities, spurred by these developments. Whereas the practical and ideological context at the time of drafting the space treaties resulted in very little attention being paid to the possibility of private space activities, history has shown this belief to be erroneous. More than ever before, private initiatives are the driving force for the development of new space technologies and activities. The applications are manifold, as are the legal questions they raise. The multiplication of actors and activities in space has left the legal framework in dire need of reinterpretation and perhaps adaptation. Acting as a catalyst for this debate is the task which this volume sets itself.

The volume is structured in three parts of comparable size and equal importance, each dealing with a separate yet related aspect of the regulatory realities of commercial spacefaring driven by private investment and technological innovation. Each chapter also raises its own questions about the applicability of the law of outer space and its interaction with air law, and the implications thereof on the boundary between the scope of application of these legal regimes.

Part I introduces the main international legal challenges posed by the changes to the spacefaring landscape in terms of both actors and types of space activities, in light of the notoriously immutable outer space legal framework comprising a limited set of treaties and instruments put to paper almost half a century ago. The chapters in this part discuss the implications of the current foray by private actors into commercial spacefaring on the 'launching State' notion as a central concept in modern space law (Chapter 1); the impact of emerging sub-orbital flight activities on the existing air and space transport liability regimes (Chapter 2); the eligibility of space tourists to receive rescue assistance in outer

[5] For the specific topic of the militarization of outer space, see R. Hansen, J. Wouters and P. De Man (eds), 'Focus Section: Military Use of Outer Space: International Legal Perspectives', 40 *Annals of Air and Space Law*, 2015, 1–218.

space as per the applicable international treaties, which focus on government astronauts and spacecraft personnel (Chapter 3); and the regulatory implications for a sustainable and effective application of existing international space law and governance structures of the increasing implementation of small satellite technology, and its application in proposed large constellations of small satellites in low Earth orbit (Chapter 4).

The chapters in this first part call for a firm but controlled approach to adapting the current international space law framework to accommodate the developments in commercial spacefaring. On the one hand, one must be heedful not to upset the carefully crafted balance of the existing international space treaty framework, where the activities of private actors are regulated through their connection with one or more States. As a result, we may prefer to integrate solutions to the challenges posed by commercial spacefaring operations into the existing space law treaties, for these new developments do not change the fact that, in an international legal environment, it remains crucial to maintain strong and predictable links with identifiable States. In this respect, some issues may simply be resolved by adopting an evolutive interpretation to the reading of key concepts and provisions in the existing treaties. Nevertheless, we must be careful not to let inaction at the international level be used as an opportunity for national regulators to turn back the clock on many hard-fought and sensible regulations, as evidenced in the current developments in some sub-orbital flight regimes.

Hence, a comprehensive analysis of the challenges posed by commercial spacefaring developments should complement the international legal appreciation with an overview of legal and regulatory responses at the national level. Part II of this volume aims to do just that, and to that effect collects contributions on a variety of topics, including the trend in national space law of the United States exempting commercial spaceflight entities from liability for a participant injury through a series of Spaceflight Liability and Immunity Acts (Chapter 5); the lessons learned and challenges exposed by the practice of the US Office of Commercial Space Transportation in regulating commercial space operations within an already existing national legal framework (Chapter 6); the development of a new safety regulatory system for commercial sub-orbital spaceplanes in the United Kingdom (Chapter 7); and the impact of commercial spacefaring pioneers such as Bigelow Aerospace on the reform of national export control restrictions on space technologies (Chapter 8).

The example of Bigelow Aerospace and the international traffic in arms regulations shows that commercial space actors are already changing the legal landscape at the national level, regardless of international efforts to comprehensively regulate commercial spacefaring. It also shows that regulatory evolution is closely connected with demands from the industry. Indeed, States are more prone to act at the national level than in the context of an international organization or committee given the close link between the space industry and the country in which it operates. National governments should therefore step up the dialogue with industry. Subsequently, such regulatory advances at the national level may be distributed to other States through the sharing of best practices and lessons learned. Such informal inter-State practices cannot replace regulatory efforts at the global international level, however. National regimes may be tempted to offer enticing regimes that may not be transposable to other countries, and may not even be sustainable in the long term. Moreover, it is clear that solutions at the national level thrive best when supported by an enabling international framework. Hence, when the sector of commercial spacefaring matures, it may be advisable to adopt a comprehensive international regime that replaces – or at least coordinates – the patchwork of national space regulations.

The means through which to do so can be of a varying nature, all of which, taken together, make up the subject of Part III of this volume, which deals with questions of space governance. In this part, a selection of authors contemplate how modern evolutions of commercial spacefaring by non-governmental actors challenge the contours and structure of a governance framework that is still predominantly tailored to governmental spacefaring actors. The ensuing analysis touches upon a variety of topics, including the elements that should be taken into consideration when making legal and policy choices for regulating sub-orbital flights carrying humans (Chapter 9); the role of the European Aviation Safety Agency in establishing a regulatory framework for the development and operations of sub-orbital and orbital aircraft in the European Union (Chapter 10); the role of UNCOPUOS in the international regulation of non-governmental space activities (Chapter 11); and the emergence of self-regulation as a possible mode of governance for a new spacefaring environment characterized by the need for specific rules in a commercial context (Chapter 12).

A targeted governance framework for addressing the challenges posed by commercial, private spacefaring may take many forms and can range from the simple extension of existing aviation regimes and institutions, over the gradual and evolutive adaptation of one or more such regimes, to the development of an entirely novel framework, either by existing or new mechanisms and actors. Whether such options will be pursued at the domestic, regional or global level depends on the ability of States to agree on the objective of such an undertaking. This future commercial space governance framework may incorporate existing actors with extended functions or altogether new institutions, replacing current organizations that are exclusively intergovernmental in nature. As such, the difficulties encountered by UNCOPUOS in maintaining its position as the central international space law-making body appear to be due, at least in part, to the lack of provision for participation by private actors. As a result, aspects of commercial spacefaring are often regulated through the actions of other institutions, typically operating in the field of aviation and telecommunication.

In the meantime, the existing governance gaps in commercial spacefaring may also be filled by self-regulation. However, the dynamics of cooperation between private actors are fundamentally different from the environment of international cooperation that characterized the initial development of the space treaties, as private self-regulation is often only considered in relation to a possible trade-off with State-sanctioned regulation. Whether the increasingly proliferating commercial spacefaring activities by private actors will ultimately benefit from such a move away from governmental regulation is doubtful, given the legal uncertainty it is likely to cause. Hence, concerted efforts must continue to work towards establishing a system of international governance that respects the role and position of private actors in the elaboration of industry-friendly regulations, while retaining the elements of the existing regime that has so far contributed to a predictable and secure international legal framework.

PART I

Legal challenges

1. The concept of the 'launching State' in commercial launch ventures

Armel Kerrest

1. INTRODUCTION

The concept of the 'launching State' is central in modern space law. It is related to the notion of liability for 'damage caused by a space object' as defined in the 1972 Liability Convention.[1] Space activities have changed a lot since the 1960s and 1970s when the legal framework for space activities was adopted. At that time few activities took place, and only States were involved. Because of the cold war, these activities were under the strict control of the two main spacefaring States. Nowadays many other States and many private entities are involved, with a plethora of new activities including commercial ventures taking place. It is often asked whether the legal framework established at the time by the Liability Convention is still efficient when commercial space ventures are concerned.

This question will be considered in the following four steps:

1. What is the importance of the notion of launching States?
2. What is the interest of holding States – the launching State(s) – liable for damage caused by a space object?
3. What happens when private entities are involved?
4. Who carries the risk created by private space activities?[2]

[1] Convention on International Liability for Damage Caused by Space Objects (hereinafter the Liability Convention), adopted by the General Assembly in its Resolution 2777 (XXVI), opened for signature on 29 March 1972, entered into force 1 September 1972.

[2] A. Kerrest and L.J. Smith, 'Article VII of the Outer Space Treaty' in S. Hobe, B. Schmidt-Tedd and K.-U. Schrogl (eds) *Cologne Commentary on*

2. THE IMPORTANCE OF THE NOTION OF THE 'LAUNCHING STATE'

From the very beginning, liability for damage caused by a space object has been linked with the notion of the launching State.[3] As early as 1959 the first report of the ad hoc Committee on the Peaceful Uses of Outer Space (UNCOPUOS) refers to the 'liability of the launching State and other States associated with it in the space activities causing injury or damage'.[4] The USA proposal concerning liability for space vehicle accidents[5] refers to the 'State or international organizations responsible for the launching of space vehicles'. During the same meeting of the UNCOPUOS Legal Subcommittee, India submitted a proposal referring to 'damage caused by space vehicles which they have launched'.[6]

The choice of the launching State as a criterion by which to apportion liability was appropriate for several reasons. First of all, it holds States liable, which is unsurprising given that only States were involved in space activities at the time. To this day this is arguably the best solution as States, especially spacefaring States, are normally rich and solvent and are able to bear significant amounts of risk without the need to purchase insurance. This is quite useful. As they do not require insurance, States may accept liability without a ceiling in either amount or time, which was exactly what happened under the Liability Convention. In light of the technical conditions at the time, these modalities were necessary to ensure the protection of potential victims. As satellites may stay in orbit for decades, it is appropriate that launching States' liability is maintained until the satellite returns to Earth.[7]

Space Law (hereinafter CoCoSL), Volume 1 (Kluwer 2009); L.J. Smith, A. Kerrest and F. Tronchetti, 'The Convention on International Liability for Damage Caused by Space Objects' in CocoSL Volume 2 (Kluwer 2013).

3 B. Cheng, 'Convention on International Liability for Damage Caused by Space Objects' in N. Jasentuliyana and R.S.K. Lee (eds), *Manual on Space Law*, Volume 1 (Oceana Publications 1979) 83.

4 Report of the Ad Hoc Committee on the Peaceful Uses of Outer Space (document A/4141, 14 July 1959) p. 23 at II B point 10; Carl Christol, 'International Liability for Damage Caused by Space Objects' (1980) 14 *American Journal of International Law* 346.

5 USA: Draft proposal on liability for space vehicle accidents 1962 (A/AC.105/C.2/L.4) http://www.unoosa.org/pdf/reports/ac105/AC105_006E-lc.pdf.

6 India: Proposal – Draft conclusions (A/AC.105/C.2/L.5) http://www.unoosa.org/pdf/reports/ac105/AC105_006E-lc.pdf.

7 The liability of the launching State has no limit in time after the launch but the victim has only one year after his learning of the damage to ask for

The very broad definition of the 'launching State' stems from its four alternative criteria: the State that launches, the State that procures the launch, the State from whose territory a space object is launched and the State from whose facility a space object is launched. As a result there is always at least one liable launching State that can be identified for any given space object. If more than one State fulfils these criteria, they are jointly and severally liable, which is testament to the victim-oriented nature of the space liability regime

The fact that more and more activities have been privatized does not change the international responsibility or liability of the States for whom the activity is a 'national activity' according to Article VI of the Outer Space Treaty,[8] or who are a launching State for the space object under Article VIII of the Outer Space Treaty. Some States would like to avoid liability through a new and narrower interpretation of the Liability Convention that would exclude liability of launching State(s) in the case of private activities. If such an interpretation were accepted, however, States could avoid being held liable simply by privatizing an activity. This would be a clear violation of the spirit of the Liability Convention, and even of the letter in the context of Article VI Outer Space Treaty whereby private activities are included in the notion of 'national activities' under the responsibility and control of the appropriate State. In the case of private activity launched from the high seas, this narrow interpretation would result in no launching State being liable, which is clearly unacceptable and contrary to the aims and purposes of the Convention.

Among the four criteria of the launching State, the most difficult to define is the one that refers to 'procuring the launch'. Its inclusion was a way to apportion some liability to States lacking autonomous launch capability and using the launch capabilities of other States. This linkage proved to be foresighted, as the State that procures the launch is currently often the only one to have a legal link to the satellite once it is in orbit.

Although for the time being there is no case of a launching State having to pay compensation to a victim under the Liability Convention

---

compensation, as per the Liability Convention Art. X/1, http://www.unoosa.org/oosa/en/SpaceLaw/gares/html/gares_26_2777.html.

[8] The Treaty on Principles Governing the Activities of States in the Exploration and Use of Outer Space, including the Moon and Other Celestial Bodies (hereinafter the Outer Space Treaty), adopted by the General Assembly in its Resolution 2222 (XXI), opened for signature 27 January 1967, entered into force 10 October 1967, http://www.unoosa.org/oosa/SpaceLaw/outerspt.html, Articles VI and VIII.

for damages caused by a privately owned or operated space object, the matter raises significant legal questions.

First, the definition of launching State(s) may pose problems when international private consortia operate satellites. The important rationale behind the fourfold criteria is to enable the victim to easily identify a liable launching State. The important side effect of the broadness of the criteria is to prevent accidents. If a State risks becoming liable for a space activity, it will efficiently control this activity and thus limit the risk of an accident. For this reason it may be useful for the various potential launching States to pass an agreement prior to the launch such as the one referred to in Article V of the Liability Convention. These agreements have no effect on the rights of the victim(s) but may ease the relationship between launching States. It may be considered unfair, for instance, that the State of the territory in which the launch takes place would be liable for damage caused by the space object in orbit, especially if a long time had passed since the launch. It may also be considered unfair that the State procuring the launch would be liable for the damage caused by the launch itself. *Dura lex sed lex.* The purpose of the Liability Convention is to protect victims. It is the launching States' responsibility to divide liability among themselves. Such agreements may be used; indeed, they should be used. For the time being, however, prelaunch agreements are rather rare, and State practice surrounding them is in need of further refinement.

A second issue relating to private space operations is the transfer of ownership of a satellite in orbit. In such cases the launching State remains liable even if it has little or no control over the satellite or the satellite operator anymore. It may be necessary to improve the way States can register space objects to reflect this evolution in economic reality. According to Article 1(c) of the Registration Convention, 'The term "State of registry" means a launching State on whose registry a space object is carried in accordance with article II'. This restricts the possibility of registering a satellite to those States which are launching States. Therefore, in the case of a transfer of ownership and control of the satellite after the launch, the launching State is liable for damage caused by the space object but can no longer control it, whereas the State of the new owner is responsible according to Article VI of the Outer Space Treaty, even if it cannot register it or exercise jurisdiction and control over it.[9] This is a point which urgently requires discussion within

[9] The Treaty on Principles Governing the Activities of States in the Exploration and Use of Outer Space, including the Moon and Other Celestial

UNCOPUOS. It is preferable that a solution is found without opening the Pandora's Box of treaty amendments.[10]

3. THE NATURE OF THE LIABILITY OF THE 'LAUNCHING STATE(S)'

As is often observed, space liability under the Liability Convention offers a high degree of protection for potential victims. When the damage is caused on Earth or to aircraft in flight, the liability is absolute. This is an exceptional provision in international law, especially when we consider that this liability has no ceiling either in amount or time, and extremely few reasons for exoneration. No act of God, nor a fault of a third party, nor a terrorist action, nor even the fault of the victim if it is short of 'gross negligence or an act or omission done with intent to cause damage' may exonerate the launching State (Liability Convention, Article VI). It should be noted that even this exceptional exoneration ceases if the activity is unlawful.

It is interesting to examine the reasons behind the acceptance by spacefaring States of such a broad and profound system of liability. It appears that at the time of the adoption of these principles, the freedom of use of outer space was not established beyond discussion. The 1919 Paris Convention on International Air Navigation recognized that the space above a State's territory was under its sovereignty. When the space era began, by contrast, it was necessary to establish beyond doubt the freedom to use outer space. The fact that such broad and unlimited liability was accepted for damage on Earth and not for damage in orbit seems to be a good indication of the deal struck between liability favouring non-spacefaring States (potential victims) and the freedom of use which they had to accept without being able to take any significant advantage of it for the time being.

In case of damage 'elsewhere than on the surface of the Earth to a space object', the Liability Convention requires that fault is proven (Article IV), which might cause difficulty even for spacefaring States. It

---

Bodies (hereinafter the Outer Space Treaty), adopted by the General Assembly in its Resolution 2222 (XXI), opened for signature 27 January 1967, entered into force 10 October 1967, http://www.unoosa.org/oosa/SpaceLaw/outerspt.html, Articles VI and VIII.

[10] A. Kerrest, 'Legal aspects of transfer of ownership and transfer of activities' (2012) IISL/ECSL Symposium Legal Subcommittee, Fifty-first session, http://www.unoosa.org/pdf/pres/lsc2012/symp-01E.pdf.

appears that States taking part in this great adventure, and their nationals, have less protection on account of their decision to participate.

For the time being there is no system of ‘space traffic management’. This compounds the problem of determining who is at fault in case of an accident. In this context it may be possible to refer to the customary rules applying on the high seas. Taking the example of a collision between two space objects, one under control and manoeuvrable, the other a piece of unresponsive and unmanoeuvrable space debris, we need to ask ourselves: would it be preferable to impose liability on the operator who does not move his satellite, provided he can, or on the operator who has dangerously left his satellite in a well-used orbit?

For damage in orbit, the Liability Convention refers to the ‘fault’ of the launching State or the ‘fault of persons for whom it is responsible’. This reference to a ‘fault’ shows that the Liability Convention does not refer to a violation of international law, as is usually the case in general international law, but to an independent concept of ‘fault’. It refers to ignorance, carelessness or negligence. If there were an accident, a judge would have to consider the behaviour of the operator with reference to the ‘normal’ or ‘right behaviour’. Experts may be consulted, but more concrete substantiations of generally accepted behaviour exist as well. These arguably carry more weight, and include ‘codes of conduct’, such as the one accepted by the Inter-Agency Space Debris Coordination Committee and summarized by UNCOPUOS. These written rules may be used as points of reference in determining whether the operator or State is at fault in the case of an accident. In this way, although the codes of conduct themselves are not compulsory, they may be considered a good indicator of the ‘right behaviour’ of a ‘careful space objects operator’.

Another point that needs interpretation in Article III of the Liability Convention is the notion of the ‘person for whom it is responsible’. Two interpretations are possible. The first refers to Article VI of the Outer Space Treaty and therefore includes the fault of any governmental agency or any non-governmental entity conducting any ‘national activity’ in outer space. The second interpretation would refer only to agents of the State, as is the case in general international law. Once more it is difficult to anticipate the possible decision of an international judge; however, the broader interpretation taking Article VI of the Outer Space Treaty into consideration is the most likely. Consequences of this choice of interpretation are significant; under the broader interpretation, private activities would expose launching States to a risk of sizeable claims for in-orbit damage as a result of faults committed by its nationals. In any case, the proof of fault for an activity conducted in outer space remains problematic. States' recognition of this issue may be seen as the basis of their acceptance of absolute/objective liability for damage on Earth.

For obvious reasons, the Liability Convention does not apply to damage caused to a national of one of the launching States. The same exclusion applies to foreigners, provided they took part in the launch – this is less obvious. This point confirms the hypothesis of a link between the protection of victims on Earth and the acceptance by every State, even non-spacefaring States, of the freedom to use outer space.

4. LIABILITY OF THE 'LAUNCHING STATE(S)' AND ACTIVITIES OF PRIVATE ENTITIES

The 1972 Liability Convention does not deal directly with the liability of private persons. It deals with the liability of launching States for space activities, including private activities, but not for the liability of the private entities themselves. Liability as set out in the Liability Convention only regulates interstate liability. When a State falls under one of the four criteria constituting a launching State, it is liable for damage caused by the space object on the ground or to aircraft in flight, and potentially liable for damage caused elsewhere.

Contrary to what was set forth in the 1969 Brussels International Convention on Civil Liability for Oil Pollution Damage, there is no channelling of the liability to operators and, therefore, no conditional exoneration of the operators. According to Article XI, paragraph 2, '[n]othing in this Convention shall prevent a State, or natural or juridical persons it might represent, from pursuing a claim in the courts or administrative tribunals or agencies of a launching State'. The victim may demand compensation directly from the private operator before a domestic judge on the basis of ordinary liability/responsibility laws. The choice belongs to the victim. He may elect to sue the operator before a domestic judge under domestic law, but he may also ask his State to enter into the diplomatic process of the Liability Convention, which does not require the prior exhaustion of local remedies. This freedom of choice implies the possibility of forum shopping. In practice, victims are likely to choose their forum case by case. If the damage is very important or if the operator is unable to pay or no longer exists, the victim benefits from asking his State to use the Liability Convention. In other cases, the route via a domestic judge might be preferable. This choice belonging to the victim is essential.

To summarize the various aspects of this choice, the positive points of using the Liability Convention are a State's liability, unlimited in amount and in time, and a joint and several liability enabling the victim to choose the liable State most able to pay compensation. The principal drawback is the State-to-State character of the procedure over which the victim has little or no control, and which holds a risk of political influence on the settlement of the dispute. As is the case with diplomatic protection, the victim State is free to select the way in which it wishes to obtain compensation and to decide whether to transfer the indemnification to its national.

On the other hand, using the domestic courts and laws enables the victim to exercise full control over his action, to receive the entire indemnification for his damage, and to avoid excessive political influence. Drawbacks of this approach include the risk of insolvency of the liable operator, the possibility of absolute liability in domestic law (in case of damage on Earth), and possible difficulties in obtaining the exequatur of the domestic judge's decision.

Until now no cases have presented themselves, either of damage caused by a space object to a private person or damage caused by a private space object. It seems that the best way for the victim to obtain compensation would be through a domestic judge. This is especially true if the liability of a private operator is recognized. Nevertheless, in some cases this would be risky or difficult. In such cases, the Liability Convention plays the role of a safety net. Therefore, although its specific mechanism for interstate liability may not be used often, it has an important role in securing space activities and in encouraging States to control activities and prevent accidents. Evidence thereof can be found in most systems of national space legislation.

There is another difficulty of interpretation in the Convention that has yet to be clarified, since no judge, either domestic or international, has ever had to apply it in a contentious case. According to Article XI the victim may use other means to obtain indemnification; the problem relates to the conjunction of the delay of one year as set out in Article IX for action under the Liability Convention and the provision of Article XI, paragraph 2, according to which the victim is 'entitled to present a claim under this Convention in respect of the same damage for which a claim is being pursued in the courts or administrative tribunals or agencies of a launching State or under another international agreement which is binding on the States concerned'. The conjunction of both articles obliges the victim to choose between action under the Liability Convention and action using domestic remedies, as it would be difficult in practice to settle the case before the domestic judge within one year. This is certainly not what the drafters of the Convention had in mind when they relieved the victim from exhausting local remedies in Article XI, paragraph 1. Nevertheless, it seems to be a consequence of the way the treaty text was drafted.

It may be argued that Article XI, paragraph 1 frees the victim from the obligation to use local remedies while not preventing him from doing so, but still the procedural rule of the one year delay applies. Here, as in many other cases, it would be useful to have a body of case law to clarify the interpretation of the Convention. This is not to say, however, that we wish to complain about a lack of space accidents.

Both the launching State(s) and the private operators may have to pay compensation. The launching State will have to do so if the victim and/or the victim's State decide to present a claim under the Liability Convention; the operator or any liable person has to pay if the victim decides to present his claim in domestic 'courts or administrative tribunals' pursuant to Article XI of the Liability Convention. Of course, the principle *non bis in idem* would apply. As was discussed above, the choice of legal forum will depend on the specific circumstances of the case, with arguments on both sides. In arguably the majority of cases it would be preferable to sue the operator directly before a domestic judge, whereas in other cases, especially if the damage in question is huge, it would be more prudent to use the diplomatic procedure offered by the Liability Convention.

5. WHO WILL BEAR THE BURDEN OF RISK IN THE CASE OF PRIVATE ACTIVITY IN OUTER SPACE?

This is the question that is at the heart of this chapter. It requires two different points to be considered. The first is the sharing of the burden of indemnification among jointly and severally liable launching States, if there is more than one, which is currently usually the case. The second is the sharing of the same burden between the State that has to pay according to the Liability Convention and the private operator whose activities lie at the origin of the State's liability.

A. Sharing the Risk Between Launching States

First of all, we have to keep in mind that the decision as to which State will be presented with a claim for compensation lies with the victim or the victim's State. Many issues may enter into consideration when choosing which launching State to claim against, including the 'deep pockets' of a given State or the practical likelihood of obtaining indemnification. Article V of the Liability Convention deals with this matter, positing that:

> A launching State which has paid compensation for damage shall have the right to present a claim for indemnification to other participants in the joint launching. The participants in a joint launching may conclude agreements regarding the apportioning among themselves of the financial obligation in respect of which they are jointly and severally liable. Such agreements shall be without prejudice to the right of a State sustaining damage to seek the entire compensation due under this Convention from any or all of the launching States which are jointly and severally liable.

UN General Assembly Resolution 59/115 on the application of the concept of the 'launching State', adopted by consensus in December 2004, 'recommends that States consider the conclusion of agreements in accordance with the Liability Convention with respect to joint launches or cooperation programmes'.[11] Article V, paragraph 2 of the Liability Convention recognizes the principle of 'a right to present a claim for indemnification to other participants in the joint launching'. The question is: who will pay? The provision is not very clear, as it does not specify the final distribution among launching States of the duty to pay indemnification. It is difficult to refer to a possible fault of one of the launching States because the system, at least for damages occurring on Earth, is based on absolute liability. Does this mean that the obligation to pay damages should be shared equally among launching States? This solution would certainly not be accepted, as the role and interest of the various launching States may differ vastly. During UNCOPUOS deliberations, the point was repeatedly raised by some States, citing for instance that it would be unfair for a State like Brazil to be solely liable on account of its territory being used for the launch.

Article IV, paragraph 2 of the Liability Convention may offer a further indication:

> In all cases of joint and several liability referred to in paragraph 1 of this article, the burden of compensation for the damage shall be apportioned between the first two States in accordance with the extent to which they were at fault; if the extent of the fault of each of these States cannot be established, the burden of compensation shall be apportioned equally between them. Such apportionment shall be without prejudice to the right of the third State to seek

[11] United Nations General Assembly Resolution 59/115 (10 December 2004), Article 2.

> the entire compensation due under this Convention from any or all of the launching States which are jointly and severally liable.

The case referred to in Article IV, paragraph 2 is different; the common damage is unexpected because it is caused by an accident. Two space objects have collided and caused damage to third parties either on Earth or in orbit. In that case the launching States of each space object are jointly and severally liable toward the third party(ies). The sharing of the risk should be done according to a potential fault of one of them. If the fault cannot be established, the sharing should be done equally. This provision may be difficult to apply given the difficulty in proving a fault in outer space. In some cases it may even produce unfair results. For example, if a nuclear power source is on board one of the space objects, the risk is considerably increased. If, in such a case, damage to a third party occurs and no fault can be proven, an equal distribution would be unfair in view of the operators' unequal contribution to the risk of damage, with the contribution of the nuclear-powered craft being much greater. One possibility might be to consider the use of a nuclear power source as a fault per se. However, this reasoning is highly uncertain.

Where damage is caused by only one space object, this mechanism of weighing respective faults is absent, although there may still be several launching States. In this case, referred to in Article V of the Liability Convention, the only practically feasible way of transferring, wholly or in part, the duty of indemnification to launching State(s) other than the one against whom the victim State chooses to claim, is for the launching States to conclude an agreement prior to launch 'regarding the apportioning among themselves of the financial obligation'. In this respect, it would be very useful if UNCOPUOS were to draft standard agreements between launching States that could be referred to before the launch in order to prevent these kinds of difficulties in the case of an accident.

B. Sharing the Risk Between Launching States and Private Operators Acting Under Their Launching State's Liability

As we have already seen, and as is commonly accepted, under the Liability Convention States are liable for damage caused by a 'national activity' conducted by a 'non-governmental entity' if they fulfil one of the criteria making them a liable launching State.

Of course, the State that has paid may look to 'get its money back' from the operator. This is one of the major functions of domestic legislation dealing with space activities. When they authorize a space activity, States do so conscious of the risk involved in being a launching

State, and they require the operator to purchase insurance to cover this risk. The fact that the majority of national space laws impose an insurance requirement lends credibility to this preoccupation among potential launching States. However, given the specific nature of space activities, especially of launch operations, it may be difficult for the operators to obtain insurance. The risk may be high, the number of launches is small, and the insurance market is narrow. In the national space laws of the USA and France, the obligation to obtain insurance may be replaced by the demonstration of financial responsibility.[12] In both cases the narrow character of the space insurance market is taken into consideration. Further, US national space law does not require the licensee or transferee to obtain insurance or demonstrate financial responsibility of more than 'the maximum liability insurance available on the world market at reasonable cost if the amount is less than the applicable amount' required by the 'Maximum Probable Loss'.[13] In the French decree the minister in charge of space may, for a limited time, exempt the operator from subscribing to an insurance policy or demonstrating financial capacities due to the state of the insurance market.[14]

Two issues should be considered here. The first is the problem of obtaining insurance without any ceiling of amount and time in order to reimburse the launching State that has paid under the Liability Convention; the second is the case of legal action against the operator before a domestic judge pursuant to Article XI of the Liability Convention.

In most domestic legislation the obligation of the operator to reimburse the launching State if it has had to indemnify a victim is limited or may be limited. This limitation in the form of a ceiling, which does not exist in respect of the State's liability under the Liability Convention, seems to be a precondition if the operator wishes to obtain insurance. No insurer can cover a risk that it cannot evaluate beforehand. This is the reason that has been put forward on many occasions to set a ceiling in respect of absolute liability.[15] To the extent that such liability is imposed on a State, this is not necessary. However, it is necessary when private operators may

---

12 51 US Code § 50914 – Liability Insurance and Financial Responsibility Requirements; Décret N° 2009-643 du 9 juin 2008, Article 16 (as modified by the Décret N° 2014-1315 of 3 November 2014 at Article 20).

13 51 US Code § 50914 – Liability Insurance and Financial Responsibility Requirements at (a) (3) (B).

14 Décret N° 2009-643 du 9 juin 2008 Article 17 (as modified by the Décret N° 2014-1315 of 3 November 2014 at Article 20).

15 It is for instance the case in the International Convention on Liability for Oil Pollution Damage 1969 (Brussels Convention) and its 1992 Protocol.

have to reimburse the State, in order to avoid the obligation for the operators to play 'Russian roulette' when a State authorizes a space operation. In their domestic legislations, most States agree to set a ceiling and thereby assume a part of the risk themselves, should damage happen that exceeds this ceiling. This is, for instance, the case of the Belgian Space Law at Article 15§3: 'the amount determined in accordance with §2 may be limited by the King, on conditions that he may determine. In such an event, the State's right of recourse against the operator may not exceed that limit.'[16]

In the US 1984/1988 Commercial Space Launch Act, the maximum probable loss (MPL) is set by NASA for the Federal Aviation Administration's Office of Commercial Space Transportation (FAA/AST) with a maximum of USD 500 million, although it is usually set much lower, from 15 to 250 million according to the financial risk of the launch.[17] In its 2008 law, France fixed a liability cap at an amount of EUR 60 million. Up to this amount the insurance covers damages; in excess of this amount the French government chips in.

C. A More Complex Problem: Action of the Victim Before a Domestic Judge

In this case it is unusual for the operator's State of nationality to intervene. Most often, it lets the operator handle the problem. If the victim, using his right under Article XI of the Liability Convention, decides to claim damages under domestic legislation where no ceiling applies, and especially if that domestic law recognizes a form of tort or objective liability, the operator may be in great difficulty if the amount of damages awarded is very large. For this reason, the USA and France, two States very much involved in space launches, have decided to support their operators by setting a liability cap, which applies not only for the reimbursement of what the State has paid as a consequence of the Liability Convention but also to the indemnification awarded by a domestic judge under domestic law. In both cases the State guarantees the amount exceeding the ceiling. This solution also benefits the victim, as he is certain to be indemnified, either by the operator or its insurance company or by the State.

---

[16] Law of 17 September 2005 on the Activities of Launching, Flight Operation or Guidance of Space Objects consolidated text as revised by the Law of 1 December 2013 (Belgian Official Journal of 15 January 2014).

[17] US Commercial Space Launch Act in US Code § 50915 – Paying Claims Exceeding Liability Insurance and Financial Responsibility Requirements.

In both cases, however, there are some limitations to this guarantee. In the USA it applies only during a launch period of 30 days.[18] It is organized under a three-tier approach. The first tier is borne by the launch company. It is limited by a maximum probable loss calculated by NASA. The maximum amount is USD 500 million, but it is usually much less. The second tier is provided by the US government, up to USD 1.5 billion (1988 value).[19] The third tier is left to the launch company, which may elect to purchase insurance,[20] and is in principle limited to an amount of USD 1500 million (1978 value).[21] This indemnification scheme is in principle limited in time, although the US Congress has repeatedly extended it. It currently runs until September 2025.[22] In France there is no limit either in time or in amount, but two phases are considered. During the launch phase itself, the guarantee applies to both damage on Earth and damage in orbit; during the in-orbit

---

[18] 14 CFR § 440.11

Duration of coverage for licensed launch, including sub-orbital launch, or permitted activities; modifications.

(a) Insurance coverage required under § 440.9, or other form of financial responsibility, shall attach when a licensed launch or permitted activity starts, and remain in full force and effect as follows:

(1) Until completion of licensed launch or permitted activities at a launch or reentry site; and
(2) For orbitai launch, until the later of—
   (i) Thirty days following payload separation, or attempted payload separation in the event of a payload separation anomaly; or
   (ii) Thirty days from ignition of the launch vehicle.

[19] Approximately $3.1 billion in 2017.

[20] Government Accountability Office (GAO), 'Commercial Space Launches: FAA's Risk Assessment Process Is Not Yet Updated' (4 February 2014) http://www.gao.gov/assets/670/660635.pdf; GAO, 'Preliminary Information on Issues to Consider for Reauthorization', http://www.gao.gov/products/GAO-12-767T (accessed 15 August 2017); J.A. Vedda, 'Study of the Liability Risk-Sharing Regime in the United States for Commercial Space Transportation' (1 August 2006) https://www.faa.gov/about/office_org/headquarters_offices/ast/reports_studies/media/Risk_Study(final).pdf (accessed 15 August 2017).

[21] Approximately $3.9 million in 2017. US Department of Transportation and FAA/AST, 'Liability Risk Sharing Regime For US Commercial Space Transportation' (2002).

[22] S. 102(d) of the Commercial Space Launch Competitiveness Act of November 2015 (H.R.2262 – 114th Congress (2015–2016)).

phase it applies only to damage on Earth.[23] Orbital mechanics and the operational aspects of space activities, however, dictate that, in the case of satellites placed in a geostationary orbit at least, the risk of damage occurring on Earth during this second phase is all but non-existent.

D. A Consequence of the Liability of the Launching State: the Authorization and Control of Private Operators

There are two reasons for States to promulgate national space legislation. The first is the obligation under Article VI of the Outer Space Treaty to authorize and supervise 'national activities' conducted in outer space by non-governmental entities. The second one, and certainly the most important at present, is the liability of the launching State under Article VII of the same Treaty and under the Liability Convention. This by itself is the reason why every piece of domestic space legislation requires a licence or authorization for 'national activities'. This requirement, although apparently based on Article VI of the Outer Space Treaty, is often implemented in a broader fashion to also cover those instances when a State may be at risk of receiving claims in its capacity as a launching State, for example when a company of its nationality launches or procures the launch, when national facilities are used or when the launch is carried out from the State's territory.

This need for control may cause difficulties when more than one launching State is involved. It is the consequence of the joint and several liability of launching States for the whole process. Every launching State is liable for any damage caused by a space object where they have, by one way or another, contributed to its launch. This creates the risk that more than one State may decide to control the launch, which in turn would impose an undue burden on operators and risk delaying the operation. In some cases control is impossible or very difficult by nature. If an American satellite is launched from China, then both States are launching States and both are liable for damages caused by the launch and by the space object after the launch without any limitation in time. We have seen that this rule, although it may be considered unfair, has proved highly effective in protecting victims.

For private companies there is a fear of redundancy of control and an important risk of delay. One State or the other may opt to control the whole process, as it is liable for every phase. The solution to this problem

---

[23] Official Journal of the French Republic No. 0129 (4 June 2008) page 9169, Law no. 2008-518 (3 June 2008) on Space Operations.

is not to modify the treaties or to reduce the liability of launching States and thus the protection of the victim; it is to support the elaboration of agreements such as those referred to in Article V of the Liability Convention. Before the launch, and while negotiating the launch agreement, launching States should negotiate a division of the burden of the risk in case of accident. At the very least they could differentiate between the launch phase itself and the in-orbit phase. Each State would assume the burden of the risk for its own phase and agree to reimburse the other State if the latter is presented with a claim for damages. In this regard it is important to remember that the victim's (or his State's) choice of respondent for a liability claim will likely be based on solvency rather than responsibility. Generalization of such agreements could avoid detrimental redundancy of control, as only one State would have to pay *in fine* for the indemnification of the victim. In launch operations, time is, of course, of the essence. Finally, as provided for in Article V of the Liability Convention, this kind of agreement has no effect on the rights of the victims.

6. CONCLUSION

The liability of launching States is an old principle that was established at the beginning of the space era. The system may have to evolve, but we need to keep in mind the necessity of maintaining a strong, efficient and fair mechanism of liability to indemnify the victims and to limit the occurrence of accidents through the preventative effects of responsibility and liability.

A wholesale modification of the treaties, the so-called 'comprehensive approach', is not desirable, as it looks certain to put the whole system at risk, given that few States would nowadays still accept responsibility and liability as they were accepted at the time of the first launches. We should keep in mind the difficulties of regulating private activities on the high seas and support efficient control by States linked with their responsibility and liability for space activities. Any difficulties, which will of course arise, may be solved individually by international, multilateral or bilateral agreements.

2. Taking a stance: managing liability for commercial space activities

Lesley Jane Smith

1. INTRODUCTION: TECHNOLOGY DRIVE AND REGULATION

Technology has been the driver of developments in transport, and aerospace flight is no exception.[1] Sub-orbital flight is the next step in revolutionizing aerospace travel, with a variety of activities underway, ranging from space tourism[2] to further plans for intercontinental passenger and cargo transport.[3] The response of the law to technological developments has been consistent; historically, pioneers of airflight accepted the risks of their travel in that very same spirit. The time lag between the initial experiments and the introduction of a licensing regime

---

[1] The Concorde, the first supersonic commercial passenger aircraft was, for example, the response of the UK and France to the space race; see http://designmuseum.org/design/concorde (accessed 14 October 2017). Following the Second World War the UK had suffered a setback at the technology level by not claiming the return of the jet plane that was licensed to the USA during the war.

[2] Dennis Tito was the first space tourist, flying in 2001 with the Russian Soyuz. See ESA Press Release No. 23–2001, 25 April 2001, at http://www.esa.int/For_Media/Press_Releases/(lang)/en/(year)/2001 (accessed 14 Oct 2017). The ISS partners agreed via the Multilateral Coordination Board (MCB) to grant an exemption allowing Tito to access the Space Station travelling on a Soyuz 2 Taxi mission that was scheduled for launch from Baikonur, Kazakhstan; see http://www.esa.int/Our_Activities/Human_Spaceflight/International_Space_Station/International_Space_Station_partners_grant_flight_exemption_for_Dennis_Tito (accessed 14 October 2017).

[3] Various spacecraft have flown since early SpaceShip One in 2004, such as XCOR's Lynx Mark I, and further companies involved in cargo and crew transportation concepts, such as Boeing, Orbital Sciences Corp. and Sierra Nevada. Virgin Galactic is currently contracting with SpaceShip Two to develop commercial sub-orbital tourism; see further, C. Chavagnac, F. Gai, T. Gharib and C. Mora, 'Astrium sub-orbital spaceplane for scientific missions' (2013) *Acta Astronautica* 92(2), 172–177.

enabled compulsory operation-based standards to develop.[4] Public interest in the aerospace sector continues to play a significant role, with safety remaining the highest priority, particularly for human space flight. Instating the most suitable licensing system requires technical planning, expertise, debate and political support for the regulatory goal.[5] As can be seen, notably in the USA, a culture of regulatory dialogue has become commonplace between developers and government agencies; in the case of sub-orbital flights, it is strongly policy-driven.[6] It has also sparked regulatory interest elsewhere, with ongoing observations of this sector by the European Aviation Safety Agency (EASA) on behalf of the European Union, as well as by select national governments.[7] Nevertheless, an appropriate, sustainable legal regime for sub-orbital operations has yet to be agreed at international level. Its absence is largely explained by open questions as to the scope of application of the existing regimes of air and space law, as well as definitional issues relating to the type of aerospace craft involved, the international authorities competent to act in this sector and, not least, the rules governing the status of space flight participants (SFP).[8]

---

4 The Wright Brothers first flew in 1903. It took until 1925 for the commercial air business to be introduced in the USA through the Air Mail Act 1925. See further, G. Nield, M. Touré, J. Sloan and D. Gerlach, 'Certification versus licensing for human spaceflight in commercial space transportation' (2013) *New Space* 1(1), 46–50.

5 R. Crowther, 'The Challenges of Human Spaceflight Safety: A Regulator's Perspective', IAF IAC-10-(D2) 9.6, http://www.iafastro.net/iac/archive/browse/IAC-10/D2/9/6385/ (accessed 14 October 2017); see further, B. Hufenbach, 'Considerations on private human access to space from an institutional point of view' (2013) *Acta Astronautica* 92(2), 131–137.

6 For a discussion of the position in the USA, see P. Sterns and L. Tennen, 'State and Municipal Regulation of the Aerospace Industry in the United States', in R. Jakhu (ed.), *National Regulation of Space Activities* (Springer 2010), 467–484.

7 See J.-B. Marciacq, Chapter 10, this volume; see further, J.-B. Marciacq, Y. Morier, F. Tomasello, Z. Erdelyi and M. Gerhard, 'Towards Regulating Sub-Orbital Flights, An Updated EASA Approach' (2010), in IAF IAC-10(D2) 9.5, http://www.iafastro.net/iac/archive/browse/IAC-10/D2/9/9148/ (accessed 14 October 2017); see further, G. Nield, M. Touré, J. Sloan and D. Gerlach, *supra* note 4.

8 'Space flight participants', though not a codified classification is the name given to non-professional tourists and volunteers who take part in sub-orbital flights but who are not technically astronauts in terms of the Rescue Agreement; see D. Webber, 'Space tourism: its history, future and importance' (2013) *Acta Astronautica* 92(2), 138–143.

Sub-orbital flights, as a fledgling sector, are also attracting increasing interest as a motor for driving innovation and spin-off technologies. Sub-orbital travel not only delivers to a distinctly high-value market, it is also currently providing an important back-up to government demand. Some governments are looking to this sector as a key element within their national infrastructure.[9] NASA has adopted commercial sub-orbital capabilities to supplement its missions, initially inspired through the Ansari X-Prize competition in 2004, and now with generous financial incentives for the commercial sector to support its own commercial crew programme.[10] Operations in the USA are authorized under experimental licence, on the basis of commercial outer space launch legislation, and are monitored by the Office for Commercial Space Transportation (AST) within the Federal Aviation Administration, (FAA), part of the Department of Transport (DoT).[11] There are trade-offs and mutual benefits for both the aviation and outer space sectors, and for government and commercial interests alike. National law is currently the driver for regulation in those jurisdictions where sub-orbital activities are, or may soon become, the subject of licensing.[12]

---

[9] See J. Stubbs, Chapter 7, this volume; for further details of the initial consultation on the development of UK spaceports and an assessment of potential locations between July and October 2014, see https://www.gov.uk/government/uploads/system/uploads/attachment_data/file/360448/spaceport-consultation.pdf (accessed 14 October 2017). The proceedings of a subsequent meeting in London now supplements the views of the UK space community; see 'Priorities for Developing the UK Space Sector – Infrastructure, Regulation, Innovation and Collaboration' (12 March 2015) http://www.westminsterforumprojects.co.uk/forums/event.php?eid=878&t=7645 (accessed 14 October 2017); see further, R. Crowther, *supra* note 5.

[10] This was first approved through the appropriations under the 2010 NASA Authorization Act 2010; NASA provides funding to SpaceX, Boeing and more recently Sierra Nevada as part of its new commercial crew space transportation system (STS) and concept for a multi-purpose crew vehicle (MPCV). See further G. Nield, M. Touré, J. Sloan and D. Gerlach, *supra* note 4.

[11] See D. Murray, Chapter 6, this volume; details of the structure of the FAA are available at https://www.faa.gov/licenses_certificates/commercial_space_transportation/ (accessed 14 October 2017). The experimental licence is available under 51 USC Ch. 509: Commercial Space Launch Activities, § 50906.

[12] The Commercial Space Launch Activities Act 2004, 51 USC 509 (USA); F. von der Dunk, 'Mixing the US and Dutch Approaches: Towards Curaçao's Legislation on Private Commercial Spaceflight' (11 December 2011), in Proceedings of the International Institute of Space Law 2012 (Eleven 2013); F. von der Dunk, 'Federal versus State: Private Commercial Spaceflight and Operator's Immunity Regulation in the US' (2013), in Proceedings of the International

2. LIABILITY REGIMES: FROM FAULT TO STRICT LIABILITY

International aviation conventions and outer space law mainly regulate aerospace safety from a public international law perspective, whereas passenger and transport law are regulated under private law aspects of international aviation law. Because outer space activities, as a matter of public international law, are only open to sovereign States, the right to participate in outer space activities is subject to national approval, whether for private commercial companies or for individual activities.[13] Currently the systems of aviation and outer space law are facing challenges through the continued growth of commercial aerospace activities, with aerospace technology pushing the borders between both regimes.

The liability rules on which these systems are based are coming under greater scrutiny, particularly as technology is narrowing the scope of what are two separate regimes. This in turn is leading to discussions about which legal regime – air law or outer space law – is the best anchor for managing the risks attached to new technology-driven aerospace vehicles which straddle their respective scope.

The general response to the management of unavoidable, or inherent, sector-specific risk has been the introduction of formal liability regimes designed to deliver a basis for restitution, or compensating loss in the event of system-related damage. In cases where the classic rule of fault or negligence liability is inappropriate in light of the type of risk, for example in the case of unavoidable sector-specific risks, systems of strict liability or statutory regimes imposing non-fault liability have taken the place of fault liability. These generally provide remedies in the form of compensation for loss and damage resulting from ultra-hazardous operations in areas where the public and private interest in safety, and fair compensation for loss, is paramount.[14] Generally, strict liability is

---

Institute of Space Law 2013 (Eleven 2014); for developments in the UK, see section 'UK case study', this chapter.

[13] Outer Space Treaty 1967, Art. VI.

[14] For a comprehensive overview of the development of the law of strict liability, and notably liability for automated manufacturing systems, see the Royal Commission on Compensation for Personal Injury (1978) London HMSO Cmnd. 7054. See further, the work of the American Law Institute, The Restatement (2nd) on Law of Torts (1964), and notably § 402A, which has been superseded by the Restatement (3rd) on Law of Torts (2008) in a 'monumental' review of the case law developed since then relating to damage resulting from

designed for high-risk sectors, with a high level of automated operations, including transport, where the individual or the environment has little to no means of controlling their inherent dangers or outcomes, including the aftermath of disasters.

Various international agreements regulate liability for loss or damage to the participants' persons and their personal environment; the interests of third parties affected by such accidents or catastrophes are invariably included.[15] Because of the high risk associated with impact and collision, strict liability regimes form the basis of international aviation and outer space operations. The community of spacefaring nations decided early on that liability for damage on Earth resulting from operations involving space objects should be subject to a special international State liability regime imposing absolute unlimited liability. The rules governing liability for outer space activities were introduced in the 1972 UN Convention on International Liability for Damage caused by Space Objects (the Liability Convention), a *lex specialis* expanding the principles originally agreed at UN level and contained in the 1967 Outer Space Treaty. The Outer Space Treaty had established the initial basis for international State-to-State responsibility for national space activities, and international State liability for space objects.[16] The Liability Convention's separation of liability for space objects from State authorization of national space activities is a distinction that continues to dominate the discussion on how to regulate the new generation of aerospace craft with launching and sub-orbital capabilities.

The exact pattern of liability chosen for allocating the risk for damage arising from hazardous operations involves calculating the probability of damage arising from the particular system, which in turn depends on the

---

dangerous products and services; see http://www.ali.org/index.cfm?fuseaction=publications.ppage&node_id=54 (accessed 14 October 2017).

[15] Convention on Damage Caused by Foreign Aircraft to Third Parties on the Surface 1952 http://www.icao.int/secretariat/legal/List%20of%20Parties/Rome 1952_EN.pdf; for a full discussion of the arguments surrounding liability for GNSS, see L.J. Smith, 'Liability for GNSS', in F. von der Dunk and F. Tronchetti (eds), *Handbook of Space Law* (Edward Elgar 2015), chapter 10. UNIDROIT is deliberating whether to adopt a new statutory system of third party liability from GNSS operations; see http://www.unidroit.org/work-in-progress-studies/studies/civil-liability/393-study-lxxix-third-party-liability-for-global-navigation-satellite-system-gnss-services (accessed 14 October 2017).

[16] See further, L.J. Smith and A. Kerrest, 'The 1972 Convention on International Liability for Damage Caused by Space Objects', in S. Hobe, B. Schmidt-Tedd and K.-U. Schrogl (eds), *Cologne Commentary on Space Law*, Vol. II (Heymann 2013).

type and level of risk involved. For commercial operations this also involves identifying the party best suited to carry the cost of liability, and this is generally the operator.[17] The operator can, in turn, arrange to limit his liability, either through contract or by virtue of the specific liability rules applicable to that sector.

Probability calculations also impact on the compulsory insurance coverage required for third party liability or for damage to those involved in the operations (employees, operators, crew, passengers and so on).[18] In return for accepting liability and any liability insurance requirement imposed, the operator is given the benefit of limiting its liability, in amount and time.[19] This type of liability regime must be anchored in the law, since no liability can arise in the absence of a legal basis. The greater industrial risks fall under specific legal or consensual regimes and are insured. One particular aspect of the liability regime for outer space activities should be recalled in the context of the law applicable to aerospace activities. In contrast to other liability regimes, international State liability for outer space activities, including those of its commercial entities, is unlimited. This is the major contrast to limited operator liability under aviation law.

---

[17] This is evidenced in large-scale industries, from oil exploration to GNSS satellite operations, where the authorities award operators a licence on the basis of a maximum insurable limited liability ceiling; see further, L.J. Smith, 'Current Challenges of Liability for Loss of Satellite Based Services' (2013), in Proceedings of the 56th Colloquium on Law of Outer Space, International Institute of Space Law (Eleven 2014), 291–304.

[18] See the use of probability calculus by the US Judge Learned Hand in *United States v. Carroll Towing Co.* (2d. Cir. 1947) 159 F.2d 169 as follows: If the burden of adequate precautions is less than the cost of injury multiplied by the probability of occurrence, then the standard of care required will not have been met, incurring liability, i.e. liability depends on whether $B < PL$. On outer space insurance see P. Monpert, 'Insurance', in L.J. Smith and I. Baumann (eds), *Contracting for Space* (Ashgate 2011), 283–289; C. Gaubert, 'Insurance in the Context of National Authorisation', in F. von der Dunk (ed.), '*National Space Legislation in Europe: Issues of Authorisation of Private Space Activities in the Light of Developments in European Space Cooperation*' (Martinus Nijhoff 2011), 163–177. Space Liability Convention 1972 Art. VII excludes the applicability of the international outer space law regime towards nationals, as the protection of this group is regulated by domestic law.

[19] The time limit for raising a direct action under the Rome Convention is six months (Art. 19), failing which only a claim against residual mass can be brought; under the Montreal Convention, claims must be raised within two years (Art. 35).

In the USA, operations licensed under the supervision of government agencies stand to benefit from a discretionary defence under the Federal Tort Claims Act against claims of negligence that result solely from the process of certifying and authorizing operations. This defence was introduced to grant government agencies the authority (and freedom of administrative discretion) to approve and license (dangerous) operations, without subsequently being held liable for accidents or loss.[20] In addition, contractors to government are protected by the so-called government contractor defence to tort claims, which can pre-empt claims under State law, even in the absence of specific legislation, where unique federal interests require this.[21] US law therefore provides a positive environment so that tort liability does not constitute an imponderable risk for government licensing agencies or commercial operators, particularly in the face of technological development.[22]

3. LIABILITY FOR AEROSPACE ACTIVITIES

Sub-orbital flight, whether vertical flight with re-entry within national airspace, point-to-point flight involving non-stop flight across foreign airspace, or inter-continental low orbital travel, requires a hybrid, reusable technology and a very reliable spacecraft. Missions may last from thirty minutes to one hour. Should orbiting take place, this may be for more than a few days, even up to a maximum of two weeks.[23] The technologies are a

---

[20] In European countries the issue of wrong or negligent certification is generally approached through technical investigations in the specific sector. It may, but need not, lead to criminal proceedings against the operator for unlawful operations. However, liability cases against the public authority are rarely brought. For the discussion in the context of GNSS certification liability, see L.J. Smith, 'Legal Aspects of Satellite Navigation', in F. von der Dunk and F. Tronchetti (eds), *supra* note 15.

[21] *Boyle v. United Techs Corp.* [1988] S. Ct 487 US 500; for the origins and operations of the Federal Tort Claims Act 1946 as amended (28 USC), see Paul F. Figley, 'Understanding the Federal Tort Claims Act: A Different Metaphor' (2009) *Tort Trial & Insurance Practice* 44(3), http://digitalcommons.wcl.american.edu/cgi/viewcontent.cgi?article=1006&context=facsch_lawrev (accessed 14 October 2017).

[22] Interestingly, discussions are now underway in the USA on a recognized need to introduce a dedicated licensing regime for unmanned vehicles in that country, to avoid losing out on competitive markets; see P. Dempsey, 'The future of international air law in the 21st century' (2015), *Zeitschrift für Luft-und Raumfahrtrecht* 64, 215–232.

[23] See G. Nield, M. Touré, J. Sloan and D. Gerlach, *supra* note 4.

mix of those known in both aviation and outer space activities, but the current craft operate neither exclusively as aeroplanes, nor as rockets.[24]

The international aviation system established by the 1944 Chicago Convention[25] is a key regulatory mechanism devised to develop international civil aviation in a safe and orderly manner.[26] Aircraft registration formalities based on nationality were introduced under Article 17, with mutual recognition of certificates of airworthiness, and licensing introduced under Article 33 alongside an ongoing framework for establishing rules and procedures, and notably recommended standards for international aviation.[27] The Chicago Convention represented the internationally recognized need for defined rules of safety, and it has continued to demonstrate a flexibility that has allowed technological development to be accommodated within its structures, and to see its rules adapted to ensure that its objectives are maintained.[28]

The conventions that preceded and followed on from the Chicago Convention provided for systems of strict but limited liability, governing carriage by air of paying passengers as well as cargo, the first of these being the Warsaw Convention of 1929.[29] This Convention has since been

---

24 On the aerodynamics involved, see R. Crowther, *supra* note 5; the hybrid aerospace craft in the USA are currently licensed as rockets. For the details of the three configurations of SpaceShip One, which is carried under the aeroplane WhiteKnight and relies on the differing aerodynamics prevailing at different altitudes, see http://science.howstuffworks.com/spaceshipone1.htm (accessed 14 October 2017).

25 Convention on Civil Aviation 1994, 15 UNTS 295.

26 Preamble to the Chicago Convention, para. 3: 'the undersigned governments having agreed on certain principles and arrangements in order that international civil aviation may be developed in a safe and orderly manner and that international air transport services may be established on the basis of equality of opportunity and operated soundly and economically.'

27 Chicago Convention Art. 26 allows for air accident investigations to be attended by experts of the State in which the aircraft is registered, while the State in which the air accident occurs is required to undertake an enquiry and share the information with the State of registry. Under the Rome Convention, the law applicable to the damage is that of the *lex loci delicti*, or place of damage – see Art. 20(1) – unless plaintiffs and defendants agree to combine actions before another jurisdiction. The rule in Art. 20(1) is to limit choices of multiple claimants and divergent fora that may otherwise be available under principles of international private law.

28 See this chapter, sections 4–5.

29 See Warsaw Convention for the Unification of Certain Rules Relating to International Carriage by Air, Warsaw Convention 1929, Rules for International Carriage by Air. The Warsaw Convention contains a presumption of liability,

updated in relation to some States, and has been supplemented by two further conventions dealing with aviation liability, notably the 1952 Rome Convention on Damage Caused by Foreign Aircraft to Third Parties on the Surface and the 1999 Montreal Convention for the Unification of Certain Rules by International Carriage. The latter regulates strict liability for death and damage to passengers. The Rome Convention provides a system of strict operator liability for third party damage from aircraft leading to passenger death and injury,[30] and a system of strict liability for loss or damage to cargo.[31] Under Article 55 the Rome Convention has priority over the Warsaw Convention for contracting States. The liability provisions of these conventions, and notably their limitations, can be overridden where there is deliberate harm, including third party or contributory negligence.[32] Whereas the Rome Convention focuses on operator liability, the Montreal Convention introduced a system of limitation of liability for passengers and goods for the carrier.

In contrast to the above, outer space law is a *sui generis* international legal regime, void of the detail contained in the Chicago Convention. The outer space regime does not dictate technical safety standards;[33] these are left to the national authorities. There are, however, initiatives towards the establishment of common regional standards, certainly at European

---

rebuttable on proof of negligence, while the Montreal Convention imposes strict liability up to the first tier of damages, and relieves the operator from further liability. On proof of third party negligence, see Montreal Convention Art. 21(2).

[30] Convention on Damage Caused by Foreign Aircraft to Third Parties on the Surface 1952, Art. 17. The European Union incorporated the Montreal Convention into its Member States' national law through the EU Regulation (EU) of the European Parliament and Council, as amended by EU Regulation 599/2014, OJ L 173/79 of 12 June 2014. The EU Regulation provides for a two-tier liability scheme, which links the levels of liability for passengers and cargo to the Montreal scheme of special drawing rights (SDR), notably by a limited first tier of liability and an unlimited second liability tier.

[31] Under Rome Convention 1952 Art. 22, liability can only be limited as from the second tier of compensation in the case of third party liability. Under Art. 35, claims must be brought within a two-year time limit.

[32] The Investigation Report of the French Authorities into the crash of a Germanwings aircraft in March 2015, published in June 2015, concludes that the pilot deliberately descended the aircraft so that the crash appears to be a deliberate action of the co-pilot; see its report available at http://www.bea.aero/docspa/2015/d-px150324.en/pdf/d-px150324.en.pdf (accessed 14 October 2017).

[33] There is authority for the position that much of the Outer Space Treaty, including Art. VI, has now acquired customary international law status; see F. Lyall and P. Larsen, *Space Law, A Treatise* (Ashgate 2009), 71.

level.[34] National authorities have duties and international responsibilities under Article VI of the Outer Space Treaty for the ongoing monitoring and authorization of their national governmental and non-governmental outer space activities.[35]

'Space law is wonderfully international.'[36] The explicit internationalization of outer space law is the first major distinction between the systems of air and space law. Air law became international to allow international flight across national airspace, almost by analogy to the right of innocent passage recognized in the law of the sea.[37] It has a strong focus on sovereign rights over national airspace, as well as a harmonized approach to regulations governing jurisdiction over aircraft, organization and air flight management.[38] Unlike air law, space law knows no national appropriation or sovereignty over outer space under Article II of the Outer Space Treaty. Space objects remain subject to the national jurisdiction of the State of registry under Article VIII of the Outer Space Treaty and Article I of the Registration Convention,[39] with ownership in outer space unaffected.

Nevertheless, the national approach taken to both air and space law is relevant given the significance of State practice as a source of international law.[40] The doctrine of State practice, whether relating to treaty or customary law, can witness and homologate a common regulatory approach by States that develop into recognized customary norms of

---

[34] See the initiatives relating to a European Cooperation on Space standardization, ECSS, information available at http://www.ecss.nl/ (accessed 14 October 2017).

[35] The Chicago system does not impose compulsory standards by virtue of their agreement at International Civil Aviation Organization (ICAO) level, unless they are contained in national law.

[36] P. Hulsroj, 'Foreword', in F. von der Dunk, *National Space Legislation in Europe* (Nijhoff 2011).

[37] S. Langton, 'Sub-orbital flights: a comparative analysis of national and international law' (2011) *Journal of Space Law* 37(2), 299–392.

[38] This applies in particular to the development of the European single sky, SES, a regulatory framework for air traffic management across the European Union that combines air traffic networks services with the interoperability of the European Air Traffic Management Network (EATN); see further, P. Dempsey, *supra* note 22.

[39] Convention on Registration of Objects Launched into Outer Space, 14 January 1975, 1023 UNTS.

[40] I. Brownlie, *Principles of Public International Law* (8th edn, Oxford University Press 2012) 24–25.

international law.[41] The current efforts to regulate hybrid aerospace should aim for a common approach and prevent national systems from developing diversely. The USA is currently at the forefront of these developments, so there is merit in observing its approach closely.[42]

4. OUT OF SCOPE: DELIMITATION AND DEFINITIONS

In the absence of dedicated national legislation as exists in the USA, there is a lack of legal certainty and agreement as to which regime (air or space) does, or should, apply to sub-orbital activities.[43] Sub-orbital vehicles operate across the confines of air and/or space law systems, which were developed prior to the emergence of this hybrid technology.

Definitional issues between air and space law derive in part from classification of flight capabilities, as well as functional differences resulting from the Earth's atmosphere and outer space orbits. The natural demarcation between air and outer space remains a recurring point of reference. Up to 120 km above sea level is generally recognized as the lower limits of orbital function, and beyond 120 km as outer space.[44] Further distinctions are made between the type of flight or mission, for example whether sounding rockets or experiments are involved. Some jurisdictions operate with 100 km as their borderline for air space, this being the famous Von Karman line.[45]

Article 3(a) of the Chicago Convention regulates aircraft that rely on aerodynamic reactions on 'surfaces under fixed conditions of flight'.[46]

---

[41] S. Hobe, 'Article I Outer Space Treaty', in S. Hobe, B. Schmidt-Tedd and K.-U. Schrogl (eds), *Cologne Commentary on Space Law* (*CoCoSL*), Vol. I, (Heymanns 2010) 61–70.

[42] P. Dempsey, *supra* note 22, at 221–223.

[43] P. Dempsey and M.C. Minheiro, 'ICAO's Legal Authority to Regulate Aerospace Vehicles', in J.N. Pelton and R. Jakhu (eds), *Space Safety Regulations and Standards* (Elsevier 2010) 245–256.

[44] The area as from 110 km above sea level is generally seen as part of outer space; see E. Vitt, 'Grundbegriffe und Grundprinzipien des Weltraumrechts', in K.-H. Böckstiegel, *Handbuch des Weltraumrechts* (Heymanns 1991) 35, 37–43; see further, UNCOPUOS, A/AC.105/C.2/2010/CRP.9 (2010), 'Concept of Sub-orbital flights: Information from the International Civil Aviation Organization'.

[45] See generally, F. Lyall and P. Larsen, *supra* note 33.

[46] UNCOPUOS, A/AC.105/C.2/2010/CRP.9 (2010). The Chicago Convention, Annex 7 defines aircraft as: 'Any machine that can derive support in the atmosphere from the reactions of the air other than the reactions of the air against

This is further explained in Amendment 2 to Annex 7 of the Convention as 'deriving lift from the reaction with the atmosphere'.[47]

Sub-orbital craft can reach their altitude in a variety of ways, whether using runway take-off, landing with their own power, through aeroplane launches or following traditional launch patterns of vertical take-off, with a variety of return techniques including gliding.[48] Whether take-off or launch is used, be this from the Earth or the air, sub-orbital vehicles go through various 'atmospheric, altitude and aerodynamic flow regimes' prior to re-entry.[49] Even if not phrased with sub-orbital flight in mind, the final leading statement on the definition of aircraft in Annex 7 of the Convention is significant: '[y]ears of considerable effort permit the classification of aircraft to be as simple as possible, and yet encompass as many types of flying machines as the human mind can devise.' Prima facie, this statement implicates an inclusive, broader and developmental interpretation of aircraft definitions.[50] There are views that the structure of the Chicago Convention is flexible enough, notably through its Article 37, which provides for additional regulation in annexe form, to encompass further rules about the vehicle type needed.[51]

Space law, on the other hand, has traditionally relied on two alternative approaches in determining its applicability, developing from discussions about the highest limit of national jurisdiction. The first is the spatial approach, by which the location of a space object when above the Earth determines the application of space law. The second is the functional approach, by which the use or function of a particular object at a particular time determines which law is applicable. In the case of a satellite, it means that space law applies. The exact approach adopted becomes particularly pertinent in the case of objects not intended for orbiting in outer space, such as sub-orbital flights, as well as those spacecraft flying or satellite activities taking place beyond the airspace

---

the earth's surface.' Aeroplane: 'A power driven heavier than air craft, deriving its lift in flight chiefly from aerodynamic reactions on surfaces which remain fixed under given conditions of flight.'

47 *Ibid.* The description in Annex 7 n.34 above appeared after an amendment made in 1967 to exclude cushion or hovercraft.

48 See G. Nield, M. Touré, J. Sloan and D. Gerlach, *supra* note 4, at 4.

49 R. Crowther, *supra* note 4.

50 Vienna Convention on the Law of Treaties (VCLT) 1969, Arts 31, 32.

51 See generally, R. Jakhu, T. Sgobba and P. Dempsey (eds), *The Need for an Integrated Regime for Air and Space ICAO for Space?* (Springer 2011).

boundary of 100 km, including in the lower Earth orbit.[52] The legal significance of a 'launch' can be seen from the provisions of Article VII of the Outer Space Treaty. It imposes liability on the launching State for damage resulting from a space object during its air passage and in outer space. Flights in sub-orbital space in the course of an Earth-to-Earth transportation remain subject to air law.[53]

Finally, and most importantly, the definition of a space object under international space law is sparse, limited to the provisions of Article VII of the Outer Space Treaty, Article I(d) of the Liability Convention and finally Article I(b) of the Registration Convention. These respectively refer to an object, including its component parts.[54] There is a presumption of intended orbital activity.[55] Hybrid spacecraft might at first sight be well accommodated under the functional approach, depending on the length and level of intended sub-orbital activity. However, the divergent systems of liability for aviation and outer space activities dictate that a choice and decision in principle must be made at some stage, both from a regulatory perspective and from one of legal certainty.

The allocation of aerospace to one or the other field through to the introduction of dedicated provisions for this sector would bridge the current gap, provide greater legal certainty and counteract concerns as to whether the law remains at the forefront of developments. Despite the tools of legal interpretation available when applying convention law, notably good faith in Article 31(1) of the Vienna Convention on the Law of Treaties (VCLT), including the specific context and intention of the parties under Article 31(2), these provisions are limited in relation to treaties which preceded its entry into force, as did both the Outer Space Treaty and the Liability Convention. What is required is an interpretation which, in line with the context of Article 32 VCLT, is void of ambiguity.

---

[52] See F. Lyall and P. Larsen, *supra* note 33, at 167–171; see further F. von der Dunk, 'Introduction to Space Law', in F. von der Dunk and F. Tronchetti (eds), *Handbook of Space Law* (Edward Elgar 2015) 87.

[53] A/AC.105/C.2/2010/CRP.9, UNCOPUOS (19 March 2010) ICAO Working Paper, para. 4.3.

[54] *Ibid.*, n.39.

[55] L.J. Smith and A. Kerrest, *supra* note 16; on Liability Convention Article I, see ns.62–64.

5. PARADIGM SHIFT: THE APPROACH OF INTERNATIONAL ADVISORY BODIES

The international authorities responsible for overseeing aviation and outer space law have been following the developments in sub-orbital flight, but with few conclusive findings about the best way forward at international level.[56] The International Civil Aviation Organization (ICAO), which has achieved a high degree of alignment of safety standards for the aviation sector,[57] initially found against including sub-orbital flights under the Chicago Convention, thereby renouncing the ICAO Council's competence to regulate these activities. This would logically mean the prima facie non-applicability of the aviation liability regime. However, as with all regulatory systems, there can be paradigm shifts, so the matter is not finally closed.[58] Indeed, there are indications that ICAO sees the introduction of an annex to the Chicago Convention as a possible approach to regulating sub-orbital flights in the future.[59]

The approach of UNCOPUOS is more flexible. UNCOPUOS, the United Nations Committee for Peaceful Use of Outer Space with its two subcommittees, has been involved in the ongoing monitoring and implementation of outer space treaty law and its subsequent development since 1959.[60] In its latest report it has heard calls from States to see the aerospace sector formally regulated. One potential regulatory approach would be to allocate the aerospace sector to the remit of one or other of

---

56 The International Civil Aviation Organization (ICAO) created by the Chicago Convention, under its governing Council (Art. 50) is responsible for international civil aviation.

57 The Chicago Convention operates with 18 separate Annexes regulating issues of airworthiness (Annex 8) to aeronautical communications telecommunications (Annex 10), and not least, nationality of aircraft and registration (Annex 7). The Convention has enabled international dialogue on standards and certification, which appear as benchmarks in the Annexes to the Convention (Standards and Recommended Practices, SARPs), enabling safety standards to develop, alongside other considerations towards economy of international flight.

58 The Air Navigation Commission is regulated by ICAO Convention Art. 56; it may make references for modifications to the Council.

59 See ICAO Working Paper C/WP-12436, concept of sub-orbital flights, at point 6.3.

60 This committee first existed as an ad hoc, then as a permanent committee from 1959. The two Subcommittees were established in March 1962; see L.J. Smith and A. Kerrest, 'Article VII Outer Space Treaty', in S. Hobe, B. Schmidt-Tedd and K.-U. Schrogl (eds), *Cologne Commentary on Space Law (CoCoSL)*, Vol. I (Heymanns 2010), 9.

the internationally competent authorities which has the greatest exposure to that main branch of the law. This would allow, for example, allocation of space sustainability and space traffic management as part of the law of outer space under the aegis of UNCOPUOS, while changes in aerospace resulting from technology such as sub-orbital flights, subsumed by air law, could be assumed under the aegis of ICAO.

As long as the international community refrains from opting for a clear regulatory approach, the national regulator alone will determine the scope and applicability of its licensing regime. If there is collaboration and consultation with other competent bodies concerned, some convergence of State practice could be achieved.[61]

6. FACILITATING MARKET ENTRY THROUGH NATIONAL LAW

National legislation is an incentive for private commercial participation in the relevant market. It enables trade-offs, with issues ranging from licensing through to questions of liability being open to discussion between commercial operators and government.

The US Commercial Space Launch Act (CSLA), amended in 2004 to become the US Commercial Space Launch Amendment Act (USCLAA), is the first federal statute dedicated to private commercial spaceflight in the USA.[62] The number of sub-orbital spacecraft developed by and launched out of the USA since its entry into force is on the increase, with SpaceX and Virgin Galactic taking the lead in California and New Mexico respectively.[63] Other more exotic locations, such as Curaçao, Netherlands Antilles, are developing newer, high-value forms of sub-orbital tourism through their advantageous climate and location.[64] There is a clear market, one which has been boosted by the US government commercial launch policy.

Various US States have also facilitated and validated the specific considerations relating to the private legal issues raised by sub-orbital travel by introducing dedicated domestic legislation. These statutes

---

[61] J.-B. Marciacq, Chapter 10, this volume.

[62] Chapter IV USC 509. Sub-orbital flights have been approved with experimental licenses by the US Federal Aviation Administration (FAA-AST) under § 50906 Commercial Space Launch Act as rocket, and not aircraft, activities.

[63] See G. Nield, M. Touré, J. Sloan and D. Gerlach, *supra* note 4; further, UNCOPUOS, Concept of Sub-orbital Flights, *supra* note 44.

[64] On the position in Curaçao, see F. von der Dunk, *supra* note 12.

validate waivers of liability towards flight participants through a doctrine of statutory informed consent. Third party liability considerations apply under the CSLA for private manned space flight based on maximum probable loss, with cross-waivers of liability for parties other than the US government.[65]

The UK is about to follow this trend by catering for sub-orbital flights. In mid-July 2014 the Secretary of State for Transport first announced the UK government's interest in developing UK sub-orbital launching capabilities by 2018. Dedicated legislation was introduced to Parliament in July 2017, as the 'Space Industry Bill to make provision for space and suborbital activities'.[66] Although it will be some time before such systems become fully operational, interest from stakeholders remains keen. The UK's stance on sub-orbital launching is even more significant since its decision to terminate its membership of the European Union; the regulatory competence for aviation certification, in terms of the main EU Regulation 216/2008, falls under the centralized competence of the EASA. While the EU Regulations stipulate the EASA's responsibility for certifying vehicles and air traffic navigation systems, this does not impinge on the original competence of Member States (including the UK) to regulate their own outer space activities. The UK's competence to authorize space activities under the Outer Space Act 1986 is being extended by this new piece of dedicated space legislation, opening up the UK to new markets as a launching state.

Not unlike sub-orbital activities in the USA, regulation of outer space activities is equally visible at national level. Since the 1990s more States have passed national space statutes, largely in response to the downsizing of technology and the opening of space capabilities for non-spacefaring nations with increased outer space technological capabilities at national level and with space being included with national strategic policy.[67] This

---

[65] For the development of State law in the USA, see P. Sterns and L. Tennen, *supra* note 6. There are domestic provisions, both municipal regulations and State law, governing space in Texas, Colorado, New Mexico, Virginia and Hawaii.

[66] As of July 2017 it had not yet passed the reading stage.

[67] The subject of national space law has been a regular item on the UNCOPUOS Legal Subcommittee agenda since UN Resolution 68/74, Recommendation on national legislation relevant to the peaceful exploration and use of outer space, A/RES/68/74, 16 December 2013. Divergences do exist in national space statutes, notably in relation to the private space sector, both as to international liability and responsibility; see latest reform of Belgian space law 2005, published 15 January 2015, Moniteur Belge.

too is being monitored and reported on centrally through the offices of UNCOPUOS.[68]

7. SIMILAR OR DISTINCT PATTERNS OF LIABILITY: WHAT ABOUT AEROSPACE?

International aviation and outer space liability systems are based on clear but distinct liability patterns for their respective sectors. The Warsaw and Montreal Conventions are applicable to aircraft and, for various reasons, are not automatically applicable to hybrid craft service operations. Whether sub-orbital flights can be subsumed within their provisions will depend on the technicalities discussed below, as well as a decision on how to extend the current concept of aircraft in flight that are used in the aviation conventions.

The first reason for their limited and divergent scope of application is that both conventions create separate schemes for limiting the strict liability of air transport undertakings operating aircraft and involved in the international carriage of persons, baggage or cargo (Article I, Montreal Convention). In contrast, outer space activities involve unlimited liability, so that the concept of a licensed commercial operator predicates a governmental rule or a national statute governing the internal apportionment and indemnification of liabilities by licensed operators; governments are deliberately designated as the first addressee of liability claims arising from outer space activities, in the light of the probable losses.[69] The second and most relevant reason is the current requirement under air convention law for a contract or 'agreement' of international carriage (whether or not gratuitous or for reward, Article 1 Montreal Convention). This gives rise to carrier conditions and documentation (passenger information, as well as cargo air waybills). Aerospace flight is not yet a truly commercial venture, so that the concept of international carriage and ticketing or air bills for a SFP are not yet seen to be relevant or to apply. These conventions would require some tweaking and several amendments to extend their current scope, notably beyond aircraft in flight. Developments here will be keenly monitored.[70]

---

68 See UNGA Resolution 68/74.

69 L.J. Smith, 'Legislating for outer space: the example of Germany: considerations on a national space law' (2012) *Zeitschrift für Luft- und Weltraumrecht* 1/2012, 6278.

70 S. Langton, *supra* note 37.

Liability provisions for outer space activities take the starting point that, although damage to the Earth is non-fault-based and in contrast to the limited liability for aircraft operators, the liability regime is unlimited and absolute. It is designed to provide a dedicated victim-oriented compensation scheme for losses caused by space operations.[71] The concept of a rebuttable presumption of fault is alien to schemes of absolute liability. In the absence of a national space law or licensing provision, a right of indemnity from commercial operators to the government does not exist. This distinction is important when governments determine their policy towards sub-orbital flight.[72]

Liability for outer space activities is therefore unique, because it is the only known system of absolute liability that is unlimited in amount and, to a certain extent, also in time.[73] In contrast, liability for aviation damage, as with other transport sectors including oil and even nuclear pollution, is a strict but limited operator liability.[74] These conventions provide predictable systems of liability for aviation and outer space, but allocate different responsibilities and liabilities on governments. Making a choice as to which liability scheme should apply – or introducing a dedicated system – requires the stakeholders to come together.

---

[71] On the victim-oriented approach, see L.J. Smith and A. Kerrest, *supra* note 16.

[72] Under the French Outer Space Law, the government assumes liability over the second tier of damage as from 60 million euro, as an incentive to commercial operations in France.

[73] L.J. Smith and A. Kerrest, *supra* note 16. Claims for compensation under the Liability Convention are subject to a one-year period from the date of the event, or the identification of the launching State; see Art. X. This does not, however, exclude claims from being raised before domestic courts and may not therefore impact on the timeframe for international State to State settlement of claims; see Art. XI, para. 2 Liability Convention.

[74] Sector-specific convention law has led to the creation of various strict and/or limited liability systems that provide for compensation on a non-fault basis, where damage is inherent to the system's own dangers. See L.J. Smith and A. Kerrest, *supra* note 16, at 124–126, especially §§ 97, 102. For the types of risks and structure as well as rationale of all the leading risk-based liability systems, see G. Brüggemeier, *Common Principles of Tort Law: a Restatement* (BIICL 2004), 267–274.

8. LEGAL TOOLS AND LEGAL CULTURE: INFORMED CONSENT AND EXCLUSION OF RISK

A. Contractual Disclaimers and the Law

Broad exclusions of liability and cross-waivers are well known among the industrials in the space market at the level of space project development and management.[75] Despite the use of contractual and statutory disclaimers for the sub-orbital markets in the USA and Curaçao, such disclaimers do not yet represent a foregone, common legal culture beyond these jurisdictions. Whether liability exclusions are tolerated by the law in respect of other hazardous operations, for example through the doctrine of *volenti non fit iniuria* (voluntary assumption of risk) as well as similarly phrased contractual disclaimers, is a recurring subject in the history of pioneering.

The validation of space flight participants' 'informed consent', a statutory equivalent to *volenti non fit iniuria*, has been made possible by State legislation in the USA.[76] In the context of experimental spaceflight the dangers are obvious, are understood and have been fully considered and accepted.

The doctrine of informed consent does not yet exist on a statutory basis in Europe for participation in dangerous activities. Informed consent grew out of the notion of defence to the crime of assault, thereafter the tort of assault and battery, and was thereafter applied symbolically to establish the limits of consent to invasive bodily 'injury' resulting from medical treatment. Other areas of law such as road traffic and the growing sector of extreme sport have produced solutions that either impose duties by criminal or administrative statutes (for example the duty to use a car passenger safety belt) or leave operators to take out liability insurance as an initial protective shield. Informed consent has also developed further into other fields of law, where breach of organizational duties such as financial or fiduciary duties of institutions

---

[75] L.J. Smith, 'The Influence of Public International Law on Space Law Contracts', in L.J. Smith and I. Baumann (eds), *Contracting for Space, Contract Practice in the European Space Sector* (Ashgate 2011).

[76] The Commercial Space Launch Act requires the operator to obtain the passenger's consent prior to flight; see § 50905(b)(5) 51 USC. On this and the position of insurance coverage, see F. Lyall and P. Larsen, *Space Law, A Treatise* (Ashgate 2011), at 493–494; see further, F. von der Dunk and F. Tronchetti, *supra* note 15, at 691.

towards clients and investors, event organization and supervision, including participation in dangerous sports, have been tested before courts of law. This has allowed a review of the common contractual disclaimer practice in cases of express or implied consent. Until the concept of informed consent is introduced on a statutory basis, courts will always be charged with clarifying its exact scope. Its limits, notably in cases of involuntary death, even in the absence of negligence, are not definitively settled and remain open to interpretation by the courts over time.

B. UK Case Study

The UK's interest in the aerospace sector makes it an appropriate forum for a case study on the law of informed consent. Under English law, legal duties of care may arise under the common law either by statute or by contract. Various statutes regulate these duties, for example in the context of property law, by imposing owner-occupier/operator duties in relation to buildings and industrial operations. These generally provide for strict liability for the owner-occupier,[77] and include complex provisions governing standards for health and safety procedures within organizations.[78]

The law of contracts, particularly the Unfair Contract Terms Clauses Act 1977 (UCTA), amended frequently over time as a result of dedicated EU-based consumer contract legislation, regulates the scope of contractual terms in regard to limiting or excluding statutory and contractual duties of care. Over the years courts have repeatedly examined the validity, and with this the limits, to so-called exclusion clauses for liability for personal death and injury under the above statute.

Clauses excluding death or personal injury in the event of negligence are invalid under UK law, as well as in the other Member States of the European Union.[79] Section 2(2) UCTA provides a further level of control, in which the reasonableness of a contract clause can be reviewed by the courts in relation to the express or implied terms of the contract in question. Schedule 2 of the statute provides a list of general tests of reasonableness. Significantly, the UCTA remains applicable, irrespective of whether exclusion or limitation clauses form part of an individually

---

[77] Occupiers Liability Act 1984, c.3, http://www.legislation.gov.uk/ukpga/1984/3 (accessed 14 October 2017).

[78] *Ibid.*

[79] Unfair Contract Terms Act (UCTA) 1977, c.50, http://www.legislation.gov.uk/ukpga/1977/50 (accessed 14 October 2017).

agreed or standard-form 'non-negotiable' contract.[80] The 1977 statute applies to all contracting partners, irrespective of whether they are commercial or business partners or conform to the European notion of 'consumer'. The aerospace sector is not excluded under the UCTA schedules from its scope, although there are restrictions relating to contracts of insurance. At first sight, and subject to the form of licensing of sub-orbital operations, whether under prerogative (sovereign) licence, defence or military authorization, the statute would appear to cover the space flight participant on a sub-orbital flight.

It is moot whether a sub-orbital flight passenger licence that included a disclaimer for death or personal injury on the basis of informed consent in the documentation ('contract'), could be struck down by English courts as being against public policy (or in Continental European terms, 'immoral') under the common law. This is, however, conceivable on the basis of the common law alone. This question would have to be examined in the light of public policy and the exact statutory context under which space flights are licensed. Sub-orbital licences can be granted by the UK government minister in statutory instrument form under Section 3 of the Outer Space Act. Under section 3(3) Outer Space Act: 'The Secretary of State may by order except other persons or activities from the requirement of a licence if he is satisfied that the requirement is not necessary to secure compliance with international obligations of the United Kingdom.'

The status of volunteer, whether in pharmaceutical trials or trials involving experimental space flight participants, calls for continued application of the law. This calls for the recognition of common law duties of care, alongside such duties as may be imposed in the course of licensing of the particular operation. These involve adherence to specified standards and, in the event of damage, some form of liability insurance. In the absence of a policy consideration to restrict common law duties of care, there is at least a rebuttable presumption that the traditional rules of duty of care would still apply.[81]

---

[80] Scottish Law Commission, 'Unfair Terms in Consumer Contracts: Advice to the Department for Business, Innovation and Skills Law Commission' https://www.scotlawcom.gov.uk/law-reform/law-reform-projects/completed-projects/unfair-terms-in-consumer-contracts/ (accessed 14 Oct 2017).

[81] On the concept of neighbour principle and duty of care in the law of negligence, see Lord Reid in *Dorset Yacht Co. v. Home Office* [1970] AC 1004 at 1027: *Donoghue v. Stevenson* may be regarded as a milestone, and the well-known passage in Lord Atkin's speech ('and who in law is my neighbour') should, I think, be regarded as a statement of principle. It is not to be treated as

Foolproof exclusions of liability, particularly where sub-orbital flight is under development, would not appear at one with general policy considerations relating to aviation and flight safety at this stage of development in sub-orbital flight technology. The history of technological progress and the law has been one of responding to risk in categories. This response was first undertaken by domestic courts that introduced techniques of imputed fault and reversal of burden of proof, notably where the costs of protection for injured parties were lower than the probable loss, so that the balance of interests fell in favour of protecting the weaker individual. This trend in jurisprudence took place at the same time in the development of technology across the world's leading jurisdictions. Automation became a category of inevitable risk recognized by the law, for which the owner-operator would bear liability. Since then the transition to systems of strict liability for owners and operators of dangerous systems, including products, has gone full circle.[82]

One additional aspect of English law sheds light on a point that may be crucial to the common law here. The English law of consideration (an essential basis for any agreement binding contractors, here in relation to services) requires, in the absence of statutory provisions, an appropriate exchange or consideration between the parties to the contract; this is generally seen as a price or payment for sub-orbital travel in return for the 'carriage'. It is conceivable that this doctrine might even be relied on to impose a requirement that SFPs be covered by compulsory liability insurance coverage, in consideration of the commitment to risk they undertake. The contractual consideration, the formal exchange of benefit and detriment generally required by the law of contract in England, could, in the absence of a statutory provision, be interpreted into the arrangement by which an individual commits or exchanges his/her safety for the greater cause of advancing technical progress. Compulsory insurance coverage would provide the necessary consideration to equal the risk that comes, or the real detriment that the SFP stands to suffer, merely through participating.

---

if it were a statutory definition. It will require qualification in new circumstances. But I think the time has come when we can and should say that it ought to apply unless there is some justification or valid explanation for its exclusion.

[82] The leading case related to car accidents, *MacPherson v. Buick Motor Co.* [1916] 217 N.Y. 382, 111 N.E. 1050; see further, the l'arrêt Jand'heur judgment by the French Supreme Civil court (Cass., Ch. Réunies, 13 février 1930), in which the court first structured the system of strict liability for dangerous property. These were followed by the introduction of statutory liability for dangerous vehicles.

'The categories of negligence are never closed.'[83] Recent case law in the UK confirms that involvement in dangerous activities, even in the legally distinct sector of the armed forces, may give rise to duties of care, breach of which can lead to civil liability.[84] With the growth of extreme sport and the offer of rewards in return for specific risks, this is a growing area of law in many jurisdictions, independent of sub-orbital flight.[85]

Finally, discussions about space flight participants have brought forth both support for and opposition to an extension, by analogy, of the principles of the Rescue and Return Agreement,[86] developed with the protection of astronauts in mind. Astronauts are seen to have contributed with their exposure to risk to the progress of humanity. Discussion on this point is divided both ways, so that a decision in principle needs to be made on a separate basis. It has, however, been pointed out that, given the geographical proximity of those States involved in the development of sub-orbital planes, notably Singapore, the United Arab Emirates and Saudi Arabia, reliance on the Rescue Agreement would be welcome, as issues of stranding and return of even SFPs could arise given the proximate geographical areas of these States and their borders.[87]

9. THE CUCKOO'S EGG: CREATING COMMON DENOMINATORS FOR LIABILITY

The aerospace sector has come a long way from initial uncertainties about the rules governing the very first operations for those accepting

---

[83] Lord McMillan in *Donoghue v. Stevenson* [1932] UKHL 100.

[84] See *Ministry of Defence v. Radcliffe* [2009] EWCA 635: liability based on instructions from a superior in the face of an obviously dangerous, but instructed activity.

[85] Extreme sport can be a lucrative market and sponsorship has provided risk funding to enable financial reward; see for example the sponsorship by Red Bull of Felix Baumgartner's Stratos jump. The UK Office of Fair Trading is involved in monitoring this sort of activity. Its powers derive from the Office of Fair Trading Act 1973, and have expanded according to legislative demand. For details, consult its 2014 report, available at https://www.gov.uk/government/uploads/system/uploads/attachment_data/file/322821/41293_HC_27_OFT_proof_2006.pdf (accessed 14 October 2017).

[86] Agreement on the Rescue of Astronauts, the Return of Astronauts and the Return of Objects Launched into Outer Space, 22 April 1968, 672 UNTS 119 (ARRA).

[87] See Chapter 3, this volume.

probable risks. Sub-orbital space flight is turning the clock back. The US State statutes governing sub-orbital operations apply the same requirements of exclusion of liability and assumption of risk as did the era of early, unregulated exploration or transport. It is paradoxical that earlier twentieth-century jurisprudence, which favoured the development of a stricter form of liability in the early days of automation, is tabled again for debate.[88] The absence of international consensus on how best to regulate these activities has been seen to constitute prevarication.[89] Sub-orbital flight is the cuckoo's egg of technological progress, and an exception within what is otherwise a homogenous, common international liability system governing safe and economic international aviation services, as well as outer space activities.

The function of public law is regulatory, triggering accountability for manufacturers and operators to standards, compliance and system safety. The separate, but equally relevant function of the private law of liability is to assess whether, in the face of the particular risk, any compensatory or protective legal norm should be put in place to cover the situations of predictable loss. Risks in society are manifold, but not all are linked to technology.[90] Many risks lie where they fall.[91]

Governments interested in this sector can develop their authorization systems on the basis of existing national space laws, while contributing to joint efforts to develop a common understanding of the requirements and risks of sub-orbital systems. Thereafter ICAO, through the competences of its Navigation Commission, can be left to review current operations and deliberate whether sub-orbitals should be preferably included within the provisions of the Chicago Convention.

As noted in his London newspaper advertisement, Shackleton, when preparing for his Arctic trips, advertised for supporters and 'ice-travel' passengers. His messages made it very clear that his fellow adventurers were more than likely to suffer tremendously as a result of the expedition, but not without honour or in vain: 'Men wanted for hazardous

---

88 L'arrêt Jand'heur (Cass., Ch. Réunies, 13 février 1930) n.68.

89 V. Nase, 'Delimitation and the sub-orbital passenger: time to end prevarication' (2012) 77 *Journal of Air Law and Commerce*, 747–767.

90 See the discussion of Lord Denning in the Court of Appeal and Lord Wilberforce on appeal to the House of Lords in *Whitehouse v. Jordan* [1980] UKHL 12 (17 December 1980) on the legal appreciation of risk and negligence ('Being born is a dangerous thing') when considering the appropriate test in determining when a breach of legal duty could arise, *in casu*, in the context of professional negligence.

91 O.W. Holmes, *The Common Law* (first published 1881, Dover 1991).

journey. Low wages, bitter cold, long months of complete darkness, constant danger. Safe return doubtful. Honour and recognition in case of success'.[92]

It is unclear whether the English doctrine of consideration would allow costly sub-orbital flights to be licensed in exchange for honour alone. The question can be solved pragmatically through a compulsory liability insurance requirement or by restricting the SFPs during trial phases to members of the armed forces who are covered by their forces' insurance schemes. This approach is known to other agencies such as NASA and the European Astronaut Corps.

The inevitable risks associated with this form of flight may otherwise see the doctrine of informed consent, in the absence of another risk-allocation model for SFPs, find itself subjected to the rigours of the common law, a century on from the original risks of automation that took place on the highways that opened up the path to strict liability.

There is merit in addressing the concept of informed consent by statute in advance, and imposing SFP liability insurance requirements.

---

[92] Shackleton (1874–1922) led a variety of heroic trips to the Arctic. It is not clear when and where this advertisement was placed in the London press, but it is attributed directly to him.

3. The scope of international obligations to extend rescue assistance to 'astronauts' and 'personnel' under the Outer Space Treaty and the Return and Rescue Agreement

Steven Wood

1. INTRODUCTION

To determine whether States must engage in search and rescue efforts for space tourists travelling in outer space, this chapter first reviews the historical contexts of the Outer Space Treaty (OST) and the Astronaut Return and Rescue Agreement (ARRA) relevant to the humanitarian purposes underlying their applicability to the rescue of 'astronauts' and 'personnel of a spacecraft' in outer space. Next, this chapter reviews applicable customary international law and the Vienna Convention on the Law of Treaties (VCLT) with respect to interpretation of Article V of the OST and Article 3 of the ARRA.

This chapter continues by reviewing the humanitarian purposes of and providing a linguistically focused interpretation of OST Article V and ARRA Article 3, specifically elucidating the ambiguity, confusion and absurd consequences resulting from these provisions. In conclusion, interpretations of OST Article V and ARRA Article 3 consistent with the humanitarian purposes of these treaties are advanced. The question of whether sub-orbital space tourists comprise 'space tourists' will remain beyond the scope of this chapter.

2. HISTORICAL CONTEXTS OF THE OUTER SPACE TREATY AND RETURN AND RESCUE AGREEMENT

The UN Ad Hoc Committee on the Peaceful Uses of Outer Space (Ad Hoc COPUOS) and its Legal Subcommittee were created by the United Nations General Assembly (UNGA) in 1958 and 1959 respectively to investigate and analyse potential legal problems in the new field of space exploration.[1] After Ad Hoc COPUOS delivered its 1959 report, the UNGA created the permanent Committee on the Peaceful Uses of Outer Space (COPUOS) to continue the work started by Ad Hoc COPUOS.[2]

The UNGA requested COPUOS 'to continue urgently its work on … assistance to and return of astronauts'[3] in December 1962, following which the USA, Soviet Union and United Arab Republic each submitted draft documents of fundamental principles for conduct and behaviour in outer space to the Legal Subcommittee.[4]

Having consensus among the Legal Subcommittee delegations, the UNGA unanimously adopted Resolution 1962 (XVIII) Declaration of Legal Principles Governing the Activities of States in the Exploration and Use of Outer Space (hereinafter Principles), including a provision concerning rescue of astronauts.[5] After completing the Principles, during its

---

1 Question of the Peaceful Use of Outer Space, GA Res 1348 (XIII), UN GAOR, 13th Sess, UN Doc A/RES/1348 (1958); P.G. Dembling and D.M. Arons, 'The Treaty on Rescue and Return of Astronauts and Space Objects' (1968) 9 WMLR 630, 632.

2 International Co-operation in the Peaceful Uses of Outer Space, GA Res 1472 (XIV), UN GAOR, 14th Sess, UN Doc A/RES/1472, (1959); Dembling and Arons, *supra* note 1, at 633.

3 International Co-operation in the Peaceful Uses of Outer Space, GA Res 1802 (XVII), UN GAOR, 17th Sess, UN Doc A/RES/1802 (1962); Dembling and Arons, *supra* note 1, at 635.

4 Report of the Legal Subcommittee to the Committee on the Peaceful Uses of Outer Space, UN GAOR, 2nd Sess, Annex I G and Annex I E, UN Doc A/AC105/12 (1963); Dembling and Arons, *supra* note 1, at 635.

5 Declaration of Legal Principles Governing the Activities of States in the Exploration and Use of Outer Space, GA Res 1962 (XVIII), UN GAOR, 18th Sess, UN Doc A/RES/1962 (1963) (hereinafter GA Res 1962). Principle 9 states that 'States shall regard astronauts as envoys of mankind in outer space, and shall render to them all possible assistance in the event of accident, distress, or emergency'; see also International Co-operation in the Peaceful Uses of Outer Space, GA Res 1721 (XVI), UN GAOR, 16th Sess, 6, UN Doc A/RES/1721 (1961); Dembling and Arons, *supra* note 1, at 635.

1964 session the Legal Subcommittee provisionally agreed to the proposed draft text of a treaty, including the preamble and three articles, for the rescue and return of spacecraft personnel.[6]

Due to the urgency expressed by the United States, the Soviet Union and many other nations, the Legal Subcommittee held two sessions in 1966, turning its attention towards establishing a treaty to set forth general legal duties and norms governing exploration and use of outer space.[7] In 1967 the UNGA unanimously approved the foundational Treaty on Principles Governing the Activities of States in the Exploration and Use of Outer Space, Including the Moon, and Other Celestial Bodies (hereinafter the Outer Space Treaty or OST),[8] which entered into force that same year.

The deaths of American and Soviet astronauts in space vehicle accidents, in January and March 1967 respectively, underscored the pressing need to create an agreement regarding rescue of astronauts.[9] In November 1967 the UNGA again appealed to COPUOS to urgently continue to work to establish 'an agreement on assistance to and return of astronauts'.[10] Shortly thereafter the UNGA unanimously adopted the Agreement on the Rescue of Astronauts, the Return of Astronauts, and the Return of Objects Launched into Outer Space (hereinafter Agreement

[6] WG 1/23, included in Report of the Legal Subcommittee to the Committee on the Peaceful Uses of Outer Space, UN GAOR, 3rd Sess, Annex III, UN Doc A/AC105/21/III (1964); Report of the Legal Subcommittee to the Committee on the Peaceful Uses of Outer Space, UN GAOR, 3rd Sess, 9, UN Doc A/5785 (1964); Report of the Committee on the Peaceful Uses of Outer Space, UN GAOR, 20th Sess, UN Doc A/6042 (1965); Dembling and Arons, *supra* note 1, at 633, 635–636.

[7] Report of the Legal Subcommittee to the Committee on the Peaceful Uses of Outer Space, UN GAOR, 5th Sess, 13, UN Doc A/6431 (1966); Dembling and Arons, *supra* note 1, at 633, 636–637.

[8] Treaty on Principles Governing the Activities of States in the Exploration and Use of Outer Space, Including the Moon and Other Celestial Bodies (entered into force 10 October 1967) Art. 6, 18 UST 2410, 610 UNTS 205 (hereinafter OST).

[9] R. Melina, 'The Fallen Heroes of Human Spaceflight', *Space.com*, 2011, http://www.space.com/11353-human-spaceflight-deaths-50-years-space-missions.html, accessed 16 March 2014; Dembling and Arons, *supra* note 1, at 638.

[10] Report of the Committee on the Peaceful Uses of Outer Space, GA Res 2260 (XXII), UN GAOR, 22nd Sess, UN Doc A/RES/2260, para. 9 (1967); Dembling and Arons, *supra* note 1, at 640.

on the Rescue and Return of Astronauts or ARRA) in Resolution 2345 (XXII), which subsequently entered into force in December 1968.[11]

3. INTERPRETATION OF TREATIES UNDER THE VIENNA CONVENTION ON THE LAW OF TREATIES

Interpreting the terminology and scope of the Outer Space Treaty[12] and the Agreement on the Rescue and Return of Astronauts[13] must be done in accordance with the Vienna Convention on the Law of Treaties[14] (hereinafter VCLT).[15]

According to the International Court of Justice (ICJ), the VCLT rules represent 'a codification of customary practice and are binding as an expression of customary international law' with respect to treaty interpretation.[16] The VCLT applies to interpreting treaties concluded prior to

---

[11] Agreement on the Rescue of Astronauts, the Return of Astronauts, and the Return of Objects Launched into Outer Space (entered into force 3 December 1968) 19 UST 7570, 672 UNTS 119 (hereinafter ARRA); Agreement on the Rescue of Astronauts, the Return of Astronauts and the Return of Objects Launched into Outer Space, GA Res 2345 (XXII), UN GAOR, 22nd Sess, UN Doc A/RES/2345 (1967); United Nations Treaties and Principles on Outer Space: Travaux Préparatoires (United Nations Office for Outer Space Affairs), http://www.unoosa.org/oosa/en/ourwork/spacelaw/treaties/travaux-preparatoires/rescue-agreement.html, accessed 19 August 2017.

[12] OST, *supra* note 8.

[13] ARRA, *supra* note 11.

[14] Vienna Convention on the Law of Treaties (entered into force 27 January 1980) Art. 31, 1155 UNTS 331, 8 ILM 679 (hereinafter VCLT).

[15] *Ibid.* Art. 31(1); R.K. Gardiner, *International Law* (Longman 2003) 67; D.S. Jonas and T.N. Saunders, 'The Object and Purpose of a Treaty: Three Interpretive Methods' (2010) 43 VJTL 565; M.J. Sundahl, 'The Duty to Rescue Space Tourists and Return Private Spacecraft' (2009) 35 JSL 163, 174.

[16] Mark J. Sundahl, *The Cape Town Convention: Its Application to Space Assets and Relation to the Law of Outer Space* (Martinus Nijhoff 2013) 118; R.K. Gardiner, *Treaty Interpretation* (Oxford International Law Library 2008) 14–16, 69 (hereinafter Gardiner 2008); *Legal Consequences of the Construction of a Wall in the Occupied Palestinian Territory* (Advisory Opinion) [2004] ICJ 136, 174, para. 94, indicating that VCLT Article 31 represents customary international law on treaty interpretation; *Armed Activities on the Territory of the Congo (Democratic Republic of the Congo v Rwanda)* (Judgement), [2006] ICJ 6, 51–52, para. 125, as States Parties to the VCLT, 'in the present case the rules

the VCLT, such as the OST and ARRA, to the extent it embodies customary international law.[17]

Under the VCLT, interpreting treaty terminology starts with the fundamental rule to interpret the 'ordinary meaning' of terms within their contexts and taking into account the objectives and purposes of the treaty.[18] The 'ordinary meaning' of a term should be the conventional meaning given to the term at the time when the treaty was concluded.[19]

4. HUMANITARIAN UNDERPINNINGS OF THE RESCUE PROVISIONS UNDER THE OST AND ARRA

The important humanitarian underpinnings of the OST and ARRA can be properly seen within the development of their appropriate historical contexts, described above.[20] Further, the preamble of the ARRA itself makes clear its humanitarian objectives by stating that it was 'prompted

contained in the Vienna Convention are not applicable, save in so far as they are declaratory of customary international law'.

[17] *Case Concerning Kasikili/Sedudu Island (Botswana/Namibia)* (Judgement), [1999] ICJ 1045, 1059, para. 18, stating that 'The Court itself has already had occasion in the past to hold that customary international law found expression in Article 31 of the Vienna Convention'. See *Territorial Dispute (Libyan Arab Jamahiriya/Chad)*, (Judgment) ICJ Reports 1994, p. 21, para. 41; *Oil Platforms (Islamic Republic of Iran v United States of America), Preliminary Objections*, (Judgment) ICJ Reports 1996 (II), p. 812, para. 23). Article 4 of the Convention, which provides that it 'applies only to treaties which are concluded by States after the entry into force of the … Convention with regard to such States' does not, therefore, prevent the Court from interpreting the 1890 Treaty in accordance with the rules reflected in Article 31 of the Convention; A. Aust, *Handbook of International Law* (2nd edn, Cambridge University Press 2010) 50; Gardiner 2003, *supra* note 15, at 68; Gardiner 2008, *supra* note 16, at 14; M.E. Villiger, *Commentary on the 1969 Vienna Convention on the Law of Treaties* (BRILL 2009) 110–112.

[18] Art. 31(1) VCLT.

[19] A. D'Amato, 'International Law, Intertemporal Problems' in R. Bernhardt (ed.), *Encyclopedia of Public International Law* (Oxford Public International Law 1992) 1234–1236; see also Art. 31(1) VCLT; Gardiner 2003, *supra* note 15, at 82–86.

[20] R.C. Beckman, '1968 Rescue Agreement – An Overview' in *Proceedings United Nations/Republic of Korea Workshop on Space Law, United Nations Treaties on Outer Space: Actions at the National Level* (United Nations Office for Outer Space Affairs 2004) 370, 370–373.

by sentiments of humanity',[21] having as its 'main principle and purpose ... the humanitarian desire to protect the life of those aboard a spacecraft'.[22]

Moreover, the ARRA further evidences its humanitarian nature through the decision not to condition the obligations to rescue or return 'personnel of a spacecraft' upon their State(s) of national origin,[23] evidencing the universal nature of these obligations and the intention to ensure the safety and safe return of all spacecraft personnel. Furthermore, the VCLT affirms that adherence by States Parties to treaties of a 'humanitarian character' cannot depend upon mutual adherence by other States Parties.[24]

Further, recognition of the fact that the ARRA was concluded so rapidly upon the heels of the OST reveals the great significance the States Parties considered this agreement to hold. Indeed, it was widely understood that practice of the principles embodied in the ARRA would be essential to make the exploration of outer space as safe as humanly possible.[25] Considering the historical contexts of the rescue provisions included in the OST and the ARRA, the many States which have ratified these treaties, the consensus of COPUOS and its Legal Subcommittee, and the unanimous adoption of these treaties by the UNGA, the great importance of these provisions is made evident.[26]

5. WHO QUALIFIES AS AN ASTRONAUT UNDER OST ARTICLE V?

According to the VCLT, the term 'astronaut' should be interpreted under the OST to have its 'ordinary meaning' at the time the OST was concluded in 1967.[27] In other words, for the purposes of OST interpretation and enforcement, modern definitions of the term astronaut, which

---

21 Preamble 4th recital ARRA.

22 Sundahl, *supra* note 15, at 187.

23 Art. 3 and Art. 4 ARRA.

24 Preamble 4th recital ARRA; Art. 60(5) VCLT.

25 Beckman, *supra* note 20 at 370–373; Dembling and Arons, *supra* note 1; Sundahl, *supra* note 15, at 167.

26 Dembling and Arons, *supra* note 1.

27 Art. 31(1) VCLT; D'Amato, *supra* note 19; Gardiner 2003, *supra* note 15, at 82–86.

often rely on training and altitude,[28] should not be taken into consideration or given weight. Leading up to the 1967 conclusion of the OST, 'astronaut' had systematically crystallized into a consistently defined term including everyone on board a spacecraft – spacecraft crew as well as private commercial passengers, that is to say space tourists.[29]

Modelled on the word 'aeronaut', which was first used in 1784 to refer to balloonists,[30] 'astronaut' derives from the root words 'astro', meaning star, and 'nautes', meaning sailor, and as combined together to form 'astronaut', meaning a sailor of the stars;[31] literally a human who travels in space.[32] The word 'astronaut' was first used as the name of a fictional spaceship in the 1880 science fiction novel *Across the Zodiac* by Percy Greg.[33] 'Astronaut' was first used in non-fiction to refer to a person travelling in space in 1929 by the British Astronomical Association.[34]

When the OST was concluded, NASA defined the term 'astronaut' primarily as 'a person who rides in a space vehicle', and secondarily as 'the test pilots selected to participate in Project Mercury, Project Gemini, Project Apollo or any other program for manned space flight'.[35] The second definition supplied in the NASA dictionary is too limited in scope to comprise the ordinary meaning of the term 'astronaut' in accordance with VCLT Article 31 because it covers only test pilots of specific early NASA manned spaceflight missions.

---

[28] F. Lyall and P.B. Larsen, *Space Law: A Treatise* (Ashgate Publishing 2009) 131; Y.A. Falait, 'Space Tourism: A Synopsis of its Legal Challenges' (2012) 1 ILJ 120, 123–124.

[29] B. Cheng, *Studies in International Space Law* (Clarendon Press Oxford 1997) 507; W.D. Reed and R.W. Norris, 'Military Use of the Space Shuttle' (1980) 13 ALR 665, 686–87; Sundahl, *supra* note 15, at 183.

[30] *Online Etymology Dictionary* (Douglas Harper ed. 2001) 'Aeronaut', http://www.etymonline.com/index.php?allowed_in_frame=0&search=aeronautics, accessed 16 March 2014; W. Grimshaw, *An Etymology Dictionary* (Lydia R. Bailey 1821) 'Aeronaut'.

[31] P. Dickson, *A Dictionary of the Space Age* (JHU Press 2010) 26–27; *Online Etymology Dictionary* (Douglas Harper ed. 2001) 'Astronaut', http://www.etymonline.com/index.php?term=astronaut, accessed 16 March 2014.

[32] Dickson, *supra* note 31, at 27.

[33] *Ibid.*; *Online Etymology Dictionary*, *supra* note 31, 'Astronaut'.

[34] Dickson, *supra* note 31, at 26–27, citing the Journal of the British Astronomical Association (1929) 331, referencing 'the first obstacle encountered by the would-be "Astronaut"'.

[35] *Dictionary of Technical Terms for Aerospace Use* (W.H. Allen ed., 1965) 21. This dictionary was used by NASA when the OST was concluded; *Dictionary of Technical Terms for Aerospace Use* (web edition, D.R. Glover, Jr ed. 2001), http://er.jsc.nasa.gov/seh/a.html, accessed 16 March 2014.

The 1967 Webster's Third New International Dictionary, which defines 'astronaut' as 'a traveler in interplanetary space', provides further support to interpret 'astronaut' under the OST as covering all persons on board a spacecraft.[36] Additionally, the 1972 Oxford English Dictionary defines 'astronaut' as 'one who travels in space, i.e. beyond the earth's atmosphere'.[37] This plurality of sources substantiates a coherent ordinary meaning of the term 'astronaut' as including all persons on board a spacecraft during the almost 40 years leading up to the OST and for at least the five years following its entry into force.[38]

On the other hand, if the term 'astronaut' is instead interpreted strictly to mean persons such as pilots, scientists, payloads and mission specialists or other crew, OST Article V search and rescue obligations would fail to apply to private commercial passengers, for example space tourists aboard a spacecraft or visiting an on-orbit hotel. However, this would result in manifestly absurd consequences, which are discussed in greater detail below, and which allow a definitive reinterpretation to challenge such a restrictive interpretation of 'astronaut'.

6. ARTICLE V OST: WHAT ASSISTANCE IS REQUIRED IN OUTER SPACE?

Article V OST stipulates that 'States Parties to the Treaty shall regard astronauts as envoys of mankind in outer space and shall render to them all possible assistance in the event of accident, distress, or emergency landing on the territory of another State Party or on the high seas'. Given the placement of the commas in this phrase, this list comprises separate distinct elements and can ultimately be read as '... in the event of accident, [or in the event of] distress, or [in the event of] emergency landing on the territory of another State Party ...'.

Further, additional support for this argument is found in the fact that, although a 'landing' may occur 'on' a territory, 'on' is not the preferred

---

36 *Webster's Third New International Dictionary of the English Language* (P.B. Gove ed. 1967).

37 *A Supplement to the Oxford English Dictionary* (R.W. Burchfield ed. 1972).

38 The ordinary meaning of 'astronaut' at the time the OST was concluded includes private passengers, and therefore 'supplementary means of interpretation can only be applied to confirm the inclusion of passengers – but not challenge it.' See Sundahl, *supra* note 15, at 183; see also Cheng, *supra* note 29; Reed and Norris, *supra* note 29.

prepositional phrase to link 'accident' and 'distress' to the noun 'territory'. Because the territory of a State is a three- dimensional volume, 'in' or 'within' would be much better prepositions for this purpose, as in 'the event of accident [or] distress [... with/in] the territory ...'. Consequently the rules of grammar demonstrate that 'accident' and 'distress' are events unrelated to the recited 'territory of another State Party' and that Article V confers an obligation upon States Parties to render 'all possible assistance' to 'astronauts' even when such 'accident' or 'distress' occurs in outer space.

Article V OST language is interpreted by doctrine as encompassing adequate breadth to include Earth-to-space search and rescue efforts for astronauts suffering accident or distress in space.[39] Accordingly, a request for Earth-to-space rescue of an astronaut crew by their State(s) of origin would be warranted if such assistance were necessary.[40] Further, Article V OST obliges States Parties to render the required assistance to astronauts, regardless of whether their State(s) of national origin are also parties to the OST.[41]

Furthermore, doctrine considers OST Article V to be customary international law[42] binding upon all States.[43] Consequently, in the event

---

[39] Dembling and Arons, *supra* note 1, at 649–650; R.C. Hall, 'Rescue and Return of Astronauts on Earth and in Outer Space' (1969) 63 AJIL 197, 205; Lyall and Larsen, *supra* note 28, at 140.

[40] Hall, *supra* note 39, at 205.

[41] Art. V, para. 1 OST states that 'States Parties ... shall render to [astronauts] all possible assistance in the event of accident, distress, or emergency landing' without placing any restriction on the national origin(s) of the astronauts.

[42] GA Res 1962; B. Cheng, 'United Nations Resolutions on Outer Space: "Instant" International Customary Law?' (1965) 5 IndianJInt'lL 36; R.J. Lee, 'Reconciling International Space Law with the Commercial Realities of the Twenty-first Century,' (2000) Singapore JInt'l&CompL 194, 203; M. Shaw, *International Law* (Cambridge University Press 2008) 123; V. Vereshchetin and G. Danilenko, 'Custom as A Source of International Law of Outer Space' (1985) 13 JSL 22.

[43] The rescue principle may have attained *erga omnes* character; see *Case concerning the Barcelona Traction, Light and Power Company, Ltd (Belg v Spain)* (Judgment), [1970] ICJ 32, para. 33; C. Cepelka and J.H.C. Gilmour, 'The Application of General International Law in Outer Space' (1970) 36 JALC 30, 48; I.A. Csabafi, *The Concept of State Jurisdiction in International Space Law* (Martinus Nijhoff 1971) 47; R. Jakhu, 'Legal Issues Relating to the Global Public Interest in Outer Space' (2006) 32 JSL 31, 38–39; M. Lachs, *The International Law of Outer Space* (Martinus Nijhoff Online 1964) 7–115, http://nijhoffonline.nl/book?id=er113_er113_001-115, accessed 16 March 2014.

of 'accident' or 'distress' in outer space, all States undertaking space activities have a fundamental duty to render 'all possible' assistance to 'astronauts', irrespective of their State of national origin.

Taking into account the humanitarian foundational basis of Article V OST and the plethora of aircraft precedents, one or both of the terms 'accident' and 'distress' should readily be interpreted to cover events including collision of spacecraft with other spacecraft or space objects, and also spacecraft malfunction, breakdown or other disablement.[44]

However, although OST Article V language may be interpreted by doctrine to encompass Earth-to-space search and rescue, the types of assistance that comprise 'possible' assistance remains a situation-dependent question, the resolution of which depends at least in part upon the discretion of the State from which such assistance is requested.[45]

7. WHO QUALIFIES AS PERSONNEL OF A SPACECRAFT UNDER ARRA ARTICLE 3?

The major difference between the rescue provisions of the OST and those of the ARRA is that ARRA Article 3 employs the term 'personnel of a spacecraft' rather than 'astronaut'.[46] Use by the ARRA of this different term means that, to fully establish the scope of search and rescue operations applicable in outer space, the ordinary meaning of the term 'personnel' must be interpreted as at the time when the ARRA was concluded.[47]

'Personnel' means 'persons employed in any work, enterprise, service, etc.' as defined in the 1968 edition of Webster's New World Dictionary.[48] This definition covers involvement of governmental representatives but does not require it, meaning that private commercial personnel are included.[49] However, although the term 'personnel' clearly includes spacecraft crew, covering the pilot and other technical staff aboard a spacecraft, the term 'astronaut' may be more amenable to inclusion of

[44] Dembling and Arons, *supra* note 1, at 646.

[45] Art. V OST; Dembling and Arons, *supra* note 1, at 649–650; Hall, *supra* note 39, at 205; Lyall and Larsen, *supra* note 28, at 140.

[46] Art. V OST; Art. 3 ARRA; Beckman, *supra* note 20, at 373; Dembling and Arons, *supra* note 1, at 643; Sundahl, *supra* note 15, at 178.

[47] Art. 31(1) VCLT; Sundahl, *supra* note 15, at 174–175.

[48] *Webster's New World Dictionary of the American Language*, (D.B. Guralink ed., 1968).

[49] Sundahl, *supra* note 15, at 185.

private passengers. This is because the requirement of employment under the conventional interpretation of 'personnel' seems too restrictive to include private paying passengers.[50]

Consequently, if a commercial spaceplane were to crash land just outside the territorial waters of a foreign country where OST Article V may be inapplicable, relying on the narrow interpretation of 'personnel' would require only the rescue of crew, permitting private passengers to be abandoned.[51] This results in a manifest absurdity and could not have been intended in view of the humanitarian purposes and objectives of the ARRA, as there is no logical reason to rescue only some of the people in danger if assistance can be given to all.[52]

The absurd consequences resulting from a narrow interpretation of 'personnel' supply the most powerful argument supporting a broader interpretation under the ARRA and VCLT.[53] Establishing these manifestly absurd consequences permits recourse to supplementary sources to support interpretation of an ordinary meaning inclusive of private passengers.[54]

The OST constitutes an excellent supplemental source to inform the interpretation of 'personnel' because it forms the basis on which the ARRA was built and because it was adopted in the same year as the ARRA. OST Article VIII stipulates that States of registration 'shall retain jurisdiction and control over [their space] object, and over any *personnel* thereof'.[55]

This provides exceptional support to the position that the ordinary meaning of the term 'personnel' included private passengers at the time the ARRA was concluded.[56] Many scholars have argued that interpreting

---

[50] B. Cheng, '"Space Objects", "Astronauts" and Related Expressions' in *Proceedings of the Thirty-Fourth Colloquium on the Law of the Outer Space* (American Institute of Aeronautics and Astronautics 1992) 17, 26, 165; S. Gorove, 'Interpreting Salient Provisions of the Agreement on the Rescue of Astronauts, the Return of Astronauts, and the Return of Objects Launched into Outer Space' in M.D. Schwartz (ed.), *Proceedings of the Eleventh Colloquium on the Law of the Outer Space* (Eleven Publishing 1969) 93; Sundahl, *supra* note 15, at 185.

[51] Sundahl, *supra* note 15, at 189.

[52] *Ibid.*; Art. 31 VCLT; Beckman, *supra* note 20, at 373; S. Freeland, 'Up, Up and … Back: The Emergence of Space Tourism and its Impact on the International Law of Outer Space' (2005) 6 CJIL 1, 10.

[53] Sundahl, *supra* note 15, at 175, 180, 189.

[54] Art. 32(b) VCLT.

[55] Art. VIII OST.

[56] *Ibid.*; Sundahl, *supra* note 15, at 186–187.

the ordinary meaning of 'personnel' in this context requires the inclusion of 'any and all people on board a spacecraft'[57] as clearly intended by the drafters.[58]

Additionally, since 'personnel' refers to all persons aboard a spacecraft under the OST, logical consistency requires that it must have the same meaning under the ARRA.[59] Further, the fact that the preamble of the ARRA affirms that the ARRA is meant to ensure the 'rendering of all possible assistance to astronauts [... and] the prompt and safe return of astronauts' also supports a broader definition for the term 'personnel'.[60]

Furthermore, a broader, more inclusive interpretation of 'personnel' is also supported by an evaluation of the historical contexts of the ARRA, including its humanitarian purpose as required under Article 31 of the VCLT.[61] Therefore, to achieve the correct humanitarian outcome intended by the drafters of the ARRA, interpreting the ordinary meaning of 'personnel' requires inclusion of all people on board a spacecraft.[62]

8. ARTICLE 3 ARRA: WHAT ASSISTANCE IS REQUIRED?

The Astronaut Rescue and Return Agreement Article 3 requires that:

> if information is received or it is discovered that the personnel of a spacecraft have alighted on the high seas *or in any other place not under the jurisdiction of any State*, those Contracting Parties which are in a position to do so shall, if necessary, extend assistance in search and rescue operations for such personnel to assure their speedy rescue.[63]

Return and Rescue Agreement Article 3 constitutes customary international law to which all States are bound.[64] Further, the ARRA does not condition the obligation to render assistance necessary to 'personnel of a

---

[57] Sundahl, *supra* note 15, at 186–187.

[58] K.-H. Bockstiegel (ed.), *Manned Space Flight: Legal Aspects in the Light of Scientific and Technical Development* (Carl Heymanns Verlag 1993), 168, 194; S. Hobe, 'Legal Aspects of Space Tourism', (2007) 86 NLR 439, 455.

[59] Sundahl, *supra* note 15, at 186–187.

[60] Preamble 1st and 2nd recitals ARRA.

[61] Art. 31 VCLT; Beckman, *supra* note 20, at 370–373.

[62] Sundahl, *supra* note 15, at 186–189.

[63] Art. 3 ARRA (emphasis added).

[64] See *supra* note 23.

spacecraft', which must be provided regardless of whether or not such personnel are nationals of a State Party to the ARRA.[65]

However, most commentators interpret the requirements of Article 3 as failing to provide any obligation to render assistance to personnel navigating outer space.[66] Among those who take this position, some interpret 'have alighted' as applying to the 'personnel'[67] or to 'the spacecraft'.[68] Further, some do not distinguish between the alighting of personnel and landing of spacecraft,[69] while others assert that Article 3 should be limited to search and rescue performed on Earth.[70]

Regardless of their respective individual positions regarding whether Article 3 applies to search and rescue of personnel in outer space, these commentators all concur that the second paragraph of OST Article V requires assistance to astronauts in outer space.[71] Most commentators also agree that Article 3 covers assistance to 'personnel of a spacecraft' who have alighted upon a celestial body such as the Moon. The phrase 'in any other place not under the jurisdiction of any State' unmistakably includes the territory of outer space.[72]

Ogunbanwo specifically states that 'Not under the jurisdiction of any State' shows that the obligation to provide assistance extends to outer space, the Moon and celestial bodies'.[73] However, to determine the full scope of search and rescue required by Article 3 ARRA, the precise meaning of the phrase 'the personnel of a spacecraft have alighted' needs to be evaluated.

---

65 ARRA.

66 Beckman, *supra* note 20, at 372–373; Dembling and Arons, *supra* note 1, at 649; Hall, *supra* note 39, at 205–206; Lyall and Larsen, *supra* note 28, at 140; O.O. Ogunbanwo, *International Law and Outer Space Activities* (Martinus Nijhoff 1975) 133; Sundahl, *supra* note 15, at 169, 194.

67 Ogunbanwo, *supra* note 66, at 133.

68 Sundahl, *supra* note 15, at 169.

69 Dembling and Arons, *supra* note 1, at 649; Hall, *supra* note 39, at 205–206; Lyall and Larsen, *supra* note 28, at 140.

70 Beckman, *supra* note 20, at 372–373.

71 *Ibid.*; Dembling and Arons, *supra* note 1, at 649; Hall, *supra* note 39, at 205–206; Lyall and Larsen, *supra* note 28, at 140; Ogunbanwo, *supra* note 66, at 133.

72 Beckman, *supra* note 20, at 372–373; Dembling and Arons, *supra* note 1, at 649; Hall, *supra* note 39, at 205–206; Lyall and Larsen, *supra* note 28, at 140; Ogunbanwo, *supra* note 66, at 133; Sundahl, *supra* note 15, at 169, 194.

73 Ogunbanwo, *supra* note 66, at 133.

A. Interpreting Use of 'Alighted' in Article 3 ARRA: Confusion Under the VCLT

By introducing the term 'alighted', the ARRA created more confusion rather than resolving the extant conceptual gaps in the OST.[74] Analysis of the ordinary meaning of 'alight' begins with its definition as an intransitive verb meaning:

1. To [spring down,] get down[,] or descend, as from on horseback or from a carriage[; to dismount].
2. To descend and settle[, lodge, rest, or stop]; as, a flying bird alights on a tree; snow alights on a roof.[75]

'Alight' was first used to describe the landing of hot air balloons, the first aeronautical vehicles, starting in 1786.[76] In accordance with the second definition,[77] most commentators interpret the Article 3 ARRA use of the term 'have alighted' as requiring the spacecraft to have made a landing.[78] However, to make the landing of a spacecraft a prerequisite to the duty to render assistance under Article 3 ARRA means that so long as the spacecraft has not landed, there is no duty to rescue 'personnel of a spacecraft' who are in orbit or navigating outer space and need assistance.[79]

An analysis of the Article 3 linguistics shows that a correct interpretation of the term 'alighted' is that 'the personnel … have alighted',

---

[74] Sundahl, *supra* note 15, at 169, 194.

[75] These definitions for 'alight' were published in the 1828 and 1913 editions of Webster's Dictionary, showing differences between the two (i.e. additions to the 1828 edition) in brackets. See *American Dictionary of the English Language* (Noah Webster ed. 1828) 'Alight', http://webstersdictionary1828.com/Dictionary/alight, accessed 18 August 2017 (hereinafter Webster) and see also *Webster's Revised Unabridged Dictionary* (Noah Porter ed. 1913) 'Alight', http://www.dict.org/bin/Dict, accessed 18 August 2017 (hereinafter Porter), both available online at The American and French Research on the Treasury of the French Language Project, http://machaut.uchicago.edu/websters, accessed 16 March 2014.

[76] *Online Etymology Dictionary* (Douglas Harper ed. 2001) 'Alight', http://www.etymonline.com/index.php?search=alight, accessed 16 March 2014.

[77] Webster and Porter, *supra* note 75, second definition.

[78] Dembling, *supra* note 1, at 649; Hall, *supra* note 39, at 205–206; Lyall and Larsen, *supra* note 28, at 140; Sundahl, *supra* note 15, at 169.

[79] C.Q. Christol, *The Modern International Law of Outer Space* (Pergamon Press 1982) 171–172; Dembling and Arons, *supra* note 1, at 649; Hall *supra* note 39, at 206; Sundahl, *supra* note 15, at 169.

meaning that the personnel, such as the pilot, crew and passengers, must have dismounted, that is to say deplaned or disembarked,[80] their spacecraft in order to be entitled to assistance. Ogunbanwo notes that '[f]or Article 3 to apply, the personnel of a spacecraft must have "alighted" … No duty to extend assistance arises until the contracting party receives information or discovers that the personnel have alighted'.[81]

This position is supported by the rules of grammar, which dictate that the plural past tense verb 'have alighted' must agree with a plural subject noun.[82] The rules of grammar also dictate that 'a spacecraft', as a singular subject noun, cannot agree with 'have alighted', it being a plural past tense verb.[83]

If instead the spacecraft was the intended subject of the verb, that is the noun or thing doing the alighting, the authors of Article 3 would have written 'alighted' as a singular past tense verb, as in 'the personnel of a spacecraft which has alighted …' or 'the personnel of a spacecraft having alighted …'. Consequently 'personnel', being the only plural subject noun present in the sentence in question, must comprise the subject of the plural past tense verb 'have alighted'.[84]

Some might argue that, just as when passengers of an aircraft touch down and pick up their cell phones to say 'hello, we've landed', similarly with 'alighted', the landing of the spacecraft is the appropriate meaning for 'the personnel … have alighted.' However, this assertion by passengers is a colloquialism impliedly referring to the landing of the aeroplane and not meant to convey that the passengers have themselves literally performed the physical landing, unassisted by the aircraft. 'Alighting' of personnel refers to their disembarkation and does not describe the landing of their spacecraft.

However, the question of whether a State Party to the ARRA may reasonably circumvent its obligations under Article 3 by relying upon the first definition of 'alight' (that personnel have disembarked) remains

---

80 Webster and Porter, supra note 75, first definition: the term 'alight' means to descend or dismount, derived from the word 'alihtan', of Old English, a combination of the word 'a', meaning 'down, aside', with the word 'lihtan', meaning 'get off, make light', and which together originally meant 'to lighten, take off, take away'. The connotation is to lighten the load of a horse or vehicle by getting down off it. See *Online Etymology Dictionary*, *supra* note 76, 'Alight'.

81 Ogunbanwo, *supra* note 66, at 138.

82 M.L. Benner, 'Self Teaching Unit: Subject – Verb Agreement' (*Online Writing Support*, Towson University 2000), http://www.towson.edu/ows/module SVAGR.htm, accessed 16 March 2014.

83 *Ibid.*

84 *Ibid.*

unresolved.[85] Recognizing these various arguments and examples supporting the opposing interpretations of 'have alighted' as alternatively referring to either the spacecraft or personnel, ambiguity and confusion exist regarding the ordinary meaning of this term.

Under the VCLT, confusion caused by ambiguous meaning calls for reconsideration of the intended ordinary meaning through consultation of the *travaux préparatoires* and other supplementary sources of interpretation.[86] Further, preconditioning the duty to render assistance on spacecraft landing or personnel disembarking contravenes the humanitarian purposes of the ARRA and results in absurd consequences, especially where a State Party is well positioned to extend assistance to those in need.

B. Absurd Results Under the VCLT: Supplementary Sources of Interpretation

The non-English official language versions of the ARRA, a primary source of authority for clarification under the ARRA and VCLT,[87] may suggest that the ordinary meaning intended for 'have alighted' was to require the landing of a spacecraft.[88] Even so, absurd consequences would result from use of either the second definition (spacecraft landing)[89] or the first definition (personnel disembarking).[90] Therefore, an alternative meaning for 'have alighted' consistent with the purposes and objectives of the ARRA must be investigated.

Several exemplary scenarios serve to inform this investigation. Take, for instance, the scenario in which a spacecraft has suffered malfunction, accident or some other damage, rendering it incapable of landing (for example, NASA Space Shuttle Columbia). Reading Article 3 to require the spacecraft to land prior to triggering the duty to render assistance results in absurd consequences because it would require the personnel to endanger their lives further by subjecting themselves to an emergency or crash landing.

---

85 Webster and Porter, supra note 75, first definition.

86 Art. 32(a) VCLT.

87 Art. 10 ARRA; Art. 33 VCLT.

88 Personal conversations with native Chinese ('外空机人员已下降于公海上'), French ('l'équipage d'un engin spatial a amerri en haute mer') and Spanish ('la tripulación de una nave espacial ha descendido en alta mar') language speakers and consultations to dictionary sources.

89 Webster and Porter, *supra* note 75, second definition.

90 Webster and Porter, *supra* note 75, first definition.

Even were the spacecraft to land with personnel still alive on board, reading Article 3 as requiring the personnel to disembark the spacecraft prior to triggering the duty to render assistance would also result in absurd consequences. For instance, the personnel may be incapable of alighting or exiting from the spacecraft for some reason, such as damage to the spacecraft rendering escape unsafe or impossible, or lack of equipment enabling survival in a surrounding hostile environment such as the high seas or Antarctica. Consequently, preconditioning assistance upon personnel alighting from a spacecraft also results in absurd consequences.

As a further example, consider a hypothetical situation wherein spacecraft regularly dock (alight or land) on a large international space station. Given a damaged spacecraft incapable of safely alighting on the space station, requiring the spacecraft to do so prior to rendering assistance would endanger both those on board the spacecraft as well as those on board the space station.

Contemplating these exemplary scenarios makes clear that reliance on either of the definitions for 'have alighted', whether advanced by the various commentators or through the linguistic and grammatical analyses set forth herein, results in manifestly absurd consequences. In accordance with the VCLT, if the ordinary meaning of treaty language results in manifestly absurd consequences, it can be challenged with the *travaux préparatoires* and other supplemental sources of interpretation.[91] Therefore, a reinterpretation of the ordinary meaning of 'have alighted' is justified.

The *travaux préparatoires* include a statement made to the UNGA by French delegate Mr Berard. In his statement, a recapitulation of previous statements before COPUOS and the Legal Subcomittee, Mr Berard indicated that the ARRA is meant to apply 'to search and rescue undertaken not only on the Earth and in its atmosphere, but also *in outer space and on celestial bodies*'.[92]

---

[91] Art. 32 VCLT: 'Recourse may be had to supplementary means of interpretation, including the preparatory work of the treaty and the circumstances of its conclusion … to determine the meaning when the interpretation according to article 31: … (*b*) leads to a result which is manifestly absurd or unreasonable'; Sundahl, *supra* note 15, at 175, 180, 188.

[92] Provisional Verbatim Record of the Sixteen Hundred and Fortieth Plenary Meeting, UN GAOR, 22d Sess, UN Doc A/PV1640, paras 77, 80 (1967), http://www.un.org/ga/search/view_doc.asp?symbol=A/PV.1640, accessed 18 August 2017 (hereinafter Provisional Verbatim Record); Sundahl, *supra* note 15, at 169, emphasis added.

Consequently the absurd results flowing from the previously reviewed definitions of 'have alighted' could not have been intended under the ARRA. Therefore, a further ordinary meaning for the term 'have alighted' in alignment with the humanitarian objectives and purposes of the ARRA needs to be explored.

To this end, 'alight' has an alternative definition, meaning to fall upon, to come upon or to chance upon by accident, or without design, as provided in the 1913 version of Webster's Revised Unabridged Dictionary and the Century Dictionary and Cyclopedia.[93] Employing this alternative definition for the Article 3 recitation of 'have alighted' circumvents the manifestly absurd results consequent to the two definitions reviewed above.

C. Article 3 ARRA: What Assistance Is Required in Outer Space?

Interpreting 'have alighted' as requiring the spacecraft to land or the personnel to disembark prior to triggering the duty to extend assistance results in manifestly absurd consequences, as shown above. However, use of the term 'alight' in the English language version of the ARRA provides an opportune means to easily resolve these unintended consequences through reinterpretation of 'alighted' in accordance with the proposed alternative ordinary meaning. Adoption of this alternative, more inclusive definition for 'have alighted' would appropriately expand the duty to rendering assistance necessary to rescue personnel of a spacecraft in orbit or navigating outer space.

Further, all States, under an interpretation of the ARRA as customary international law, would be required to extend assistance necessary to assure the speedy rescue of the personnel of a spacecraft who 'have alighted' in, that is to say have fallen upon, come upon or happened upon, 'any other place not under the jurisdiction of any State' (any location in outer space). Consequently, in accordance with the alternative definition of 'alighted', as proposed herein, the duty to extend assistance necessary to the speedy rescue of personnel would be extended to all areas of outer space, not just to celestial bodies.

---

[93] Porter, *supra* note 75, third definition: 'to come or chance (*upon*)'. See also, *Century Dictionary and Cyclopedia* (William Dwight Whitney ed., The Century Co. 1891): 'to fall (upon); come (upon) accidentally, or without design; light: as, to *alight* on a particular passage in a book, or on a particular fact; to *alight* on a rare plant', http://www.global-language.com/CENTURY/, accessed 16 March 2014 (all emphasis in original references).

However, the recitations which say that States from whom assistance is requested must be 'in a position to do so' in order to have a duty to render 'necessary assistance' presents something of an equivocation. What assistance precisely qualifies as 'necessary assistance' is a likely point of argumentation and whether a State is 'in a position' to render such assistance remains a situation-dependent question, depending at least in part upon the discretion of the State from which such assistance has been requested.[94]

9. CONCLUSIONS

Employing more expansive definitions for the terms 'astronaut', 'personnel' and 'have alighted' in line with the humanitarian objectives of Article V OST and Article 3 ARRA enables them to provide for rescue assistance to astronauts and personnel in outer space. More limited definitions of 'astronaut' generally require criteria including training and the attainment of specific altitudes, while more constrained definitions of 'personnel' typically require such persons to be employed or acting as service providers. Further, more restrictive definitions for the Article 3 term 'have alighted' require either the landing of a spacecraft or the disembarking of personnel.

However, in accordance with the VCLT, the ordinary meanings for these terms need to be interpreted as at the time their respective Treaties were concluded, that is 1967 and 1968 respectively. Consequently, definitions for these terms accepted in modern times should not be accorded primary authority in such interpretations. Evidence reviewed herein demonstrates that, as used in the OST and ARRA, the respective ordinary meanings of 'astronaut' and 'personnel' were intended to include all persons on board a spacecraft located in outer space and in need of assistance.

However, even if authorities were to ignore this evidence and attempt to adopt more restrictive definitions for 'astronaut' and 'personnel', such a position would be thwarted by the fact that these more restrictive definitions result in manifestly absurd consequences, because a rescuing party would be required to rescue only those persons who satisfy the criteria of 'astronaut' or 'personnel', and private commercial passengers could be left to fend for themselves.

---

94 Art. 3 ARRA; Dembling and Arons, *supra* note 1, at 649–650; Hall, *supra* note 39, at 205; Lyall and Larsen, *supra* note 28, at 140.

This absurdity could not have been intended by the drafters of the OST and ARRA, therefore justifying recourse to the *travaux préparatoires* and other supplemental sources supporting a more expansive reading of the ordinary meanings for 'astronaut' and 'personnel', inclusive of private passengers. Consequently, for the purposes of the OST and ARRA, 'astronaut' and 'personnel' can respectively be interpreted to include all persons on board a spacecraft including private commercial passengers.

Similarly, in the interpretation of 'have alighted' the meanings for this term respectively advanced by most commentators (spacecraft landing) and suggested by the linguistic and grammatical analysis provided herein (personnel disembarking) result in manifestly absurd consequences and justify recourse to the *travaux préparatoires* and supplementary sources of interpretation.

Adoption of an alternative definition for the term 'alight' in accordance with VCLT Article 32(b) is therefore required. As suggested herein, a satisfactory tertiary definition for 'alight' exists, meaning to fall upon, to come upon or to happen upon by chance or accident. Adopting this alternate definition of 'alight' will appropriately extend the duty to provide rescue assistance under Article 3, as compared to the conventionally advanced interpretations.

This result is in proper alignment with the humanitarian intent of the drafters, as evidenced by the statement of the French Delegate Mr Berard included in the *travaux préparatoires*. In accordance with the alternative definition of 'have alighted' proposed herein, the assistance necessary to quickly rescue personnel of a spacecraft in orbit or navigating outer space would be ensured.

However, a great deal of discretion is ultimately at the disposal of States from which rescue assistance is requested. In the case of OST Article V, the requirement is to provide 'all possible' assistance, but what assistance is in fact 'possible' is determined first and foremost by the State that is to provide such assistance. Similarly, for Article 3 ARRA the assistance must be 'necessary' to ensure the speedy rescue of personnel, and the State from which such assistance is requested must be 'in a position' to extend such assistance. Whether such State is 'in a position' to provide this assistance remains largely at the discretion of that State.

4. Small satellites, large constellations and space debris: *in dubio pro LEO*?

Ward Munters

1. INTRODUCTION

At the dawn of the space age, an unidentified, 'dark' space object circling the Earth caused uneasiness and headlines in American media.[1] Relief came just weeks later when the object was identified. 'It was a space derelict, the remains of an Air Force Discoverer satellite that had gone astray.'[2] Although a degree of trepidation concerning unknown space objects during the Cold War is understandable, that historical example perhaps serves best to illustrate how current preoccupations in the space domain might yet be eclipsed by emerging developments. Once the innocuous resolution to a news story – a curiosity – today 'space debris is considered the single most important threat to the safety of operations as well as peace and security in outer space.'[3]

Inspired by that perspective, the present chapter addresses two currently emerging and interrelated developments in space activities that appear similarly poised to have far-reaching effects on international space law and governance, as well as on their efficacy *vis-à-vis* the sustainable use of outer space: (i) the rapidly growing implementation of small

[1] J.W. Finney, 'A Mystery Polar Satellite, Believed Russian, Spotted; POLAR SATELLITE DETECTED BY D.S.', Special to The New York Times, *New York Times* 1960, 11 February 1960. Available online at http://query.nytimes.com/gst/abstract.html?res=9D03EED81138E333A25752C1A9649C9461 91D6CF&legacy=true (last accessed 15 February 2017).

[2] Editors, 'Science: Space Watch's First Catch', *TIME Magazine* 1960, 7 March 1960. Available online at http://content.time.com/time/magazine/article/0,9171,894745-2,00.html (last accessed 15 February 2017).

[3] T. Brisibe, 'Parliamentary diplomacy in the United Nations and progressive development of space law', *European Journal of Law Reform* 2016, Vol. 18, 26.

satellite technology, and its application in (ii) proposed large constellations of small satellites in low Earth orbit (LEO). Although those developments are complex and multifaceted, here they are discussed primarily with regard to the potential effects of constellations on the proliferation of space debris.

To that end, the first section provides a brief outline on the emerging activities at hand, while the second contains an overview of the awareness of those two emerging phenomena within relevant international forums. A third part presents a number of regulatory considerations apropos the developments and the international legal framework on space. Particular attention is paid to international debris mitigation guidelines as well as international liability for space activities in concert with scientific studies to date on the effects of planned large constellations. Inspiration is also taken from the field of international environmental law, that is, obligations regarding the prevention of transboundary harm and the precautionary principle.

In its legal analysis, the chapter aims primarily to take a broad and explicitly normative view. Instead of providing a particular set of clear-cut answers, it attempts to raise questions that could – or should? – merit reflection, in light of the future development of international space law and governance. Herein lies the crucial question this chapter seeks to emphasize with regard to megaconstellations: are we *certain* – and if so, to what degree – that, even with the best intentions and the utmost diligence, the risks to the amenity of the space environment represented by large constellations may presently not be too great?[4]

2. SMALL SATELLITE TECHNOLOGY AND LARGE CONSTELLATIONS: A PRACTICAL OVERVIEW

A. Small Satellite Technology

Although satellites may be classified in a number ways for example by cost, size, mission, lifetime or orbit – the term 'small satellites' is largely

[4] Perhaps an even more important and corollary consideration, that cannot be addressed here, is: who gets to decide and on what grounds? E.g., see: R.S. Jakhu, 'Legal issues relating to the global public interest in outer space', *Journal of Space Law* 2006, Vol. 32, 31–110, and S. Cogolati, L. Hamid and N. Vanstappen, 'Global public goods and democracy in international legal scholarship', *Cambridge Journal of International and Comparative Law* 2016, Vol. 5, 4–30.

used today[5] to denote a general class of spacecraft with a launch mass of less than one thousand kilograms (1000 kg).[6] Table 4.1 provides an overview of a general mass-based classification scheme for satellites.

Table 4.1 Classification of satellites by mass[7]

Satellite class		Launch mass
Large satellite		>1000 kg
Minisatellite	'Small' satellite class	100–1000 kg
Microsatellite		10–100 kg
Nanosatellite		1–10 kg
Picosatellite		0.1–1 kg
Femtosatellite	'Satellite-on-a-chip'	1–100 g

Several advantages make small satellite technology an exciting prospect for established space actors and entrants alike, thereby driving growth in the use and number of small satellites in recent decades.[8] The overarching drivers can be said to be the significant reduction of cost and complexity in the planning, design, manufacture, launch and utilization of small satellites when compared to their larger, integrated and more expensive counterparts. These are coupled with the resulting operational flexibility of using smaller, cheaper and focused platforms.

[5] For another overview of different classification schemes, the history of development and uptake of small satellite technology, and other relevant aspects, see O. Koudelka, 'Micro/nano/picosatellite-activities: challenges towards space education and utilisation' in I. Marboe (ed.), *Small Satellites: Regulatory Challenges and Chances*, Leiden, Brill Nijhoff, 2016, 7–27.

[6] H.J. Kramer and A.P. Cracknell, 'An overview of small satellites in remote sensing', *International Journal of Remote Sensing* 2008, Vol. 29, 4289–4290.

[7] *Ibid.*, 4290.

[8] For an extensive overview of the technical history of small satellite development, see: *ibid.*, 4285–4337. For an additional overview of the evolution of small satellite technology, its rationale and drivers, and its applications, see R.S. Jakhu and J.N. Pelton, *Small Satellites and Their Regulation*, New York, Springer, 2014, 4–41.

On the one hand, costs are reduced by the growing commercial availability of affordable off-the-shelf components – which could ultimately allow for these small spacecraft to be mass-produced[9] – and ongoing advancements in the miniaturization of electronics and sensors.[10] As a consequence, development times are generally shorter[11] and projects are often less cumbersome to plan than those involving larger spacecraft. Moreover, small satellites are generally not as technologically complex, often adhering to a 'design-to-capability' philosophy and carrying only a single instrument for a specific mission objective.[12] This drives down costs even more and adds to the flexibility and rapid turnaround times of small satellite operations for experimental and technology test-bed purposes.[13]

On the other hand, the decrease in launch mass that characterizes these satellites translates directly into lower launching costs per unit.[14] Depending on the mission, smaller and cheaper launch vehicles are often equally effective. Multiple small spacecraft can also share launch costs when they are launched collectively by a single launch vehicle[15] or when they 'piggy-back' as secondary payloads[16] on launches for larger spacecraft.[17] Concurrently, the present unfolding of an internationally competitive and

---

[9] M. Yakovlev, M. Kulik and V. Agapov, 'Small satellites and space debris issues' in H. Sawaya-Lacoste (ed.), *Proceedings of the Third European Conference on Space Debris, 19–21 March 2001, Darmstadt, Germany*, ESA SP-473, Vol. 2, Noordwijk, ESA Publications Division, 2001, 733.

[10] N.H. Crisp, K. Smith and P. Hollingsworth, 'Launch and deployment of distributed small satellite systems', *Acta Astronautica* 2015, Vol. 114, 65.

[11] International Academy of Astronautics Space Sciences Committee, IAA Position Paper, 'The case for small satellites', *Acta Astronautica* 1993, Vol. 31, 103.

[12] H.J. Kramer and A.P. Cracknell, *supra* note 6, 4285.

[13] The European Space Agency and small satellite activities, Conference Room Paper submitted by the European Space Agency (ESA) to the fifty-fifth session of the Legal Subcommittee (8 April 2016), UN Doc. A/AC.105/C.2/2016/CRP.19 (2016).

[14] H.J. Kramer and A.P. Cracknell, *supra* note 6, 4290.

[15] M. Yakovlev, M. Kulik and V. Agapov, *supra* note 9, 733.

[16] Nevertheless, small satellite deployment can also be hampered by their status as 'secondary payload' by having to adhere to primary launch schedules and orbit destinations. This is particularly prohibitive for missions that require small satellites to be deployed as a constellation in accurate and coordinated orbits or multiple planes, for which a dedicated launch is required. See N.H. Crisp, K. Smith and P. Hollingsworth, *supra* note 10, 66.

[17] N.H. Crisp, K. Smith and P. Hollingsworth, *supra* note 10, 65. Consider the recent launch by the Indian Space Research Organisation (ISRO) of 104 satellites of seven States on a single Polar Satellite Launch Vehicle (PSLV-C37). ISRO, 'PSLV-C37 Successfully Launches 104 Satellites in a Single Flight', 15 February

commercial launching market provides interested operators with increasingly flexible solutions – that is, dedicated and accurate launches of small spacecraft – to obstacles under the current launching paradigm.[18]

Other perceived advantages of these relatively inexpensive spacecraft over larger versions can include: more varied and more frequent mission opportunities; a larger involvement of local industry; an increase in capacity building; greater accumulation of scientific and technical knowledge; and a wider array of potential users.[19]

Indeed, whereas satellite capabilities were once the exclusive domain of a limited number of developed spacefaring States, today an ever-growing variety of space activities progressively falls within the reasonable scope of ambition of any entity with space-based aspirations. Small or developing countries, research groups, universities and students or commercial and private actors, among others, are no longer precluded from operating a satellite of their own and benefitting therefrom. In sum, '[s]mall satellites seem to embody both the potential benefits and problems associated with the expansion of space activities, which were, until recently, restricted to governments and a few companies with the resources required to build and launch a full-size satellite.'[20]

B. Large Constellations

The idea of large groups, or constellations, of numerous satellites under single control and performing a number of shared and collective functions is not new, nor particularly recent. Notable examples of existing commercial constellations of small satellites in LEO include, for instance, the Iridium (66 satellites), Globalstar (40 satellites) and Orbcomm (up to 36 satellites) telecommunication systems.[21] Those systems are not novel, but circumstances today can be said to be different than during the time when they were originally conceived.

2017. Available online at http://www.isro.gov.in/update/15-feb-2017/pslv-c37-successfully-launches-104-satellites-single-flight (last accessed 20 February 2017).

[18] N.H. Crisp, K. Smith and P. Hollingsworth, *supra* note 10, 66–67.

[19] International Academy of Astronautics Space Sciences Committee, *supra* note 11, 109.

[20] Proposal for a Single Issue/Item for discussion at the fifty-fifth session of the Legal Subcommittee in 2016 on: 'Exchange of views on the application of international law to small satellite activities', Conference Room Paper submitted by the Group of Latin American and Caribbean States (GRULAC) to the fifty-fourth session of the Legal Subcommittee (22 April 2015), UN Doc. A/AC.105/C.2/2015/CRP.23/Rev.1 (2015), para. 1, *in fine*.

[21] R.S. Jakhu and J.N. Pelton, *supra* note 8, 16–17.

The abovementioned advantages and capabilities of small satellites have not only already made them uniquely suited for use in constellations or distributed systems, particularly in LEO. Continuous modern technological advancements are also bound to have a profound impact on both the capabilities and the number of constellations of small spacecraft in the near future. Indeed, statements in the press and media on the planned OneWeb megaconstellation in LEO confirm the importance of this synergy:

> Greg Wyler, OneWeb's chief executive, said the new venture's satellites would each weigh less than 300 pounds and cost $350,000 each. He said they would be manufactured quickly, using more of an assembly line approach. Current satellites are manufactured one at a time – a process that can take years – and can cost tens of millions of dollars.[22]

Small satellite technology and large constellations thus appear to be a match made in heaven. Although that technology is, of course, not limited in its use to large constellations, this chapter will continue with a focus on the latter. The expected rise in the number of space objects and spacefaring actors engendered by small satellites in general will, however, only further complicate the issues identified hereinafter.

3. INCREASED AWARENESS IN INTERNATIONAL FORUMS

A. The United Nations Committee on the Peaceful Uses of Outer Space

Having agreed to a proposal[23] tabled at its fifty-fourth session[24] and pursuant to UN General Assembly Resolution 70/82,[25] the Legal

---

[22] M. Petersen, 'Elon Musk and Richard Branson invest in satellite-Internet ventures', *Los Angeles Times* 2015, 16 January 2015. Available online at http://www.latimes.com/business/la-fi-satellite-entrepreneurs-20150117-story.html (last accessed 10 February 2017).

[23] UN Doc. A/AC.105/C.2/2015/CRP.23/Rev.1 (2015), *supra* note 20.

[24] Report of the Legal Subcommittee on its fifty-fourth session, held in Vienna from 13 to 24 April 2015 (30 April 2015), UN Doc. A/AC.105/1090 (2015), para. 222.

[25] Resolution 70/82 of the UN General Assembly – International cooperation in the peaceful uses of outer space (9 December 2015), UN Doc. A/RES/70/82 (2015), 3, para. 4.

Subcommittee (LSC) of the UN Committee on the Peaceful Uses of Outer Space (UNCOPUOS) considered for the first time, at its fifty-fifth session in 2016, a new single issue agenda item entitled 'General exchange of views on the application of international law to small satellite activities'.[26] Recognizing the potential of small satellites for empowering aspiring and developing space actors with 'limited funds' and for 'fostering the general technological progress in the area of space activities',[27] the Subcommittee noted that the new agenda item 'would provide valuable opportunities for addressing a number of topical issues relating to international and national policy and regulation measures regarding the use of small satellites by various actors'.[28] The item was retained on the agenda for discussion at the Subcommittee's fifty-sixth session.[29]

Moreover, the LSC acknowledged the existence of a number of unspecified legal challenges and, with regard to the safe and responsible use of outer space in the future, that 'it was important to include small satellite missions appropriately in the scope of application of international and national regulatory frameworks.'[30] In response:

> [s]ome delegations expressed the view that all international rights and obligations of States with respect to large satellites were equally relevant for the conduct of space activities with the use of small satellites, and thus the United Nations treaties and principles on outer space, the ITU Constitution and Convention and Radio Regulations, as well as certain non-binding instruments, such as the Space Debris Mitigation Guidelines, were providing the legal framework to be applied to various space objects, including small satellites.[31]

Notwithstanding legal questions that currently remain unanswered, this view is undoubtedly correct from an international law standpoint. As the space treaties do not contain a legal criterion for distinguishing between different classes of spacecraft on the basis of size or any other technical

---

[26] Report of the Legal Subcommittee on its fifty-fifth session, held in Vienna from 4 to 15 April 2016 (27 April 2016), UN Doc. A/AC.105/1113 (2016), paras 217–231.

[27] *Ibid.*, paras 221 and 222.

[28] *Ibid.*, para. 220.

[29] *Ibid.*, para. 249.

[30] *Ibid.*, paras 224–225.

[31] *Ibid.*, para. 226.

characteristic, there can be no doubt that a satellite, whether 'small' or 'large', is to be qualified as a 'space object' pursuant to, for example, Article VIII of the Outer Space Treaty (OST)[32] or Article I(d) of the Liability Convention.[33] Through this qualification important rights and obligations with regard to such objects are established, both for States Parties to the space treaties and space actors in general.

The topic of small satellites also appears under the LSC's agenda item for a '[g]eneral exchange of information and views on legal mechanisms relating to space debris mitigation measures, taking into account the work of the Scientific and Technical Subcommittee (STSC)'. Some delegations proposed the growing number of small satellite deployments should be considered in connection with the problem of space debris,[34] while '[t]he view was expressed that space debris mitigation should be effectively implemented, regardless of the size and constellation of space objects, and that special attention should be given to the new concept of megaconstellations.'[35]

In turn, these prospective concerns are echoed and elaborated upon in discussions of the STSC under the item of 'Space debris', where, reflecting also the topic of the current chapter, '[t]he view was expressed that recent developments in the field of small satellites, as well as announcements by private companies to launch large constellations of satellites, had raised questions about their potential impact on the space debris environment in the short and long terms.'[36] Similarly, at the fifty-ninth session of UNCOPUOS in 2016, under the agenda item of 'Space debris', '[t]he view was expressed that it was necessary to analyse

---

[32] Treaty on Principles Governing the Activities of States in the Exploration and Use of Outer Space, including the Moon and Other Celestial Bodies, Adopted on 19 December 1966 and entered into force 10 October 1967, UN General Assembly Resolution 2222(XXI), Annex, UNTS 1967, Vol. 610, 205.

[33] Convention on International Liability for Damage Caused by Space Objects, adopted on 29 November 1971 and entered into force 1 September 1972, UN General Assembly Resolution 2777(XXVI), Annex, UNTS 1975, Vol. 961, 187.

[34] Report of the Legal Subcommittee on its fifty-fifth session, *supra* note 26, para. 183.

[35] *Ibid.*, para. 184.

[36] Report of the Scientific and Technical Subcommittee on its fifty-third session, held in Vienna from 15 to 26 February 2016 (9 March 2016), UN Doc. A/AC.105/1109 (2016), para. 111.

the possible impact of the deployment of large constellations of satellites in low-Earth orbit and to investigate the end-of-life disposal of constellation members.'[37]

In addition to a growing awareness evidenced by these remarks, the Committee has been kept abreast of the emerging developments by a number of relevant observers, international organizations and other space actors. As an example, in considering the new agenda item, the LSC[38] was provided with a conference room paper by the European Space Agency (ESA) entitled 'The European Space Agency and small satellite activities.' The paper sets out details on the general notion of 'small satellites' and the Agency's activities, as well as elaborating on several issues that might benefit from legal clarification in the future – for example, the tentative need for diversified regulatory approaches for large and small satellites.[39] Also worth noting, finally, are presentations held in the context of the Subcommittees' sessions on, inter alia, the planned OneWeb megaconstellation consisting of hundreds of small satellites in LEO,[40] or on the 'Legal and practical considerations of registering mega-constellations and space debris.'[41]

B. The International Telecommunication Union

In an effort to boost discussions on the newly agreed agenda item, the LSC concurrently decided to invite updates by the International Telecommunication Union (ITU) on 'relevant developments and issues

---

[37] Report of the Committee on the Peaceful Uses of Outer Space on its fifty-ninth session, held in Vienna from 8 to 17 June 2016 (28 June 2016), UN Doc. A/71/20 (2016), para. 90.

[38] Report of the Legal Subcommittee on its fifty-fifth session, *supra* note 26, para. 219.

[39] The European Space Agency and small satellite activities, *supra* note 13, 3.

[40] M. Lindsay, 'OneWeb', Scientific and Technical Presentation at the fifty-third session of the Scientific and Technical Subcommittee (hereinafter OneWeb Presentation). Available online at the website of UNOOSA, http://www.unoosa.org/oosa/en/ourwork/copuos/stsc/technical-presentations.html#stsc2016 (last accessed 10 February 2017). See Report of the Scientific and Technical Subcommittee on its fifty-third session, *supra* note 36, para. 94(c).

[41] A. Soucek, 'Legal and practical considerations of registering mega-constellations and space debris', Technical Presentation at the fifty-fifth session of the Legal Subcommittee. Available online at the website of UNOOSA, http://www.unoosa.org/oosa/en/ourwork/copuos/lsc/2016/symposium.html (last accessed 10 February 2017). See Report of the Legal Subcommittee on its fifty-fifth session, *supra* note 26, para. 11.

regarding ITU procedures and regulations applicable to small satellites'.[42] Notable, in that regard, are the 2015 Prague Declaration[43] and 2016 Santiago Declaration[44] on Small Satellite Regulation and Communication Systems by the participants in the ITU's symposia on the same topic.

Both declarations recognize there is an urgent need for adherence to international laws, regulations and procedures as well as for compliance with space debris mitigation guidelines in small satellite activities. In particular, they underscore 'the importance for small satellite community [*sic*] to be prepared for implementing existing and newly developing recommendations and practices supporting the long-term sustainability of outer space activities'.[45] As a consequence, the declarations urge 'the small satellite community to comply with the applicable international and national laws, regulations and procedures, indispensable to guarantee the long-term sustainability of small satellite projects, the avoidance of harmful interference and proper management of space debris', and thereafter recommend ways in which to facilitate these efforts.[46]

C. The Inter-Agency Space Debris Coordination Committee

In February 2016 the Steering Group of the Inter-Agency Space Debris Coordination Committee (IADC), uniting the world's predominant governmental space agencies in cooperative space debris research, issued the IADC Statement on Large Constellations of Satellites in Low Earth Orbit.[47] The focus of the statement is on 'the emerging plans for large

---

42 Report of the Legal Subcommittee on its fifty-fourth session, *supra* note 24, para. 222.

43 Prague Declaration on Small Satellite Regulation and Communication Systems, 3 March, 2015 (hereinafter Prague Declaration). Available online at http://www.itu.int/en/ITU-R/space/AmateurDoc/PragueDeclaration.pdf (last accessed 10 February 2017).

44 Santiago Declaration on Small Satellite Regulation and Communication Systems, 8 November 2016 (hereinafter Santiago Declaration). Available online at https://www.itu.int/en/ITU-R/space/workshops/2016-small-sat/Documents/Santiago-Declaration-DRAFT.pdf (last accessed 10 February 2017).

45 Prague Declaration, 1 and Santiago Declaration, 1.

46 Prague Declaration, 1. The Santiago Declaration uses the term 'invites' instead. Santiago Declaration, 2.

47 Inter-Agency Space Debris Coordination Committee (IADC), Statement on Large Constellations of Satellites in Low Earth Orbit, IADC-15-03, February 2016. Available online at http://www.iadc-online.org/Documents/IADC-15-03%20Megaconstellation%20Statement.pdf (last accessed 10 February 2017).

constellations of satellites in Low Earth Orbit (LEO) and … the potential for such systems to have an important influence on the evolution of the space debris environment and consequent impact on the population of man-made satellites orbiting the Earth'.[48]

The following paragraph is especially relevant with respect to the following sections:

> [r]ecognising that a number of constellation architectures are already in the process of development, and that timely guidance could help to inform the design and operation of such systems, the IADC resolves to follow a two-track approach to address this issue. First, on the basis of the outcome of initial reflections, it offers a number of preliminary qualitative observations that operators could consider in their conceptual design, and subsequently, as more substantive and comprehensive modelling data becomes available from the coordinated international studies, offer more detailed, quantitative guidance to inform the development and operation of such constellation systems.[49]

Apart from the high-level recognition of the imminent and acute implementation of several constellation concepts and of their potential effects, the preliminary qualitative approach by the IADC provides inspiration for novel methods to strengthen the relevance of current international space debris mitigation measures, with regard to large constellations (see section on Post-mission lifetime below). It also holds clues as to the a priori (non-)fulfilment by authorizing States of obligations to prevent transboundary harm (see section on Application to space activities and large constellations below). In addition, three preliminary considerations by the IADC, in particular, will imbue and bolster the analysis in the current chapter.[50]

First, it is clear *ab initio* that 'the planned constellation architectures represent a step change in the number of satellites operating in the low Earth orbit regime, and may question the validity of the assumptions used to derive the existing space debris mitigation guidelines (e.g. launch traffic models and the numbers of objects in orbit).' A paradigm shift in space activities might, thus, bring unforeseen consequences and carries with it scientific uncertainty – for example, as to the efficacy of current mitigation measures (see section 4.A).

---

48 *Ibid.*, 4.

49 *Ibid.*

50 Unless otherwise stated, all quotes, citations and considerations in the following paragraphs of the current section are with sole reference to IADC, *supra* note 47, 6.

Second, there is a need to focus on the reliability of spacecraft systems pertaining to, inter alia, post-mission disposal that depends in part on the level of technological maturity, on the choices in the design of constellations and on the concepts and practices that are put into actual operation. Along the same lines, the Committee continues with the proviso that '[i]t is clear that significant improvements in the reliability of the disposal function at end of life will be needed for the new constellations compared with that currently demonstrated by space systems on orbit' (see section on A temporary moratorium on large constellations? below).

Third, as a result of the large number of anticipated spacecraft needed to sustain megaconstellations, the greatest short-term effects will be on the constellations themselves 'in terms of possible close approaches, and the consequent potential burden in terms of conjunction assessment and the possible need for avoidance manoeuvres, which will ultimately affect fuel margins and platform lifetimes.' Therefore, where this chapter refers to notions of self-interest and self-regulation, this caveat should remind us not only that implementing large constellations in a sustainable manner will prove burdensome on operators, but also that there are reasonable underlying concerns with regard to cost–benefit analysis. Short-term individual benefits and decreased costs for operators deriving from a more relaxed application of mitigation measures – for example, preserving fuel and augmenting lifetimes – may conflict with longer-term collective costs and benefits relating to the amenity of the near-Earth orbits for other users and future generations.

4. LEGAL CONSIDERATIONS RELATING TO LARGE CONSTELLATIONS AND THE LONG-TERM SUSTAINABILITY OF SPACE ACTIVITIES

The unprecedented introduction of large constellations into the already congested, contested and competitive near-Earth orbits,[51] compounded by an increase in space users enabled by small satellite technology, clearly

[51] 'The result of the increase in space actors and space users is that the space environment, especially key Earth orbits, has become increasingly utilized over the past few decades. As a consequence, the outer space environment is becoming increasingly congested, contested and competitive.' Report of the

raises concerns for an already deteriorating space debris environment. This chapter, therefore, advances what might be considered a 'radical' view, by asserting that there is a pressing need for the international community to step out of the murky waters of endless deliberations and navel-gazing and to move towards concerted and 'radical' legal action – or, more precisely, temporary *inaction* – to preserve the amenity of the outer space environment for future activities.

The author contends that the rationale to this approach is twofold. On the one hand, a focus on worst-case scenarios related to mega-constellations might be admittedly pessimistic, but should nevertheless function as a call for action and for a healthy dose of scepticism with respect to emerging developments, lest the international community be faced with a *fait accompli*. On the other hand, the 'radical' perspective is perhaps the only tenable position, inasmuch as a 'step change' in the number of space objects constitutes a paradigm shift that, by definition, will inevitably put an end to the status quo.

To make its normative case, the legal analysis touches upon a non-exhaustive appraisal of legal questions relating to, inter alia, the efficacy of liability and reparation schemes to safeguard sustainability, the protection of the 'space environment' per se, international obligations to prevent 'transboundary harm' and the application of the 'precautionary principle' to outer space activities. It is important to emphasize that the example of the OneWeb constellation is used in this analysis, not to single out any particular entity or entities, but for the simple reason that it exemplifies the superlative nature of megaconstellations. It also provides a link with real-world developments, underscoring the acute nature of the following considerations.[52]

Group of Governmental Experts on Transparency and Confidence-Building Measures in Outer Space Activities (29 July 2013), UN Doc. A/68/189 (2013), 9, para. 6.

[52] OneWeb has obtained substantial financial backing from a number of high-profile commercial space actors, including Intelsat and Airbus, as well as securing a billion-dollar launch contract with Arianespace and the Russian space agency ROSCOSMOS, among others. The first constellation satellites are slated for launch as early as 2018. See S. Clark, 'OneWeb launch deal called largest commercial rocket buy in history', *Spaceflightnow.com* 2015, 1 July 2015. Available online at http://spaceflightnow.com/2015/07/01/oneweb-launch-deal-called-largest-commercial-rocket-buy-in-history/ (last accessed 15 February 2017). See also OneWeb's official website, available online at http://oneweb.world (last accessed 15 February 2017).

A. The Limits of Mitigation Guidelines and Liability

As noted by the IADC in its preliminary statement, announcements with regard to plans for large constellations of hundreds to thousands of small satellites in LEO represents a 'step change' in the number of space objects. This unprecedented development may render invalid the underlying assumptions and challenge the efficacy of current international efforts to mitigate proliferation of hazardous space debris – that is, various guidelines on space debris mitigation. In fact, '[t]here is also a question regarding the robustness of the existing debris mitigation guidelines to effectively manage the new constellations and their impact on the orbital environment in a sustainable manner (e.g. limit residence times in orbit).'[53] It is instructive, then, to turn our attention to two sets of widely recognized international guidelines.

The 2007 UNCOPUOS Space Debris Mitigation Guidelines (UNCOPUOS SDM Guidelines) are a set of 'high-level qualitative guidelines',[54] distilled through political process from the more technical and more *quantitative* mitigation guidelines, developed under the auspices of the IADC (IADC SDM Guidelines).[55] That distinction in character between those sets of guidelines acquires heightened relevance with respect to their application to large constellations in LEO. That is to say, the qualitative character of the UNCOPUOS SDM Guidelines may offer a relatively straightforward medium through which to deal with the potentially disruptive effect of megaconstellations on the robustness of existing guidelines.

(i) Post-mission disposal

In particular, the albeit limited number of scientific studies to date on the effects of megaconstellations on the proliferation of space debris in LEO have uniformly stressed the importance of one mitigative measure that impacts the long-term debris environment, that is, 'post-mission disposal' (PMD):[56] the timely de-orbiting or re-orbiting of spacecraft to ultimately

---

[53] IADC, *supra* note 47, 6.

[54] Report of the Committee on the Peaceful Uses of Outer Space (UNCOPUOS) on its fiftieth session (15 June 2007), UN Doc. A/62/20 (2007), Annex, 47–48 (hereinafter UNCOPUOS SDM Guidelines).

[55] IADC, *Space Debris Mitigation Guidelines*, IADC-02-01, Rev. 1, 2007 (hereinafter IADC SDM Guidelines). Available online at http://www.iadc-online.org/index.cgi?item=docs_pub (last accessed 10 February 2017).

[56] These considerations have also been made part of the recent discussions on space debris at UNCOPUOS: '[t]he view was expressed that it was necessary

remove them from outer space. This takes place at the end of their mission phase or end-of-life (EOL) with the aim of making them re-enter Earth's atmosphere, either directly or later in time, by repositioning them to an orbit with a reduced orbital lifetime due to, inter alia, atmospheric drag.[57] Two important variables of PMD emerge for our purposes in this context. The first is the amount of time over which the space object is ultimately de-orbited. The second concerns the rate of compliance by space actors with this crucial guideline.

Post-mission lifetime As to the first element, the IADC SDM Guidelines state that the 'IADC and some other studies and a number of existing national guidelines have found 25 years to be a reasonable and appropriate lifetime limit' for objects that have the potential to interfere with the LEO region.[58] The explicit reference to '25 years' contributes to the more quantitative character of those guidelines. In contrast, the corresponding UNCOPUOS SDM Guideline, Guideline 6: Limit the long-term presence of spacecraft and launch vehicle orbital stages in the low-Earth orbit (LEO) region after the end of their mission, does not expressly specify any time requirement for what 'long-term presence' might entail, for proffering 'a numerical value would have conflicted with the character of the text as representing high-level qualitative guidelines.'[59]

---

to analyse the possible impact of the deployment of large constellations of satellites in low-Earth orbit and to investigate the end-of-life disposal of constellation members.' Report of the UN Committee on the Peaceful Uses of Outer Space (UNCOPUOS) on its fifty-ninth session, *supra* note 37, 15, para. 90.

[57] Alternatively, spacecraft reaching EOL in orbits where de-orbiting is prohibitively difficult may be repositioned, re-orbited, to 'graveyard orbits' from where they will not necessarily re-enter Earth's atmosphere but will no longer pose a significant threat to objects in more congested orbits.

[58] IADC SDM Guidelines, 9–10 ('5.3.2 Objects Passing Through the LEO Region').

[59] The inclusion of an explicit time requirement was controversial, such that no consensus arose from discussions within UNCOPUOS's STSC. For the quoted text as well as the controversial discussions, see A. Soucek, 'SDM Guideline 6' in S. Hobe, B. Schmidt-Tedd, K.-U. Schrogl (eds) and P. Stubbe (Assist. ed.), *Cologne Commentary on Space Law: Vol. III*, Cologne, Carl Heymanns Verlag, 2015, 637 and note 155.

Even so, the ubiquitous nature of the '25-year rule' has led it to form a common, though perhaps optimistic,[60] baseline for scientific studies on PMD apropos debris in general, and the effects of megaconstellations in particular. However, as one study notes, 'several of the announced large constellations aim to have shorter post-mission lifetimes, so to understand how these efforts will affect the evolution of the object population, post-mission lifetime requirements of 0 (direct re-entry), 10 and 25 years were considered.'[61] This is especially relevant as OneWeb plans to limit the post-mission presence of its spacecraft in LEO to a more stringent five years, in an effort to substantially alleviate the impact on the debris population.[62] Significant uncertainties remain nonetheless, as exemplified by another study, wherein a number of variables 'were chosen to account for an ambiguity in the current OneWeb statements, which emerge from the disposal to a 1100x200km orbit and a disposal lifetime of about 5 years'.[63]

What becomes clear on the basis of the IADC's statements and the cited studies, furthermore, is that the 25-year rule is not a *one-size-fits-all* measure – that is, there is likely a need for lifetime requirements to dynamically adapt according to the particularities of any space activity. Mindful also of the IADC's recent reliance on a qualitative approach in its preliminary statement, the explicit choice in UNCOPUOS to promulgate qualitative measures may here prove to be opportune. Qualitatively circumscribed criteria can provide a flexible means of integrating dynamically tailored lifetime requirements into the existing quasi-legal guidelines. Notably, the conspicuous absence of any quantified measure of time arguably allows for a reading of UNCOPUOS SDM Guideline 6 wherein the period of time corresponding to 'long-term presence' can be interpreted from the *telos*[64] of the guideline, according to relevant circumstances.

---

[60] B. Bastida Virgili, J.C. Dolado, H.G. Lewis, J. Radtke, H. Krag, B. Revelin, C. Cazaux, C. Colombo, R. Crowther and M. Metz, 'Risk to space sustainability from large constellations of satellites', *Acta Astronautica* 2016, Vol. 126, 155.

[61] *Ibid.*, 158.

[62] OneWeb Presentation, *supra* note 40, 17.

[63] J. Radtke, C. Kebschull and E. Stoll, 'Interactions of the space debris environment with mega constellations – using the example of the OneWeb constellation', *Acta Astronautica* 2017, Vol. 131, 56.

[64] See: A. Soucek, *supra* note 59, 637.

If the overarching goal of the UNCOPUOS SDM Guidelines is indeed 'preserving the outer space environment for future generations',[65] then what constitutes a 'long-term presence' of space objects should be interpreted from, and judged with reference to the obstacles particular (constellations of) spacecraft pose to achieving that goal. In the unprecedented and exceptional case of very large constellations it seems warranted, therefore, to expect from space actors and the corresponding authorizing States full compliance with significantly more restrictive limits on post-mission lifetime, as opposed to a period of 25 years – to be progressively enumerated on the basis of scientific studies – even under the current manifestation of the UNCOPUOS SDM Guidelines.

The teleological approach has the additional benefit of circumventing, to a degree, the update mechanism in the UNCOPUOS SDM Guidelines, while nevertheless strengthening their efficacy with regards to evolving technologies and emerging developments. That is to say, '[t]his document will be reviewed and may be revised, as warranted, in the light of new findings.'[66] Even though the recognition of the need for future updates is commendable, it should not be forgotten that it has become increasingly difficult to reach consensus on substantial issues within UNCOPUOS, leading some to question the Committee's productivity and efficiency.[67] Indeed, the present UNCOPUOS SDM Guidelines were issued only after protracted deliberations.[68] They are likely to remain unaltered for some time – likely, at least, until the IADC has concluded its multi-year studies on the effects of large constellations and only after a significant number of additional spacecraft have been placed in orbit.[69] Coincidentally, the mere possibility of a teleological-interpretative method can, *de lege ferenda*, serve to invalidate arguments holding that '[g]iven the evolving technical aspects of debris mitigation, and the practical, economic reality

---

[65] UNCOPUOS SDM Guidelines, *supra* note 54, 47.

[66] *Ibid.*, 50.

[67] E.g., see X. Yu, 'UNCOPUOS 50 years on: Assessing current dynamics and exploring its future role', *Space Policy* 2012, 147; T. Brisibe, *supra* note 3, 22–23.

[68] A report detailing space debris and potential mitigating measures was put before the UNCOPUOS's STSC in 1979–1980, while the IADC's SDM Guidelines were put before it in 2003 and only formalized in the UNCOPUOS STSC's SDM Guidelines in 2007. See Mutual relations between space missions (7 December 1979), UN Doc. A/AC.105/261 (1979); Report of the Scientific and Technical Subcommittee on its forty-fourth session, held in Vienna from 12 to 23 February 2007 (6 March 2007), UN Doc. A/AC.105/890 (2007), para. 99 and Annex IV.

[69] IADC Statement, *supra* note 47, 4–5.

that existing platforms cannot be replaced overnight, we do not see the wisdom in ossifying debris mitigation standards into international law at this time.'[70]

Finally, although the cited studies underscore in certain terms the paramount importance of reducing post-mission lifetime, in order to curb the impact of megaconstellations on the debris environment, the impact of shorter lifetimes should not be overstated either. The beneficial effects will likely be limited over the long term as 'such efforts may still have an effect on the LEO object population.'[71] While the qualitative nature of the UNCOPUOS SDM Guidelines might allow for interpretation to accommodate or, when seen as a potential standard of care for ascertaining liability, even necessitate adapted lifetimes, we should turn our attention to the second dimension mentioned above: compliance with and success rates of PMD measures. As one authoritative study puts it, the longer-term consequences 'are likely to be dominated by the success, or otherwise, of post-mission disposal activities within the constellations'.[72]

Post-mission disposal compliance: not a panacea To be sure, there seems to be agreement in the currently scarce scientific literature 'that the most critical parameter of a sustained environment is to limit the number of objects that stay on orbit over the very long-term [*sic*], for example by performing an end-of-life disposal at a very high success rate'.[73] That conclusion corroborates similar concerns that moderate compliance with, or success of, PMD at a rate of 50 per cent for a constellation in LEO could, nevertheless, lead to an accumulation of space objects and could provide a catalyst for a detrimental growth of the debris population. Even a 90 per cent success rate may see an average increase in the number of pieces of debris of at least a quarter, although these levels may also be sufficient to stabilize the future debris environment. As the use of the term 'may' implies, significant uncertainty remains in this area as well, and is compounded by the consideration that 'the results … are likely to be optimistic, especially if more than one large constellation is deployed in LEO as anticipated.'[74]

---

70 Statement by the United States of America at the 52nd session of the Legal Subcommittee of UNCOPUOS in 2013, as cited in T. Brisibe, *supra* note 3, 29.

71 B. Bastida Virgili *et al.*, *supra* note 60, 158.

72 *Ibid.*, 158.

73 J. Radtke, C. Kebschull and E. Stoll, *supra* note 63, 56.

74 B. Bastida Virgili *et al.*, *supra* note 60, 158.

More particularly, compliance with PMD is likely to be of special importance for the specific altitude of 1200 kilometres at which OneWeb or other planned and even larger megaconstellations[75] plan to operate. A lower atmospheric density at these chosen orbits leads to only a slow orbital decay over millennia of defective satellites that are not otherwise disposed of.[76] Consequently, a lower level of adherence to PMD will lead to higher, possibly prohibitive collision probabilities – for example, between derelict spacecraft or fragmentation debris and constellation members – and of concomitant conjunctions and avoidance manoeuvres.[77] In fact, for just a single constellation, '[e]ven when performing post-mission disposal with 90% success, the collision probability against constellation objects increases from 22% after 10 years of operation to over 42% after 20 years of operations for the 1200 km constellation.'[78]

Facing these emerging developments, a very high or near-perfect PMD compliance rate is thus a *conditio sine qua non*[79] – though, in general, not a sufficient condition[80] – for the sustainability of activities in the

---

[75] E.g., Boeing is planning to operate a constellation of between 1396 and 2956 satellites at an altitude of 1200 km. See: P.B. De Selding, 'Boeing proposes big satellite constellations in V- and C-bands', *Spacenews* 2016, 23 June 2016. Available online at http://spacenews.com/boeing-proposes-big-satellite-constellations-in-v-and-c-bands/ (last accessed 10 February 2017).

[76] F. Alby, 'The space debris environment and its impacts' in W. Rathgeber, K.-U. Schrogl and R.A. Williamson (eds), *The Fair and Responsible Use of Space: An International Perspective*, Vienna, Springer-Verlag, 2010, 63.

[77] 'In further analyses, over a few generations of OneWeb satellites, it becomes apparent that the chosen orbital region is more susceptible to failed post mission disposal (PMD) manoeuvres than other regions. The lower the PMD rate is, the higher will the number of inter-constellation encounters be. Failing to achieve a PMD success rate of at least 50% will lead to a significantly higher number of avoidance manoeuvres that need to be performed. After 10 years the 1500 manoeuvres per year mark will be reached. After 20 years 2500 manoeuvres will be necessary. The effect of this drastic increase in inter-constellation encounters is accounted to the lower atmospheric density, which leads to a slow decay of defective satellites': J. Radtke, C. Kebschull and E. Stoll, *supra* note 63, 67.

[78] *Ibid.*

[79] *Ibid.* ('on higher orbits are also fewer satellites in use and a lot more volume of space available per satellites [*sic*] the chance of encounters is also reduced. In the light of the chosen 1200 km altitude a very high PMD rate will help to keep the orbital region as usable for future satellite generations as it is today.')

[80] Apart from the consideration that even a 90 per cent PMD compliance rate will significantly augment collision probabilities over the long term, general active debris removal and on-orbit servicing activities will be necessary to

affected orbits and for the amenity of the LEO space environment itself. A sobering assessment of present reality shows, however, that current compliance with and success rates of PMD in LEO are *many times lower* than those ostensibly needed to even allow for a chance to stabilize the debris population over the long term. Only approximately 20 per cent of spacecraft are currently being de- or re-orbited in LEO over the more generous period of 25 years, which is unlikely to be adequate in the future.[81]

Although OneWeb statements point towards an intent to self-regulate and adhere to PMD over a significantly shorter period of five years, it is uncertain, inter alia: whether this constitutes the fulfilment of an international legal 'obligation' with any 'teeth' that could be enforced in cases of non-compliance; whether private and voluntary industrial standards offer any solace in this regard;[82] whether the national regulatory and licensing mechanisms of States that must authorize and will be responsible for the activities of OneWeb pursuant to Article VI of the Outer Space Treaty are sufficiently tailored to the unprecedented scale and accompanying risks of the activities; or whether such strict adherence is, at least currently, even a technological possibility.[83]

---

stabilize the long-term debris environment. E.g., see Active Debris Removal – An Essential Mechanism for Ensuring the Safety and Sustainability of Outer Space, A Report of the International Interdisciplinary Congress on Space Debris Remediation and On-Orbit Satellite Servicing (27 January 2012), UNCOPUOS, UN Doc. A/AC.105/C.1/2012/CRP.16 (2012), 39.

[81] B. Bastida Virgili *et al.*, *supra* note 60, 155. See also V. Morand, J.C. Dolado-Perez, T. Philippe and D.A. Handschuh, 'Mitigation rules compliance in low Earth orbit', *Journal of Space Safety Engineering* 2014, Vol. 1, 84–92.

[82] On ISO Standard 24113, see A. Soucek, *supra* note 59, 637 and notes 151–152. See also L. Viikari, 'Environmental aspects of space activities' in F. von der Dunk and F. Tronchetti (eds), *Handbook of Space Law*, Cheltenham and Northampton, MA, Edward Elgar Publishing, 2015, 752–753. For a more general analysis of advantages and concerns with regard to the ISO's environmental standards, see D.A. Wirth, 'The International Organization for Standardization: private voluntary standards as swords and shields' in G. Van Calster and D. Provost (eds), *Research Handbook on Environment, Health and the WTO*, Cheltenham and Northampton, MA, Edward Elgar Publishing, 2013, 139–153.

[83] J. Radtke, C. Kebschull and E. Stoll, *supra* note 63, 56 ('OneWeb understands the high responsibility that comes with introducing hundreds of satellites into the LEO environment, which furthermore will cross the highly populated regions between 700 and 900 km twice. While this is a very positive signal, still many open questions remain on how OneWeb will be able to fulfill

It is, moreover, far from certain that operators of similar large constellations, possibly under the authorization of other States, will act in a similar fashion. If compliance, bringing with it additional costs for operators, is not universal, domestic space industries could be at a competitive disadvantage on the global markets.[84] Purported reasons of self-interest[85] so far also do not seem to have engendered widespread compliance with undeniably vital PMD measures. This casts a worrisome shadow on the sustainability of space activities, *a fortiori* when multiple heretofore unseen megaconstellations may interact with one another or cause collisions in shared orbital regimes.[86] Of course, myriad other concerns arise with regard to, inter alia, the undermining of international security and stability, the regulatory frameworks for active debris removal or to any of the other complex issues in space governance that have long since been intrinsically linked to the space debris conundrum, as well as to a growing number of spacefaring actors.

The lack of an enforceable international legal framework or, alternatively, of sufficiently harmonized national legislation on matters involving debris or large constellations should, therefore, give significant pause for thought in view of these novel developments. This is especially so, when even single collision events can drastically affect the debris population[87] and considering that even during timeframes of historically

---

these self-imposed requirements' and '[o]ne very large unknown of the final satellites is the electrical engine to be used').

[84] D. Rex, 'Will space run out of space? The orbital debris problem and its mitigation', *Space Policy* 1998, 104–105; T.K. Percy and D.B. Landrum, 'Investigation of national policy shifts to impact orbital debris environments', *Space Policy* 2014, 32.

[85] 'Based on our experience, we believe Sates [*sic*] are motivated first and foremost by enlightened self-interest in the safety and sustainability of space activities. We do not believe that the force of legal obligation is necessary for States to take measures to mitigate debris.' Statement by the United States of America at the 52nd session of the Legal Subcommittee of UNCOPUOS in 2013, as cited in T. Brisibe, *supra* note 3, 29.

[86] See, e.g., *supra* note 75.

[87] Towards Long-Term Sustainability of Space Activities: Overcoming the Challenges of Space Debris, A Report of the International Interdisciplinary Congress on Space Debris (3 February 2011), UNCOPUOS, UN Doc. A/AC.105/C.1/2011/CRP.14 (2011), 29.

and comparatively low traffic in and towards LEO, there was a doubling in the spatial density of debris.[88]

As an example, if something were to go wrong – such as a collision or break-up event – in the implementation of a megaconstellation operating, by its very nature, in a sphere surface around the Earth and at an altitude exhibiting only moderate or little natural decay, such as the 1200 kilometres envisaged by OneWeb,[89] the resulting debris would *irreversibly* affect the orbits at the operational altitude for several centuries or millennia and beyond.[90] Moreover, whereas the exact motion of fragments after a break-up is uncertain, the resulting debris cloud would be uniformly distributed over the same surface through 'chaotic motion', essentially creating a sphere of debris indiscriminately affecting any future activities: not only at the entire given altitude, but, in the long run, also above and below it.[91] Evidently, if more than one megaconstellation

---

[88] J. Radtke and E. Stoll, 'Comparing long-term projections of the space debris environment to real world data – looking back to 1990', *Acta Astronautica* 2016, Vol. 127, 489.

[89] It must be granted, however, that the choice of an orbital altitude of 1200 kilometres does have beneficial effects and is supported by scientific recommendations. The altitude exhibits a comparably low spatial density of existing debris, limiting the number of interactions between the constellation and the background debris population. Statements by OneWeb show that these recommendations have been taken into account. See OneWeb Presentation, *supra* note 40, 16. See also J. Radtke, C. Kebschull and E. Stoll, *supra* note 63, 67 ('During this analysis it was found that orbit of 1200 km altitude chosen for OneWeb is recommendable as it lies within a local minimum of the number of required collision avoidance manoeuvres for the first generation of OneWeb satellites'). See further S.-Y. Su, 'Impact of satellite constellation arrangement on the future orbital debris environment', *Advances in Space Research* 1997, Vol. 19, 354; R. Walker, R. Crowther and G.G. Swinerd, 'The long-term implications of operating satellite constellations in the low Earth orbit debris environment', *Advances in Space Research* 1997, Vol. 19, 355–358.

[90] F. Alby, *supra* note 76, 63.

[91] Z. Binbin, W. Zhaokui and Z. Yulin, 'An analytic method of space debris cloud evolution and its collision evaluation for constellation satellites', *Advances in Space Research* 2016, Vol. 58, 903–904 and 908–909. 'The injection of many satellites into a single altitude and inclination regime could provide the initiating conditions for a collision-induced cascading effect. This could be generated by interactions between the constellation and the background debris population. It would cause the debris population growth to accelerate and pose an increased hazard to future systems. The collision-induced fragmentation of a constellation satellite generates a debris cloud that expands to envelop all other satellites in the region, including satellites both internal and external to the constellation': R. Walker, R. Crowther and G.G. Swinerd, *supra* note 89, 355. E.g., see also

is operating at the same altitude or in close proximity to another,[92] the risks of collision and debris proliferation are likely to increase by orders of magnitude.[93]

Although a mitigative approach by itself is insufficient to curtail the proliferation of debris over the longer term, the above considerations evidently raise questions with respect to the swift implementation by States of mitigation measures in space activities under their authority and responsibility. We enter here into the murky legal waters pertaining to the non-binding nature of the international mitigation guidelines and to the *corpus iuris spatialis* itself.

(ii) The Liability Convention and its inadequacy to preserve the environmental amenity of outer space

As stated in the UNCOPUOS SDM Guidelines, 'they are not legally binding under international law' and rely on voluntary implementation by Member States and international organizations, through national or other mechanisms, for their application.[94] Whether it is opportune to have non-legal and explicitly non-binding international SDM Guidelines continues to be an issue of debate in the literature and at the UN,[95] but is all the more important in light of the preceding remarks. It must be conceded that making the guidelines non-binding at the international level may contribute to the limited compliance rates witnessed in regard to LEO to date. If the sustainable development of outer space is to be assured, we might look, therefore, at other legal methods of ensuring that the UNCOPUOS SDM Guidelines in general, and PMD measures in particular, are implemented to the greatest extent possible.

Mitigation guidelines as standards of care? In this respect, Stubbe and Schrogl argue that international SDM Guidelines nevertheless hold legal significance, since 'they can be regarded … as defining the standard of

---

S.P. Barrows, G.G. Swinerd and R. Crowther, 'The cascade fragmentation of a satellite constellation', *Advances in Space Research* 1995, Vol. 16, 119–122; J. Radtke, C. Kebschull and E. Stoll, *supra* note 63, 65–66.

92 See, e.g., *supra* note 75.

93 See, e.g., *supra* note 91.

94 UNCOPUOS SDM Guidelines, 48.

95 Compare, e.g., '[s]ome delegations expressed the view that the Legal Subcommittee should develop legally binding rules for space debris mitigation' with '[t]he view was expressed … that it was not necessary to establish debris mitigation standards in international law at present': Report of the Legal Subcommittee on its fifty-fifth session, *supra* note 26, paras 167 and 169.

reasonable conduct, which is relevant to the assessment of fault within the context of the Liability Convention.'[96] They go on to note that, for the determination of fault-based liability for damage caused in outer space in accordance with Article III of the Liability Convention, fault is established when there is either negligent conduct (*culpa*) or conduct with an intent to harm (*dolus*). In turn, the existence of negligence, and hence fault, is assessed objectively by comparison with the conduct that can be reasonably expected under given circumstances and that can be defined, inter alia, by established standards and practices.[97]

As the UNCOPUOS SDM Guidelines are substantially based on the IADC SDM Guidelines, which comprise input from space agencies of all major and other spacefaring States:

> there is very good reason to regard the mitigation norms contained therein as those objective standards and practices that define the reasonable conduct with respect to the avoidance of space debris. If non-compliance with these norms leads to the causation of damage in the scope of Art. III LIAB, the respective launching State of the damage causing object can be held to be at fault.[98]

It is submitted that the same holds for a teleological interpretation of UNCOPUOS SDM Guideline 6, on the basis of its qualitative character. The corresponding standard of reasonable conduct for megaconstellations should, thus, be understood to comprise a post-mission lifetime of five years or less.

On the one hand, the prospect of sizeable claims for compensation under the Liability Convention might well prove to be an incentive for actors, and their sponsoring States, to observe the UNCOPUOS SDM Guidelines and to adopt necessary preventive or mitigative measures in order to pre-emptively exculpate themselves.[99] On the other hand, current State practice points towards low compliance with PMD measures in LEO and should raise at least some doubts as to the incentivizing effect deriving from the retrospective application of the Liability Convention or, indeed, as to the efficacy of any scheme of reparations.

---

96 P. Stubbe and K.-U. Schrogl, 'The Legal Significance of the UNCOPUOS SDM Guidelines' in S. Hobe, B. Schmidt-Tedd, K.-U. Schrogl (eds) and P. Stubbe (Assist. ed.), *supra* note 59, 643.

97 *Ibid.*, 648. See also L. Viikari, *supra* note 82, 763.

98 P. Stubbe and K.-U. Schrogl, *supra* note 96, 648.

99 See C. Cinelli and K. Pogorzelska, 'The current international legal setting for the protection of the outer space environment: the precautionary principle *avant la lettre*', *Review of European, Comparative & International Environmental Law* 2013, 190–191.

A healthy dose of scepticism is especially warranted from the current chapter's perspective on the need for safeguarding the long-term sustainability of space activities, and the amenity of the near-Earth space environment, before the potentially *irreversible* consequences of large constellations come to pass. Of course, much will depend on if and how the concept of negligence and its standards for assessment will evolve proportionally to the increased impact that megaconstellations may have on the space environment.

Likelihood of controversy, discord and augmented tensions Hypothetical claims for compensation under the Liability Convention could be very large as a result of the costly nature of space infrastructure. Were an incident to occur, any assessment of fault-based liability with reference to the UNCOPUOS or IADC SDM Guidelines is, hence, likely to be vehemently challenged by potentially liable launching States on the basis of their non-binding status. The same holds true for the more restrictive teleological interpretation of the UNCOPUOS SDM Guidelines proposed above. The potential for disputes is exacerbated by a lack of legal certainty due to the extreme scarcity of precedents and the absence of any case law pertaining to the Liability Convention on which a case, in one way or another, could otherwise be based.[100] In turn, and paradoxically so for any presumed incentive-creating effect, this legal uncertainty 'undermines the value of international law as a guide to future State conduct'.[101]

Moreover, as large constellations will represent a 'step change' in mankind's activities in outer space, it is not hard to envision the argument being put forward that the threshold for 'reasonable conduct' under these unique circumstances should be lower. Mindful of the nascent and

---

[100] The Liability Convention has never been invoked before a Court. The only precedent to date is the claim by Canada against the Soviet Union invoking, inter alia, absolute liability in the Liability Convention to claim compensation for damage when nuclear material was scattered over its territory by the re-entry of the uncontrollable Cosmos 954 satellite. Canada Claim Against the USSR for Damage Caused By Soviet Cosmos 954, 23 January 1979, *International Legal Materials* 1979, Vol. 18, 899–908. See also E. Galloway, 'Nuclear powered satellites: The U.S.S.R. Cosmos 954 and the Canadian claim', *Akron Law Review* 1979, Vol. 12, 401–415; T. Masson-Zwaan and R. Crowther, 'Legal and regulatory issues' in M. MacDonald and V. Badescu (eds), *The International Handbook of Space Technology*, Berlin, Springer Praxis Books, 2014, 661.

[101] A. Ovchar, 'Estoppel in the jurisprudence of the ICJ. A principle promoting stability threatens to undermine it', *Bond Law Review* 2009, Vol. 21, 33.

uncertain nature of the activities, negligence might be precluded. In fact, the IADC's admission that the robustness of the current guidelines is shrouded in doubt may in some sense serve – or be seen to serve – as a justification for derogating from the measures prescribed therein.

Incidentally, with regard to the mere production of debris, a number of delegations have invoked the 'principle of common but differentiated responsibilities'[102] in UNCOPUOS. Even so, in contrast with any intention to protect the space environment, the underlying political motive appears to be both to shift the burden of cleaning up the near-Earth orbits to those States that have historically contributed most to the proliferation of debris, and to justify a less restrictive application of mitigative measures to their space activities proper.[103] Any perceived negligence, actionable or not, with regard to mitigation in the implementation of megaconstellations, can only serve to strengthen their resolve, thereby potentially instigating a divisive race to the bottom that puts the space environment at increased risk.

Damage to the 'space environment'? The Liability Convention sets out a rather narrowly delineated category of 'damage', more precisely defined in Article I(a) of the Convention as 'loss of life, personal injury or other impairment of health; or loss of or damage to property of States or of persons, natural or juridical, or property of international intergovernmental organizations'. Even if the mere production of debris by itself could somehow be held to be negligent in reference to the SDM Guidelines, this would not give rise to a valid claim in light of the lack of 'damage' and of any injured State Party. That is to say, the *sole* fact that debris is created does not engage the liability of the launching State(s) under the Convention. 'Damage' to what can be termed the 'space environment' per se – or, alternatively, 'damage' consisting solely of the degradation of the environmental amenity or utility of LEO – seems to

102 E.g., see T. Honkonen, *The Common but Differentiated Responsibility Principle in Multilateral Environmental Agreements: Regulatory and Policy Aspects*, Alphen Aan Den Rijn, Kluwer Law International, 2009, 1.

103 Although the exact wording has evolved, the underlying motives appear unchanged, e.g. Report of the Legal Subcommittee on its fifty-fifth session, *supra* note 26, para. 172 ('Some delegations expressed the view that the issue of space debris should not be treated in a way that limited access to outer space or impaired the development of space capabilities by the least developed or developing countries, and that it was necessary to take into account the principle of proportional responsibility for space debris removal').

fall outside the limited definition set out in Article I(a) of the Liability Convention.[104]

One author sums up the legal predicament, already salient in 1992:

> Strikingly, outer space as an area – vital to the continued existence of the Earth – is excluded from the applicability of this definition. … There is the practical consideration that in the area called 'outer space', if damaged, it will become next to impossible to proceed with scientific and other space missions hitherto considered essential to a better insight into the structure and dynamics of outer space. Also, a number of space applications will become difficult if not impossible at all. … In view of the present largescale pollution of outer space, inclusive of the near-Earth environment and the Earth itself by space debris, … it is submitted that the term 'damage' as contained in the [Liability Convention] is insufficiently covering the activities of States in outer space.[105]

Furthermore, where damage to the space environment per se is concerned, there cannot be said to exist an injured State Party with *locus standi* for the purposes of the Convention. It is, moreover, unlikely that any entity could hope to pursue on the international stage a claim similar to an *actio popularis*[106] that does not represent the specific interests of an injured State Party but, instead, a more diffuse public interest in the preservation of a hospitable or amenable orbital environment.[107] To date, there has been no instance of a State bringing a claim against another State, under the Convention or on other grounds, involving either 'damage' to the space environment per se or otherwise predicated on a loss of its amenity for activities by another due to space debris – that is, decades of State practice runs counter to any view to the contrary.

Hanqin sheds light on this circumstance:

---

104 As one author suggests, an interpretative method might here bring solace, but the approach is far from a legal certainty: '[a]lthough the definition does not refer to "environmental" harm, it can be interpreted to allow compensation claims for the "property of states" which are environmental assets or other natural resources', P. Sands, *Principles of International Environmental Law*, Cambridge, Cambridge University Press, 2003, 897.

105 B.C.M. Reijnen, *The United Nations Space Treaties Analysed*, Gif-sur-Yvette, Editions Frontières, 1992, 180–181, original emphasis.

106 E.g., see W.J. Aceves, 'Actio popularis – the class action in international law', *University of Chicago Legal Forum* 2003, 356–359;

107 E.g., see G.I. Hernandez, 'A reluctant guardian: the International Court of Justice and the concept of "international community"', *British Yearbook of International Law* 2013, Vol. 83, 13–60.

> [I]in accordance with the principles of State responsibility and liability, remediable damage must be connected with a legal right or interest of a State, i.e. an entity with plenary legal personality in international law. ... However, in the case of damage to the global commons – namely areas situated beyond natural jurisdiction and control (e.g. polar areas, the high seas or outer space) – it has traditionally been thought that no State can claim damage on behalf of the international community under international law if its own legal rights or interests are not directly affected.[108]

De lege ferenda, it will be interesting to see how state practice and opinion evolve. Under growing pressure by the steady rise in the number of spacefaring actors and space objects, current inertia in the international community must inevitably give way to more advanced legal frameworks to modulate the increasingly complex interactions in the space domain. New instruments will hopefully include more detailed measures for the protection of the space environment as a goal in and of itself.

Consider, in addition, the hypothesis of only one active constellation at a given orbital altitude and containing only a few or no space objects of States other than the launching States. No international liability under the Convention with respect to third States Parties should be expected to arise in the near future since the main effects of or any 'damage' caused by debris – necessitating, for example, additional avoidance manoeuvres or incapacitating other satellites in the constellation – initially will mainly impact the constellation and its launching State(s). Although the constellation's operator(s) or launching State(s) might hence implement mitigative measures out of self-interest or even out of a sense of moral obligation, these reasons are not tied to liability per se and do not have foundations in certain and enforceable legal grounds.

Finally, a likely scenario is that of a typical case concerning uncontentious damage to space objects of third States Parties being brought forward. The conclusion, on what might be valid and tenable legal arguments, could very well be that the launching States' efforts were sufficient to preclude negligence and, thus, to insulate them from any claims for compensation. Under the current legal framework there might simply not exist any wrongful act upon which to base an obligation to repair or compensate. In the meantime, debris will nevertheless have

---

[108] X. Hanqin, *Transboundary Damage in International Law*, Cambridge, Cambridge University Press, 2003, 6–7.

been created that, wrongful or otherwise, will continue to impede the sustainable use of the affected orbit for a long time.

Compensation and reparation after the fact To be sure, in accordance with its provisions, the mechanism for reparation under the Liability Convention for any damage is compensation – that is, monetary or financial payment to be determined as set out in Article XII. From a sustainability standpoint, a compensatory measure falls short of protecting the space environment or its amenity from proliferating debris, for it must necessarily be quantified *ex post facto* when an unspecified number of hazardous debris have already been created – a *fait accompli*. Moving beyond the Liability Convention's limited scope, it can be argued that the entire concept of reparation in international law is unlikely to be an effective means with which to preserve the amenity of the near-Earth orbits.[109] The International Law Commission (ILC) clarifies the problem: '[p]revention should be a preferred policy because compensation in case of harm often cannot restore the situation prevailing prior to the event or accident.'[110]

With regard to reparations following a wrongful act,[111] as elucidated by the International Court of Justice (ICJ), 'customary international law provides for restitution as one form of reparation for injury, restitution being the re-establishment of the situation which existed before occurrence of the wrongful act' and 'where restitution is materially impossible

---

[109] For a more detailed look at reparations in the context of international environmental law, see P. Sands, *supra* note 104, 883–894; M. Burkett, 'Climate reparations', *Melbourne Journal of International Law* 2009, Vol. 10, 509–542. More generally, see also C. Gray, 'The choice between restitution and compensation', *European Journal of International Law* 1999, Vol. 10, 413–423.

[110] Draft Articles on Prevention of Transboundary Harm from Hazardous Activities, with commentaries, in Report of the International Law Commission on its fifty-third session held from 23 April to 1 June and 2 July to 10 August 2001, Official Records of the General Assembly on its fifty-sixth session, Supplement No. 10, UN Doc. A/56/10 (2001), Yearbook of the International Law Commission 2001, Vol. II, Part Two, UN Doc. A/CN.4/SER.A/2001/Add.1 (Part 2) (2001), 148, para. 2 (hereinafter ILC Draft Articles on Transboundary Harm with commentaries).

[111] E.g., see D. Shelton, 'Righting wrongs: reparations in the Articles on State Responsibility', *American Journal of International Law* 2002, Vol. 96, 833856.

or involves a burden out of all proportion to the benefit deriving from it, reparation takes the form of compensation or satisfaction, or even both.'[112]

For the time being there are very good reasons to assume, however, that restitution to the *status quo ante* – that is, an orbit once again unencumbered by, at least, the most dangerous debris – is 'materially impossible' since '[t]here is no way to clean up orbits by eliminating debris.'[113] Myriad proposals in scientific literature notwithstanding,[114] so-called active debris removal (ADR) technology is not yet sufficiently advanced to allow an entity to first manoeuver a spacecraft into close proximity of even a *single* uncooperative, uncontrollable piece of debris that may be exhibiting an erratic attitude and to subsequently capture as well as remove it from orbit, let alone to make this a routine practice.

Despite statements by OneWeb touting the ability of its satellites to perform routine removal of constellation members that are uncooperative or otherwise unable to enact the crucial PMD phase by themselves,[115] substantial doubt remains as to how this highly complicated capacity, seemingly still evading many of the world's top space agencies, is integrated with the spacecraft or how it will be enacted in a routine, reliable and safe manner. This technology is untested and unverified to the author's best knowledge. In other words, the capability is mired in uncertainty and carries potential consequences that might, ultimately, do more harm than good, by creating additional debris in the process. It is, moreover, far from certain that spacecraft in megaconstellations envisioned by other actors will feature similar capabilities.

Even if such technology were to exist today, the potentially large number of unresponsive satellites or, after a collision, of debris fragments to be individually targeted would make cleaning up prohibitively difficult and complex for any actor. It would become progressively more so, due to the inescapable physical realities of orbits at relatively higher and more difficult-to-reach altitudes, such as the 1200 kilometres envisioned by OneWeb. This is not to mention that 'the cost of such a mission to

---

112 *Pulp Mills on the River Uruguay (Argentina v. Uruguay)*, Judgment, ICJ Reports 2010, 103–104, para. 273. For an earlier dictum by the Court in this regard, see *Gabčíkovo-Nagymaros Project (Hungary/Slovakia)*, Judgment, ICJ Reports 1997, 81, para. 152.

113 F. Alby, *supra* note 76, 65.

114 For an extensive overview, see M. Shan, J. Guo and E. Gill, 'Review and comparison of active space debris capturing and removal methods', *Progress in Aerospace Sciences* 2016, Vol. 80, 18–32.

115 OneWeb Presentation, *supra* note 40, 15.

retrieve a single piece of debris, or a few pieces of debris located in adjacent orbits, would seem exorbitant.'[116]

Therefore, at present and for the foreseeable future, restitution with respect to any affected orbits, in the unlikely circumstance that the creation of debris is held to be a wrongful act, is both 'materially impossible' and almost certainly 'a burden out of all proportion to the benefit deriving from it', due to the astronomical investment required and the technological obstacles yet to be surmounted. That leaves compensation and satisfaction as the forms of possible reparation. Compensation is inadequate, as already mentioned, because it cannot fundamentally alleviate the problem of any created debris or physically restore the orbits to a state more amenable to future activities. The same is evidently true for reparation in the form of mere 'satisfaction' of seeing a claim validated or substantiated. At best, both would prove to be a pyrrhic victory for any party wishing to preserve the environmental amenity of near-Earth orbits for future activities.

The preceding considerations were succinctly summarized, *mutatis mutandis*, in the joint dissenting opinion of judges Al-Khasawneh and Simma to the *Pulp Mills* judgment of the ICJ. In the context of transboundary harm in the field of international environmental law, the judges argue to take 'a forward-looking, prospective approach, engage in a comprehensive risk assessment and embrace a preventive rather than compensatory logic'.[117] Indeed, '[t]his logic carries with it particular cogency in the realm of environmental law'[118] and dovetails wholly with the earlier proclamation by the Court that 'in the field of environmental protection, vigilance and prevention are required on account of the often irreversible character of damage to the environment and of the limitations inherent in the very mechanism of reparation of this type of damage.'[119]

Prescribing to the Court's logic and considering the highly novel and superlative nature of large constellations, as well as of the resulting future environmental uncertainties and potentially destabilizing effects on the long-term space debris population, it is submitted that valid *present-day*

---

[116] F. Alby, *supra* note 76, 66.

[117] Joint Dissenting Opinion of Judges Al-Khasawneh and Simma, *Pulp Mills on the River Uruguay (Argentina v. Uruguay)*, Judgment, ICJ Reports 2010, 118, para. 22.

[118] *Ibid.*

[119] *Gabčíkovo-Nagymaros Project (Hungary/Slovakia)*, Judgment, ICJ Reports 1997, 78, para. 140.

concerns must engender a more future-oriented and prospective manner in which to regulate this novel activity. Prevention is better than the cure.[120]

B. Large Constellations from a Preventive and Precautious Perspective

To avoid being confronted with a *fait accompli* in the management of LEO, it can be helpful to ascertain whether the relevant space actors, spacefaring States and the international community as a whole can be legally compelled to steer clear of the risks of large constellations, both certain and uncertain.

As Brisibe argues:

> [T]here are laws and practices from which immediate and reliable analogies can be drawn, such as those concerning the responsibility of states for breaches of international law and appropriate remedies, with specific reference to regimes on liability deriving from other environmental agreements and international law principles. Such laws and practices inspire and illustrate the benefits of adopting binding rules of conduct concerning outer space activities, against which legal obligations can be established to take appropriate measures preventing harm.[121]

The following sections hence take inspiration from analogous developments[122] in the field of international environmental law with respect to transboundary harm and the precautionary principle.

(i) Obligations to prevent transboundary harm: an inadequate alternative?

Scope of the obligations The body of interrelated obligations on the prevention of transboundary harm from hazardous activities might offer a suitably complementary legal framework accommodating some of the above concerns. To be sure, '[i]nternational law, on the one hand,

---

[120] ILC Draft Articles on Transboundary Harm with commentaries, *supra* note 110, 148, para. 2.

[121] T. Brisibe, *supra* note 3, 27–28.

[122] For a comprehensive analysis of the concept of transboundary harm and the ICJ's potential role, in the context of GHG emissions, offering useful analogies for its application with regard to space activities, see J.C. Glickenhaus, 'Potential ICJ Advisory Opinion: duties to prevent transboundary harm from GHG emissions', *New York University Environmental Law Journal* 2015, Vol. 22, 117–155.

provides for a series of autonomous prevention obligations and, on the other, a due diligence obligation to prevent transboundary harm where the prevention obligations play an important role.'[123]

At first glance, slated activities with regard to large constellations of satellites fall within the scope set out in Article 1 of the ILC's Draft Articles on Prevention of Transboundary Harm from Hazardous Activities,[124] which is limited to activities that are 'not prohibited by international law and which involve a risk of causing significant transboundary harm through their physical consequences'.[125] Article 2(d) further delineates the scope as pertaining to 'those activities carried out in the territory or otherwise under the jurisdiction or control of a State'.[126] Not only can the no-harm obligations be held to be applicable to space activities pursuant to Article III of the Outer Space Treaty's general invocation of international law, they can also be seamlessly read together with the State of registry's exclusive jurisdiction and control over space objects set out in Article VIII of that treaty, as well as with the non-wrongful character, in principle, of the implementation of large constellations.

A closer look at the scope and definitions set out in Article 2 of the ILC's Draft Articles supports this view. Importantly, Article 2(b) defines 'harm' not only as harm caused to persons or property, but also explicitly includes harm to the environment per se. Moreover, for activities to be covered by the obligations, the required threshold is that of a 'risk of causing significant transboundary harm': this is defined in Article 2(a) as 'risks taking the form of a high probability of causing significant transboundary harm and a low probability of causing disastrous transboundary harm'.[127] It is submitted that, on the basis of the many aforementioned risks to the space environment or its amenity inherent in highly novel large constellations, this threshold is reached across the

---

123 I. Plakokefalos, 'Prevention obligations in international environmental law', *Yearbook of International Environmental Law* 2012, Vol. 23, 43.

124 The Draft Articles may serve in our analysis since, inter alia, the Court made explicit reference to them in its *Pulp Mills* judgment: 'which apparently indicates an increasing trust placed by the Court on the [International Law] Commission.' S. Villalpando, 'On the International Court of Justice and the determination of rules of law', *Leiden Journal of International Law* 2013, Vol. 26, 247. See also *Pulp Mills on the River Uruguay (Argentina v. Uruguay)*, Judgment, ICJ Reports 2010, 68 para. 152, 82 para. 203 and 87 para. 215.

125 ILC Draft Articles on Transboundary Harm with commentaries, *supra* note 110.

126 *Ibid.*, 149, Article 1, para. 1.

127 *Ibid.*, 151–152, Article 2(a).

entire spectrum of probabilities and corresponding magnitudes of risk, for both significant as well as disastrous transboundary harm.[128] It would then appear incumbent, under international law, on the authorizing States of the OneWeb constellation to meet the no-harm obligation, as elaborated in more detail by the ICJ (see below).

In particular, the no-harm obligation states that '[t]he State of origin shall take all appropriate measures to prevent significant transboundary harm or at any event to minimize the risk thereof,' as set out in Article 3 of the Draft Articles. This 'primary' no-harm obligation reflects the fundamental principle of *sic utere tuo et alienum non laedas*[129] – requiring States, in international law, to use assets under their control in a manner that does not injure other States[130] – and is supported by a series of secondary obligations, the fulfilment of which substantially contributes to the satisfactory exercise by States of the due diligence requirement under the primary rule.[131]

Notable examples of secondary obligations in the Draft Articles include, inter alia: the obligation to cooperate internationally in good faith and to seek assistance of one or more competent international organizations (Article 4); the obligation for the responsible State to confer prior authorization for an activity carrying sufficient risk (Article 6); the obligation to assess the risks of possible transboundary harm resulting from an activity, including an environmental impact assessment (EIA) (Article 7); the obligation to notify and inform likely affected States (Article 8); or the obligation to consult with all concerned States with a view to achieving mutually acceptable solutions as to possible preventive measures (Article 9).

---

[128] *Ibid.*, 152, Article 2, para. 1–4: 'The definition … allows for a spectrum of relationships between "risk" and "harm", all of which would reach the level of "significant".'

[129] *Ibid.*, 153, Article 3, paras 1–3. See also V. Gupta, 'Critique of the International Law on Protection of the Outer Space Environment', *Astropolitics* 2016, Vol. 14, 33.

[130] For an assessment of the principle and its legal history, see L. Viikari, *The Environmental Element in Space Law: Assessing the Present and Charting the Future*, Leiden, Martinus Nijhoff Publishers, 2008, 150–157; F. Weiss and B. Scherzer, '(Existence of) common or universal principles for resource management (?)' in M. Bungenberg and S. Hobe (eds), *Permanent Sovereignty over Natural Resources*, Cham, Springer, 2015, 45–47.

[131] For difficulties with respect to secondary obligations acting as 'objective criteria' for the due diligence no-harm obligation and to the due diligence standard itself, see I. Plakokefalos, *supra* note 123, 32–36.

The no-harm obligation and the International Court of Justice An exhaustive assessment of the obligations falls beyond the scope of the current chapter. The exact contours, with regard to their application to large constellations and concomitant scientific uncertainty on the environmental effects, will need to be identified through further legal analysis: this should take into account, for example, the ICJ's recognition of and elaborations upon[132] the substance and procedural requirements of the differentiated secondary obligations and of the primary due diligence obligation.

Notably, the Court concluded in its 2010 *Pulp Mills* judgment that the general no-harm obligation, identified earlier in its Advisory Opinion on Nuclear Weapons,[133] requires a State 'to use all means at its disposal in order to avoid activities which take place in its territory, or in any area under its jurisdiction, causing significant damage to the environment of another State'.[134] It went on to state that 'it may now be considered a requirement under general international law to undertake an environmental impact assessment where there is a risk that the proposed industrial activity may have a significant adverse impact in a transboundary context, in particular, on a shared resource.'[135]

More recently, the ICJ applied its conclusions in *Pulp Mills* to a general category of 'proposed activities which may have a significant

---

[132] There is, however, a relative scarcity of environmental case law and the Court's contribution to the development of international environmental law has been called into question. It has been noted, e.g., that '[p]art of the reason for the lack of environmental cases may be the Court's notorious difficulty dealing with complex facts': R.J. Bettauer, 'International environmental law-making and the International Court of Justice', *Proceedings of the Annual Meeting of the American Society of International Law* 2011, Vol. 105, 63. See also J.E. Viñuales, 'Legal techniques for dealing with scientific uncertainty in international law', *Vanderbilt Journal of Transnational Law* 2010, Vol. 43, 437–503; P. Sands, 'Climate change and the rule of law: adjudicating the future in international law', *Journal of Environmental Law* 2016, Vol. 18, 19–35.

[133] The Court held that '[t]he existence of the general obligation of States to ensure that activities within their jurisdiction and control respect the environment of other States or of areas beyond national control is now part of the corpus of international law relating to the environment': Legality of the Threat or Use of Nuclear Weapons, Advisory Opinion, ICJ Reports 1996, 241–242, para. 29. See also O.W. Pedersen, 'From abundance to indeterminacy: the precautionary principle and its two camps of custom', *Transnational Environmental Law* 2014, Vol. 3, 330.

[134] *Pulp Mills on the River Uruguay (Argentina v. Uruguay)*, Judgment, ICJ Reports 2010, 55–56, para. 101.

[135] *Ibid.*, 82–83, para. 204.

adverse impact in a transboundary context'.[136] The Court set out procedural aspects of the no-harm obligation, dividing the latter into three distinct and sequentially cascading obligations whereby the activation of each progressive obligation is conditional on the results of the preceding one.[137] As a first step, 'to fulfil its obligation to exercise due diligence in preventing significant transboundary environmental harm, a State must, before embarking on an activity having the potential adversely to affect the environment of another State, ascertain if there is a risk of significant transboundary harm'[138] (ascertainment of risk). Second, if the State uncovers a risk, it must perform an EIA,[139] the specific content of which is to be determined in its domestic legislation or in the authorization process for the proposed activity (assessment of risk).[140] If, in turn, the EIA identifies a risk of significant transboundary harm, the third step requires the State(s) planning the activity 'to notify and consult in good faith with the potentially affected State, where that is necessary to determine the appropriate measures to prevent or mitigate that risk'[141] (notification and consultation).

Application to space activities and large constellations The above obligations are congruent with – or elaborate upon – similar provisions in the *corpus iuris spatialis*, adding to the likelihood of their being applicable, or needing to be applicable, to large constellations.

Given the status of outer space as a global commons, that is an area beyond national jurisdiction pursuant to Article II of the Outer Space Treaty, and considering the freedom of exploration and use of outer space by all States on the basis of equality and in accordance with international law pursuant to Article I OST, all States in general, and *a fortiori* all States conducting or planning to conduct space activities, can likely be held to have an interest in the prevention of significant transboundary

---

[136] *Certain Activities carried out by Nicaragua in the Border Area (Costa Rica v. Nicaragua) proceedings joined with Construction of a Road in Costa Rica along the San Juan River (Nicaragua v. Costa Rica)*, Judgment, ICJ, 16 December 2015, http://www.icj-cij.org/docket/files/152/18848.pdf, 45, para. 104 (hereinafter *San Juan River*).

[137] J.K. Cogan, 'Certain Activities carried out by Nicaragua in the Border Area; Construction of a Road in Costa Rica along the San Juan River', *American Journal of International Law* 2016, Vol. 110, 323 and 325–326.

[138] *San Juan River, supra* note 136, 45, para. 104.

[139] E.g., see L.E. Viikari, 'Environmental impact assessment and space activities', *Advances in Space Research* 2004, Vol. 34, 2363–2367.

[140] *San Juan River, supra* note 136, 45, para. 104.

[141] *Ibid.*

harm resulting from constellations. Moreover, in addition to the space treaties' general insistence upon international cooperation – for example, in Articles I or III OST, *in fine* – international responsibility of and prior authorization by States for national space activities by non-governmental actors under their supervision is a fundamental fixture of the Outer Space Treaty pursuant to Article VI.

Furthermore, apart from the general argument for the protection of the space environment as a goal in itself, to be derived from the spirit of its text and deliberations, the recent first set of consensual Guidelines for the Long-Term Sustainability of Outer Space Activities (LTSSA Guidelines)[142] should induce the adoption of a preventive approach. The guidelines expressly call for States to 'adopt, revise or amend regulatory frameworks to ensure the effective application of relevant, generally accepted international norms, standards and practices for the safe conduct of outer space activities' (LTSSA Guideline 1.2) and to '[a]ddress, to the extent practicable, risks to people, property, public health and the environment associated with the launch, in-orbit operation and re-entry of space objects' (LTSSA Guideline 2.2(c)). Finally, LTSSA Guideline 27.4 makes clear that States and international organizations 'should consider appropriate safety measures to protect the Earth and the space environment from harmful contamination, taking advantage of existing measures, practices and guidelines that may apply to those activities, and developing new measures as appropriate.'

In the same vein, Article IX OST[143] takes an expressly future-oriented perspective with regard to 'harmful interference', namely that:

> [i]f a State Party to the Treaty has reason to believe that an activity or experiment planned by it or its nationals in outer space, including the Moon and other celestial bodies, would cause potentially harmful interference with activities of other States Parties in the peaceful exploration and use of outer space … it shall undertake appropriate international consultations before proceeding with any such activity or experiment.

For interference with a more general and contingent category of activities in the peaceful exploration and use of outer space, Article IX, *in fine*, is less stringent and merely confers a right instead of prescribing an obligation: a State Party which has reason to believe that an activity

---

[142] Guidelines for the long-term sustainability of outer space activities: first set, Report of the Committee on the Peaceful Uses of Outer Space on its fifty-ninth session, *supra* note 37, Annex, 56–67.

[143] For a critical analysis of Art. IX OST apropos environmental protection, e.g., see V. Gupta, *supra* note 129, 24–29.

planned by another State Party would potentially cause harmful interference to activities in general 'may request consultation concerning the activity or experiment'.

Consider the following statement by Reijnen in 1992:

> [T]here is neither in general international law – *lex generalis* – nor in space law – *lex specialis* – to be found any norm more specific than the one of article IX [of the Outer Space Treaty], which obliges States to consult each other prior to the planning of activities in space of a potentially harmful nature.[144]

As shown, the international legal landscape has since changed. The three-pronged methodology advanced by the ICJ under the no-harm rule is not only more precise procedurally, laying out a clear series of progressive obligations of process. Its scope also captures a different and broader set of planned activities, because it does not require potentially harmful interference with regard to only a limited category of 'space activities of other States Parties to the Outer Space Treaty'. Instead, the due diligence obligations are triggered by a significant risk of transboundary harm, including harm to the environment of outer space, ultimately engaging of their own accord an obligation to notify and consult. Alternatively, both regimes do point to the a priori nature of their respective obligations: States must act preventively 'before embarking upon' or 'before proceeding with' any such proposed or planned activity. Although they bear a striking prima facie resemblance, additional legal analysis should demarcate the precise boundaries between both sets of obligations.

Considering those corroborating elements, proposed space activities of States and commercial actors relating to large constellations and their potentially significant effects on the 'shared resource' of the near-Earth orbits appear to fall within the scope of the no-harm rule in international law. The actors can be expected to comply outright with the procedural course set out by the ICJ.

It is presently unclear to what extent the authorizing States of OneWeb, be it domestically or internationally, have acted in accordance with the cascade of obligations put forth by the ICJ, thereby satisfying the due diligence requirement, if at all. One might ask, for example, whether the IADC's ongoing studies conform to the obligation to conduct a comprehensive and prior environmental impact assessment with regard to the effects of large constellations on the space environment. It should

---

144 B.C.M. Reijnen, *supra* note 105, 131.

also not be forgotten that the IADC felt compelled to issue a preliminary statement on the matter, '[r]ecognising that a number of constellation architectures are already in the process of development'.[145] This statement – or admission? – leads to the consideration that any current or future ascertainment or assessment by the IADC is not a prior one – *before embarking upon* or *before proceeding with* – but will at best be contemporaneous to activities already underway.

Considering the ICJ's three-step procedure, it might be legally more precise to qualify the IADC's statement as only *potentially* fulfilling the first step – *if*, indeed, the IADC could be held to act on account of the authorizing States – consisting of the preliminary ascertainment of risk of significant transboundary harm that would require, under international law, a comprehensive EIA to be conducted as the obligatory second step. In turn, and dependent on its results, the EIA can trigger the third a priori step, involving the obligation to notify and consult with all potentially affected States, that is the international community at large. As to the latter, it is doubtful whether the presentation given by OneWeb on its planned activities in the context of UNCOPUOS's STSC is sufficient to discharge the participating State(s) of their duties.[146]

Insofar as the sequence of steps set out by the ICJ has not yet been allowed to run its due procedural course, it appears OneWeb's authorizing State(s) would be acting in contravention to the due diligence obligation in international law to prevent transboundary harm, and increasingly so, should they persist in the process of deploying the constellation.

Fundamental inadequacy Nonetheless, as Plakokefalos rightly points out, the exact content of the entirety of obligations in different contexts remains vague, their normative weight is limited, they leave a wide margin of discretion to the State(s) of origin – for example, in deciding on the specificities of an EIA – and '[i]t is this lack of strength and clarity of the prevention obligations that renders the content of the general no-harm obligation equally weak.'[147] Although the Court's development has certainly contributed to delimiting more clearly dimensions of the due diligence obligation, it remains one of process instead of result,[148] calling into question the degree to which the rule can adequately protect the near-Earth orbits.

---

145 IADC, *supra* note 47, 4.
146 OneWeb Presentation, *supra* note 40.
147 I. Plakokefalos, *supra* note 123, 43.
148 O.W. Pedersen, *supra* note 133, 338.

Indeed, the efficacy of the obligations is equally hampered by the same problem already identified in earlier sections: '[t]he situation where a state has fulfilled its due diligence obligation and nevertheless transboundary environmental harm does occur is the most troubling aspect of the whole matrix of obligations.'[149] In simple terms, debris will have been created and irreversible harm to the space environment will have been done.

Plakokefalos concludes, *de lege ferenda*:

> It seems that the task international law faces is twofold. On the one hand, strengthening the obligations to prevent transboundary harm must become a priority. On the other hand, an effective international system of liability must be developed so as to cover the instances where damage occurs and the engagement of the rules on state responsibility is not an option since the primary obligation [to prevent transboundary harm] has not been breached. Only if these two goals are met will the states develop more nuanced, detailed, and, more importantly, efficient rules governing the prevention of transboundary harm.[150]

Despite their fundamentally more prospective approach, no-harm obligations still appear to fall short. The next section therefore shifts the focus from 'prevention' to 'precaution' as an envisioned ultimate means to safeguard LEO from possibly deleterious effects of large constellations.

(ii) The precautionary principle and large constellations

The fractured principle The status of the much-discussed precautionary principle and its many relevant aspects with regard to general international law are unequivocally problematic.[151] Although – or because – numerous and divergent incarnations of the principle have been

149 I. Plakokefalos, *supra* note 123, 43.

150 *Ibid.*

151 'Whether in spite or because of the vast attention it has been receiving, the precautionary principle continues to be prone to considerable misgivings. In particular, its legal status and consequences are subject to a confusion of tongues of Babylonian proportions', nurturing in turn 'another peculiarity of the principle: the strikingly polemic nature of the academic debate on the subject': A. Trouwborst, 'The precautionary principle in general international law: combating the Babylonian confusion', *Review of European, Comparative & International Environmental Law* 2007, Vol. 16, 185–186.

codified in international treaties, instruments and declarations,[152] debate between critics[153] and proponents[154] on the principle or its application continues on multiple fronts. Foremost, perhaps, is the discussion on whether the principle has entered into the realm of customary international law and on all of the consequences such a status would entail.[155] In a similarly controversial fashion, it has inspired varying levels of endorsement or reticence regarding its application in proceedings before international tribunals and judicial bodies,[156] further adding to the fractured legal landscape.

Suffice it to say, significant complexity surrounds any argument basing itself on a precautionary approach in international law. For the sake of brevity, the current section casts only a cursory and superficial glance at the principle in the *lex lata*, referring for a more comprehensive enquiry to the ample attention the topic has received in literature. Instead, the general principle and salient considerations carried in its wake are used as a lens through which to view the current topic.

General scope and effects of precaution and large constellations Following its introduction to the intergovernmental plane in 1987,[157] the

---

152 *Ibid.*, 187; D.L. Vanderzwaag, 'The ICJ, ITLOS and the precautionary approach: paltry progressions, jurisprudential jousting', *University of Hawai'i Law Review* 2013, Vol. 35, 617, note 4.

153 Inter alia, critics argue that an application of the precautionary principle would make any progress or innovation impossible or cumbersome. For an overview of common criticisms and counterarguments, see P. Sandin, M. Peterson, S.O. Hansson, C. Ruden and A. Juthe, 'Five charges against the precautionary principle', *Journal of Risk Research* 2002, Vol. 5, 287–299. See also C.R. Sunstein, 'Cost-benefit analysis and the environment', *Ethics* 2005, Vol. 115, 351–385.

154 E.g., see N.M. Sachs, 'Rescuing the strong precautionary principle from its critics', *University of Illinois Law Review* 2011, 1285–1338.

155 For an extensive assessment of the customary law character of the precautionary principle, see: O.W. Pedersen, *supra* note 133, 323–339.

156 E.g., see D. Kazhdan, 'Precautionary pulp: Pulp Mills and the evolving dispute between International Tribunals over the reach of the precautionary principle', *Ecology Law Quarterly* 2011, Vol. 38, 527–552; D.L. Vanderzwaag, *supra* note 152, 617–632; A.H. Ansari and S. Wartini, 'Application of precautionary principle in international trade law and international environmental law: A comparative assessment', *Journal of International Trade Law and Policy* 2014, Vol. 13, 19–43; K. Garnett and D.J. Parsons, 'Multi-Case review of the application of the precautionary principle in European Union law and case law', *Risk Analysis 2016*, DOI: 10.1111/risa.12633.

157 See A. Trouwborst, *supra* note 151, 187.

precautionary principle rapidly evolved into a central tenet of international environmental law.[158] A short period thereafter it was included in the 1992 UN Rio Declaration on Environment and Development, the instrument most cited with respect to the principle,[159] and became enshrined in its principle 15: '[i]n order to protect the environment, the precautionary approach shall be widely applied by States according to their capabilities. Where there are threats of serious or irreversible damage, lack of full scientific certainty shall not be used as a reason for postponing cost-effective measures to prevent environmental degradation.'[160]

Ever since, it has reappeared in a constant state of flux, throughout a variety of permutations and wordings, each oftentimes differing from the others and adding to or opting to omit parts of its substance – for example, the reference to cost-effective measures is far from universal. This definitional indeterminacy[161] makes the principle difficult both to operationalize and to apply.[162] Despite this fact, Sandin attempts to distil its fundamental essence into a single expression with four dimensions: '[i]f there is (1) a threat, which is (2) uncertain, then (3) some kind of action (4) is mandatory.'[163] Leaving aside (4), the controversial mandatory nature of precaution in international law, the three former elements require clarification in general, and in relation to large constellations in particular.

Incidentally, the application of the precautionary principle to the space environment has featured in scholarly literature.[164] Notably, Larsen seeks

---

158 *Ibid.*, 187; D.L. Vanderzwaag, *supra* note 152, 617.

159 D. Kazhdan, *supra* note 156, 532–533 and note 41.

160 Rio Declaration on Environment and Development, Report of the UN Conference on Environment and Development (14 June 1992), UN Doc. A/CONF.151/26/Rev.1 (1992), Vol. 1, Annex, principle 15. See also F. Weiss and B. Scherzer, *supra* note 130, 47–49.

161 Definitional indeterminacy would not appear, prima facie, to prevent the precautionary principle from achieving the status of customary international law. For a comparison with similarly indeterminate norms, such as the no-harm rule and the EIA-rule, see O.W. Pedersen, *supra* note 133, 338.

162 P. Sandin, 'Dimensions of the precautionary principle', *Human and Ecological Risk Assessment: An International Journal* 1999, Vol. 5, 890.

163 *Ibid.*, 898.

164 E.g., see L. Viikari, *supra* note 130, 157–178; C. Cinelli and K. Pogorzelska, *supra* note 99, 186–201.

to underpin the application of the principle to (activities in) the lunar environment by invoking the latter's fragility.[165] He argues specifically that:

> All astronauts agree that outer space, including the Moon, is a very dangerous place that does not permit any mistakes to be made. Its fragile environment makes the outcome of lunar activities uncertain. It is particularly important to focus on uncertainties in making decisions about the Moon's environment because erroneous decisions about the Moon cannot be reversed ... [T]he Moon is fragile ... it is unable to 'heal' itself.[166]

In an almost prescient manner, his arguments can today, *mutatis mutandis*, just as well be extended to the near-Earth orbits. Hence, in the current context, considerable weight should be given to the precise wording in the UN General Assembly's Resolution 70/82, stating that the Assembly is '[d]eeply concerned about the fragility of the space environment and the challenges to the long-term sustainability of outer space activities, in particular the impact of space debris which is an issue of concern to all nations'.[167] Interpreted together with the recent promulgation in UNCOPUOS's STSC of a first set of LTSSA Guidelines,[168] and in combination with the various statements of concern in UNCOPUOS, ITU and IADC on the effects of small satellites and large constellations (see above), this is indisputable evidence of the 'delegates' (and thus the world's) awareness of the fragility' of the space environment as well as of a 'general admonition against disrupting the existing environmental balance'.[169] That is to say, if '[t]he rise of the precautionary principle in international law can, to a large degree, be ascribed to a growing awareness of the serious and/or irreversible nature of much environmental harm,'[170] supporting the precautious protection under international law of the space environment per se should merit considerable attention.

---

165 P.B. Larsen, 'Application of the precautionary principle to the Moon', *Journal of Air Law and Commerce* 2006, Vol. 71, 295–306.

166 *Ibid.*, 298.

167 Resolution 70/82 of the UN General Assembly, *supra* note 25, 2.

168 Guidelines for the long-term sustainability of outer space activities: first set, Report of the Committee on the Peaceful Uses of Outer Space on its fifty-ninth session, *supra* note 37, Annex, 56–67.

169 For both quotes, originally pertaining to the lunar environment and applied here to the general context of the space environment and the near-Earth orbits, see P.B. Larsen, *supra* note 165, 301.

170 A. Trouwborst, *supra* note 151, 187.

Returning to the aforementioned three elements distilled from the principle, and considering the emphasis in the previous sections on the many risks and uncertainties inherent in a shift to a new paradigm of large constellations, the author submits that (1) a threat, or the possibility of a threat, of space debris to the amenity of the fragile space environment of LEO, as a result of such an unprecedented development, is self-evident. The same holds true for (2), the uncertainties with regard to necessary technological capabilities – for example, removal capability, conjunction detection capability, collision avoidance capability – and uncertainty in scientific studies as to the environmental effects of single, or especially multiple constellations. This is compounded, in addition, by degrees of scientific uncertainty on the evolution of the space debris population in general.[171]

Confronted by such a lack of certainty, (3) the action(s) to be taken, under a general interpretation of the precautionary principle, should be *effective* and *proportional.* In this regard, the action should effectively safeguard the endangered environment while corresponding to the cumulative probability and gravity of the risk involved, thereby progressively limiting the margin of discretion left for authorizing States as the severity of risks and/or their probabilities increases. Examples of actions to be taken range from additional research and environmental impact assessments, to increased safety margins and protocols or shifting the burden of proof with respect to potential negative effects to the instigator, to the *ultimum remedium* of instituting a moratorium on the activity.[172]

A temporary moratorium on large constellations? In light of the hitherto often reiterated unprecedented nature of large constellations and their superlative size – of both the constellations and their potential effects – it is reasonable to look at a temporary moratorium on their implementation, to ascertain whether it is a precautionary measure for authorizing States that is both proportional and effective in preventing what could be irreversible harm to the near-Earth orbits. The moratorium might then stretch in time beyond the temporary stand-still already incumbent on those States while they complete the process of due diligence under the no-harm rule. It could last until the combination of probability of and gravity of the risk from constellations has sufficiently decreased – that is,

---

171 J. Radtke and E. Stoll, *supra* note 88, 482–490.
172 A. Trouwborst, *supra* note 151, 189.

when technological capabilities and more sophisticated debris environment modelling techniques have been confirmed and less stringent measures eventually become proportional and effective.[173]

This position will certainly prove contentious. It might be expected, for instance, that some may argue that such a moratorium would stifle innovation in space technologies, would represent a loss of socio-economic benefits associated with novel globalized telecommunications or would put an economic burden on projects already beyond the mere planning phase. These arguments should be rejected, at least to start with.

On the one hand, the paramount importance of near-perfect compliance to PMD measures should move relevant States to proffer strict mitigation regulations, linked directly to any authorization for large constellations. Included in such a regulatory framework should be a burden of prior proof[174] on operators to demonstrate *in practice* and *in space*, through test missions with a limited number of satellites, that they are able to fulfil at least two criteria. Their spacecraft should have the reliable capability of autonomously complying to a post-mission lifetime of 5 years or less and the operators should be capable of managing the increasingly complicated collision avoidance protocols[175] that will surely have to be enacted.[176]

---

[173] The proposed temporary moratorium on large constellations draws inspiration from an application of the 'strong precautionary principle' that 'clearly places a default "hold" on certain activities that may pose serious risks, until the proponent can prove that the activity can be conducted safely, or at least with acceptable risk': N.M. Sachs, *supra* note 154, 1296.

[174] For an assessment of the allocation and reversal of the burden of proof in the application of the precautionary principle, see M. Ambrus, 'The precautionary principle and a fair allocation of the burden of proof in international environmental law', *Review of European, Comparative & International Environmental Law* 2012, Vol. 21, 259–270.

[175] 'Even if OneWeb and other mega-constellation operators take strong efforts to avoid collisions and other debris-generating events, their development will still pose challenges for efforts to avoid satellite collisions in general, as the number of operations satellites [*sic*] dramatically increase': J. Foust, 'Mega-constellations and mega-debris', *The Space Review* 2016, 10 October 2016. Available online at http://www.thespacereview.com/article/3078/1 (last accessed 10 February 2017).

[176] E.g., significant uncertainty as to the OneWeb constellation's satellites' capabilities remain, *a fortiori* in an operational context. See J. Radtke, C. Kebschull and E. Stoll, *supra* note 63, 56 ('One very large unknown of the final satellites is the electrical engine to be used'). OneWeb appears to be aware of these requirements, but, as shown, uncertainties remain. 'OneWeb also plans to swiftly deorbit its satellites at the end of their lives. "At OneWeb, the ability to

In addition, and perhaps more importantly, it is not unwarranted, in the context of effective and proportional measures, that the touted capability of constellation spacecraft to perform removal of uncooperative craft unable to enact the PMD phase should be shown to work in a routine, safe and reliable fashion *before* the moratorium can give way to lesser measures: thus, *before* the constellation is put into orbit. Alternatively, if applicable routine removal services are developed by another actor, the constellation's operators should be required to furnish prior proof of having contracted for those services.

Such an approach appears particularly justifiable in light of the IADC's statement that '[i]t is clear that significant improvements in the reliability of the disposal function at end of life will be needed for the new constellations compared with that currently demonstrated by space systems on orbit.'[177] Rather than stifling innovation, the adoption of a temporary precautionary stand-still would ostensibly incentivize widespread technological development in active debris removal, on-orbit servicing and other mitigative technological measures as necessary stepping stones to implementation, while enabling a future wherein cleaning up the space environment becomes a feasible prospect.

On the other hand, the socioeconomic concerns must be framed in their proper long-term context. While a temporary pause in any claimed benefits deriving from constellations will result from a moratorium by definition, this decrease should not serve as an argument to temporally displace[178] the potentially grave and long-lasting effects. Any social or economic benefits derived from rapid implementation in the short term must be weighed against the temporally perpetual costs, mostly accruing over the long term,[179] of potential harm to the amenity of the near-Earth orbits by the nominal operation of constellations or by the materialization of the graver risks associated therewith – costs, moreover, that

deorbit is specified as the highest reliability function in the spacecraft, even above the payload itself," MacLay said. Each spacecraft will have enough fuel to lower its perigee "all the way to re-entry" at the end of their lives, "and are committed to being out of orbit within five years of decommissioning"': J. Foust, *supra* note 175.

[177] IADC, *supra* note 47, 6. See also B. Bastida Virgili *et al.*, *supra* note 60, 158 ('[t]he greatest challenge will be in delivering the reliability of disposal over the lifetimes of the satellites and the constellations envisaged').

[178] T.S. Aagaard, 'Environmental law as a legal field: an inquiry in legal taxonomy', *Cornell Law Review* 2010, Vol. 95, 270 ('A cost-benefit disjunction arises when, as is often true, the benefits of using the environment are experienced immediately, but the costs are not experienced until much later').

[179] E.g., see L. Viikari, *supra* note 82, 767.

must be borne by all future generations of mankind and in the creation of which they will have had no say. All actors, including currently authorizing States, now and forever into the future, will reap the benefits of safer, cleaner and sustainable, albeit delayed, implementation of large constellations.

The need to act – or *not* act – is now. In the words of Trouwborst:

> [p]recaution means taking action at a moment which is early enough to prevent unacceptable environmental damage effectively. In many cases that moment arrives well before conclusive scientific evidence concerning the risk in question becomes available – in a time of doubt, therefore. Under the precautionary principle, the benefit of any such doubt is to go to the environment. *In dubio pro natura.*[180]

Mindful of doubts expressed by the IADC in particular, the international community should consider applying the same precautionary logic and make the concerted choice to err on the side of the space environment and the near-Earth orbits. *In dubio pro LEO.*

5. CONCLUSION

Developments in small satellite technology not only render space activities a feasible prospect for an ever-increasing number of space actors, but also lend themselves to implementation in large constellations of hundreds to thousands of spacecraft performing collective functions. The international space governance community is waking up to these unprecedented developments, as well as their potential impacts. Discussions in UNCOPUOS, ITU and IADC, in particular, highlight the potential risks large constellations may pose in relation to the creation of space debris.

From the perspective of protecting the environment of the near-Earth orbits themselves, or at least their environmental amenity, the current chapter sets out a number of legal considerations apropos large constellations and the proliferation of debris. Near-perfect compliance with the mitigative measure of post-mission disposal over a relatively short post-mission lifetime will likely be of paramount importance if the effects of megaconstellations on LEO are to be sustainable. There is good reason to assume international debris mitigation guidelines, post-mission disposal among them, constitute those standards of conduct that States

---

[180] A. Trouwborst, *supra* note 151, 187.

must follow, lest they are found to be negligent and thus internationally liable under the Liability Convention should damage occur to another State. However, liability, and in its wake the system of reparations and compensation for any damages, are not well suited to protecting the space environment itself, because they do not or cannot protect the space environment per se and, in addition, are necessarily assessed retrospectively – after environmental harm has already occurred.

The ICJ's three-step series of obligations under the primary obligation to prevent transboundary harm in international law might be better equipped to legally accommodate the environmental challenges of large constellations. A risk of significant or disastrous harm to the space environment arising from the implementation of such constellations falls within the scope of the obligations. The authorizing State(s) must, hence, ascertain if there is a risk and if so, conduct an environmental impact assessment. Should the assessment find evidence of potential risks, the State is obliged to notify and consult with all affected States, a group that could be said to be the international community as a whole. Nevertheless, the due diligence obligation to prevent transboundary harm is an obligation of process and not of result. Were space debris to be created in LEO after the due diligence requirement was met, no further recourse would be available for those wishing to protect the environment itself from significant harm.

In an effort to prevent the risks to the space environment relating to the emergence of large constellations, the precautionary principle might offer inspiration to act prospectively: *before* the constellations are put into space. The large degree of uncertainty about crucial technological capabilities, and the exact consequences of constellations with regard to the debris environment, should compel authorizing States and the international community to pause and reflect on those doubts. As long as that significant uncertainty remains, they should opt to err on the side of the near-Earth environment and choose to institute the effective and proportional measure of a temporary moratorium on the implementation of constellations – at least until much-needed technological capabilities and more detailed environmental impact assessments coalesce and together diminish the doubts. Rather than impeding progress, a precautious course will ensure that the fruits of progress, once achieved, do not acquire a bitter aftertaste in the decades to come.

PART II

Regulatory responses

5. US State Spaceflight Liability and Immunity Acts in context

Andrea J. Harrington

1. INTRODUCTION

'The realization of affordable and safe commercial private human access to outer space may be as significant to mankind as the Wright brothers' Flyer, Gagarin's first spaceflight and the Saturn V "Moon rocket" that put Aldrin and Armstrong on the Moon.'[1] With a wide range of companies in the marketplace seeking funds and developing new and innovative space technologies, it would seem that we are well on our way to becoming a society with access to space. Unfortunately, '[d]espite tantalizing commercial possibilities, the long-term technological and commercial viability of commercial human spaceflight remains to be seen. Among the factors contributing to the industry's ultimate success or failure will be the application of laws and the formulation of regulations governing the carriage of human beings into space.'[2]

Given the current state of development of space tourism, effective national regulations which can both foster development of the sector without undue hindrance and simultaneously foster public trust are critical for the functioning of this industry. It is important to note that, in the short term:

> [m]any of the currently planned space tourism projects will operate from one and the same territory. As long as the intended vehicles will 'take off' and 'land' in that territory, the likelihood of cross-border damage is limited, and

[1] T. Masson-Zwaan and S. Freeland, 'Between Heaven and Earth: The Legal Challenges of Human Space Travel' (2010) 66 *Acta Astronautica* 1597, 1597.

[2] T. Hughes and E. Rosenburg, 'Space Travel Law (and Politics): The Evolution of the Commercial Space Launch Amendments Act of 2004' (2005) 31:2 *J Space L* 12, 3.

> in principle that State's national law will apply, whether it concerns orbital or sub-orbital flight.[3]

While there is a plethora of sub-orbital travel endeavours planned, none that have progressed significantly in their development are prepared to serve as point-to-point transportation between two States. Thus, for the time being these activities are more likely to be governed by national laws than bilateral or multilateral agreements, acknowledging of course that these activities must be performed in conformity with governing treaty law and customary international law. Questions of legal liability will rise in prominence and complexity as we move closer to cross-border space transportation.

The necessity for a liability regime that can both promote the growth of the space industry and adequately protect consumers cannot be overestimated. Though regulations may only govern activities with regard to a particular State, it is still necessary for such regulations to be carefully drafted and easily comparable to the regimes of other States. Unfortunately slight drafting differences in the space flight liability and immunity acts being promulgated by US States could cause confusion. Additionally, these liability regimes are unlikely to be acceptable in some other jurisdictions and thus may only be viable while the space travel industry remains in its infancy.

2. DEFINITIONS

While '[t]he term "space tourism" has been defined as "any commercial activity offering customers direct or indirect experience with space travel",'[4] this definition seems to imply that such activities would be carried out solely for the purposes of tourism. It is but one of many problematic definitions in this area of legal discourse. In that vein Professor Crowther states that '[s]pace tourism is the term commonly used to refer to ordinary members of the public buying tickets to travel to space and back.'[5] It is important to note that for the foreseeable future

---

[3] Masson-Zwaan, *supra* note 1, at 1600.

[4] S. Freeland, 'Up, Up and … Back: The Emergence of Space Tourism and Its Impact on the International Law of Outer Space' (2005) 6 *Chi J Int'l L* 1, 6.

[5] R. Crowther, 'The Regulatory Challenges of Ensuring Commercial Human Spaceflight Safety' (2011) 27 *Space Pol'y* 74, 74.

tourist activities in space are not likely to be large-scale tourist operations, but rather small-scale activities with two to six seats per flight.[6] The term 'tourism' is more evocative of large cruise ships and sprawling resorts that are filled with tour groups and sightseers, and thus the usefulness of the term fails in that regard. Privately funded scientific study, exploration or human spaceflight for the purposes of asteroid mining or lunar colonization would not be activities that one would readily classify as tourism.

> [O]perators certainly are not only looking for clients among people interested in travelling into or in outer space for fun, but also for space agencies willing to pay for the training of astronauts or others willing to pay for small experiments – not touristic activities in any normal sense of the word.[7]

The European Space Agency (ESA) has defined space tourism as encompassing 'the execution of sub-orbital flights by privately funded and/or privately operated vehicles and the associated technology development driven by the space tourism market';[8] the term in this context is inclusive rather than exclusive, though it does not take non-touristic sub-orbital flights into account. The functional overlap with tourism may actually come in the form of liability waivers, disclaimers and informed consent forms, which have previously been used in national and local jurisdictions to deal with the high-risk adventure tourism activities that have preceded space tourism.[9] Certainly, at least for the time being, human spaceflight, including commercial human spaceflight, will continue to be a high-risk activity.

It is necessary to clarify the status of persons engaging in space tourism, so as to avoid confusion. Though the term 'astronauts' literally means 'persons who sail among the stars,' and 'cosmonauts' means 'those who navigate the universe,' these terms are too broad, in that they

---

6 Masson-Zwaan and Freeland, *supra* note 1, at 1599. The XCOR Lynx vehicle only has two seats available per flight: see 'About Lynx' (*XCOR Aerospace*), http://xcor.com/lynx/, accessed 8 April 2015.

7 F. von der Dunk, 'Space Tourism, Private Spaceflight and the Law: Key Aspects' (2011) 27 *Space Pol'y* 146, 147.

8 'ESA's Position on Privately-Funded Sub-orbital Spaceflight' (10 April 2008) http://esamultimedia.esa.int/docs/gsp/Suborbital_Spaceflight_ESA_Position_Paper_14April08.pdf.

9 F. von der Dunk, 'The Integrated Approach – Regulating Private Human Spaceflight as Space Activity, Aircraft Operation, and High-Risk Adventure Tourism' (2012) *Acta Astronautica*, http://dx.doi.org/10.1016/j.actaastro.2012.05.020 at 8.

include those persons travelling into space as employees or contractors for their respective governments who have extensive specialized training.[10] Those persons who might be considered career astronauts are outside the scope of this chapter. The term 'space tourist,' which has been defined as 'someone who tours or travels into, to, or through space or to a celestial body for pleasure and/or recreation,'[11] however, is too narrow for the reasons discussed above, namely that space tourism may be undertaken for purposes other than pleasure and/or recreation. For the purposes of the liability discussion here it is to be assumed that any governing treaty provisions that are applicable to astronauts or personnel of a spacecraft would apply to spaceflight participants, and thus that such participants would, for example, be entitled to the same return and rescue efforts as any other human space traveller, trained or otherwise.[12]

The USA uses the term 'spaceflight participant' and defines the term as follows: 'an individual, who is not crew, carried within a launch vehicle or reentry vehicle,'[13] while 'crew means any employee of a licensee or transferee, or of a contractor or subcontractor of a licensee or transferee, who performs activities in the course of that employment directly relating to the launch, reentry, or other operation of or in a launch vehicle or reentry vehicle that carries human beings.'[14] Of course, the American definition of spaceflight participant is, like astronaut or cosmonaut, too broad from an academic perspective in the sense that it may encompass those persons who may not be career astronauts, but who are still not commercially engaging in such space travel. The distinction from crew is not necessarily helpful, particularly in light of the potentially confusing difference in the definition of the same term used with regard to the International Space Station: 'Spaceflight participants are individuals (e.g. commercial, scientific and other programs; crewmembers of non-partner space agencies, engineers, scientists, teachers, journalists, filmmakers or tourists) sponsored by one or more partner(s). Normally, this is a

---

10 B. Cheng, *Studies in International Space Law* (Clarendon Press 1997) 507.

11 Freeland, *supra* note 4, at 6.

12 See Cheng, *supra* note 10, at 509; M. Lachs, *The Law of Outer Space* (Sijthoff 1972) 71; V.S. Mani, 'The Agreement on the Rescue of Astronauts, the Return of Astronauts and the Return of Space Objects Launched into Space 1968' (UN Office of Outer Space Affairs, 25 September 2003), http://www.unoosa.org/pdf/publications/st_space_22E.pdf, at 224.

13 'Commercial Space Launch Activities', 51 USC § 50902 (20) (2010).

14 *Ibid.* at (2).

temporary assignment that is covered under a short-term contract.'[15] Of course, spaceflight participant is the term that must be used from a practical perspective with regard to the USA. Likewise, the State Spaceflight Liability and Immunity Acts apply the US definition of spaceflight participant, either wholesale or with minor modifications. As the USA is the focus of this chapter, the term spaceflight participant will be applied here.

3. ORBITAL AND SUB-ORBITAL TRAVEL

Whether sub-orbital travel is considered to be space travel and thus a space activity within the meaning of the international space law regime and national space regulations is a critical issue that must be resolved in order to understand which liability rules may apply. The key difference between orbital and sub-orbital space travel is that an orbital velocity is not achieved during sub-orbital spaceflights, which typically climb to an altitude around 100 km.[16] To achieve orbital spaceflight, a velocity must be achieved that will follow a path consistent with the curvature of the Earth, thus preventing the vehicle from being pulled back to Earth.[17] The USA defines a sub-orbital trajectory as follows: 'the intentional flight path of a launch vehicle, re-entry vehicle, or any portion thereof, whose vacuum instantaneous impact point does not leave the surface of the Earth.'[18] A vehicle would therefore be an orbital vehicle if, allowed to continue on its launch trajectory, it would not strike the surface of the Earth.

While orbital spaceflight is more technically complex and expensive than sub-orbital flight, that has not prevented some private companies, for example Excalibur Almaz or SpaceX, from venturing into this business.[19] The majority of the planned commercial human spaceflight activities, at least those that are further along in their respective development processes, are, however, sub-orbital.

---

15 'Principles Regarding Processes and Criteria for Selection, Assignment, Training and Certification of ISS (Expedition and Visiting) Crewmembers' (Multilateral Crew Operations Panel, November 2001), http://esamultimedia.esa.int/docs/isscrewcriteria.pdf, accessed 3 August 2017.

16 Masson-Zwaan and Freeland, *supra* note 1, at 1599.

17 *Ibid.*

18 'Commercial Space Launch Activities', 51 USC § 50902 (23) (2010).

19 Masson-Zwaan and Freeland, *supra* note 1, at 1599.

With regard to sub-orbital flight, '[t]he main issues relate to the definition and delimitation of airspace and outer space, the status of astronauts, and the issues of liability, registration and jurisdiction.'[20] Of these, the only one that is more or less unique to sub-orbital flight is the problem of the definition and delimitation of airspace and outer space.

The problem of whether sub-orbital flight should be defined primarily as fitting within the space law regime or the air law regime is tangential to the focus of this chapter. As was pointed out by the International Civil Aviation Organization (ICAO), the international aviation regulatory regime could not and would not apply unless the space tourism involved transportation between two States.[21] Likewise, the Warsaw or Montreal Convention liability regimes would not apply for the same reason.[22] That said, the discussion of the policies implemented by the individual States mentioned often gives an indication of how such States view the issue. Additionally, a disharmonious series of national regulations will create a confusing regulatory environment for operators.

For this purpose it is useful to see how these flights can be considered to fit, or not, within the current treaty regimes. While the space law treaties do not define a space object, other than to state that it includes launch vehicles and component parts,[23] Annex 1 to the Chicago Convention does define an aircraft as 'any machine that can derive support in the atmosphere from the reactions of the air other than the reactions of the air against the earth's surface.'[24] Many reusable launch vehicle concepts, particularly with regard to sub-orbital vehicles, can derive such support from the atmosphere, though they may only do so at certain points during their flight.

Professor von der Dunk indicates that actually deriving such support for any portion of the flight would in fact not be required under this definition.[25] He points out, however, that:

---

[20] S. Gorove, 'Aerospace Object – Legal and Policy Issues for Air and Space Law' (1997) 25:2 *J Space L* 102, 105.

[21] UN COPUOS, 49th Session, C-WP/12436 of 30/05/05 (2010).

[22] Convention for the Unification of Certain Rules Relating to International Carriage by Air 1929 (Warsaw Convention), 137 LNTS 11; Montreal Convention for the Unification of Certain Rules for International Carriage by Air 1999 (ICAO), 2242 UNTS 309.

[23] Convention on Registration of Objects Launched into Outer Space 1975, 1023 UNTS 15, Art. I; Convention on International Liability for Damage Caused by Space Objects 1972, 961 UNTS 187, Art. I.

[24] Convention on Civil Aviation 1994 (Chicago Convention), 15 UNTS 295, Annex 1.

[25] Von der Dunk, *supra* note 9, at 6.

> At the national level, in many cases the application of elaborate and specially crafted sets of rules is made contingent upon an aircraft belonging to a *specific category of* aircraft, referring to such criteria as size, use and operational characteristics, or to *specific types of operations* regardless of the craft used.[26]

Generally, space tourism vehicles are not well suited to such established requirements[27] due to 'distinctions in physics between "lift" and "thrust"' needed to accomplish a sub-orbital flight,' so that 'the law classifies commercial space tourism ships as rockets, not aircraft.'[28]

Professor Jakhu and Dr Nyampong suggest that, for the time being, the applicable space law responsibility and liability regime is most appropriate for commercial human spaceflight, due to the fact that such activities are still ultra-hazardous activities and such classification would promote the appropriate regulation and supervision of those activities and increase the likelihood that compensation would be paid to victims of any damage thereunder.[29] The scope of the Liability Convention includes an attempted launch within the definition of a launch, and therefore would be relevant to any incidents that may occur in the course of operation before the vehicle achieves substantial altitude.[30]

'Robert Heinlein famously remarked that once you've reached low-Earth orbit, you're halfway to anywhere in the solar system … at one hundred miles up, you're above the thickest of the atmosphere and the strongest gravitational pull; thus you have spent the lion's share of the fuel needed to travel' to any one of a number of relatively nearby space destinations.[31] Though one hundred miles is higher than some sub-orbital flights will travel, the idea that escaping the atmosphere is the most important factor in space travel remains, and sub-orbital flights certainly reach that mark. With regard to sub-orbital flights, the USA has labelled this form of transportation as a space activity rather than an aviation

---

26 *Ibid.* at 16, original emphasis.

27 *Ibid.*

28 J.I. Gabrynowicz, 'One Half Century and Counting: The Evolution of US National Space Law and Three Long-Term Emerging Issues' (2010) 4(2) *Harv L & Pol'y Rev* 405, 418; 'Commercial Space Transportation: Sub-orbital Rocket Launch Notice', 68 Fed.Reg. 59977 (20 October 2003).

29 R.S. Jakhu and Y.M. Nyampong, 'International Regulation of Emerging Modes of Space Transportation' in J.N. Pelton and R.S. Jakhu (eds), *Space Safety Regulations and Standards* (Elsevier 2010) 215, 233.

30 Liability Convention, *supra* note 23, Art. I(b).

31 G. Klerkx, *Lost in Space: The Fall of NASA and the Dream of a New Space Age* (Pantheon Books 2004) 91.

activity.[32] As an example, the 2004 launch of SpaceShipOne 'was considered a space launch and treated as such.'[33] Though the USA did not register SpaceShipOne in its national register or the UN Registry, this is theoretically permissible under the Registration Convention in that the Convention only calls for the registration of those objects launched into Earth orbit or beyond.[34] Though Professor von der Dunk calls for an alternative interpretation,[35] the US interpretation of beyond orbit is reasonable absent contrary established guidance or practice, and thus should not be considered a breach of international obligations under the Convention. Perhaps one reason why the USA has utilized the option not to register sub-orbital vehicles is the fact that they would be subject to a cumbersome registration system under national law. US regulations require reusable launch vehicles to be reregistered every time the vehicle is launched.[36] Interestingly, the space shuttles were registered a total of 125 times, in compliance with the regime being applied to orbital missions.[37]

It is noteworthy in consideration of the demarcation issue with regard to sub-orbital flights that 'the US DOT [Department of Transportation] awards commercial astronaut wings to pilots and flight crew on board a licensed launch vehicle on a flight that exceeds 80.45 kilometers as a recognition for having reached outer space.'[38]

---

32 'Final Rule on Experimental Permits for Reusable Sub-orbital Rockets', 72 Fed.Reg. 17001 (2007).

33 P. van Fenema, 'Sub-orbital Flights and ICAO' (2005) 30 *Air & Space Law* 396, 708.

34 Convention on Registration of Objects Launched into Outer Space 1975, 1023 UNTS 15, Art. II.

35 F. von der Dunk, 'Beyond What? Beyond Earth Orbit? ...! The Applicability of the Registration Convention to Private Commercial Manned Sub-Orbital Spaceflight' (2013) 43(2) *California Western International Law Journal* 269.

36 'Commercial Space Transportation Regulations', 14 C.F.R. §431.85 (2008).

37 'Register of Space Objects' (UN Office of Outer Space Affairs, Online Index of Objects Launched into Outer Space), http://www.oosa.unvienna.org/oosa/en/osoindex.html, accessed 3 August 2017.

38 M.C. Mineiro, 'Assessing the Risks: Tort Liability and Risk Management in the Event of a Commercial Human Spaceflight Vehicle Accident' (2009) 74 *Air L & Com* 371, 373.

4. SAFETY ISSUES

Safety regulation is one target area to be addressed by States with active space tourism industries. It is of particular importance given that physical dangers exist not only to third parties on the ground and in the air, but also to persons aboard the craft. The goal of safety policies for manned spaceflight is a 'reliability of 0.999 against a catastrophic accident defined as one in which human lives are lost'; these are better than odds of 1 in 1000. However, from the beginning of manned spaceflight to 1993 these odds were not maintained, as there were three fatal accidents among 139 missions.[39] Such odds, though, are still clearly far from optimal. Some of the dangers of human space travel have been articulated as follows:

> radiation, fire, solar flares, and equipment degradation or failure … the presence of toxic chemicals on board space craft and difficulty in providing swift and proper medical treatment … psychological stress caused by long-term isolation which will have to be dealt with and countered, crimes in outer space.[40]

Of course, these dangers exist in conjunction with the dangers posed by the vacuum of space itself, such as extreme temperatures and lack of oxygen.

Safety regulation for space tourism vehicles is a critical component of a fully functional national space regulatory regime and predictable, reliable risk management.

> [O]ne of the primary hazards or risks associated with this young industry is that there are no accepted standards guiding the industry regarding critical concerns like the physical condition of the SFP [space flight participant], what gear the SFP should be required to wear, what safety equipment should be in the vehicle, what is required in a safety briefing, what type of vehicle is capable of routinely travelling to sub-orbital space, or even what specific categories of aircraft or specific instrument ratings a pilot must have.[41]

---

[39] J.H. Carver, 'Factual Issues' in K.-H. Böckstiegel (ed.), *Manned Space Flight: Legal Aspects in the Light of Scientific and Technical Development* (Carl Heymanns Verlag 1993) 149, 150.

[40] N.M. Matte, 'Safety and Rescue: Introduction' in K.-H. Böckstiegel (ed.), *Manned Space Flight: Legal Aspects in the Light of Scientific and Technical Development* (Carl Heymanns Verlag 1993) 145, 147.

[41] T. Knutson, 'What is "Informed Consent" for Space-Flight Participants in the Soon-to-Launch Space Tourism Industry?' (2007) 33 *J Space L* 105, 114.

Thus, this issue must be considered and addressed if the successful development of this industry is to be achieved.

5. US FEDERAL SPACE LAW

'The US Government has passed quite a substantive body of rules governing private human spaceflight, mainly designed to enable entrepreneurs to go ahead with offering sub-orbital flights under conditions which are less stringent than for classical transport.'[42] With regard to understanding US federal space law, Dr Stephen Doyle provides a helpful overview of the law-making process in the USA:

> Under authority granted by the Constitution, the US Congress promulgates national legislation, published, after executive approval, in the form of Public Laws, and consolidated in the US Code and in the Statutes … In its turn, Congress authorizes federal government agencies to adopt necessary rules, regulations, and policies to implement the roles and functions given to the agencies by the Constitution and by Congress. Rules and regulations of the agencies are collected and published as the Code of Federal Regulations … In addition to adoption of rules and regulations, federal agencies may adopt and implement internal agency policies to guide and manage administrative practices of agencies.[43]

The Reagan era saw emphasis on the conviction that the private sector could accomplish spaceflight more efficiently and more appropriately than the government, and thus the Commercial Space Launch Act was passed in 1984 to encourage commercial space launches and implement appropriate licensing requirements for such launches.[44] The private space sector in the USA has been perceived as integral to the use of space in terms of economic viability as well as international prestige and competitiveness, to the extent that NASA has been required to utilize commercial services where possible.[45] In 2005 NASA initiated its Commercial Crew and Cargo Program that, in part, has the intent to establish a market

---

42 Masson-Zwaan and Freeland, *supra* note 1, at 1600.

43 S.E. Doyle, 'Astronauts and Cosmonauts in International Cooperation: A View of the American Experience' in K.-H. Böckstiegel (ed.), *Manned Space Flight: Legal Aspects in the Light of Scientific and Technical Development* (Carl Heymanns Verlag 1993) 43, 43–44.

44 Gabrynowicz, *supra* note 28, at 410–411.

45 'National Aeronautics and Space Program', 51 USC § 20102 (2010); P.S. Dempsey, 'The Evolution of US Space Policy' (2008) 33 *Ann Air & Sp L* 325, 340.

environment such that the space transportation services will be available to both public and private sector customers; in accordance with this objective, NASA has awarded Space Act Agreements, which fundamentally are flexible contracts, to private sector entities.[46] Given the USA's intent to foster the growth of the commercial space industry, there has been a wide array of legislation and regulation promulgated in this area.

The relevant federal legislation on National and Commercial Space Programs has been consolidated into Title 51 of the US Code.[47] In addition to this legislation, the executive branch issues regulations in the form of National Space Policy and Space Transportation Directives.[48]

6. LIABILITY IN US FEDERAL SPACE LAW

'US aerospace companies continue to cite the commercial activities of foreign governments and their use of industrial policy as reasons for continuing the favorable US government-industry risk-sharing regime contained in US launch law.'[49] The American regime includes mandatory cross-waivers of liability, insurance and financial responsibility requirements, and additional indemnification above required insurance amounts.[50]

The cross-waiver of liability provisions specifically excludes spaceflight participants from waivers with private entities (though they are required with respect to the federal government).[51] 'This exclusion leaves open the possibility of CHSF (commercial human space flight) operators suing manufacturers for indemnification of damages paid to injured SFPs. Manufacturers may also be sued for indemnification of damages paid to injured third parties.'[52] Though they have the authority to do so,[53] the Federal Aviation Administration (FAA) has not implemented any

---

46 D. Stone *et al.*, 'NASA's Approach to Commercial Cargo and Crew Transportation' (2008) 63 *Acta Astronautica* 192, 192–193.

47 'National and Commercial Space Programs', 51 § USC (2010).

48 M. Mineiro, 'Commercial Human Spaceflight in the United States: Federal Licensing and Tort Liability' (LLM Thesis, McGill University Institute of Air and Space Law, 2008) (unpublished), 11.

49 Gabrynowicz, *supra* note 28, at 410–412.

50 Mineiro, *supra* note 38, at 392.

51 M. Schaefer, 'The Need for Federal Preemption and International Negotiations Regarding Liability Caps and Waivers of Liability in the US Commercial Space Industry', (2015) 33(1) *Berkeley Journal of International Law* 223.

52 Mineiro, *supra* note 38, at 397.

53 'Commercial Space Launch Activities', 51 USC § 50905 (2010).

specific medical requirements for spaceflight participants; however, the FAA strongly recommends that spaceflight participants seek medical advice before engaging in more taxing orbital missions.[54]

> The holder of a license or a permit under this chapter may launch or reenter a spaceflight participant only if—
>
> (A) in accordance with regulations promulgated by the Secretary, the holder of the license or permit has informed the spaceflight participant in writing about the risks of the launch and reentry, including the safety record of the launch or reentry vehicle type, and the Secretary has informed the spaceflight participant in writing of any relevant information related to risk or probable loss during each phase of flight gathered by the Secretary in making the determination required by section 50914(a)(2) and (c);
>
> (B) the holder of the license or permit has informed any spaceflight participant in writing, prior to receiving any compensation from that spaceflight participant or (in the case of spaceflight participant not providing compensation) otherwise concluding any agreement to fly that spaceflight participant, that the United States Government has not certified the launch vehicle as safe for carrying crew or spaceflight participants;
>
> (C) in accordance with regulations promulgated by the Secretary, the spaceflight participant has provided written informed consent to participate in the launch and reentry and written certification of compliance with any regulations promulgated under paragraph (6)(A).[55]

It is useful to note a few facts that are not directly related to the liability provisions with regard to space tourism, but are relevant to understanding the overall context. The holder of an experimental sub-orbital permit may not carry passengers for compensation; this type of permit may be used for research, development, testing and demonstrating compliance only.[56] Licence holders must satisfy the Human Spaceflight Requirements for Crew and Spaceflight Participants in order to be allowed to carry spaceflight participants.[57] These requirements 'apply to all applicants for a license or permit who propose to have a SFP on board a vehicle, all

---

[54] K. Wong, 'Developing Commercial Human Space-Flight Regulations' in J.N. Pelton and R.S. Jakhu (eds), *Space Safety Regulations and Standards* (Elsevier 2010) 149, 155.

[55] 'Commercial Space Launch Activities', 51 USC § 50905 (2010).

[56] 'Commercial Space Launch Activities', 51 USC § 50906 (2010).

[57] 'Final Rule on Human Spaceflight Requirements for Crew and Spaceflight Participants', 71 Fed.Reg. 75616 (15 December 2006).

operators licensed or permitted who have a SFP on board, and all SFP engaged in an activity authorized under the Act.'[58]

It should be noted that, 'rather than imposing certification, the US regime is one of non-certification, leading to an absence of contractual liability as long as the "informed consent" requirement is complied with.'[59] Generally speaking, the substantial case law regarding pre-recreational warnings is fairly consistent in holding that legally supportable warnings are 'specific, obvious and direct, unambiguous, easy to understand, simple and complete.'[60] Thus, such warnings and their sufficiency can be adequately addressed within the context of the domestic US legal system where there is no international component.

7. US STATE SPACE LAW

'While States are prohibited from having laws inconsistent with federal law, the [Commercial Space Launch Act] specifically grants States the authority to implement law in addition to or more stringent than a requirement of, or regulation prescribed under, the Act.'[61] As regulation of the space industry by individual States has not been pre-empted, State and local legislation is permitted to the extent that it does not conflict with federal regulation.[62] Several US States have undertaken legislative activity with the intention to attract space tourism. Such State law incentives include spaceport incentives intended to leverage existing facilities, establishment of space authorities and industry-favourable liability regimes dealing with the federal informed consent rules.[63] State Spaceflight Liability and Immunity Acts will be the focus for the purposes of this section.

A. State Spaceflight Liability and Immunity Acts

Virginia pioneered Spaceflight Liability and Immunity Acts for spaceflight participants (or simply participants as these acts call them) in

[58] Mineiro, *supra* note 48, at 51.

[59] Von der Dunk, *supra* note 7, at 152.

[60] Knutson, *supra* note 41, at 118.

[61] 'Commercial Space Launch Activities', 51 USC. § 50919 (2010); Mineiro, *supra* note 38, at 381.

[62] P. Sterns and L. Tennen, 'State and Municipal Regulation of the Aerospace Industry in the United States' in R.S. Jakhu (ed.), *National Regulation of Space Activities* (Springer 2010) 467, 468.

[63] Gabrynowicz, *supra* note 28, at 420.

2007.[64] Since then, Florida, California, Texas, New Mexico and Oklahoma have followed suit.[65] The content of these Acts is remarkably similar, though there are a few notable differences of which we need to be aware. All of the Acts specify that, if the procedures of the Act are followed, a spaceflight entity will not be liable for a participant injury resulting from the risks of spaceflight activities.[66] Florida, New Mexico and California use the word 'inherent' before risks of spaceflight; the provision is thus less favourable to spaceflight entities in those States as it only protects against inherent risks, not all risks generally.[67] The Virginia, Oklahoma and Texas Acts contain broad definitions of a spaceflight entity which clearly protect entities such as spacecraft manufacturers that provide vehicles for use by space tourism companies. New Mexico and California do not include suppliers and contractors in the waiver scheme.[68] The clarity of these broad definitions is decidedly favourable to providers of space transportation, protecting the other entities in the chain of service.

In all cases the participant must be presented with warning language and return a signed form evidencing that they are providing their informed consent. All but New Mexico include minimally acceptable sample language to be used. Of the sample language provided, Florida's is the least robust.[69] It does not reference voluntary participation or informed consent. It does, however, include language demonstrating that the participant had an opportunity to consult with an attorney before signing the form.

For one example, this is the language provided by the California statute:

---

64 Spaceflight Liability and Immunity Act 2007, Va H.B. 3184, §8.01-227.8 & §8.01-227.9 (VA Spaceflight Act).

65 Spaceflight Informed Consent Bill 2008, Fla S.B. 2438 (FL Informed Consent); Spaceflight Liability and Immunity Act 2012, 5 Ca Civil C tit 7 § 2210 (CA Spaceflight Act); Limited Liability for Space Flight Activities Act 2011, Tex Civ Prac C tit 4 Ch 100A (TX Spaceflight Act); Spaceflight Informed Consent Act 2013, N Mex S.B. 240 (NM Informed Consent); Spaceflight Liability and Immunity Act 2013, Okla Stat tit 3 § 351 (OK Spaceflight Act).

66 VA Spaceflight Act, *supra* note 65; FL Informed Consent, *supra* note 65; CA Spaceflight Act, *supra* note 65; NM Informed Consent, *supra* note 65; OK Spaceflight Act, *supra* note 65.

67 FL Informed Consent, *supra* note 65; CA Spaceflight Act, *supra* note 65; NM Informed Consent, *supra* note 65.

68 Schaefer, *supra* note 51, at 19–20.

69 VA Spaceflight Act, *supra* note 65; FL Informed Consent, *supra* note 65; CA Spaceflight Act, *supra* note 65; OK Spaceflight Act, *supra* note 65.

> WARNING AND ACKNOWLEDGMENT: I understand and acknowledge that, under California law, there is limited civil liability for bodily injury, including death, emotional injury, or property damage, sustained by a participant as a result of the inherent risks associated with spaceflight activities provided by spaceflight entity. I have given my informed consent to participate in spaceflight activities after receiving a description of the inherent risks associated with spaceflight activities, as required by federal law pursuant to Section 50905 of Title 51 of the United States Code and Section 460.45 of Title 14 of the Code of Federal Regulations. The consent that I have given acknowledges that the inherent risks associated with spaceflight activities include, but are not limited to, risk of bodily injury, including death, emotional injury, and property damage. I understand and acknowledge that I am participating in spaceflight activities at my own risk. I have been given the opportunity to consult with an attorney before signing this statement.[70]

In all cases, liability is not waived if the appropriate warning language is not provided and signed.

All of the Acts contain a list of circumstances under which a spaceflight entity will not be protected from liability. Gross negligence evidencing wilful or wanton disregard for the safety of the participant is included in all acts.[71] All six States likewise include intentionally injuring the participant or causing intentional injury to the participant.[72] The third exception, however, is only applied by Florida, New Mexico and California: when the spaceflight entity had actual knowledge or reasonably should have known of a dangerous condition.[73] This standard is more closely aligned with standard negligence. Thus, those three States are more favourable to the spaceflight participant in that respect.

Virginia, California and Oklahoma make clear direct reference to the relevant federal law on informed consent and liability waivers,[74] while Florida and New Mexico do not.[75] Texas mentions only 'language required by federal law.'[76] Those statutes that include reference to the

---

[70] CA Spaceflight Act, *supra* note 65.

[71] NM Informed Consent, *supra* note 65.

[72] VA Spaceflight Act, *supra* note 65; FL Informed Consent, *supra* note 65; CA Spaceflight Act, *supra* note 65; NM Informed Consent, *supra* note 65; OK Spaceflight Act, *supra* note 65.

[73] FL Informed Consent, *supra* note 65; CA Spaceflight Act, *supra* note 65; NM Informed Consent, *supra* note 65.

[74] VA Spaceflight Act, *supra* note 65; CA Spaceflight Act, *supra* note 65; OK Spaceflight Act, *supra* note 65.

[75] FL Informed Consent, *supra* note 65; NM Informed Consent, *supra* note 65.

[76] TX Spaceflight Act, *supra* note 65.

federal requirements are better drafted, in the sense that they are less likely to be found to conflict with the existing federal provisions if challenged in court.

Generally speaking, an informed consent regime is one in which something is done to an individual (for example, a medical procedure) rather than a situation in which a participant is engaging in an activity. It appears that, in the context of spaceflight participants, the US Congress had instead intended to impose what would normally be considered a duty to warn.[77] The State Acts, however, go much farther toward protecting spaceflight entities.

Table 5.1 Spaceflight Liability and Immunity Act comparison chart

State	'Inherent' risk	Sample warning language	Excludes gross negligence/ wanton or wilful disregard	Excludes 'knew or reasonably should have known' of dangerous condition	Reference to federal law on informed consent	Broad definition of space-flight entity
New Mexico	Y	N	Y	Y	N	N
Florida	Y	Y	Y	Y	N	N
California	Y	Y	Y	Y	Y	N
Virginia	N	Y	Y	N	Y	Y
Oklahoma	N	Y	Y	N	Y	Y
Texas	N	Y	Y	N	Y (vague)	Y

B. Enforceability of Liability Waivers

Liability waivers are based on the principle of *volenti non fit injuria*: there is no injury to one who consents.[78] A liability waiver is a contract modifying the rights of parties under tort law and is generally upheld in the USA with regard to adventure activities in circumstances where it has

[77] Knutson, *supra* note 41, at 109.

[78] S.M. Grieshop Corrada, 'Liability Waivers in the United States Travel and Adventure Sports Industry' (2006) 2006 *Int'l Travel LJ* 156, 156.

been properly drafted and consented to by a participant, though some States will not enforce these contracts on public policy grounds.[79] '[I]t is generally agreed that the liability waiver: (1) must not violate public policy; (2) must have been procured through adequate consideration; (3) must contain clear and unambiguous language; and (4) the signatory must have the capacity to contract.'[80] Generally speaking, these waivers cannot include gross negligence or recklessness.[81] Some courts have held such waivers to be against public policy where a public duty is involved,[82] which would not be the case with regard to space tourism.

These forms, however, are not always accepted or enforceable in other jurisdictions, and thus may not provide a useful model moving forward with regard to the development of national or international space regulation. They are commonly referred to as assumption of risk forms in Australia, where they are still at an earlier stage in their development. Only when it can be demonstrated that the participant had a personal responsibility for their own safety are these waivers likely to be successful in Australia. In a space context, where accidents are unlikely to be caused due to any actions of the participants themselves, this manner of assumption of risk is less likely to be successful than in the United States.[83]

These waivers are likely to be enforceable in Canada in a space context, however, where there is appropriate advance notice and the terms are clear and easy to understand.[84] In Canadian law contracts are upheld unless they are unconscionable, which is generally only an issue where there is a problem with unequal bargaining positions.[85] In this case, an individual would have the option to not agree to the terms and not travel on the spaceflight, thus there would be no such issue. It is worth noting that not all Canadian jurisdictions permit the waiver of gross negligence, and thus caution must be used with regard to waivers that may attempt to exclude such liability.[86]

---

79 *Ibid.* at 156–157.

80 *Ibid.* at 157.

81 *Ibid.* at 158.

82 J.O. Spengler and B.B. Hronek, *Legal Liability in Recreation, Sports, and Tourism* (Sagamore Publishing 2011) 69.

83 S. Beeton, 'Horseback Tourism in Victoria, Australia: Cooperative, Proactive Crisis Management' (2001) 4(5) *Current Issues in Tourism* 422, 433.

84 R. Cloutier *et al.*, *Legal Liability and Risk Management in Adventure Tourism* (Hignell Printing 2000) 56–57.

85 *Ibid.* at 61–63.

86 *Ibid.* at 66–70.

In the European Union, Directive 93/13 regulates such contractual waivers, including waivers with regard to space tourism. Under this Directive a term will be regarded as unfair if 'it causes significant imbalance in the contractual rights and obligations of the parties to the detriment of the consumer.'[87] The directive includes a list of potentially unfair terms, which includes such terms as excluding or limiting the liability of an entity in the event of death or personal injury of a consumer that results from an act or omission of that entity.[88] It is important to note that this provision only applies with respect to standardized boilerplate language, and not where such contracts are drafted on an individualized basis.

The enforceability of liability waiver provisions with regard to spaceflight participants in European jurisdictions will depend upon the implementing legislation in the relevant Member State.[89] Legislation varies from those disallowing any exclusion or limitation on personal injury or death liability[90] on the one hand, to those that do not have any provisions addressing this issue either way.[91] Where implementing legislation permits a case-by-case analysis, the Directive's rules regarding unfairness must be applied, taking into account the good or service and all circumstances surrounding the contract's conclusion.[92] Thus, where a case-by-case analysis applies, circumstances particular to space travel will favour the acceptability of the waiver.[93]

8. CONCLUSION

In conclusion, while Spaceflight Liability and Immunity Acts and their ilk are likely to succeed at the US level in limiting liability for space tourism operators, they are less likely to be successful in a European

---

87 M.P. Chatzipanagiotis, 'Regulating Sub-orbital Flights in Europe: Selected Issues' (19 November 2012), available at SSRN: http://ssrn.com/abstract=2177671 at 60–61; citing Council Directive 93/13/EEC of 5 April 1003 OJ L 095, 21/04/1993 at 29–34.

88 Council Directive 93/13/EEC *supra* note 87, Annex 1.

89 Chatzipanagiotis *supra* note 87, at 63.

90 For example, § 309 Nr. 7(a) German Civil Code; Arts 33(2)(a) and 36(2)(a) Italian Consumer Code; Art. 10 bis (1) combined with the First Additional Disposition, part II, Nr. 10 of the Spanish Law 26/1984 on Consumer Protection.

91 Chatzipanagiotis *supra* note 87, at 62.

92 Council Directive 93/13/EEC *supra* note 87, Art. 4(1).

93 Chatzipanagiotis *supra* note 87, at 62.

environment. Thus, certain US States are providing a more enticing limited liability regime for such operators while their activities are in their infancy. As space tourism transforms into an activity that is more likely to cross State lines, however, a comprehensive international regime may be necessary to reduce confusion. Given that this industry has yet to truly begin carrying spaceflight participants with any regularity, this Spaceflight Liability and Immunity Act regime is untested. It will be interesting to see how the courts apply it moving forward. Regardless, now that the sub-orbital industry has seen its first fatality in the form of the October 2014 Virgin Galactic tragedy,[94] it is apparent that the 'inherent risks' of space flight are here to stay, at least for a while.

[94] T. Malik, 'Virgin Galactic SpaceShipTwo Crash: Full Coverage and Investigation' (*Space.com*, 14 December 2014), http://www.space.com/27629-virgin-galactic-spaceshiptwo-crash-full-coverage.html, accessed 8 April 2015.

6. Commercial space operations within an existing national legal framework: some lessons learned and the challenges ahead

Daniel P. Murray[1]

1. INTRODUCTION

The US space program today has three sectors: civil, military and commercial. The commercial sector had its official start in 1984 with the passage of the Commercial Space Launch Act (CSLA).[2] Congress assigned regulatory oversight for the commercial sector to the Office of Commercial Space Transportation, which was originally a staff office within the US Department of Transportation. Today, the Office of Commercial Space Transportation (AST) is one of four lines of business within the Federal Aviation Administration (FAA). The AST's mission is to ensure the protection of the public, property and the national security and foreign policy interests of the United States during commercial launch and reentry activities, and to encourage, facilitate and promote US commercial space transportation.

The CSLA requires US citizens to obtain a licence before conducting the commercial launch of a launch vehicle or the commercial reentry of a reentry vehicle. The only exception is for missions conducted by and for the US government, such as launches by the National Aeronautics and Space Administration (NASA) or the US Air Force. The operators of commercial reusable sub-orbital rockets may otherwise obtain an experimental permit from the AST, authorizing them to conduct sub-orbital

[2] 'National and Commercial Space Programs,' Title 51 U.S. Code, Sec. 50901, 2011 ed. Available: http://www.gpo.gov/, accessed 2 September 2013.

launches for research and development, crew training or demonstrating compliance with other regulations. Altogether there have been over 270 licensed or permitted launches, as well as eight reentries from orbit, all conducted without any injuries to the public or damage to public property. The AST has also licensed the operations of nine launch sites, commonly referred to as spaceports.

The CSLA also directs the AST to encourage, facilitate and promote the US commercial space transportation industry. As such, the AST develops products for the industry, such as launch forecasts, market studies and economic analyses, and publishes them on its website.[3] Further, the AST promotes research in and development of commercial space transportation in partnership with academia and industry through grants provided by its Commercial Space Transportation Center of Excellence (COE).[4] The AST also provides Space Transportation Infrastructure Matching grants to State agencies to promote the development of launch and reentry sites through industry partnerships.

Since the late 1990s the AST has gained experience in regulating US commercial space transportation that may be relevant to other nations contemplating their own national legal frameworks. To that end, this chapter provides a brief overview of the AST's regulatory approach. It then proceeds to highlight some of the challenges posed by commercial space operations that the AST anticipates in the near future.

2. OVERVIEW OF THE AST'S REGULATORY APPROACH

The AST's public safety mission and its mission to encourage, facilitate and promote the industry are purposely separated, as safety is the AST's and the FAA's top priority. As established in the AST's annual Business Plan,[5] the AST's licence and permit application evaluation functions directly and distinctly support the FAA's goal of moving the agency to the next level of safety. However, the developing state of the industry and the direction of national law and policy continue to guide the AST's regulatory approach. For example, the CSLA, as amended, encourages

[3] http://www.faa.gov/about/office_org/headquarters_offices/ast/, accessed 4 August 2017.

[4] http://www.coe-cst.org/, accessed 4 August 2017.

[5] AST FY2013 Business Plan, Office of Commercial Space Transportation, Federal Aviation Administration http://www.faa.gov/about/plans_reports/media/2013/AST_FY2013_Business-Plan_12-12-10.pdf, accessed 4 August 2017.

the simplified and expedited issuance and transfer of commercial licences for launch vehicles, reentry vehicles and associated services. Likewise, the President's 2010 National Space Policy[6] reaffirmed the US government's commitment to the growth of commercial space transportation by instructing departments and agencies to encourage an innovative and entrepreneurial commercial space sector, to minimize regulatory burden and to ensure that the regulatory environment is timely and responsive.

This guidance is notably present in the structure and approach of Chapter III in Title 14 of the Code of Federal Regulations[7] (14 CFR Chapter III), which contains the FAA's regulations for commercial space transportation. For example, the regulations for expendable launch vehicles (ELVs) include a number of prescriptive requirements developed by the US Air Force and NASA over several decades through operations conducted at the US Federal Ranges. The FAA worked closely with the Air Force and NASA to develop these common safety requirements. Common requirements capture best safety practices, but they also provide an ELV operator with a consistent, predictable path to approval, regardless of whether the operation is conducted for commercial, military or civil purposes. Prescriptive regulations for ELVs reduce the regulatory burden on the operator and facilitate the AST's timely evaluation of an application. Numerous examples exist within 14 CFR part 417 of specific, quantitative limits for minimum reliability, performance, capacity and margin of particular ELV components and systems. This is particularly true with regard to flight termination systems.

Nevertheless, the AST does not certify vehicles or their designs. A certification approach, such as the approach that the FAA uses to certify aeroplanes, requires an extensive amount of flight history. Currently most launch and reentry vehicles are considerably more expensive to operate than aeroplanes, and as such launch and reentry vehicle developers and operators cannot demonstrate the reliability of their vehicles in the same manner as most aeroplane developers and operators. Aeroplane developers and operators conduct hundreds of hours of flight tests that thoroughly investigate the envelope of operations, both nominal and contingency, before introducing their designs into commercial service. Due to the higher costs associated with operating their vehicles, launch

---

[6] National Space Policy of the United States of America, Executive Office of the President, 28 June 2010, https://obamawhitehouse.archives.gov/sites/default/files/national_space_policy_6-28-10.pdf, accessed 4 August 2017.

[7] Commercial Space Transportation, Federal Aviation Administration, Department of Transportation, Title 14 Code of Federal Regulations, Chapter III, 2013 ed.

and reentry vehicle developers and operators must use reliability estimates based on analysis and subscale testing to supplement their limited flight history.

Further, a certification approach may require intensive regulatory insight and oversight of the design, manufacturing and testing of vehicles and their components. With such an approach, regulators or their designees would be present and engaged with vehicle developers throughout the design and manufacturing lifecycle to ensure that safety standards are met. While the AST engages with prospective licensees and permittees as early as possible through its pre-application consultation process and remains engaged throughout the licensing and compliance monitoring phases, it has not adopted an embedded regulatory approach. Currently specific safety standards for designing and manufacturing the variety of launch and reentry vehicles under development and in use today do not exist. Indeed, vehicle developers use a variety of innovative methods, techniques and technologies to design and manufacture launch and reentry vehicles. Accordingly, the AST licenses and permits the safe operation of vehicles rather than certifying the vehicles themselves.

Moreover, the AST takes a performance-based approach to regulation in respect to reusable launch vehicles (RLVs) and reentry vehicles. This approach provides the emerging industry with the needed flexibility to develop commercially viable reentry and RLV systems, while the level of experience associated with ELVs, and the flight history associated with aeroplanes, is gained over time. In practice, this approach allows applicants to use a methodology of their choosing to demonstrate safety, so long as that method satisfies the safety performance goals stated in the regulation. At the same time the AST provides the industry with advisory circulars, which demonstrate to the industry one acceptable means of meeting a performance-based regulation, as well as guides, templates and checklists. In general, the AST believes that performance-based regulation furthers the public interest by encouraging innovation and technology development. Likewise, performance-based regulations allow the AST's safety oversight to keep pace with a dynamic industry, as it is difficult to anticipate industry trends and the development of new regulations can be a time-consuming, resource-intensive undertaking.

In developing its RLV regulations, the AST created a 'three-pronged approach', consisting of quantitative risk, system safety and operating restrictions. Title 14 CFR part 431 requires an applicant for a licence to launch an RLV to submit a quantitative risk analysis demonstrating that the public safety risk threshold defined in part 431 is not exceeded. It also requires an applicant to develop and implement a system safety process to identify hazards and mitigate risks, and to impose operating requirements that ensure compliance with the assumptions used in the quantitative risk analysis and the mitigations identified in the system safety process.

A multi-pronged approach is commonly used in safety regulation to capitalize on the strengths of interrelated safety elements in a manner that compensates for their limitations. The strength of quantitative risk analyses lies in their ability to identify risk drivers and provide useful indications of relative risk. Through quantitative risk analysis, a vehicle operator can explore options for accomplishing a given mission, comparing the risk to public safety associated with each option and choosing less risky alternatives from among those options.

Further, quantitative analyses, as generally applied to launch and reentry safety assessments, focus on failure responses rather than failure mechanisms. While a launch or reentry vehicle could suffer a myriad of failures that could pose a hazard to public safety, the outcomes can be categorized into a limited list of applicable responses. For instance, the vehicle can fail in a way that causes it to break into pieces that fall to the surface, or the vehicle could fail in a way that causes it to fall intact to the surface. When the vehicle or its debris impacts the surface, it may or may not explode. A quantitative analysis explores each of these applicable failure responses for a given vehicle on a given mission. In that regard it simplifies the safety assessment by computing the risks to public safety at locations and times that span the full duration of the launch or reentry operation, but without dissecting each failure scenario at each location and time to determine the various sequences of events that must take place in order for the failure to occur.

However, the usefulness of the results of quantitative risk analyses is limited by the quality of the data used as inputs. Uncertainty in key parameters, including the overall probability of the vehicle failing in a manner that creates a hazard to the public, requires careful scrutiny in order to discern their influence on the results of the analysis.

Likewise, system safety processes are designed to be systematic and thorough, but they can omit key risks when hazard sources and mechanisms are poorly understood. System safety engineers may examine a vehicle using a function-by-function or system-by-system approach, identifying the sources, mechanisms and outcomes of failure associated with each function, component, subsystem and system of a vehicle, as well as the interfaces between systems. The ability of system safety engineers to identify the various ways in which complex, interacting systems can fail is often based on their experience when working with similar systems. Where such experience may be lacking, hazards may elude identification. This is particularly true of new, innovative systems like those that the sub-orbital RLV industry is developing.

Thus, an applicant's use of both quantitative and qualitative assessments forms a complementary means of assessing public safety. Where a system safety analysis may fail to identify a particular hazard or one of its causes, an analyst may still capture the vehicle response to that hazard in the quantitative analysis. The analyst may then assess the effect on public safety and identify operational adjustments that may manage the risk. For example, a change in vehicle trajectory may reduce the likelihood of a vehicle impact within a populated area, even in instances where the chain of events that leads to that impact has not been specifically identified. Likewise, a system safety analysis may identify the potential for a critical vehicle system to fail as well as opportunities for the vehicle developer to make adjustments in the design of that system to reduce that system's likelihood of failure. These adjustments can reduce the overall likelihood of a vehicle failure that leads to an impact. As stated in the Final Rule, just as system redundancy compensates for failure or flawed design or performance, interrelated safety elements assure that actual hazards from vehicle operation, whether anticipated in analytical assessments or unforeseen, will not increase risk to the public beyond an acceptable level.

3. CHALLENGES

Performance-based regulation can present challenges to commercial space vehicle developers and operators, and particularly so for new entrants, as it produces a level of uncertainty in the resources, cost and schedule required to obtain approval. For this reason the AST provides a pre-application consultation service to the industry as a means of familiarizing new entrants with the content and purpose of the regulations. This service also provides the AST with initial insights into a prospective applicant's plans. The process typically consists of a series of meetings in which information is shared and feedback is provided. At least one meeting is required by regulation before the AST can accept an application. Early consultation helps both an applicant and the AST to identify possible regulatory issues at the planning stage, when changes to an application or to proposed licensed or permitted activities are less likely to result in significant delays or costs to the applicant.

The performance-based approach also challenges the AST to ensure consistency across evaluations, as the acceptability of a demonstration of safety can be based on qualitative factors, at least in part, and it is therefore to some extent subjective. This is particularly challenging in the present state of the industry, where each mission is unique. For the flight tests of sub-orbital reusable rockets, the AST designed the experimental permit to allow vehicle operators to more readily make changes to their vehicles and operational concepts as they demonstrate technologies and develop new techniques. However, even for more routine missions, such as placing a satellite into geosynchronous orbit or sending cargo to the International Space Station, vehicle operators can make changes to their vehicles or trajectories to increase performance or reusability, or accommodate a secondary payload. These changes often preclude the AST from simply approving a given mission under the same pretences as a previous mission.

Accordingly, the path to public safety differs from vehicle to vehicle and from mission to mission. In these instances the AST must consider the full case for safety, ensuring that the combination of safety approaches a vehicle developer or operator employs sufficiently addresses the hazards. In this manner the AST achieves consistency through the careful execution of its evaluation processes. The AST's safety engineers scrutinize licence and permit applications to ensure that they identify and address all reasonably foreseeable hazards. They then ensure that the applicants have identified the means for controlling the hazards and that the applicants have verified the effectiveness of those means.

The AST will continue to be challenged to keep pace as the commercial space transportation industry expands in new directions. For example, a number of companies are developing vehicles and technologies that will allow them to begin offering transportation services for crew and other occupants to orbital destinations or flights of space flight participants to sub-orbital space. The US Congress directed the AST not to implement additional regulations on crew or space flight participant safety until 2015, barring an accident or close call, so as to allow the industry an opportunity to develop the necessary technologies, approaches and standards that will provide for safe commercial human space flight. In 2015 the US Congress extended this date to 2023. The AST will eventually develop commercial human space flight regulations when the need for such regulations arises. In the meantime, however, the AST will continue to monitor safety trends and gather data that it can share across the industry to promote best practices. In 2014 the AST developed and issued a set of recommended practices for human space flight occupant safety.[8] This document provides

---

8 Recommended Practices for Human Space Flight Occupant Safety, Federal Aviation Administration, 27 August 2014, https://www.faa.gov/about/office_org/

a compilation of practices that the AST believes are important and recommends for commercial human space flight occupant safety. It is intended to enable a dialogue among, and perhaps consensus between, government, industry and academia on practices that will support the continuous improvement of the safety of launch and reentry vehicles designed to carry humans. The document can also be used to help identify subject areas that could benefit from industry consensus standards and it may serve as a starting point for a future rule-making project, should there be a need for such an effort at some point in the future. However, it is not a regulation and it has no regulatory effect.

4. THE WAY FORWARD

The AST may eventually certify launch and reentry vehicles and designs in a manner more similar to the way the FAA certifies aeroplanes. Such an approach to regulation would look to capitalize on a convergence of vehicle designs and development, and manufacturing and operational approaches that has not yet occurred. In the meantime, anticipated increases in flight rates and the expansion of the industry to new markets will provide the AST with opportunities to gain more regulatory experience as the commercial space transportation industry gains more technical experience and benefit from the use of best practices.

The AST will look to the Commercial Space Transportation COE to conduct the research and development work necessary to support such an approach. This work includes identifying indicators and trends that will confirm such a convergence. It also extends into capturing and disseminating best practices and developing new safety techniques. Some of these techniques may hold similarities to those used in aviation, but others will most likely be unique to the physical and financial dynamics of the commercial space transportation industry.

5. CONCLUSION

The AST continues to improve its existing regulations as it identifies any need for additional regulations. Fortunately, the AST's performance-based regulations have generally demonstrated the flexibility and broad applicability to various vehicle concepts. For this reason the AST

headquarters_offices/ast/media/Recommended_Practices_for_HSF_Occupant_Safety-Version_1-TC14-0037.pdf, accessed 4 August 2017.

strongly encourages other nations who are currently contemplating developing a regulatory framework for space transportation to consider a performance-based regulatory approach similar to the approach embodied in 14 CFR Chapter III. Over the course of the past several years AST representatives have conducted a number of bilateral meetings with representatives of foreign governments to introduce them to its regulations and share some of its experiences and lessons learned. In 2014 the AST signed memoranda of cooperation with Italy[9] and the United Kingdom.[10] The AST regularly participates in international space safety forums, such as those hosted by the International Association for the Advancement of Space Safety (IAASS) and the International Astronautical Congress (IAC), and it recently accepted an invitation from the International Civil Aviation Organization (ICAO) to chair a Space Learning Group. The Space Learning Group is composed of representatives of ICAO Member States, and it serves as a forum to share information and provide best practices on commercial space activities for other States interested in emulating existing regulations or established practices. Through these interactions and other opportunities to engage the global safety community, the AST will continue to lead efforts to ensure safety and expand the use of best practices.

[9] Memorandum of Cooperation in the Development of Commercial Space Transportation between the Federal Aviation Administration, Department of Transportation, United States of America, and the Italian Ente Nazionale Per L'Aviazione Civile, Italian Republic', Federal Aviation Administration, 12 March 2014, http://www.faa.gov/about/office_org/headquarters_offices/ast/programs/international_affairs/media/Memorandum_of_Cooperation_FAA_and_Italy_ENAC_signed_March-12-2014.pdf, accessed 4 August 2017.

[10] Memorandum of Cooperation in the Development of Commercial Space Transportation between the Federal Aviation Administration, Department of Transportation, United States of America, the Department of Transportation, United Kingdom of Great Britain and Northern Ireland, The United Kingdom Space Agency, the Secretary of State for Business Innovation and Skills, United Kingdom of Great Britain and Northern Ireland, and the Civil Aviation Authority, United Kingdom of Great Britain and Northern Ireland, Federal Aviation Administration, 15 July 2014, http://www.faa.gov/about/office_org/headquarters_offices/ast/programs/international_affairs/media/Memorandum_of_Cooperation_FAA_and_UK_AST_DFT-UKSA-CAA_signed_July-2014.pdf, accessed 4 August 2017.

7. Developing a new safety regulatory system for commercial sub-orbital spaceplanes in the UK – the challenge!

Jeremy Stubbs

1. INTRODUCTION

For the past 60 years access to space has mainly been provided through very expensive national spaceflight programmes. On 28 October 1971 a UK designed and built Black Arrow Rocket successfully inserted a UK satellite called Prospero into orbit. Unfortunately the Black Arrow programme was subsequently cancelled and Prospero is still the only satellite launched into orbit by a UK rocket system. Since 1971 the UK has built up a strong national space presence, but it relies on other nation States for satellite launch services.

The dawn of the twenty-first century has ushered in a new era in spaceflight where the commercial sector is increasingly able to provide a space launch capability that is considerably less expensive than in the past. However, there is still a desire to further reduce launch costs and achieve true low-cost access to space; reusability is seen to have the potential to substantially reduce those launch costs.

Several commercial organisations are developing spaceplanes for both sub-orbital and orbital operations. They are completely reusable and are seen as a game-changing technology that has the potential to both reduce costs and provide increased operational flexibility. The emerging market for personal space flight experience, or space tourism as it is more commonly called, is helping to drive forward development of sub-orbital spaceplanes. Some designs are being test-flown, with commercial operations expected to follow in the near future. Sub-orbital spaceflight experience is seen as an enabler to the development of wider market opportunities such as scientific research and satellite launch.

The joint UK government/industry Space Growth Action Plan (2011)[1] set out a programme to grow the UK's world share of the estimated £400 billion global space-enabled market to 10 per cent by 2030. As part of this growth the UK government aspires to take advantage of the potential for commercial low-cost access to space and have a functioning spaceport hosting commercial spaceplane operations from the UK by 2018.

In August 2012 the UK Department for Transport (DfT) and UK Space Agency (UKSA) tasked the UK Civil Aviation Authority (CAA) to undertake a detailed review of what would be required from an operational and regulatory perspective to enable spaceplanes to operate from the UK by 2018, pending demonstration of feasibility and a decision to do so. This chapter concentrates on the thought processes and challenges that were faced in carrying out the review.

2. GOVERNMENT TASK

The CAA is the UK's specialist aviation regulator and is widely respected throughout the world. It was established in its current form in 1972 and regulates aviation activities by powers granted under the UK Civil Aviation Act 1982[2] and through the articles of the UK Air Navigation Order 2016 (ANO).[3] However, whilst the CAA has technical and regulatory expertise in all aspects of aviation, spaceflight was not part of its normal activity.

The government task was to carry out a review to:

- Assess the extent to which the UK can support safe spaceplane operations.
- Develop options for the certification of spaceplanes, engines and associated systems.
- Identify the key characteristics and potential locations of a spaceport.
- Develop an understanding of the future market for spaceplane operations.

---

[1] https://www.gov.uk/government/uploads/system/uploads/attachment_data/file/298362/igs-action-plan.pdf, accessed 4 August 2017.

[2] http://www.legislation.gov.uk/ukpga/1982/16/contents, accessed 4 August 2017.

[3] http://www.legislation.gov.uk/uksi/2016/765/contents/made, accessed 4 August 2017.

The review was to be carried out in partnership with the UKSA and with the support of industry.

3. SCOPE OF THE REVIEW

It was evident from the government request that the review should concentrate on spaceplanes and not on the more conventional vertical launch rocket systems. However, that decision in itself posed a problem, as there are many different spaceplane designs and no internationally agreed definition of what constitutes a 'spaceplane'. Therefore, more clarity had to be obtained as to the scope of the review.

Sub-orbital spaceplanes were already in development, particularly in the USA. In particular, Virgin Galactic[4] and Xcor Aerospace[5] were seen as market leaders in this field. These sub-orbital spaceplanes were intended not only for the emerging market for space tourism but also for scientific payloads and the launch of orbital satellites. The UKSA was very interested in scientific research and the availability of a UK satellite launch system; therefore, sub-orbital spaceplane systems appeared to be an ideal starting point for UK space launch operations. Hence, sub-orbital spaceplanes were seen as a key part of the review.

There were also spaceplane designs intended for orbital use. For example, the NASA Shuttle could be defined as a spaceplane, but it was launched vertically as a conventional rocket system and returned as a glider. Similarly the US Company Sierra Nevada[6] is developing the Dream Chaser, which is launched to orbit vertically by a conventional rocket system. These systems would require large vertical launch facilities and were considered unlikely to be viable from the UK.

However, one type of system was in development in the UK. A British company, Reaction Engines Limited (REL),[7] was developing a revolutionary new air-breathing rocket engine called Sabre. This engine had the potential of enabling a spaceplane to fly directly to orbit, that is, a single-stage-to-orbit spaceplane. As there was government and European Space Agency (ESA)[8] interest in the company, it was decided that spaceplane single-stage-to-orbit operations would also be within the scope of the review. Sabre also showed promise as a potential engine for

---

4 http://www.virgingalactic.com/, accessed 4 August 2017.
5 http://www.xcor.com/, accessed 4 August 2017.
6 http://www.sncspace.com/, accessed 4 August 2017.
7 https://www.reactionengines.co.uk/, accessed 4 August 2017.
8 http://www.esa.int/ESA, accessed 4 August 2017.

a hypersonic aircraft. Therefore, it was decided that options for regulating long-range hypersonic transport aircraft would also be included in the review.

As to the final scope of the review, discussions with the UKSA and DfT resulted in an agreement that it should consider only spaceplanes that used a runway for take-off and landing in a similar way to conventional aircraft. These would be sub-orbital and orbital spaceplanes and long-range, high-speed transport.

It was also important to gain clarity on what would be needed for a UK 'spaceport' and where would be the best location for this. Again, there is no internationally agreed definition of a spaceport and therefore agreement had to be reached as to the scope of such a facility. To ensure compatibility with horizontally launched spaceplanes, it was agreed that a UK spaceport would need to be similar in nature to a conventional aerodrome. The requirements for a vertical launch site would not form part of the review, apart from where best to locate such a site.

As the UK has no experience of spaceplane operations, it was important that the review fully examined the processes worldwide that were currently supporting such operations. It was quickly evident that the most mature system was in the USA, where spaceplanes were actively being developed, and indeed test-flown. Therefore, the review would need to assess the regulatory frameworks available in the USA and the way in which they were being applied.

Finally, the business case for UK commercial spaceplane operations had to be examined by the review team.

The agreed final scope of the review was as follows:

- a description and analysis of actual or anticipated key spaceplane operations and their requirements;
- an analysis of the applicability of the procedures and requirements utilised by the US Federal Aviation Administration Office of Commercial Space Transportation (FAA AST) to the UK;
- recommendations for the appropriate regulatory framework for commercial spaceplane operations in the UK, to include:
  - spaceplane airworthiness;
  - airspace requirements;
  - air traffic management;
  - flight operations;
  - flight crew licensing; and
  - flight crew and participant medical requirements;
- an analysis of and recommendations regarding the appropriate regulatory requirements for spaceport operations;

- recommendations as to the most suitable locations for a spaceport in the UK;
- consideration of the likely environmental impacts peculiar to spaceplane and spaceport operations;
- an assessment of the potential for the growth of the spaceplane industry beyond sub-orbital space tourism and satellite launches; and
- an assessment of the value to the UK of commercial spaceplanes and related technologies.

4. EDUCATION AND CONSULTATION

The CAA has a great deal of expertise in all aspects of aviation; however, this did not extend to spaceflight operations. The only regulations in the ANO for rockets are those relating to airspace requirements for vertical launch rockets, and in the past the majority of these have been small and sub-orbital. In order to carry out the review, the CAA assembled a small team of experts from across the main aviation technical disciplines. Before starting to address the many questions that would need to be answered for the review it was immediately evident that this team would need to take time to learn about the emerging spaceplane industry and spaceflight in general.

Initially the review team used as much information as was available from discussions with the UKSA, in publications and on the Internet in general. It was quickly apparent that substantive information on spaceplanes was scarce and there were very few publications that considered options for regulation. The only developed regulatory framework for space launch operations in Europe was that of the French regulator CNES,[9] and that was designed for unmanned vertical launch using expendable rockets. Elsewhere there were very few frameworks in place outside of the USA that could be assessed and possibly used as a basis for a UK regulatory framework.

The UKSA was already carrying out regulatory workshops with the spaceplane industry. Support and learning from industry would be essential in achieving a balanced approach to regulation. The addition of the CAA to these workshops quickly enabled the team to establish very good working relationships with the industry representatives present, which continued throughout the review.

---

9 https://cnes.fr/en, accessed 4 August 2017.

Discussions in the various workshops were open and frank. It was obvious that there was some disagreement between European industry and that of the USA. The USA had an established regulatory framework that treated spaceplanes as rockets. It only regulated the safety of the uninvolved general public and did not call for any form of vehicle certification. European industry, on the other hand, had no clear regulatory system, but thought that spaceplanes should be treated more as aircraft and that it should be possible to establish a form of vehicle certification by setting safety targets and using aviation methodology.

The industry group also discussed the impact of EU legislation. Who should regulate, both in the short term and the longer term? Industry representatives were very keen to establish clear and concise legal and regulatory frameworks without too much regulatory burden. The position of most parties present was that, in Europe, nation States should develop their own frameworks for the short term. However, a unified European framework could be developed once sufficient experience of operations had been gained.

With the scarcity of detailed information available in the UK and given the short timescale, it was decided that the team needed to look elsewhere. Therefore, a request was made to government to carry out a formal fact-finding visit to the USA. This was to include opportunities to meet with all the appropriate government departments involved in the space industry as well as the spaceplane industry itself. The DfT agreed that this would be essential to help shape the review and the USA agreed to host a cross-UK government/CAA fact-finding team over a period of ten days.

The visit was very successful and covered Washington DC, New Mexico Spaceport and Mojave Air and Spaceport. All the stakeholders visited were very supportive and the knowledge gained by the team was invaluable in helping shape their understanding of the issues that needed to be addressed. Through this visit the UK team established and maintained a very good working relationship with FAA AST, trade bodies and US industry. Several of the key issues that were discussed are expanded on below.

5. BARRIERS

International export controls exist to restrict the distribution of arms and proliferation of strategic missile systems. There are four major global agreements, two of which are relevant to spaceplanes: the Missile

Technology Control Regime (MTCR)[10] and the Wassenaar Arrangement (WA).[11] The UK and the USA are signatories to both agreements.

The MTCR was established in 1987 by the USA, Canada, France, Germany, Italy, Japan and the UK, but now there are a total of 34 signatories. The MTCR aims to limit the proliferation of strategic missile systems capable of delivering weapons of mass destruction. The WA was established to contribute to regional and international security and stability, by ensuring greater responsibility in transfers of conventional arms and related dual-use goods and technologies.

During the team's exploratory visit to the USA in June 2013, a meeting was held with members of the US Department of State. They confirmed that manned sub-orbital and orbital spacecraft are on the US Munitions List (USML)[12] and therefore subject to the US International Traffic in Arms Regulations (ITAR).[13] Even with the changes to the classification of satellites introduced into US Law in 2013, these systems would still be covered by ITAR.

It is the view of the US Department of State that, irrespective of the commercial aspirations of companies developing sub-orbital and orbital vehicles, the potential for spaceplanes to deliver weapons of mass destruction is the critical issue. The US Department of State therefore begins from a position of a strong presumption to deny any export licence applications for spaceplane technology.

US export controls would have a significant impact on industry's desire to use their spaceplanes worldwide. If the UK wished to encourage the import of US spaceplane systems, then the British government would have to work with the US government and the US sub-orbital spaceplane industry to obtain the necessary export licence. This would also include the necessary Technical Assistance Agreements on the information that the UK regulatory authority would have access to in order to put in place a suitable regulatory regime. Such an outcome would enable operators to start planning their activities, and the regulatory process to continue. However, formal discussions could not start until an operator signalled its intent to apply for an export licence.

---

10 mtcr.info/?lang=en, accessed 4 August 2017.

11 http://www.wassenaar.org, accessed 4 August 2017.

12 http://www.ecfr.gov/cgi-bin/text-idx?SID=86008bdffd1fb2e79cc5df41a180750a&node=22:1.0.1.13.58&rgn=div5, accessed 4 August 2017.

13 https://www.pmddtc.state.gov/regulations_laws/itar.html, accessed 4 August 2017.

6. LEGAL CONSIDERATIONS

There were many legal issues that needed resolution or on which policy decisions needed to be taken. This section discusses some of the most interesting issues.

A. Aviation or Space Legislation

Spaceplanes are a new class of vehicle that are not formally defined or even mentioned within existing UK or European aviation or space legislation. Spaceplanes could be assessed as vehicles that act as an aircraft while in the atmosphere and as a spacecraft while in space; therefore, both space law and aviation law may be applicable. This is further complicated by the fact that the limits of the atmosphere and space are not legally defined.

The International Civil Aviation Organization (ICAO)[14] defines an aircraft as 'any machine that can derive support in the atmosphere from the reactions of the air other than the reactions of the air against the earth's surface' (ICAO Annex 1 or Annex 6 Part I). This would suggest that aviation law should apply.

The legal frameworks for aviation in the UK are the Civil Aviation Act and the Air Navigation Order. In addition, the UK is a member of the EU, and as such the EU Aviation Law and its associated regulatory frameworks, through the European Aviation Safety Agency (EASA),[15] apply to most UK aviation activity. In informal discussions with EASA it became quickly apparent that they considered spaceplanes as aircraft under the ICAO definition and that EU Aviation legislation and regulations should apply.

As stated above, following discussions with potential European spaceplane manufacturers and operators, it was apparent that they favoured an aviation approach to the design, manufacture, operation and regulation of spaceplanes. They stated that obtaining some form of certificate of airworthiness would give more certainty as to the safety performance of the spaceplane to potential operators, passengers, regulators and the insurance industry.

---

14 http://www.icao.int/Pages/default.aspx, accessed 4 August 2017.

15 https://www.easa.europa.eu/, accessed 4 August 2017. Note that EASA currently has no clear competency for regulating space matters. For more on the EASA, see Chapter 10 of this volume.

Even though the decision could be made to apply aviation law to spaceplane operations, would the UK Outer Space Act (OSA)[16] also apply to spaceplanes? Clearly the OSA would apply to single-stage-to-orbit spaceplanes and to orbital satellites launched from a spaceplane (or conventional aircraft), but its applicability to sub-orbital spaceplanes is less clear. As there is no formal international definition of where space starts, it could be argued that a sub-orbital spaceplane operating below the minimum levels for maintaining an orbit would not be considered a space object, and therefore the OSA would not apply. However, this would be subject to a UK policy decision. Therefore, there is the potential for two legal and regulatory regimes to apply to a single flight.

The situation in the USA is different, however. The USA rules define sub-orbital rocket as 'a vehicle, rocket-propelled in whole or in part, intended for flight on a sub-orbital trajectory, and the thrust of which is greater than its lift for the majority of the rocket-powered portion of its ascent'.[17] As a result, FAA aviation legislation does not apply to spaceplanes. All spaceplane activity will be regulated by FAA AST through the US Code of Federal Regulations Part 400 series.[18]

As for the UK, a way needed to be found to provide a suitable legal basis on which to regulate UK spaceplane operations. The preferred government option was to keep everything as simple as possible and find a way to accommodate spaceplane operations within existing UK legislation without the need for an amendment. A further complication was that any US system that wished to operate from the UK would also need to be licensed by FAA AST and therefore comply with both the US and UK legal and regulatory systems. Finally, it would be helpful if agreement could be reached to regulate spaceplanes nationally by the UK, even if this were to be for the short term pending future European rule-making.

Following discussions between the CAA, DfT and EASA it was agreed that if the UK wished to operate spaceplanes within UK airspace only, a solution could be to consider the use of Annex II of the EASA Basic

---

16 http://www.legislation.gov.uk/ukpga/1986/38/contents, accessed 4 September 2017.

17 http://www.ecfr.gov/cgi-bin/text-idx?SID=ac1f56a04cfaa0ce8ba3283961011b82&mc=true&node=pt14.4.401&rgn=div5#se14.4.401_15, accessed 4 August 2017.

18 http://www.ecfr.gov/cgi-bin/retrieveECFR?gp=&SID=ac1f56a04cfaa0ce8ba3283961011b82&mc=true&r=PART&n=pt14.4.400, accessed 4 August 2017.

Regulation.[19] This Regulation allows certain categories of aircraft (specifically designed or modified for research, experimental or scientific purposes, and likely to be produced in very limited numbers) to be regulated under national rules.

Spaceplanes could easily be considered as 'experimental'. However, there would be difficulties with this methodology. In particular, could fare-paying passengers be carried on 'experimental aircraft'? Any US spaceplane operation in the UK would be operating under the terms of a US licence and would not be 'experimental'. There was also the question of industry acceptance of the use of 'experimental'. Interestingly the EASA regulations are 'silent' about the carriage of fare-paying passengers on experimental aircraft, and following consultation, industry indicated that it understood the situation and was willing to adopt this approach.

B. Passengers or Participants and Informed Consent

Using an EASA Basic Regulation Annex II opt-out to ensure national regulation seemed to offer an acceptable solution for the UK. However, the issue of commercial passengers on aircraft that did not hold a certificate of airworthiness still needed to be addressed.

The US regulatory and legal system for commercial spaceflight does not recognise paying customers as passengers; instead it calls them participants. These participants would not have the same 'passenger rights' as established in the aviation system and would have to accept that flying as a participant on a spaceflight carries a greater degree of risk than conventional aviation does.

The US Code of Federal Regulations Part 460 includes a requirement that the operator must 'inform each spaceflight participant in writing about the risks of the launch and re-entry, including the safety record of the vehicle'. In the absence of any form of FAA 'Safety Certification' of the vehicle, the operator must also inform the participant that the US government has not certified the vehicle as safe for participants or crew. Following receipt of this information, the participant and crew must provide written consent to 'participate' in the flight.

There are several issues that need to be addressed when considering informed consent:

---

19 http://eur-lex.europa.eu/legal-content/EN/TXT/?qid=1474978980580&uri=CELEX:32008R0216, accessed 4 August 2017.

- The question must be asked: 'is the participant capable of fully understanding the risks?' In order to answer this question, should the potential participant be required to have a form of medical assessment as to his or her mental fitness?
- Should the regulator be required to set out the required content for the operator's informed consent?
- Should an age limit for acceptance of these risks be set? Would it be legal for parents or guardians of minors to sign on their behalf?
- Following a pre-flight medical check, should the potential participant also sign as informed consent as to their own personal medical risk?
- When should this informed consent be signed? If it were to be just before the flight and the potential participant refused to sign, should there be an automatic refund of deposit or any other payments made to the operator?

In the UK, the principle of informed consent is used by the aviation industry for activities such as wing-walking and parachuting. Therefore, in principle, informed consent could be used for spaceflight operations in the UK; but again, the same issues outlined above would have to be addressed. Also, assuming that the flight is carried out by a US operator, the participant may need two informed consent processes, one to satisfy US regulatory authorities and one to satisfy the UK authorities.

C. Liability

It must be made clear that informed consent itself does not absolve the operator of liability claims brought by crew or participants or their families in the event of death or serious injury following an accident or serious incident. It would also not affect the operators' basic responsibilities for 'duty of care' to comply with the relevant health and safety legislation.

In the event of sub-orbital spaceplane operations from the UK, what would be the impact on the liability of government?

If we are to assume that spaceplanes are aircraft, then section 76 of the UK Civil Aviation Act 1982 would apply. This states that the owner or operator of an aircraft cannot be sued for trespass or nuisance simply because the aircraft has flown over any property at a reasonable height and in accordance with the ANO. However, where damage is caused by anything falling from an aircraft, the owner is liable and the person suffering damage does not have to prove negligence.

The UK Outer Space Act could also apply to the spaceflight operation. This treats liability in a different way and gives the government the ability to cap liability for the majority of missions to 60 million euros. However, if the operation is licensed by the USA, it would be subject to liability under the Commercial Space Launch Act (CSLA).[20] This uses the principle of maximum probable loss to determine liability.

Having two different systems applying to the vehicle would give rise to considerable uncertainty and may make securing adequate insurance a difficult proposition. Before operations commence, it would be sensible for all parties concerned with the spaceflight operation to work together to establish a better understanding of liability and therefore insurance requirements.

From the above, it is clear that using an EASA Annex 2 approach to bring regulation under national control could enable the UK to establish a legal framework for the operation of sub-orbital spaceplanes. However, there is still a need to further clarify the use of informed consent together with the liability and insurance requirements.

7. RISK

In order to effectively evaluate the options for safety regulation, the spaceplane review team had to understand the safety risks associated with the sub-orbital spaceplane industry.

Aircraft are designed, certificated and operated against standards that are the result of over one hundred years of experience. The aviation regulatory system has also developed to protect the safety of all those on board aircraft and is coordinated worldwide through the ICAO. For commercial aircraft operations, adhering to these standards, combined with the ICAO regulatory system, results in catastrophic failures occurring only once in every 10 million flight hours.

The statement that 'Spaceflight is Risky!' is generally accepted by industry and regulators worldwide; however, this generally refers to orbital operations using vertical launch rocket systems. Would sub-orbital spaceplanes also be considered to be as risky? A typical flight profile of taking off from a runway, using a relatively low-powered rocket motor to climb and gliding back to a landing on a runway seemed to indicate that these operations may be a lower risk. Sub-orbital operations are also not as exposed to the dangers of orbital debris as are orbital activities, and a

---

[20] http://uscode.house.gov/view.xhtml?path=/prelim@title51/subtitle5/chapter509&edition=prelim, accessed 4 September 2017.

short flight time means radiation exposure is also considerably reduced. Orbital spaceplanes, however, would be subject to many of the same risks as current rockets; in fact, the continual and rapid turnaround that may be possible with such vehicles may increase the overall cumulative system risk.

Commercial sub-orbital spaceplanes are new, not many have flown and there are very few statistics available to build up a picture of the risks associated with commercial spaceplane operations. If commercial operations with fare-paying passengers on board were to become commonplace, then overall safety would probably need to be orders of magnitude better than currently experienced by the space industry in general. The challenge is therefore to design a suitable regulatory system that contributes to improving the safety of the operation without being so burdensome that the operation becomes no longer viable.

The US CSLA requires the FAA AST to regulate commercial spaceflight in terms of third party risk only, and provides maximum risk totals for groups of people as well as individual risk. The sub-orbital spaceplane operator is required to carry out a casualty analysis for every planned flight profile from every spaceport/launch location and has to demonstrate that it meets the requirements as laid down in the CSLA. In general, the space industry achieves this by locating its launch sites in areas of limited population density, for example, at desert or coastal locations, and operating in segregated airspace away from other aircraft.

In the UK there are no desert locations; however, on careful examination there were several UK coastal aerodrome locations that looked likely to provide operators with a good chance of meeting the US casualty analysis requirements. These locations were already established UK danger areas that could be adapted to provide suitable segregated airspace. However, it was acknowledged that more work had to be done to understand the impact of flying over the whole of the UK's coastal areas, the potential for operations close to sites of large national infrastructure such as power stations and hospitals, and how a large influx of tourists to the launch site would impact the casualty analysis results.

Developing a regulatory framework using third party risk only may not be the only option in the UK. Those personnel on the ground working on the operation are classed as second parties and are regulated under UK health and safety legislation. The challenge would therefore be to see if there were options available to provide some safety assurance for the occupants of the spaceplane, both crew and participants. One of the key methods of enhancing occupant safety would be to look at the art of the possible in terms of the safety assurance of the spaceplane itself and the establishment of safe operating procedures.

In the USA, whose legal and regulatory system only addresses third party risk, industry and the FAA AST have worked together and published their 'Recommended Practices for Human Space Flight Occupant Safety'.[21] This document is not legally binding but effectively outlines 'draft standards' that could be used to enhance occupant safety. The International Association for the Advancement of Space Safety (IAASS)[22] is also working to establish principles for sub-orbital spaceflight safety. These could form the basis of a limited level of safety regulation for crew and participants.

8. THE 'CERTIFICATION' WORD

The majority of current space launch regulatory frameworks were developed to provide protection for the uninvolved general public and do not seek to regulate the safety of spacecraft occupants. They include frameworks such as that of the USA and State limits on the overall third party risk, as well as individual third party risk.

The US regulatory framework does not impose safety standards for the design, build or operation of the spacecraft other than to provide for the safety of third parties. The FAA AST does, however, require a System Safety Review assessment of the vehicle that identifies all safety-critical systems, including a hazard analysis and risk assessment, and a validation and verification analysis. The verification process aims to produce tangible evidence that the design meets safety requirements and could be considered as a step towards a certification process.

Aviation regulation seeks to protect the safety of aircraft occupants; if the aircraft and its operation are safe, then third parties will be protected by default. This is achieved by a system of 'certification' of the aircraft against set standards and by establishing standard operating procedures for the operation. Aviation certification involves the regulator's technical experts being involved from the very beginning of the design process through to the completion of the flight test programme. It is not carried out in retrospect and cannot be applied to sub-orbital spaceplanes that are already established. Therefore, if the aviation model of regulation were to

---

21 https://www.faa.gov/about/office_org/headquarters_offices/ast/media/Recommended_Practices_for_HSF_Occupant_Safety-Version_1-TC14-0037.pdf, accessed 4 August 2017.

22 http://iaass.space-safety.org, accessed 4 August 2017.

be applied to existing designs, then an alternative to traditional certification would need to be developed.

Following consultation with industry, it was immediately apparent that there was a major difference of opinion between established sub-orbital spaceplane manufacturers in the USA and Europe.

US industry was firmly opposed to the establishment of any form of aviation-style certification for their sub-orbital spaceplanes. They argued that the industry was too new, with few standards available, and certification would place too high a burden on the new industry.

As stated earlier, European industry was keen to establish a certification approach for sub-orbital spaceplanes using current aviation-style standards and developing new ones where no appropriate standards existed. It argued that certification would provide greater safety assurance for occupants and would make securing insurance easier.

9. A PROPOSED UK SOLUTION

There are several different regulatory options for, or approaches to, the regulation of sub-orbital spaceplane operations. Any regulatory framework must also take into account the complex legal environment present in the UK and the EU. This, combined with the different operators and operating models currently envisaged, indicates that a 'one size fits all' regulatory framework may be difficult to achieve.

However, in the interests of clarity and transparency, there is a strong case to find as much common ground as possible, not only across the regulatory frameworks proposed for different operations, but also within existing regulatory frameworks.

Given sufficient demand from Member States, it is probable that the EU will introduce a regulatory framework for commercial sub-orbital spaceplane operations at some time in the future. However, European legislative development takes time and such development is unlikely to occur quickly enough for the proposed start of operations in the UK. Therefore, it would make sense to establish an initial national solution that would be broadly in line with European thinking; this solution could then form the basis for future European development.

Looking at the legal arguments discussed in section 6, it was clear that a sub-orbital spaceplane framework based on aviation regulation was possible and the EASA Annex II solution would allow for operations to be regulated under national law by the CAA. In the short term this would give the freedom to allow operations using a sub-orbital spaceplane that holds a current FAA AST operating licence by using exemptions and special conditions to the ANO to cover those areas of UK aviation legislation with which the design and manufacture of the spaceplane and its operation could not comply.

These special conditions would aim to help mitigate the risks to the uninvolved general public and, where possible, identify and mitigate the risks to spaceplane occupants. They should be based on industry best practice and derived from aviation regulation and suitable space safety regulation, where appropriate and available.

But whom should the framework seek to protect? It was agreed that third party safety had to be assured within bounds established under UK law. Whilst these bounds had yet to be established, it was assumed that they would also satisfy US casualty analysis requirements. Second party safety would most likely be a UK health and safety responsibility. First party safety, that is, the safety of spaceplane occupants, could not match that offered to conventional commercial aircraft operations. However, it would be essential that the regulator worked closely with industry to provide a level of safety assurance that was as robust as reasonably practical.

In the absence of any airworthiness certification of the vehicle, a short-term pragmatic approach to airworthiness approval needs to be found. A solution may be to use an enhanced version of the FAA AST System Safety Review; this would require the operator to demonstrate that it is satisfied that the vehicle is 'fit for the intended purpose' under an overall safety management system framework.

In the longer term, however, the aim must be to start the development of a 'certification process' that can be applied to new sub-orbital spaceplane designs. With sufficient successful operating experience, the safety performance of systems such as rocket engines, reaction control systems and other space-related systems will become much better understood. This could lead to the development of standards and recommended practices for the design, manufacture and operation of such vehicles, leading the way to substantive improvements in overall safety.

Applying elements of aviation law to a US licensed system (under space law) would, however, still present issues that would need careful consideration. These include differing requirements for liability and insurance together with the use of informed consent.

10. CONCLUSIONS

The regulation of sub-orbital spaceplane operations is a challenge for any nation that does not have an established space launch capability. The following are a series of questions, thoughts and advice that should be taken into consideration before developing any regulatory framework:

- Developing new legislation and regulatory frameworks takes time and expense. What is your ambition? Do you wish to establish an industry that is purely sub-orbital or do you wish for orbital operations including vertical launch?
- Look at your existing legislation; do you have an Outer Space Act already established? Is it sufficient or is new legislation required?
- Where does your planned launch system originate? Is it home-grown or is it simply brought in from another nation?
- If you only plan a few demonstration launches, then it may be sensible to temporarily accept the regulation established by the State of the operator.
- Consider other established regulatory systems; one may be suitable, so why not adopt that system?
- Consult with industry.
- Be prepared to adapt and evolve your framework as operations become established.

For the UK, there are many options available for the development of a sub-orbital spaceplane regulatory framework. However, it should be supported in law and compatible with the wishes of European space industry, whilst at the same time be capable of allowing for a US licensed operation from the UK.

The review[23] therefore sets out a comprehensive series of recommendations to government highlighting how this could be achieved. It proposes a hybrid framework that is aviation-based, but introduces elements of space-based regulation to cover those areas where pure aviation regulation would not be appropriate.

However, before the UK could accept applications from operators, the UK government would have to be satisfied that the current UK legislation was sufficient. For a US-licensed operation, agreement would need to be reached with the USA over the division of responsibilities together with issues surrounding liability. Government would also need to be satisfied that the regulatory framework was sufficiently robust to manage the activity.

---

[23] https://www.gov.uk/government/uploads/system/uploads/attachment_data/file/329756/spaceplanes-summary.pdf and https://www.gov.uk/government/uploads/system/uploads/attachment_data/file/329758/spaceplanes-tech.pdf, both accessed 4 August 2017.

11. THOUGHTS FOR THE FUTURE

This chapter has concentrated on exploring the options for the regulation of UK national sub-orbital spaceplane operations. However, designs are already being proposed that will allow spaceplanes to achieve low Earth orbit. This could result in a dramatic reduction in operating costs, bringing about a big increase in demand, with space hotels, space power stations, spacecraft refuelling stations and support for outer space activity such as asteroid mining and planetary exploration. It will also be much easier for current non-spacefaring nations to acquire such a space launch capability: all that will be required is a suitable runway and a spaceplane.

To understand the potential impact that a single-stage-to-orbit spaceplane may have on infrastructure requirements, as well as legal and regulatory systems, it is important to examine the flexibility and performance that is being planned. For example, the Reaction Engines Skylon is designed to be fully reusable, place more than 10 tonnes into orbit and could be adapted to carry passengers. It will take off and land from a conventional runway; it will be able to self-ferry to a new departure location and could operate daily. A worldwide fleet of 30 such vehicles would completely transform the low Earth orbit environment and the operation will more closely resemble normal aviation rather than the current spaceflight model.

We must not of course forget the more conventional way of accessing space. Spaceplanes may eventually replace vertical launch as the preferred method of reaching space, but current activity points to the possibility of completely reusable vertical launch systems. These systems could also substantially lower launch costs and lead to a marked increase in orbital activity, and would have to be factored into any new legal and regulatory system.

Successful operation of such single-stage-to-orbit spaceplanes and reusable rockets may require a complete change in infrastructure requirements. To manage large numbers of flights to orbit and return, together with transfer flights between orbits, a unified system of space traffic management will be essential. This would need to provide tracking, communications, debris avoidance and data links, and be a global solution rather than a national responsibility.

Although single-stage-to-orbit spaceplanes and reusable vertical launch systems should reduce the generation of space debris, the large increase in the number of flights will result in a large overall increase in risk of damage from space debris. A space traffic management system should also be responsible for space debris mitigation, but eventually a method of debris removal will be essential.

Such a future system will require a complete rethink of global space law and regulation. It is likely that the current UN treaties that underpin current space law and space operations will be too cumbersome to support the anticipated level of activity. Issues surrounding national sovereignty and the need for a global solution would suggest that the United Nations should recognise the need for change and establish a global space legal system that is fit for the future.

This legal system should call for the establishment of a space traffic management system and eventually make the deliberate release of space debris illegal. It should also call for the establishment of a global safety regulatory system that would cover all aspects of operation, including access to and from the Earth, Earth orbit and eventually outer space operations.

Developing such a legal system would, of course, be difficult and take time, but if it is to be in place for such future reusable spaceplane operations, the time to start is now.

8. Red tape in the final frontier: Bigelow Aerospace's adventures in export control

Mike N. Gold and Christopher M. Hearsey

1. INTRODUCTION

While science fiction involves many frightening villains, Klingons, angry Martians, and various creatures from the great beyond, real space projects have faced a foe that has been far more pernicious than what even the most creative writer's imagination can conjure: specifically, the International Traffic in Arms Regulation (ITAR).[1] The overly broad implementation of the ITAR has crippled commercial space projects while failing in its primary mission to bolster national security and foreign policy interests.[2] Like in any good sci-fi epic, victory against this threat was eventually achieved, but the battle has been both lengthy and costly.

In July 2006 and June 2007 Bigelow Aerospace, LLC ('Bigelow Aerospace' or 'BA') successfully launched two pathfinder expandable

[1] 22 CFR §§ 120–130 (2016).

[2] See discussion below. See for example M.N. Gold, 'Thomas Jefferson, We Have a Problem: The Unconstitutional Nature of the U.S.'s Aerospace Export Control Regime as Supposed by Bernstein v. U.S. Department of Justice' (2009) 57 *Clev. St. L. Rev.* 629; United States Government Accountability Office, 'Export Controls: Vulnerabilities and Inefficiencies Undermine System's Ability to Protect U.S. Interests' GAO-07-1135T (2007); and Center for Strategic & International Studies, 'Briefing of the Working Group on the Health of the US Space Industrial Base and the Impact of Export Controls (February 2008), available at http://csis.org/files/media/csis/pubs/021908_csis_spaceindustryitar_final.pdf.

habitat prototypes, designated *Genesis I* and *Genesis II*.[3] The goal of the *Genesis* programme was to demonstrate the benefits of Bigelow Aerospace's innovative expandable habitat technology in an actual orbital environment. Bigelow Aerospace launched both *Genesis* spacecraft from the Russian Federation using ISC Kosmotras, a Ukrainian–Russian–Kazakhstani joint stock company, as the transportation provider for the spacecraft.[4] The *Genesis* programme, designed as a first step toward expanding the paradigm of commercial opportunities and capabilities for human spaceflight, quickly became embroiled in a constant struggle over legal obstacles, namely the reach and application of the ITAR.[5]

While the ITAR experiences of the *Genesis* programme have been discussed at length elsewhere,[6] this chapter instead seeks to provide some history and background regarding BA's uphill but ultimately successful battle to reform the export control regime surrounding the 'passenger experience'.[7] The publication of the final rule for the Amendment to the

---

3 See Bigelow Aerospace, 'Genesis I' (*BigelowAerospace.com*), available at http://www.bigelowaerospace.com/genesis-1.php; L. David, 'Bigelow Aerospace's Genesis-1 Performing Well' (*Space.com*, 21 July 2006), available at http://www.space.com/2649-bigelow-aerospace-genesis-1-performing.html; Bigelow Aerospace, 'Genesis II', available at http://www.bigelowaerospace.com/genesis-2.php; T. Malik and L. David, 'Bigelow's Second Orbital Module Launches into Space' (*Space.com*, 28 June 2007), available at http://www.space.com/4007-bigelow-orbital-module-launches-space.html.

4 See M.N. Gold, 'The Wrong Stuff: America's Aerospace Export Control Crisis' (2008) 87 *Nebraska L. Rev.*, 521, 523. See also L. David, 'Bigelow's Space Module Plans in Overdrive' (*NBC News*, 22 November 2006), available at http://www.nbcnews.com/id/15851693/ns/technology_and_science-space/t/bigelows-space-module-plans-overdrive/#.U6jJDfldXh4.

5 See L. David, 'Genesis-1: Reaching Escape Velocity from Red Tape' (*USA Today*, 26 July 2006), available at http://usatoday30.usatoday.com/tech/science/space/2006-07-26-genesis-1_x.htm.

6 See M.N. Gold, commentary, 'Lost in Space: A Practitioner's First-Hand Perspective on Reforming the US's Obsolete, Arrogant, and Counterproductive Export Control Regime for Space-Related Systems and Technologies' (2008) 34 *Journal of Space Law* 163.

7 As Gold noted in 2009, the 'passenger experience has been ruled non-licensable, [i.e., Bigelow Aerospace does not] need to file paperwork for foreign passengers to receive training on how to live and function aboard a Bigelow station. [This] includes being exposed to the hardware externally and internally, learning how to operate systems such as communications, how to connect hardware for water and air, ingress and egress procedures, and just being in the habitat. All of this would have required licenses previously.' Interview by Amy Klamper with Mike Gold, Corp. Counsel and Director of Washington Operations, Bigelow Aerospace, in Washington DC (31 August 2009), available

International Traffic in Arms Regulations: Revision of U.S. Munitions List Category XV[8] promulgates for the first time a general passenger experience exemption under the ITAR, which exempts from export licensing the activities that potential spaceflight clients are likely to experience.[9] The passenger experience exemption resulted from BA's commodity jurisdiction (CJ) request determination of 2009 issued by the Department of State (DOS) regarding the company's expandable space habitat technology.[10] Without such a determination, seeking and servicing international clientele would have been significantly more costly, burdensome, and a major barrier to accessing the global private sector and sovereign markets.[11]

This chapter reviews the legal, policy, and historical factors that impacted the current reforms as well as how, subsequent to the *Genesis* campaigns, Bigelow Aerospace pursued changes to how the ITAR was applied to its commercial operations.[12] Moreover, the issues discussed

---

at http://www.spacenews.com/article/mike-gold-corporate-counsel-amd-director-washington-operations-bigelow-aerospace. See 'Space Technology: Freedom to Fly' (*The Economist*, 22 April 2009), available at http://www.economist.com/node/13525115.

[8] 79 Fed. Reg. 66,608 (10 November 2014) ('final rule, correction'). The Passenger Experience and space habitat technology rules technically came into effect on 10 November 2014 even though the Department of State published the language in the final interim rule of May 2014. See 79 Fed. Reg. 27,180, 27,188 (13 May 2014) ('final interim rule').

[9] The amendments also established the transfer of certain commercial communication satellites and 'in-space habitat' technologies from the 'United States Munitions List' to the 'Commerce Control List', except in certain situations where defence articles and their associated technical data may still be subject to the United States Munitions List (USML). See discussion below.

[10] See C. Clark, 'Bigelow Aerospace Petitions State for Export Jurisdiction Change' (*SpaceNews*, 3 March 2008), available at http://www.spacenews.com/article/bigelow-petitions-state-export-jurisdiction-change; and P.J. Blount, 'Bigelow Aerospace to File ITAR Challenge' (*Res Communis*, 12 March 2008), http://rescommunis.wordpress.com/2008/03/12/bigelow-to-file-itar-challenge/. See also M.J. Sundahl, 'Bigelow Aerospace's Commodity Jurisdiction Request Under ITAR and Its Impact on the Future of Private Spaceflight' (2009) Proceedings of the 52nd Colloquium on the Laws of Outer Space, IAC-09-E8.6.13, and interview by Amy Klamper with Mike Gold, *supra* note 7.

[11] See L. David, *supra* note 5, and M.N. Gold, 'OpEd: Common Sense' (*SpaceNews*, 12 March 2008), http://www.spacenews.com/article/oped-common-sense.

[12] The evolution in the menagerie of export control laws and regulations that has led to an *in rem* jurisdiction regime filled with ambiguity, complexity, and strong enforcement. For an example of the issues of ambiguity and complexity

herein are scoped to the current promulgation of the final rule to the amendments to Category XV of the United States Munitions List (USML).[13] Finally, this chapter ends with a description of the lessons learned from Bigelow Aerospace's singular effort.

2. A BRIEF HISTORY OF THE EVOLUTION OF UNITED STATES EXPORT CONTROL LAWS

A. The Menagerie of Export Control Laws

Since the constitutional establishment of the United States, Congress has had the power to regulate the import and export of goods and services with foreign entities.[14] The continual evolution of technology and international relations has led to various temporary restrictions on the export of domestic goods and services.[15] Restrictions almost always coincided with armed conflict or emergency situations where important national security issues or maintaining peaceful foreign relations were in play. The rationales for such restrictions were usually reasonable in light of the need to control domestic goods and services designated by law as munitions of armed conflict or singled out as important, scarce domestic

---

under the ITAR, see C.M. Hearsey, 'The Foreign Space Flight Participant Problem: Can a Space Flight Operator Balance Satisfaction of FAA Informed Consent Information Requirements with ITAR?' (2013) 6 *Phoenix L. Rev.* 303, 325 (2013), available at http://papers.ssrn.com/sol3/papers.cfm?abstract_id=2072479. On enforcement, see for example Arent Fox, 'ITAR Criminal Violations – Where the Less You Know Is Not Necessarily the Better' (*Arent Fox*, 26 February 2014), available at http://www.arentfox.com/newsroom/alerts/itar-criminal-violations-where-less-you-know-not-necessarily-better#.VE-0EvnF_h4; Boeing Company, DOS Bureau of Political-Military Affairs (28 March 2006) (Consent Agreement), available at http://www.pmddtc.state.gov/compliance/consent_agreements/pdf/Boeing_ConsentAgreement_06.pdf; and US DOS Directorate of Defense Trade Controls, Consent Agreements, available at https://www.pmddtc.state.gov/compliance/consent_agreements.html.

[13] 22 CFR § 121.1, Category XV (2016).

[14] See US Constitution Art. I, § 8, Cl. 3 (Commerce Clause) and US Constitution Art. I, § 10, Cl. 2 (Import-Export Clause).

[15] See, for example *US v. Barber* [1815] 13 US 243 (first Supreme Court case establishing 'fat cattle' as munitions of war and therefore as controlled goods under the export control laws of the USA), and *US v. Sheldon* [1817] 15 US 119 (establishing the scope of the meaning of the transportation of munitions and therefore establishing liability under the export laws of the USA).

material resources.[16] Moreover, the United States had developed a sizeable commercial arms export industry during the early twentieth century.[17] The outbreak of various armed conflicts just prior to the Second World War required the enactment of a new regulatory regime to maintain the balance between the export of munitions and the neutral foreign policy in vogue prior to the outbreak of the Second World War.[18]

The reactionary rhythm of temporary export restrictions from the nineteenth century continued on through to the mid-twentieth century with a treble increase in the delegated authority to the president to regulate goods considered munitions of armed conflict, that is, arms, ammunition and implements of war (collectively, 'controlled goods').[19] Congress addressed the need for a defined, modern 'temporary'[20] federal export control system over controlled goods and important domestic resources beginning with the enactment of the Neutrality Act of 1935.[21] The Neutrality Act of 1935 vested regulation of controlled goods in the DOS in concert with an inter-agency National Munitions Control Board, and also gave the president expansive authority to implement the legislation by proclamation.[22] The broad jurisdictional reach of the Neutrality Act of 1935 included, inter alia, the enumeration of the scope

---

[16] See M.C. Mineiro, *Space Technology Export Controls and International Cooperation in Outer Space* (Springer 2012) 43–45. See also *supra* note 15.

[17] Mineiro, *supra* note 16, 43–45.

[18] See Neutrality Act 1935 § 1, 22 USC § 441, 49 Stat. 1081 (1935).

[19] See for example Trading with the Enemy Act 1917, 40 Stat 411, 6 October 1917 (codified at 12 USC §§ 95a–b and 50 USC App. §§ 1–44), and Neutrality Act 1935 § 1, 22 USC § 441, 49 Stat. 1081 (1935). See also Proclamation, Neutrality Act 1939, Pub. Res. 54, H.J. Res. 306, 54 Stat. 4, 4 November 1939 (acting under Section 12 of the Neutrality Act 1939, the DOS issued the ITAR for the first time. See 4 Fed. Reg. 4,510, 4,512–4,522 8 November 1939).

[20] Historically, Congress tended to put time limits on the effect of such export controls, usually ending when certain hostilities ceased. As Berman and Garson note, '[w]ith the end of World War II, however, drastic wartime controls over exports were continued from year to year until February 28, 1949, when the Export Control Act was enacted, the first comprehensive system of export controls ever adopted by the Congress in peace time. Even that Act was initially conceived as a temporary measure, and might well have been allowed to lapse in 1951 but for the Korean War' (internal citations omitted): H.J. Berman and J.R. Garson, 'United States Export Controls – Past, Present, and Future' (1967) 67 *Colum. L. Rev.* 791, 791–92 (analysing the legal history of the Export Control Act 1949).

[21] Neutrality Act 1935 § 1, 22 USC § 441, 49 Stat. 1081 (1935).

[22] *Ibid.* at § 2.

of controlled goods prohibited under the Act, prohibition of the export and transportation of controlled goods to belligerent countries, the establishment of a registration and licensing regime of persons engaged in the business of manufacturing, exporting or importing controlled goods, and the restriction on travel by American citizens aboard belligerent ships during war.[23] An amendment to the Neutrality Act of 1935 passed by Congress in 1937 extended presidential power to include the ability to restrict the 'shipment of certain articles or materials in addition to arms, ammunition and implements of war' where such restrictions were 'necessary to promote the security or preserve the peace of the United States or to protect the lives of citizens of the United States'.[24] Congress further broadened presidential export control powers with the enactment of the Neutrality Act of 1939,[25] which granted the president the 'power to restrict the export of any articles or materials to any country involved in a declared state of war'.[26]

Several issues inherently affected the application of both Neutrality Acts. First, Congress did not provide a general licence provision and therefore registered entities had to obtain an export licence for each export transaction.[27] Second, neither Neutrality Acts addressed the terms 'technology or technical data in any of its provisions regarding export control' and 'neither term was mentioned in the list of arms, ammunitions and implements of war subject to export control pursuant to' the Neutrality Act of 1935 or its amendment.[28] However, pursuant to Section 2 of the Neutrality Act of 1935, Congress authorized the National Munitions Control Board to develop regulations over controlled goods.[29] In April 1936 President Franklin D. Roosevelt issued a Proclamation that for the first time classified controlled goods into six general categories: the beginning of the USML.[30] The subsequent DOS publication in

---

23 *Ibid.* at § 3.

24 S.J. Res. 51, Ch. 146, § 2(a), 50 Stat. 122 (1937) (superseded by the Neutrality Act 1939, 4 November 1939, 54 Stat. 4). See P.J. Monahan, 'The Regulation of Technical Data under the Arms Export Control Act of 1976 [AECA] and the Export Administration Act of 1979: A Matter of Executive Discretion' (1983) 6 *B.C. International & Comparative L. Rev.* 169, 172. See discussion below on technical data.

25 Neutrality Act 1939, 54 Stat. 4, 4 November 1939.

26 Monahan, *supra* note 24, at 172. See *supra* note 19, Neutrality Act of 1939, at § 1(a)–(b).

27 Monahan, *supra* note 24, at 173.

28 *Ibid.*, at 172. See discussion below on technical data.

29 See Neutrality Act of 1935 at § 2, and 1 Fed. Reg. 314 (5 May 1936).

30 1 Fed. Reg. 314, 315 (5 May 1936).

volume one of the Federal Register in May 1936 listed the categories of controlled goods under the title 'International Traffic In Arms'.[31] Although Congress eventually replaced the Neutrality Acts, the general framework of and rationale for, as well as many of the underlying issues presented by, the modern export control regime arose from the Neutrality Acts and persist to this day.[32]

In combination with the Neutrality Acts, Congress enacted An Act to Expedite the Strengthening of the National Defense[33] (Export Control Act of 1940) to establish additional peacetime controls over commercial goods considered to have military significance under the rationale of 'national defense'. The intent of the Export Control Act of 1940 was to ensure and expedite the build-up of resources and goods for the national defence as the threat of international armed conflict loomed on the horizon of foreign relations.[34] Again, the president was granted broad and unilateral authority to regulate and prohibit the export 'of military equipment or munitions or component parts thereof, or machinery, tools, or material, or supplies necessary for the manufacture, servicing, or operation thereof'.[35] However, the implementation of the Export Control Act of 1940 soon evolved into an exercise to achieve foreign policy rather than purely national defence objectives when President Roosevelt prohibited the export without a licence of petroleum, petroleum products and scrap metal from the United States.[36] Not until the enactment of the Export Control Act of 1949[37] were foreign policy objectives enumerated as congressional policy independent of national security objectives. Under the rationale of foreign policy, the origins of the export control of 'non-military goods' and the concept of 'dual-use goods' materialized.[38]

---

31 *Ibid.*

32 See for example Neutrality Act 1939, Pub, Res. 54, H.J. Res. 306, Nov. 4, 1939; Mutual Security Act 1954 (MSA) § 414, Pub. L. 83-665, §414, 68 Stat. 832 (1954); Exec. Order No. 10575, § 103(b), reprinted in 19 Fed. Reg. 7,249, 7,251 (9 November 1954).

33 Export Control Act 1940, 54 Stat. 714, Pub. L. 76-703, 2 July 1940.

34 Mineiro, *supra* note 16, at 44.

35 Export Control Act 1940 at § 6.

36 Mineiro, *supra* note 16, at 44 (noting a historical catalyst for economic warfare that contributed to the Japanese bombing of Pearl Harbor in December 1941).

37 Export Control Act 1949, Pub. L. 81-11, 63 Stat. 7, 26 February 1949, available at http://legisworks.org/sal/63/stats/STATUTE-63-Pg7.pdf.

38 Mineiro *supra* note 16, at 45.

The beginning of the Cold War with the Union of Soviet Socialist Republics (Soviet Union) led to major changes in export controls including the formal recognition of Soviet States as the new post-war security threat to the USA.[39] The need to control commercial goods of potential military significance was subsequently addressed with the enactment of the Export Control Act of 1949,[40] which became the first extensive export control regime implemented by the USA to address the existential threats posed by communism.[41] Much of the common statutory and regulatory language related to the modern export control system of goods and services arose from the Export Control Act of 1949. For example, the legislation retained considerable delegated authority granted to the president in previous legislation as well as expanded export controls to all communist States, including the authority to unilaterally deny any person 'export privileges', limited procedural rights under the Administrative Procedures Act (APA),[42] the authority to 'cut off the entire export trade of the United States', and the authority to issue heavy administrative and criminal sanctions for violations.[43] As Berman and Garson noted in 1967:

> [p]robably no single piece of legislation gives more power to the President to control American commerce. Subject to only the vaguest standards of 'foreign policy' and 'national security and welfare', he has authority to cut off the entire export trade of the United States, or any part of it, or to deny 'export privileges' to any or all citizens. Moreover, the procedures for implementing this power are left almost entirely to his discretion, and at the same time heavy administrative and criminal sanctions may be imposed for violation of any export regulations he may introduce. Under the Export Control Act, the Executive regulates all exports from the United States regardless of destination.[44]

---

[39] See I.F. Fergusson, 'The Export Administration Act: Evolution, Provisions, and Debate' (15 July 2009) Congressional Research Service Report for Congress, at 1–2, RL31832, available at https://www.fas.org/sgp/crs/secrecy/RL31832.pdf.

[40] Export Control Act 1949, P.L. 81-11, 63 Stat. 7, § 2, 26 February 1949, available at http://legisworks.org/sal/63/publaws/publaw-11.pdf.

[41] Berman and Garson, *supra* note 20, at 791–792 (analysing the legal history of the Export Control Act 1949).

[42] Administrative Procedure Act 1946 § 4, Pub. L. 79-404, 60 Stat. 237, 11 June 1946 (codified and amended at 5 USC § 500 *et seq.*). See *supra* note 37, at § 7.

[43] Berman and Garson, *supra* note 20, at 792. See generally, Fergusson, *supra* note 39, at 1–2.

[44] Berman and Garson, *supra* note 39, at 791, 792.

This observation still resonates today in export control literature and domestic corporate criticisms regarding the ITAR.

Furthermore, just as in subsequent export control laws and regulations, the Export Control Act of 1949 enumerated three rationales for imposing export controls: supply controls, foreign policy controls, and national security controls.[45] Despite the Act being a peacetime undertaking, Congress clearly intended that the rationales underpinning the Act should focus on opposing and marginalizing the military capabilities of communist regimes.[46] Moreover, the USA sought to achieve greater policy coordination with Western bloc States through the Coordinating Committee for Multilateral Export Controls (CoCom).[47] CoCom was a voluntary multilateral organization that pursued the goal of coordinating national controls over the export of technologies and strategic resources to communist States.[48] The passage of the Export Control Act of 1949 functioned coextensively with CoCom measures, the practical effect being a total embargo of communist States and related entities.[49]

By the early 1950s the US commercial industry became more vocal about the impact the various new export control regulations were having on international business.[50] Concurrently, administrative problems that stemmed from the fast-growing export control system became a greater issue for the US government to address.[51] As a result, Congress repealed the Neutrality Acts and their amendments and implemented reforms through a variety of legislation. One solution Congress devised was the enactment of the Mutual Security Act of 1954 (MSA).[52] Section 414 of the MSA captured, for the first time, the control of technical data as a part of the control of munitions. As Section 414 states:

---

[45] See *supra* note 37, at §§ 2–3.

[46] See *supra* note 37, at §§ 1–2.

[47] See 30 Fed. Reg. 9,034 (17 July 1965). See also, Mineiro, *supra* note 16, at 54–56, and 'Technology and East-West Trade' (November 1979) Chapter VIII, Multilateral Export Control Policy: The Coordinating Committee (CoCom) 153–170, available at https://www.princeton.edu/~ota/disk3/1979/7918/7918.PDF.

[48] 'Technology and East-West Trade'.

[49] See Fergusson, *supra* note 39, at 2–3.

[50] Monahan, *supra* note 24, at 173.

[51] *Ibid.*

[52] Mutual Security Act 1954 (MSA), Pub. L. 665, 68 Stat. 832, 26 August 1954.

> [t]he President is authorized to control, in furtherance of world peace and the security and foreign policy of the United States, the export and import of arms, ammunition, and implements of war, including technical data relating thereto, other than by a United States Government agency. The President is authorized to designate those articles which shall be considered as arms, ammunition, and implements of war, including technical data relating thereto, for the purposes of this section.[53]

Of course, these changes occurred prior to the development of spaceflight capabilities, but the underlying impact persists to this day via the emphasis of the ITAR and the Export Administration Regulations (EAR)[54] on and control over technical data.[55]

In concert with the Export Control Act of 1949 and its amendments, the legal effect of the enactment of the MSA was essentially the establishment of controls over all goods and services for export as well as all technical information related to such goods and services. However, the MSA never defined the meaning or scope of the term 'technical data'. The Department of State, and subsequently the Department of Justice in prosecutions over violations of technical data controls, were given wide latitude by the courts to define the meaning and scope of the term 'technical data'.[56] Eventually, despite the ITAR's exemption from the APA, the DOS promulgated a definition of the meaning and scope of the term 'technical data' in the 1965 reissue of the ITAR.[57] The legacy and effect of the reauthorization of the Export Control Act of 1949 and the Mutual Security Act of 1954 and its progeny continue to this day.

As policy rationales for and implementation of the modern export control regime evolved from the post-war period through to the end of the Cold War, Congress eventually replaced many of the old export control laws with two new pieces of legislation. Specifically, to renovate the US export control system, Congress enacted the Arms Export Control

---

53 *Ibid.* at § 414.

54 Export Administration Regulations, 15 CFR §§ 730–774 (2014).

55 See below note 136, Cox Report.

56 Monahan, *supra* note 24, at 172–173. For relatively modern case law on US court interpretations of export-controlled technical data, see *Bernstein v. US* [9th Cir. 1999] 176 F.3d 1132, *US v. Elder Industries, Inc.* [9th Cir. 1978] 579 F.2d 516, and *US v. Van Hee* [6th Cir. 1976] 531 F.2d 532.

57 See *infra* note 108.

Act of 1976[58] (AECA) and the Export Administration Act of 1979[59] (EAA) (see Table 8.1), which:

> [function] by capturing all goods, services, and information about goods and services [for export] under two jurisdictions. First, the [AECA] authorizes the Department of State, through the Directorate of Defense Trade Controls ('DDTC'), to administer the ITAR, which covers items on the [USML]. Second, the [EAA] authorizes the Department of Commerce, through the Bureau of Industry and Security ('BIS'), to administer the [EAR], which covers items on the Commerce Control List ('CCL').[60]

As authorized by these statutes, 'the EAR controls goods and technologies generally, with special consideration given to domestic economic impact of export controls, [while] ITAR controls defense articles and services.' Under the EAR system, the Bureau of Industry and Security (BIS) gives presumptive approval to export licence applicants. Conversely, under the ITAR, the Directorate of Defense Trade Controls (DDTC) gives 'presumptive denial' to export licence applicants,[61] which created a regime wherein commercial companies are constructively guilty until proven innocent of export control violations. Moreover, each statute retained much of the authority granted to the president in previous export control laws, while subsequent legislation gave the president greater flexibility to designate the control of goods and services under two different licensing regimes administered by two different Federal departments and governed, to some degree, by an inter-agency process.[62]

---

[58] Arms Export Control Act 1976 [AECA], P.L. 94-329, 90 Stat. 729, 30 June 1976 (codified at 22 USC § 2751 *et seq.*) (as amended by the National Defense Authorization Act for Fiscal Year 2013 § 1265, Pub. L. 112-239, 126 Stat. 1632, 2 January 2013). The AECA repealed and replaced the Mutual Security Act 1954 (MSA).

[59] Export Administration Act 1979, Pub. L. 96-72, 93 Stat. 503, 29 September 1979 (codified at 50 USC § 2402(c) 2012). The policy of the EAA is to encourage international trade and minimize the uncertainties in export control policy with full consideration of its impact on the US economy. See also *ibid.* at § 2402(a)(2)–(3).

[60] Hearsey, *supra* note 12, at 325, internal citations omitted.

[61] *Ibid.* at 325 (citing Mineiro, *supra* note 16, at 47).

[62] See discussion below. See generally, I.F. Fergusson and P.K. Kerr, 'Congressional Research Service, The U.S. Export Control System and the President's Reform Initiative' (13 January 2014) R41916.

Table 8.1 The modern US export control system[63]

Enacting legislation	Arms Export Act of 1976 (AECA)	Export Administration Act of 1979 (EAA)
Statutory delegation of administration	Department of State (DOS)	Department of Commerce (DOC)
Agency of jurisdiction	Directorate of Defense Trade Controls (DDTC)	Bureau of Industry and Security (BIS)
Implementing regulation	International Traffic in Arms Regulations (ITAR)	Export Administration Regulations (EAR)
In rem articles regulated (control list)	United States Munitions List (USML)	Commerce Control List (CCL)
Scope of congressional intent	National security, foreign policy, destination, end-user controls	Economic growth, supporting international trade objectives
Regulatory posture	Presumptive denial	Presumptive approval

It is worth noting that the EAA expired in 2001 and Congress has yet to reauthorize the statute. Nonetheless, the EAA and EAR continue in force as authorized by presidential invocation of the National Emergencies Act[64] through an Executive Order.[65]

[63] I am grateful to Christopher Johnson from the Secure World Foundation for assistance with this table.

[64] See 80 Fed. Reg. 48,233 (11 August 2015), Presidential Notice – Continuation of the National Emergency With Respect to Export Control Regulations (4 August 2016) which extended the EAA until 17 August 2017 under Section 202(d) of the National Emergencies Act (codified at 50 USC § 1622(d)), available at https://obamawhitehouse.archives.gov/the-press-office/2016/08/04/notice-continuation-national-emergency-respect-export-control. A new presidential memorandum extends EAR through 17 August 2018: https://www.whitehouse.gov/the-press-office/2017/08/15/notice-regarding-continuation-national-emergency-respect-export-control.

[65] Beginning in 2001, the periodic executive reauthorization of the EAA had been originally continued pursuant to the International Emergency Economic Powers Act, Pub. L. 95-223, 91 Stat. 1626, 28 October 1977 (codified at 50 USC §§ 1701–1707), beginning with Exec. Order 13,222 (17 August 2001).

B. AECA and ITAR: The Basis of Space-Related Export Controls

Although much literature has been written on the EAA and AECA and their respective regulations, the rest of this section will focus on the ITAR, specifically the ITAR provisions implicated in Bigelow Aerospace's CJ request.

ITAR is a well-known 'four-letter word' within the commercial aerospace industry because of the costs it imposes on commercial companies with and/or seeking international clientele. Many export control practitioners will generally frame the ITAR as pernicious or imposing onerous regulatory burdens on commercial companies because of the seemingly arbitrary technical interpretations made by the DDTC with respect to specific classifications of new technologies, the integration or use of known, older technologies for new purposes, or the degree to which information relating to controlled items may be communicated, to whom, and how. Nevertheless, the DDTC is generally concerned about the use and performance of controlled items as well as who the end-user(s) may be and for what purpose(s) end-users may utilize the exported item(s). On the other hand, while practitioners may argue amongst themselves or with the US government about the very nature of the ITAR, problems may arise not because of the control over export of technologies, but because of how the DDTC implements the rules of the ITAR. Thus, while many DDTC interpretations may be viewed as generally reasonable, determining the appropriate classifications of defence articles requires constant diligence on the part of export control practitioners and that diligence could be the difference in implementing an international business strategy or not. A review of the relevant law provides a sense of the scope of issues involved and the frustrations faced by US commercial entities conducting or seeking to conduct international business.

Generally, the purpose of the ITAR is 'to control the export of commercial goods, services, and technologies categorized as munitions'.[66] Section 38 of the AECA states that:

> In furtherance of world peace and the security and foreign policy of the United States, the President is authorized to control the import and the export of defense articles and defense services and to provide foreign policy guidance to persons of the United States involved in the export and import of such articles and services. The President is authorized to designate those items which shall be considered as defense articles and defense services for the purposes of this section and to promulgate regulations for the import and

[66] Mineiro, *supra* note 16, at 54. See AECA §§ 1–2.

> export of such articles and services. The items so designated shall constitute the United States Munitions List.[67]

Moreover, Section 47(7) of the AECA also states that '"defense articles and defense services" means, with respect to commercial exports subject to the provisions of section 38 of [the AECA], those items designated by the President pursuant to [22 U.S.C. § 2778(a)(1)]'.[68] Thus, all items on the USML are designated pursuant to sections 38 and 47(7) of the AECA.

The complexity and ambiguity that arises under the AECA relates to the status and scope of defence articles, defence services, and technical data under the ITAR. The AECA defines a defence article as including:

> (3)(A) Any weapon, weapons system, munition, aircraft, vessel, boat, or other implement of war,
>
> (B) Any property, installation, commodity, material, equipment, supply, or goods used for the purposes of making military sales,
>
> (C) Any machinery, facility, tool, material, supply, or other item necessary for the manufacture, production, processing, repair, servicing, storage, construction, transportation, operation, or use of any article listed in this paragraph, and
>
> (D) Any component or part of any article listed [on the USML].[69]

Next, under the ITAR, a defence article also includes associated 'technical data' defined as:

> (a)(1) Information … which is required for the design, development, production, manufacture, assembly, operation, repair, testing, maintenance or modification of defense articles. This includes information in the form of blueprints, drawings, photographs, plans, instructions or documentation.
>
> (2) Classified information relating to defense articles and defense services on the [USML] …
>
> (3) Information covered by an invention secrecy order; or
>
> (4) Software … directly related to defense articles.
>
> (b) The definition in paragraph (a) of this section does not include information concerning general scientific, mathematical, or engineering principles

---

[67] AECA at § 38 (codified at 22 USC § 2778(a)(1)). Compare AECA § 38 with MSA § 414.

[68] AECA § 47(7) (codified at 22 USC § 2794(7)). See also 22 CFR § 121.1(a)–(e) (2016).

[69] 22 USC § 2794(3) (2012).

> commonly taught in schools, colleges, and universities or information in the public domain as defined in §120.11 of this subchapter. It also does not include basic marketing information on function or purpose or general system descriptions of defense articles.[70]

Additionally, the AECA covers defence services, which 'includes any service, test, inspection, repair, training, publication, technical or other assistance, or defense information … used for the purpose of making military sales, but does not include design and construction services'.[71] The ITAR also expands the definition of defence services to include:

> (a)(1) furnishing of assistance (including training) to foreign persons, whether in the United States or abroad in the design, development, engineering, manufacture, production, assembly, testing, repair, maintenance, modification, operation, demilitarization, destruction, processing or use of defense articles;
>
> (2) The furnishing to foreign persons of any technical data controlled under this subchapter … whether in the United States or abroad.[72]

The AECA statutorily defines defence information as including 'any document, writing, sketch, photograph, plan, model, specification, design, prototype, or other recorded or oral information relating to any defense article or service'.[73] Hence, almost every conceivable aspect of a commercial good designated as a dual use item or a munition potentially falls under the licensing regime of the ITAR.

The general logic underlying the AECA and ITAR is that every defence article that could be exported or imported is categorically controlled except those which are specifically exempted. How then does the DDTC classify technologies designated as defence articles? The DDTC classifies defence articles, services, and information by technical category and publishes these classifications on the USML. As of March 2016, the USML is divided into 21 categories. The USML categorizes defence articles by specific technical scope, which includes various levels of technical detail and generality of items captured by each category.[74] Because technology is constantly evolving, it may not always be clear

---

70 22 CFR § 120.10 (2016).

71 22 USC § 2794(4) (2012).

72 22 CFR § 120.9 (2016).

73 22 USC § 2403(e) (2012). See Foreign Assistance Act 1961, Pub. L. 87-195, § 644(e), 75 Stat. 424, 30 August 1961.

74 See 22 CFR § 121.1 (2016).

what types of technologies fit within a particular category or whether a technology would be controlled under the USML or the CCL.[75]

To illustrate the types of export control issues commercial companies face under the ITAR, consider how a company would need to manage technologies under USML Category XV, which is titled 'Spacecraft Systems and Associated Equipment'.[76] Category XV has two relevant subsections that specifically pertain to the issues addressed in this chapter. First, subsection (a) lists '[s]pacecraft, including communication satellites, remote sensing satellites, scientific satellites, research satellites, navigation satellites, experimental and multi-mission satellites' as technologies controlled under the USML.[77] Second, subsection (f) lists the technical data and defence services associated with technologies controlled under Category XV(a):

> as well as detailed design, development, manufacturing or production data for all spacecraft and specifically designed or modified components for all spacecraft systems. This … includes all technical data, without exception, for all launch support activities (e.g., technical data provided to the launch provider on form, fit, function, mass, electrical, mechanical, dynamic, environmental, telemetry, safety, facility, launch pad access, and launch parameters, as well as interfaces for mating and parameters for launch).[78]

---

[75] See below commodity jurisdiction request. See also M. Goldstein, 'The Impact of Export Control Reform on Emerging Technology' (*Defense Trade Law*, 30 January 2014), available at http://defensetradelaw.com/2014/01/30/1146/.

[76] 22 CFR § 121.1 (2016), Category XV. See for example 22 CFR § 121.1 (2016), Category IV.

[77] 22 CFR § 121.1, Category XV (a) (2016) (Note to paragraph (a): 'Commercial communications satellites, scientific satellites, research satellites and experimental satellites are designated as SME [Significant Military Equipment] only when the equipment is intended for use by the armed forces of any foreign country').

[78] 22 CFR § 121.1, Category XV(f) (2016) ('Technical data directly related to the manufacture or production of any article described elsewhere in [Category XV] that is designated as Significant Military Equipment (SME) shall itself be designated SME. Further, technical data directly related to the manufacture or production of all spacecraft, notwithstanding the nature of the intended end-use [e.g., even where the hardware is not SME], is designated SME'.)

Note to paragraph (f): 'The special export controls contained in §124.15 of this subchapter are always required before a U.S. person may participate in a launch failure investigation or analysis and before the export of any article or defense service in this category for launch in, or by nationals of, a country that is not a member of the North Atlantic Treaty Organization or a major non-NATO ally of the United States. Such special export controls also may be imposed with

As repeatedly noted, each type of defence article and service, as well as all technical data, is subject to export control. The ITAR defines an export as:

> (1) Sending or taking a defense article out of the United States in any manner, except by mere travel outside of the United States by a person whose personal knowledge includes technical data; or
> (2) Transferring registration, control or ownership to a foreign person of any aircraft, vessel, or satellite covered by the U.S. Munitions List, whether in the United States or abroad; or
> (3) Disclosing (including oral or visual disclosure) or transferring in the United States any defense article to an embassy, any agency or subdivision of a foreign government (e.g., diplomatic missions); or
> (4) Disclosing (including oral or visual disclosure) or transferring technical data to a foreign person, whether in the United States or abroad; or
> (5) Performing a defense service on behalf of, or for the benefit of, a foreign person, whether in the United States or abroad.
> (6) A launch vehicle or payload shall not, by reason of the launching of such vehicle, be considered an export for purposes of this subchapter. However, for certain limited purposes (see §126.1 of this subchapter), the controls of this subchapter may apply to any sale, transfer or proposal to sell or transfer defense articles or defense services.[79]

In addition, the transmission of protected items or data can also take the form of a 'deemed export' or 're-export' under the ITAR.[80] A deemed export under the ITAR means 'the discourse or transference of any defense article or defense service (including technical data) to a foreign person in the United States'.[81] A re-export or retransfer means 'the transfer of defense articles or defense services to an end-use, end-user, or destination not previously authorized by license, written approval, or exemption pursuant to this subchapter'.[82]

The AECA licensing regime follows much of the foundational legal structures adopted since the enactment of the Neutrality Acts. The basic structure consists of a licensing regime that permits a person[83] who is empowered to transact with the US government[84] and registered with the

---

respect to any destination as deemed appropriate in furtherance of the security and foreign policy of the United States.'

79 22 CFR § 120.17 (2016).
80 22 CFR §§ 120.17(a)(2)–(5) & 120.19 (2016).
81 22 CFR § 120.17(a)(1)–(6) (2016).
82 22 CFR § 120.19 (2016).
83 22 USC § 2778(b) (2012).
84 22 USC § 2778(b) (2012).

DDTC[85] to enter into several types of agreements that may permit the export of technologies and their associated technical data listed on the USML. Moreover, statutory and regulatory controls on end-users, destinations, and how items are used by end-users continue to serve as necessary metrics when the DDTC makes a determination on a registered entity's application for export. However, one critical difference relative to the Neutrality Acts is that while a technology may be removed from the USML, that technology's underlying technical data may still be captured under the USML and subject to control under the ITAR's licensing requirements.[86] Thus, awareness of potential export liabilities and due diligence become integral in the management of technologies such as launch vehicles and 'in-space habitats'.

In many, if not most, cases the DDTC may require specific technological management agreements to limit any potential unauthorized export of controlled technologies. Common legal vehicles utilized under the modern US export control regime are Technical Assistance Agreements (TAAs) and Technology Transfer Control Plans (TTCPs). A TAA serves as the broad, basic legal vehicle in order to transfer technical data for USML items. TAAs are:

> [a]n agreement (e.g., contract) for the performance of a defense service(s) or the disclosure of technical data, as opposed to an agreement granting a right or license to manufacture defense articles. Assembly of defense articles is included … , provided production rights or manufacturing know how are not conveyed.[87]

Moreover, due to the complex nature of TAAs, it can often take months to process and gain approval from the DDTC, because TAAs generally include provisos that must be agreed to by all parties: for example, '24-hour monitoring of all hardware, mandatory presence of U.S.

---

[85] *Ibid.*

[86] See, for example, Cox Report infra note 136. Compare CCL 9A004 (ISS EAR control designation) and CCL 9A515 (general EAR control designation), 15 CFR §774, Supp. No. 1 (2016) with USML Category XV; Compare CCL 9E001/-002 (and 15 CFR §774, Supp. No. 1 (2016)) with USML Category XV(f).

[87] 22 CFR § 120.22 (2016).

Government officers during any "technical" conversations'[88] by Defense Technology Security Administration (DTSA) personnel.[89]

Generally, a US commercial company may be required to enter into a TTCP as a proviso of a TAA if the USML items implicated in a potential transaction between a US commercial company and a foreign entity fall under, for example, Category XV.[90] A TTCP is a document by which a US commercial company provides a detailed and specific plan to the DTSA for ensuring secure communications and handling of items and their associated technical data with foreign entities and personnel. A TTCP requires:

> any U.S. person or entity involved in the export to notify the Department of Defense in advance of all meetings and interactions with any foreign person or entity that is a party to the export and require[s] such U.S. person or entity to certify that it has complied with this notification requirement within 30 days after launch.[91]

This can be an expensive and burdensome requirement for commercial companies and it can take months for TTCPs to gain final approval from the DTSA.[92]

Occasionally, a US company may not be able to determine whether a particular technology under its ownership or control falls within the USML or the CCL. When this situation occurs, the ITAR provides a means by which to obtain a determination through a commodity jurisdiction (CJ) request. When confusion arises regarding which export control list is appropriate, companies file a CJ request to resolve any questions that they may have regarding a 'commodity's jurisdiction', that is, DOS or DOC jurisdiction, and the DOS institutes an inter-agency process to resolve the matter.[93] Details of CJ determinations are provided

---

88 M.N. Gold, 'The Wrong Stuff: America's Aerospace Export Control Crisis' (2008) 87 *Nebraska L. Rev.* 521, 524.

89 See P.J. Blount, 'Informed Consent v. ITAR: Regulatory Conflicts that Could Constrain Commercial Human Space Flight' (2010) 66 *Acta Astronautica* 1608, 1611 (discussing the impact of DTSA monitors on commercial human spacecraft, like BA habitat systems).

90 See 22 CFR § 124.15(a) (2016).

91 22 CFR § 124.15(a)(1) (2016).

92 Gold, *supra* note 88, at 524–525.

93 See 22 CFR § 120.4 (2016).

only to the company seeking a resolution and are not disseminated publically by the DOS.[94]

C. A Brief Review of the Policy and Historical Changes that Affected the Evolution of the Modern US Export Control Regime

While the menagerie of export control laws continues to grow and limit how and to where US companies may export goods and services of controlled technologies, the current export control regime has suffered and continues to suffer from implementation issues and periodic discordant policy changes. For example, the ITAR's concentric administration fluctuated several times between the late 1980s and the early 2000s.[95] The roots of these fluctuations began in the privatization of commercial telecommunication satellites in the 1960s.[96] The contentious and complex implementation of the Commercial Satellite Act of 1962[97] began the integral path toward a commercial space market.[98] Not until the launch of the Early Bird Communication Satellite (Intelsat I) in 1965 did the USA officially start to regulate commercial satellite technologies.[99]

Soon after NASA's *Apollo* program achieved its goal of placing Americans on the Moon before the end of 1969, a movement began to privatize the launch vehicle industry. The 1970s saw increasing advocacy

---

[94] Except for the data listed on the DDTC website listing final determinations of commodity jurisdiction requests. See US Department of State, Directorate of Defense Trade Controls, 'Commodity Jurisdiction Final Determinations' http://pmddtc.state.gov/commodity_jurisdiction/determinationAll.html.

[95] See Mineiro, *supra* note 16, at 54 (Mineiro frames the evolution of export controls over commercial space technologies as representing three distinct eras. The first era consists of the Cold War period from 1949 to 1991. The second era consists of the post-Cold War liberalization era from 1991 to 1999. The third era consists of the Strom Thurmond era from 1999 to the present. With the current changes to export control policy and regulations, a question arises about whether we are or will soon be in a new fourth era dominated by the need to provide flexibility to commercial space companies to engage in international business bounded by the traditional national security and foreign policy constraints of US export control laws and regulations).

[96] See R. Launius, 'The First Commercial Activity: Communication Satellites' (*Roger Launius' Blog*, 13 June 2014), available at http://launiusr.wordpress.com/2014/06/13/the-first-commercial-space-activity-communications-satellites/.

[97] Communications Satellite Act 1962, Pub. L. 87-624, 76 Stat. 419, 31 August 1962.

[98] Mineiro, *supra* note 16, at 54–55.

[99] *Ibid.*

for commercial space capabilities both in the United States and internationally.[100] By 1988 several events had occurred, the roots of which transformed the commercial space industry, including the first launches of vehicles not wholly controlled by the US government,[101] the signing of a Presidential Executive Order that established the Office of Commercial Space Transportation within the Department of Transportation,[102] congressional enactment of the Commercial Space Launch Act of 1984 and subsequent amendments,[103] and robust growth in the demand for commercial launch services and telecommunication satellites.[104] Consequently, recognizing lucrative global business opportunities for the American commercial space sector, the Reagan Administration began a process to open up the export of US commercial telecommunication satellites for international launches.[105] Thus began a period of consonant liberalization of the US export control regime for commercial telecommunication satellites and their associated technologies. However, less than a decade later, a launch failure in China would lead to a dissonance that provided the necessary catalyst for a contraction in export control liberalization.

The evolution of the regulation of commercial telecommunication satellites and their associated technologies began in 1965 when the DOS reissued the ITAR and the USML.[106] The recodification of Title 22 of the

---

[100] *Ibid.* at 54–56.

[101] See, for example, J.C. Abell, 'Sept. 9, 1982: 3–2–1– … Liftoff! The First Private Rocket Launch' (*Wired*, 9 September 2009), available at http://www.wired.com/2009/09/dayintech0909privaterocket/.

[102] See Exec. Order No. 12,465, 49 Fed. Reg. 7,211 (24 February 1984).

[103] See Commercial Space Launch Act 1984, Pub. L. 98-575, 98 Stat. 3055, 30 October 1984; Commercial Space Launch Act 1988, Pub. L. 100-657, 102 Stat. 657, 15 November 1988; and Commercial Space Launch Act 2004, Pub. L. 108-492, 118 Stat. 492, 23 December 2004.

[104] See J.L. Reed, 'The Commercial Space Launch Market and Bilateral Trade Agreements in Space Launch Services' (1999) 13 *American University International Review* 157, available at http://digitalcommons.wcl.american.edu/cgi/viewcontent.cgi?article=1349&context=auilr.

[105] See United State Space Launch Strategy, NSDD-254 (27 December 1986), available at http://marshall.wpengine.com/wp-content/uploads/2013/09/NSDD-254-U.S.-Space-Launch-Strategy-27-Dec-1986.pdf; National Space Policy, NSDD-293 (5 January 1988); Mineiro *supra* note 16, at 54–62; and Reed, *supra* note 104, at 164–169.

[106] See 30 Fed. Reg. 9,034 (17 July 1965) pursuant to Section 414 of the Mutual Security Act 1954 (MSA), Pub. L. 83-665, §414, 68 Stat. 832 (1954). Exec. Order No. 10,575, § 103(b), reprinted in 19 Fed. Reg. 7,249, 7,251 (9 November 1954).

Code of Federal Regulations introduced for the first time the term 'spacecraft' and its 'associated equipment' on the USML.[107] These technologies were placed under Category VIII ('Aircraft, Spacecraft, and Associated Equipment') of the USML and listed as '[s]pacecraft including manned and unmanned, active and passive satellites' and the '[l]aunching, arresting, and recovery equipment for' such spacecraft.[108] Subsequent changes following the enactment of the AECA have expanded and moved spacecraft and their associated technologies into a separate USML category.[109]

To protect and foster US commercial space-based telecommunications and launch services, the USA sought to achieve greater policy coordination with Western bloc States through the CoCom.[110] The reorganization of the ITAR and the USML in 1965 was a direct result of integrating export control of spacecraft technologies under the ITAR into the CoCom to ensure US technological and market supremacy.[111] Consequently, 'a U.S. government monopolized launch services industry virtually assured all western satellites would be launched in the United States, the practical result being that no U.S. satellites would be exported for launch services.'[112]

This status quo persisted until the early 1980s when a series of events started to significantly affect the US monopoly over launch services and satellite manufacturing. Initially, the Reagan Administration and the US commercial aerospace industry took notice of the European development of indigenous launch vehicles and satellite manufacturing.[113] With the US monopoly over international satellite sales and launches threatened by international competition, it became exceedingly difficult to maintain policy coordination within the CoCom for spacecraft technologies, inter alia, as member interests diverged.[114] Subsequently, after an ascent

---

107 30 Fed. Reg. 9,034 (17 July 1965).

108 *Ibid.*

109 By 1993 the DOS had reissued the USML, which did not include the classification of 'manned satellite' specifically in Category XV. Not until the reissue of the ITAR and the USML in May 2014 did an analogous term for a 'manned satellite' ('in-space habitat') return to the USML. See discussion below.

110 See 30 Fed. Reg. 9,034 (17 July 1965); Mineiro *supra* note 16, at 54–56; and 'Technology and East-West Trade', *supra* note 48.

111 See 30 Fed. Reg. 9,034 (17 July 1965), and 'Technology and East-West Trade', *supra* note 48.

112 Mineiro, *supra* note 16, at 55.

113 *Ibid.* at 54–55.

114 See R. Cupitt and S. Grillot 'COCOM is Dead, Long Live COCOM: Persistence and Change in Multilateral Security Institutions' (1997) 27 *British*

anomaly destroyed the Space Transportation System Challenger in 1986, the Reagan Administration adopted an outline of a new policy that would transition the US government from a sponsor of launches to the establishment of a domestic private commercial launch industry capable of competing for international launch services.[115] Finally, the fall of the Soviet Union in 1991 gave way to the possibility of increased trade between Western and former soviet States, as well as the need to protect soviet rocket and nuclear technologies from global proliferation.[116] Consequently, the United States entered into bilateral technology and launch services agreements with the People's Republic of China (PRC) and the Russian Federation in 1989 and 1993 respectively,[117] and by 1996 CoCom was replaced with the voluntary, multilateral Wassenaar Agreement on Export Controls for Conventional Arms and Dual-Use Goods and Technologies.[118]

The PRC was (and the Russian Federation may still be)[119] an attractive destination for US commercial space companies seeking satellite launch services. Because China offers a burgeoning market and could provide subsidized commercial launch services, it became an attractive alternative for US satellite manufacturers due to the significant cost advantages of mass-to-orbit over most US and European launch service providers.

Journal of Political Science 361; J. Johnson-Freese, 'Alice in Licenseland: U.S. Satellite Export Control Since 1990' (2000) 16 *Space Policy* 195; and M. Mastanduno, *Economic Containment: Cocom and the Politics of East-West Trade* (Cornell University Press 1992).

[115] See United States Space Launch Strategy, NSDD-254 (27 December 1986), available at http://marshall.wpengine.com/wp-content/uploads/2013/09/NSDD-254-U.S.-Space-Launch-Strategy-27-Dec-1986.pdf, and Mineiro, *supra* note 16, at 54–60.

[116] For a concise history of the US–Russian partnership over the International Space Station, see Marcia Smith, 'Testimony before the Subcommittee on Science and Space, Committee on Commerce, Science, and Transportation before the United States Senate' (20 April 2005), available at http://history.nasa.gov/isstestimonysmith.pdf.

[117] See, for example, Reed, *supra* note 104, and Mineiro, *supra* note 16, at 54–60.

[118] The Wassenaar Arrangement on Export Controls for Conventional Arms and Dual-Use Goods and Technologies (12 July 1996), see 15 CFR 743.1 (2016).

[119] But see P.B. de Selding, 'Iridium, frustrated by Russian red tape, to launch first 10 Iridium Next satellites with SpaceX in July' (*SpaceNews*, 25 February 2016), available at http://spacenews.com/iridium-frustrated-by-russian-red-tape-to-launch-first-10-iridium-next-satellites-with-spacex-in-july/?utm_content=buffer7f62d&utm_medium=social&utm_source=facebook.com&utm_campaign=buffer.

Moreover, Russia also provides significant cost advantages over other alternatives because of subsidized commercial launch services as well as past investments made in launch systems and infrastructure. Additionally, in the 1990s concern over the fate of highly skilled Russian engineers and the Russian aerospace sector as a whole presented the USA with a foreign policy rationale to bring the Russians into the International Space Station (ISS) partnership[120] and grant export approvals for launches out of current Russian and former soviet spaceports.[121]

The dramatic policy changes in the United States between 1984 and 1996 led to a cultural shift in the control of space-related technologies. The prospect of cheaper launch vehicles in the international launch market signalled a significant change from just a decade earlier. As noted by Zelnio, 'all items related to satellite manufacturing were controlled by the State Department as munitions and were subject to ITAR approval. In 1988, increased demand for launch services led President Ronald Reagan to sign a deal with China to launch US built comsats.'[122] However, in response to the Tiananmen Square incident in 1989, in February 1990 Congress enacted Section 902 of the Foreign Relations Authorization

---

[120] See Smith, *supra* note 116.

[121] However, the current Ukrainian situation has strained US-Russian relations and has affected, to some degree, the availability of US commercial access to Russian spaceports and technologies. See, for example, C.R. Wall *et al.*, 'U.S. and UK Take Modest Steps to Restrict Exports to Russia' (*Pillsbury Law*, 1 April 2014), available at http://www.pillsburylaw.com/publications/us-and-uk-take-modest-steps-to-restrict-exports-to-russia; Reuters, 'Russia to ban US from using Space Station over Ukraine sanctions' (*The Telegraph*, 13 May 2014), available at http://www.telegraph.co.uk/news/worldnews/europe/russia/10828964/Russia-to-ban-US-from-using-Space-Station-over-Ukraine-sanctions.html; and D. Leone, 'Notwithstanding Sanctions, ULA Standing by for RD-180 Deliveries through 2017' (*SpaceNews*, 6 August 2014), available at http://www.spacenews.com/article/military-space/41507notwithstanding-sanctions-ula-standing-by-for-rd-180-deliveries-through. Russia has threatened to end its relationship with the USA in relation to the ISS multiple times in the last several years. But so far the threats have not materialized into significant policy changes. See https://arstechnica.com/science/2017/07/russian-official-on-new-us-sanctions-and-nasa-nothing-lasts-for ever/ and https://sputniknews.com/politics/201707281055965009-us-sanctions-russia-response/.

[122] See R. Zelnio, 'A Short History of Export Control Policy' (*The Space Review*, 9 January 2006), available at http://www.thespacereview.com/article/528/1.

Act, Fiscal Years 1990 and 1991[123] (FRAA), which suspended the export of US satellites for launch as payloads on Chinese launch vehicles as well as imposing strict sanctions on the PRC government. Additionally, President George H.W. Bush imposed sanctions on the PRC government for humanitarian violations resulting from the crackdown on protesters through an Executive Order in concert with the FRAA.[124]

Eight months after the enactment of the FRAA, Congress passed HR 4653, the Export Facilitation Act of 1990,[125] which reauthorized the Export Administration Act of 1979. Congress intended the Act to help liberalize and facilitate the export of controlled technologies internationally.[126] Included in the Act were provisions that would have significantly affected international trade, provided for an increase in congressional oversight on export controls, and enabled Congress to micromanage the control of technologies for export.[127] However, President George H.W. Bush pocket vetoed the Act and issued a Memorandum of Disapproval.[128] In the memorandum President Bush stated:

> I am withholding my approval of HR 4653 [because,] [a]lthough this legislation contains constructive provisions, it would severely constrain Presidential authority in carrying out foreign policy … Other sections of HR 4653 contain useful provisions that will be implemented as soon as possible. However, additional legal authority is not required to make our export control system reflect the economic and national security realities of today's world.[129]

Furthermore, in temporal assignment the memorandum directed that 'the United States will remove from the [USML] all items contained on the CoCom dual-use list unless significant US national security interests would be jeopardized [as well as implement] new interagency procedures

---

[123] Foreign Relations Authorization Act, Fiscal Years 1990 and 1991 §902, Pub. L. 101-246, 112 Stat. 2177, 6 February 1990 (codified at 22 USC § 2778 note).

[124] See Exec. Order 12,711 (11 April 1990).

[125] HR 4653, Omnibus Export Amendments Act 1990 (pocket vetoed 17 November 1990), available at https://www.congress.gov/bill/101st-congress/house-bill/4653/actions.

[126] See A. Riviezzo Dineen, 'An Iron Curtain to Free Trade: An Evaluation of H.R. 4653, The Export Administration Act Amendments' (1991) 22(4) *Law and Policy in International Business* 857.

[127] *Ibid.*

[128] *Ibid.*

[129] Memorandum of Disapproval for the Omnibus Export Amendments Act of 1990 (16 November 1990), available at http://bushlibrary.tamu.edu/research/public_papers.php?id=2456&year=1990&month=11.

... to make dual-use export license decisions more predictable and timely'.[130] With this directive the US government instituted several inter-agency groups to begin the process of moving commercial communication satellites and other designated dual-use items off of the USML and onto the CCL.

As the inter-agency process moved forward slowly, the DOS and the DOC began to issue advance notices of proposed rule-making to address the directives given by President Bush to reform export control regulations. Beginning in September 1991, the DOS proposed to 'create a new category of defense articles and related technical data involving spacecraft by combining articles from several existing categories into a single, new Category XV'.[131] After a complicated two-year process of review, comment and revisions, the DOS published its final rule on amendments to the ITAR by:

> consolidating all spacecraft to be controlled under the USML, as well as all specifically designed or modified components, parts, accessories, and attachments of such satellites, and their directly related technical data and defense services, into a single category of the USML ... At the same time, this rule initiate[d] the movement off the USML to the [DOC's CCL] ... of ... all components, parts, accessories, attachments and associated equipment of spacecraft which have not been specifically designed or modified to provide the characteristics of capabilities described in Category XV which cause a spacecraft to remain under the control of the USML.[132]

Throughout the rest of the 1990s the DOS and the DOC updated, amended, and transferred various items covered under Category XV from the USML to the CCL. However, by 1995 the US government still had not made a decision to transfer commercial communication satellites off of the USML and on to the CCL. Not until after the DOS decided to narrow the list of militarily sensitive technologies on the USML did the DOC make an appeal to President Clinton.[133] In March 1996 President Clinton finally ordered the transfer of commercial communication satellites to the CCL, but with a caveat that an inter-agency group comprising the DOS and the Departments of Defense and Energy, as well as the Arms Control and Disarmament Agency, would make the final licensing

---

130 *Ibid.*

131 56 Fed. Reg. 43,894 (5 September 1991).

132 58 Fed. Reg. 47, 636 (10 September 1993). See 58 Fed. Reg. 47,322 (8 September 1993) (corresponding DOC regulations).

133 Zelnio, *supra* note 122.

decisions.[134] Six months later the DOC would finally receive full jurisdiction over commercial communication satellites.[135] Unfortunately, the liberalization of the export control regime over commercial satellites and their associated technologies had achieved its peak, but substantial and dramatic changes lurked on the horizon.

D. The Strom Thurmond National Defense Authorization Act of 1999

As the inter-agency processes to reform export controls soldiered on through the mid 1990s, two failures of PRC Long March launch vehicles would frame that era of export controls in the USA. First, in January 1995 a Long March 2E with a Hughes-built Apstar-2 satellite as its payload exploded.[136] Subsequently, in February 1996 a Long March 3B exploded along with its payload of an Intelsat satellite constructed by Loral.[137] In an effort to discover the cause of the launch failures and recover insurance claims, the DOC reviewed the failure analysis presented by Hughes and Loral and determined that the export of the analysis fell under the DOC-issued licence of February 1994.[138] The DOC cleared the export of the failure analysis to the PRC, so that Hughes and Loral could work with PRC engineers to analyse the engineering data to determine the cause of the respective launch failures and the destruction of the payloads.[139]

However, the export of the failure analysis to the PRC ignited a dramatic reaction from the US Congress. In the wake of the discovery of the export of the technical data included in the failure analysis, Congress reviewed the information provided by the commercial companies and approved by the DOC for release and, as a result, published the Cox Committee Report in 1998.[140] The report's major findings, inter alia, included the decision that the DOC did not have the authority to

---

[134] See 61 Fed. Reg. 12,714, 12,928 (25 March 1996).

[135] See 61 Fed. Reg. 54,540 (21 October 1996) and 62 Fed. Reg. 50,865 (29 September 1997).

[136] See 'Report of the Select Committee on US National Security and Military/Commercial Concerns with the People's Republic of China' (Cox Report) (US Congress, 1999). See also, China Great Wall Industry Corporation, 'A Review of China's Launch Services for US-Made Satellites' (*FAS*, July 1998), available at http://fas.org/news/china/1998/980700-prc-gwic.htm.

[137] *Ibid.*

[138] *Ibid.*

[139] *Ibid.*

[140] *Ibid.*

authorize the export of the failure analysis because the data provided in the analysis was subject to DOS licensing.[141] Moreover, Congress found that the export of the technical data in the report constituted a violation of the ITAR.[142] These findings soured relations with the aerospace industry and served as the catalyst for the changes that followed, including the termination of presidential authority to move certain items between the USML and the CCL that President Bush had sought to defeat with his pocket veto just eight years earlier.[143]

The movement to lock down the ITAR and eliminate previously held presidential discretion from export control decisions manifested itself in the passage of the Strom Thurmond National Defense Authorization Act of 1999[144] (1999 NDAA), which created greater congressional oversight and explicitly prohibited the export, inter alia, of commercial satellites to China.[145] Acting under Section 38 of the AECA, Congress included

---

[141] *Ibid.*

[142] *Ibid.*

[143] As a practical and unforeseen consequence, the Department of Justice sought to prosecute ITAR violations arising from the unauthorized disclosure of technical data to the PRC. See, for example, Office of the Spokesman – Washington, DC, 'U.S. Department of State Reaches Settlement with Boeing and Hughes' (5 March 2003), available at http://2001-2009.state.gov/r/pa/prs/ps/2003/18275.htm. See also http://www.pmddtc.state.gov/compliance/consent_agreements/pdf/Boeing_ConsentAgreement_06.pdf.

[144] Strom Thurmond National Defense Authorization Act for Fiscal Year 1999, Pub. L. 105-261, 112 Stat. 1995 (17 October 1998).

[145] *Ibid.* at §§ 1511–1516. See 'Statement on Signing the Strom Thurmond National Defense Authorization Act for Fiscal Year 1999' (17 October 1998), available at http://www.presidency.ucsb.edu/ws/?pid=55105 ('I am strongly opposed to a provision that, effective March 1999, will transfer the jurisdiction over satellite exports from the Department of Commerce to the Department of State. This change is not necessary to ensure effective control of U.S. exports of satellites and could hamper the U.S. satellite industry. The Congress repeatedly supported the transfer of satellite licensing jurisdiction to the Department of Commerce long before I ordered the transfer in 1996. I strongly urge the Congress to demonstrate its support for a strong domestic satellite industry by passing remedial legislation to halt this transfer of jurisdiction prior to its effective date. In the meantime, I will take action to minimize the potential damage to U.S. interests that could arise from the Act's export control related requirements. I will direct the appropriate agencies to implement these provisions, subject to appropriate law and regulation, in a manner that supports legitimate commercial communications satellite exports while ensuring that the extensive safeguards needed to protect our national security remain in effect. I will also direct all concerned agencies, subject to appropriate law, regulation, and U.S. national security interests, to employ, to the extent appropriate, time-lines

Section 1513 in the 1999 NDAA that transferred back control of all satellites and related technologies to DOS jurisdiction and placed the technologies under the USML.[146] Congress also enacted a provision in the 1999 NDAA that required the president to notify Congress of any export of satellite technology to the PRC within 15 days and to certify that the export of such technology did not harm the US launch industry and/or assist the PRC in missile technology development.[147]

These dramatic changes sent shockwaves through the US aerospace industry. Congress had shut down over sixty years of presidential authority and closed off China as an option for the launch of US satellites.

As these events played out in the early 2000s, Bigelow Aerospace began to consider whether its technologies would fall under DOS or DOC jurisdiction, since habitats were neither communication satellites nor launch vehicles.[148] Commercial expandable space habitats were a new type of technology, one which had not been previously contemplated[149] either for commercial development or international sale under US export control laws and regulations. The next section briefly describes Bigelow Aerospace's export control reform efforts.

---

and transparent licensing practices for satellites and related items described in section 1513(a) of the Act in a manner consistent with current dual-use export license processing. I note that H.R. 3616 also requires that I make certain certifications to the Congress in advance of any export of missile technology or equipment to the People's Republic of China (PRC). Specifically, I must certify that such exports will not be detrimental to the U.S. space launch industry and will not measurable [*sic*] improve the PRC's missile or space launch capabilities. In making this certification, I will be guided by the conference report that notes that "this certification is not, and is not intended to be, a prohibition on the export of U.S. satellites to be launched by the PRC, but is intended to ensure that U.S. national security would not be jeopardized by any such export". I agree with this objective. Further, I take note of the bill's legislative history with respect to the export of U.S.-made items in connection with emergency repair or replacement for commercial aircraft, and I will exercise the certification authority consistent with that view'). President Clinton formally transferred satellites back to the USML in March 1999. See 64 Fed. Reg. 13,679 (22 March 1999).

146 Strom Thurmond National Defense Authorization Act for Fiscal Year 1999 at § 1513, Pub. L. 105-261, 112 Stat. 1995 (17 October 1998).

147 *Ibid.* at § 1515.

148 Gold, *supra* note 11.

149 But see 30 Fed. Reg. 9,034 (17 July 1965) ('spacecraft including manned and unmanned, active and passive satellites', Category VIII(b). This reference disappears in the subsequent reissue of the ITAR).

3. BIGELOW'S SPACE-RELATED EXPORT CONTROL REFORMS

A. Bigelow Aerospace's Commodity Jurisdiction Request

After suffering from the financial and substantive burdens of the ITAR during the *Genesis* programme, Bigelow Aerospace became acutely aware that any business case involving international astronautics could not be successfully implemented under the then existing regime. The prospect of flying DTSA monitors alongside passengers and stationing them aboard a Bigelow habitat in orbit seemed both costly and slightly ludicrous, but such requirements remained a possibility, if not a certainty, under the existing regulations of the time. Moreover, the paperwork involved to train, transport, and support a foreign Bigelow customer also seemed to be problematic. Drafting, reviewing, filing, and gaining approval for all of the necessary TAAs and TTCPs to support international customers would have been nearly impossible to accomplish in a timely and consistent manner.

Therefore, in the autumn of 2007, with the *Genesis* launches complete, BA turned its attention to reforming the system to allow its own business and the commercial space field in general to survive and thrive.[150] Specifically, on 21 December 2007 Bigelow Aerospace filed a CJ request for its expandable space habitats with the DDTC.[151] Once word spread of this blasphemous action, many commentators ridiculed BA's decision to seek a determination for its expandable habitats.[152]

The Bigelow CJ request filed in 2007 took an aggressive position by requesting that the DOS confirm that all Bigelow habitat technologies belong on the CCL. Although BA had concluded that, according to the letter of the law, its habitats fell under the auspices of the USML, and therefore the ITAR, the hope was to use the CJ request as a means to at least open up a broader conversation regarding the manner in which the rules were being implemented and the pernicious impact of the ITAR on Bigelow Aerospace's ability to execute its plans. After the request was reviewed by the DOS, a dialogue began between Bigelow Aerospace and

---

150 Sundahl, *supra* note 10.

151 Clark, *supra* note 10; and Blount, *supra* note 10.

152 See, for example, C. Burns, 'Best Acronym Ever: "POOF"' (*Export Law*, 12 March 2008), available at http://www.exportlawblog.com/archives/305, derogatorily defining BA's habitat technology as POOFs, citing T. Dinerman, 'Inflatable POOFs' (*The Space Review*, 19 July 2004), available at http://www.thespacereview.com/article/187/1; and see Gold, *supra* note 11.

Ann Ganzer, the Director of the Office of Defense Trade Controls Policy (ODTCP) within the DDTC. While the ODTCP did not believe that it was viable to transfer all of Bigelow's habitat technology onto the CCL, Ms Ganzer still hoped to find a way to assist the company in overcoming the barriers to implementing its business case.

In 2008 these discussions resulted in a letter from Bigelow Aerospace to the ODTCP articulating the relief that it needed to support the passenger experience.[153] Within the letter BA described the activities that potential foreign customers would be involved in, such as terrestrial training, operating communications equipment, working with their payloads, and emergency protocols. Of equal if not greater importance, the letter also stated which activities foreign customers would not be involved in, specifically space station operations and maintenance. This letter established the path for providing Bigelow Aerospace with relief for its customer base and was the model for future CJ requests and amendments to export control regulations addressing this issue.[154]

Despite the common ground that was established between Bigelow Aerospace and the ODTCP, producing a response to the Bigelow CJ request still proved challenging. Resistance to any change to the existing export control regime remained strong on Capitol Hill. However, Bigelow Aerospace's CJ request presented a compelling and easily understood narrative. The analogy that was often used by both the DDTC and BA was that just because a foreign national is a passenger on a military cargo plane it does not mean that a foreign national can build one.[155] This very analogy was introduced and used by the Managing Director of the DDTC, Robert Kovac. In 2008 Mr Kovac was the highest career level official at the DDTC. A well-respected and spectacularly effective leader, Mr Kovac was a critical and early advocate for export control reform. Moreover, Mr Kovac was a staunch supporter of small businesses, and without his support, Bigelow Aerospace's CJ request may have failed and all of the reforms that it would later engender would have been delayed by years, or potentially never have occurred at all.

Ultimately, in 2009 under Mr Kovac's stewardship, the DDTC responded to Bigelow Aerospace's CJ request, and the results were better

---

153 Clark, *supra* note 10; and Blount, *supra* note 10.

154 See, for example, Commodity Jurisdiction Final Determinations, U.S. Department of State, Directorate of Defense Trade Controls, 2012 list, available at https://www.pmddtc.state.gov/commodity_jurisdiction/determination2012.html.

155 See M.J. Sundahl, 'Space Tourism and Export Controls: A Prayer for Relief' (2010) 75 *Journal of Air Law and Commerce* 581, 595–596.

than could have been expected. All of the activities representing the passenger experience articulated in BA's correspondence with Ann Ganzer were deemed 'non-licensable' by the DOS, including, inter alia, those related to ingress and egress procedures, emergency and medical training procedures, zero gravity training, use of the galley, onboard exercise equipment, civil communication devices, and the use of personal hygiene and sanitary facilities. This immediately freed Bigelow Aerospace from the costly and onerous burden of filing TAAs, TTCPs, and similar paperwork that would have otherwise been involved in supporting foreign astronauts and spaceflight participants, and it also eliminated the potential for crippling DTSA monitoring requirements. At the same time, the CJ response protected national security and foreign policy interests by excluding foreign nationals from operating or maintaining Bigelow habitats, and the DDTC's response also included a caveat under which its relief would not apply to foreign nationals under the ITAR's Section 126.1[156] list of prohibited countries (including Iran, North Korea, Sudan, and China).

Bigelow Aerospace publicly announced the success of its CJ request in the spring of 2009, and the results were hailed as a significant step forward by Virgin Galactic, SpaceX, and other commercial space companies.[157] Once the precedent had been set for the passenger experience, commercial human spaceflight companies began to understand the utility of this new ruling. Many export control practitioners and trade associations, some of whom had previously opposed BA's efforts, eventually embraced the need for export control reform in the following years.[158] Additionally, Virgin Galactic, leveraging the precedent set by Bigelow Aerospace, filed its own CJ request.[159] The response that Virgin Galactic received in January 2012 represented the continuing progress of the export control movement, since its CJ response did not include ITAR's Section 126.1 limitation.

---

[156] 22 CFR § 126.1 (2016) (the list currently includes: Côte d'Ivoire, Democratic Republic of Congo, Eritrea, Iraq, Iran, Lebanon, Liberia, Libya, North Korea, Somalia, and the Republic of the Sudan. See 22 CFR § 126.1[c][1]–[11]).

[157] See 'Space Technology: Freedom to Fly' (*The Economist*, 22 April 2009), available at http://www.economist.com/node/13525115.

[158] *Ibid.*

[159] See Commodity Jurisdiction Final Determinations, *supra* note 154.

B. Working Towards a New Paradigm Within the COMSTAC

In 2008 the Commercial Space Transportation Advisory Committee (COMSTAC)[160] established the Export Controls Working Group (ECWG) and appointed Mike Gold as the new Working Group's first Chair.[161] Subsequently, at the autumn 2009 meeting of the COMSTAC, the ECWG put forward recommendations related to HR 2410, Foreign Relations Authorization Act, Fiscal Years 2010 and 2011:[162]

1. The COMSTAC supports export control reform in general and the goals and objectives of HR 2410 in regard to, among others, reviewing and revising the USML. We urge the Senate to move as quickly as possible in taking up and passing comparable legislation.
2. Obtaining a broad array of industry input is vital for the U.S. Department of State and other federal entities to effectively implement positive changes to the nation's export control regime. Therefore, future export control reform legislation in the Senate and House should encourage the Department of State to share significant planned rules and policy shifts with the Defense Trade Advisory Group ('DTAG'), the COMSTAC, and any other relevant Federal Advisory Committees that express a desire to become involved.
3. We support HR 2410's call for the public disclosure of properly redacted Commodity Jurisdiction Request determinations via the Internet, and would encourage the Department of State to take such action immediately.[163]

---

[160] President Ronald Reagan initially created the COMSTAC in 1984 to help facilitate commercial input for the commercialization of the launch services sector in the United States. See 49 Fed. Reg. 14,621 (12 April 1984). COMSTAC operates under the Federal Advisory Committee Act 1972 (FACA) by 'provid[ing] information, advice, and recommendations to the Department of Transportation and the Administrator of the Federal Aviation Administration (FAA) on the critical matters facing the US commercial space transportation industry'. See Federal Advisory Committee Act 1972, Pub. L. 92-463, 86 Stat. 770 (codified at 5 USC App. 2). Per the FACA, the Federal Aviation Administration must authorize the renewal of the COMSTAC every two years. See, for example, 79 Fed. Reg. 72,057 (2 December 2013).

[161] Commercial Space Transportation Advisory Committee (COMSTAC) Meeting Minutes, 30 October 2008, at 12, available at http://www.faa.gov/about/office_org/headquarters_offices/ast/advisory_committee/meeting_news/media/2008/Oct%2030%202008.pdf.

[162] Foreign Relations Authorization Act, Fiscal Years 2010 and 2011 §§ 801–832, HR 2410, (111th Congress 2009). See David Damast, 'Export Control Reform and the Space Industry' (2010) 42 *Georgetown Journal of International Law* 211.

[163] COMSTAC Meeting Minutes, 29 October 2009, at 5–7, available at http://www.faa.gov/about/office_org/headquarters_offices/ast/advisory_committee/meeting_news/media/comstac_minutes_102909.pdf.

The COMSTAC adopted these recommendations unanimously and set the precedent of coordinating export control reform positions of the commercial space industry within the Committee. Moreover, starting in 2010 the DOS partially adopted the third recommendation. The DOS now publicly posts the names of companies that have received a final CJ determination, the date of the determination, the result, and a brief description of the determination.[164]

In 2013 the COMSTAC transitioned the equities of the ECWG to the International Space Policy Working Group (ISPWG). The ISPWG continues to provide observations, findings, and recommendations on, inter alia, export control issues important to the US commercial space industry. The ISPWG's observations, findings, and recommendations to the Federal Aviation Administration's Office of Commercial Space Transportation (FAA AST) provide FAA AST leadership with guidance relative to how export controls are impacting the commercial space transportation industry. Moreover, the COMSTAC continues to serve as a forum for collegial discussions of and consensus-building around the various export control topics affecting the commercial space industry on a regular basis. The COMSTAC's involvement in export control reform efforts have been and continue to be an important tool for the commercial space industry in finding solutions to streamlining the export control processes of the DDTC and BIS to the benefit of all parties.

C. The National Defense Authorization Acts of 2010 and 2013

In August 2009 President Barack Obama ordered a comprehensive review of the US export control system.[165] Following this review, in April 2010

---

[164] See, for example, Commodity Jurisdiction Final Determinations, *supra* note 154. Additionally, at this meeting the ECWG issued a second set of recommendations for COMSTAC adoption. Unfortunately, due to confusion over COMSTAC bylaws, several of the recommendations were blocked from being adopted by the ECWG for a full COMSTAC vote. Compare Bylaws for the COMSTAC, at 3, available at http://www.faa.gov/about/office_org/headquarters_offices/ast/media/COMSTAC_Bylaws.pdf with COMSTAC Meeting Minutes, *supra* note 163, at 6–7. If COMSTAC had adopted these recommendations, then it would have gone on the record early in support of, inter alia, an inter-agency effort to review and revise the USML and CCL with private sector input, promoting the Office of Science and Technology Policy to coordinate with other federal agencies to review technologies on the export control lists for removal and placement, and publishing redacted CJ requests.

[165] See Fergusson and Kerr, *supra* note 62 (synopsis of President Obama's export control reform initiative).

the Department of Defense proposed an outline of a new system for export controls, which included a single export licensing agency, a unified control list, a single enforcement coordination agency, and an integrated information technology (IT) system for users.[166] While the Obama Administration took steps to implement an integrated IT system for users and to establish a single enforcement agency, no specific legislative or regulatory proposals have been put forward to consolidate the export licensing agencies or technology control lists.[167]

At the end of October 2009 President Obama signed the National Defense Authorization Act for Fiscal Year 2010[168] (2010 NDAA), and the legislation explicitly established the next steps for the export control reform process.[169] Specifically, Section 1248 of the 2010 NDAA 'directed the Secretaries of Defense and State to conduct a review of US space export control policy, including a risk assessment of removing satellite and related components from the USML'.[170] In May 2011 the Secretaries of Defense and State transmitted the interim Section 1248 report to Congress.[171] The interim report concluded that commercial communication satellites (CCS), their 'related components, and integration and launch information "with certain exceptions, conditions and limitations" could be removed from the USML and transferred to the CCL "without posing an unacceptable security risk".'[172] After significant review, the president delivered the final Section 1248 report to Congress in April 2012 with a 'recommendation that Congress should return export control jurisdiction for CCS to Presidential discretion, as well as authorize the Department of Defense to determine the need for special export control monitoring and oversight services for CCS'.[173]

Ultimately, subsequent to much deliberation and congressional discussions with the Departments of Defense and State, on 2 January 2013 President Obama signed the National Defense Authorization Act for

---

166 *Ibid.*

167 *Ibid.*

168 National Defense Authorization Act for Fiscal Year 2010, Pub. L. 111-84, 123 Stat. 2190 (28 October 2009).

169 See Damast, *supra* note 162.

170 Fergusson and Kerr, *supra* note 62; see National Defense Authorization Act for Fiscal Year 2010, *supra* note 168.

171 See Departments of Defense and State, Report to Congress Section 1248 of the National Defense Authorization Act for Fiscal Year 2010 (*Public Law* 111–184, 15 March 2012), available at http://archive.defense.gov/home/features/2011/0111_nsss/docs/1248_Report_Space_Export_Control.pdf.

172 Fergusson and Kerr, *supra* note 62.

173 *Ibid.*

Fiscal Year 2013[174] (2013 NDAA) that included Section 1261, which amends Section 1513(a) of the 1999 NDAA and authorizes the transfer of:

> satellites and related items on the CCL with certain country restrictions and reporting requirements. It would continue the prohibition on satellite sales to or launches by China, or to countries designated as State sponsors of terrorism … and North Korea. In addition, a license application to export satellites and related items to a country in which the [US] maintains a comprehensive arms embargo will face a presumption of denial, although not an outright prohibition.[175]

Thus, Congress returned to the president many of the discretionary powers eliminated by the 1999 NDAA to remove space technologies from the USML and transfer them back to the CCL, although with a new and strict requirement for congressional notification.[176]

The next section describes two of the many substantive changes the DOS and the DOC incorporated into their final rule that amends Category XV of the USML that Bigelow Aerospace, individually, and COMSTAC, collectively, had sought to achieve.

D. Amendment to the International Traffic in Arms Regulations: Revision of US Munitions List Category XV[177]

After the DDTC issued determinations on BA and Virgin Galactic's CJ requests, discussions arose within the DOS and with commercial human spaceflight companies on the next steps for export control reform. In 2014 the DOS issued the Amendment to the International Traffic in Arms Regulations: Revision of U.S. Munitions List Category XV, first as a

---

[174] National Defense Authorization Act for Fiscal Year 2013, Pub. L. 112-239, 126 Stat. 1632 (2 January 2013). See 78 Fed. Reg. 16,129 (13 March 2013) (directing the executive branch on how to execute the administration of export control reform).

[175] Fergusson and Kerr, *supra* note 62. See National Defense Authorization Act for Fiscal Year 2013 § 1261, Pub. L. 112-239, 126 Stat. 1632 (2 January 2013).

[176] See National Defense Authorization Act for Fiscal Year 2013 §§ 1262–1263, Pub. L. 112-239, 126 Stat. 1632 (2 January 2013).

[177] The DOS promulgated two final rules to the Amendments to Category XV of the USML. See 79 Fed. Reg. 27,180 (12 May 2014) ('final interim rule'), and 79 Fed. Reg. 66,608 (10 November 2014) ('final rule, correction').

final interim rule[178] in May and as a final rule (corrected)[179] in November. Two major outcomes resulted. First, the DOS and the staff of the DDTC facilitated the adoption of a broad passenger experience exemption into the final interim rule. Second, the DOS reintroduced what had been previously designated a 'manned satellite' under the pre-1993 USML categorization as an 'in-space habitat' by classifying the technology for the first time on the CCL, provided the system did not include an integrated propulsion system except for attitude control.[180]

Although the DOS placed the passenger experience into the interim final rule as part of broader amendments to Category XV of the USML, the Passenger experience exemption did not come into effect until 10 November 2014.[181] In a note to Category XV(f), which covers technical data, the DOS states that the passenger experience represents:

> [a]ctivities and technology/technical data directly related to or required for the spaceflight (e.g., sub-orbital, orbital, lunar, interplanetary, or otherwise beyond Earth orbit) passenger or participant experience, regardless of whether the passenger or participant experience is for space tourism, scientific or commercial research, commercial manufacturing/production activities, educational, media, or commercial transportation purposes, are not subject to the ITAR or the EAR. Such activities and technology/technical data include those directly related to or required for: (a) spacecraft access, ingress, and egress, including the operation of all spacecraft doors, hatches, and airlocks; (b) physiological training (e.g., human-rated centrifuge training or parabolic flights, pressure suit or spacesuit training/operation); (c) medical evaluation or assessment of the spaceflight passenger or participant; (d) training for and operation by the passenger or participant of health and safety related hardware (e.g., seating, environmental control and life support, hygiene facilities, food preparation, exercise equipment, fire suppression, communications equipment, safety related clothing or headgear) or emergency procedures; (e) viewing of the interior and exterior of the spacecraft or terrestrial mock-ups; (f) observing spacecraft operations (e.g., pre-flight checks, landing, in-flight status); (g) training in spacecraft or terrestrial mock-ups for connecting to or operating passenger or participant equipment used for purposes other than operating the spacecraft; or (h) donning, wearing, or utilizing the passenger's or participant's flight suit, pressure suit, or spacesuit, and personal equipment.[182]

---

178 79 Fed. Reg. 27,180 (13 May 2014) ('final interim rule').

179 79 Fed. Reg. 66,608 (10 November 2014) ('final rule, correction').

180 79 Fed. Reg. 27,180 (13 May 2014) ('final interim rule').

181 *Ibid.*

182 *Ibid.* at 27, 188.

In addition to taking the substance of the Bigelow Aerospace CJ request and transforming it into an industry-wide standard, the changes to Category XV also included the reclassification of controlled technologies under the USML. Specifically, spacecraft listed under Category XV(a) will only remain under the USML if such spacecraft include integrated propulsion systems.[183] Per the DOS note in the interim final rule:

> [s]pacecraft specially designed for human space flight that have integrated propulsion present another security concern, for such capabilities may be used for the purposes of weapons targeting from space. So, although these technologies and capabilities are used in commercial endeavors, they continue to merit control on the USML. As technologies develop, and as there may come to be a greater differentiation between military-critical and commercial technologies, their licensing jurisdiction will be reassessed, as provided for in section 38(f) of the AECA.[184]

Moreover, the DOS and the DOC have split the Category XV reforms for habitat technologies between the USML and CCL.[185] The DOS and the DOC made the distinction based upon whether such habitats have integrated propulsion.[186] The DOS will continue to leave habitats with integrated propulsion systems, except for attitude control, on the USML.[187] Additionally, the DOC will now license space habitat technologies enumerated under the CCL's ISS export control regulations or not identified under Category XV(a) of the USML.[188] However, the DOS continues to keep the technical data related to such habitat technologies split between the USML and the CCL, except when such technology is subject to or associated with the ISS.[189] When CCL Category 9 hardware and associated technical data relate to the ISS as 'specially designed' or subject to the direction of and control by the National Aeronautics and Space Administration, such hardware and technical data will generally remain under the CCL.[190]

---

[183] *Ibid.* at 27, 182.

[184] *Ibid.*

[185] See 79 Fed. Reg. 27,418 (13 May 2014), and 79 Fed. Reg. 27,180 (24 May 2014).

[186] 79 Fed. Reg. 27,418 (13 May 2014).

[187] 79 Fed. Reg. 27,180, 27,182 (24 May 2014).

[188] 79 Fed. Reg. 27,418, 27,425 & 27,440 (13 May 2014). 15 CFR § 774, Supp. No. 1 (2016).

[189] See Category 9: 9A004 & 9E001 & -002 (2016); 15 CFR § 774, Supp. No. 1 (2016). See also USML Category XV(f), notes 1–3.

[190] *Ibid.*

4. CONCLUSIONS AND OBSERVATIONS

Bigelow Aerospace's successful legal challenge to the ITAR over the passenger experience represents a model for opposing counterproductive regulations. It is possible to fight city hall, and much of the blame for the sluggish nature of ITAR reform in the last decade can be directed not at the DDTC or DTSA, but directly at industry itself. Industry has done a great deal of complaining relative to the pernicious nature of the ITAR, but such complaints were rarely combined with action. The Bigelow CJ request, particularly at the time it was filed, was all too unique in that no other commercial human spaceflight companies were willing to dedicate time and resources to challenging the law. As this chapter makes clear, obtaining the successful CJ determination was neither quick nor easy, but it was successful and demonstrated that progress was possible, opening the door for additional changes and eventually the dramatic shift represented by the 2013 NDAA.

Prior to the 2013 NDAA, when industry asked the question as to why change was not occurring, the answer could quite easily be found in a mirror. Companies were either satisfied with the status quo or were contributing to a culture of non-compliance wherein the ITAR was ignored.[191] Complaining about the US export control regime or simply ignoring the ITAR entirely is a large part of why even common-sense reforms took a decade or more to come to fruition. As the Bigelow Aerospace CJ request process and the 2013 NDAA development demonstrate, there are many innovative, experienced, and dedicated civil servants who are determined to establish a regime that supports domestic business interests while also bolstering national security and foreign policy objectives. Moreover, members of the defence and intelligence community played a critical role in pushing the 2013 NDAA forward and supporting the overall reform movement.[192] The US government can and often is eager to play a role in effectuating positive change, but industry must show the courage to initiate the process and the persistence to see it through. When this occurs, and industry associations such as the Satellite

---

[191] This is a situation which also occurs all too often in government agencies. See, for example, United States Government Accountability Office, 'Export Controls – NASA Management Action and Improved Oversight Needed to Reduce the Risk of Unauthorized Access to Its Technologies' (April 2014) GAO-14-315, available at http://www.gao.gov/products/GAO-14-315.

[192] Such military and intelligence support was due largely to concerns regarding the deterioration of the US industrial base for space hardware and components.

Industry Association,[193] the Aerospace Industries Association, the American Institute of Aeronautics and Astronautics, and the Commercial Spaceflight Federation become engaged, anything is possible.

However, technology is not static, and while the current US export control regime for space-related hardware represents a dramatic step forward, the fight is far from over. Industry should take the hard-earned lessons learned from Bigelow Aerospace's work to heart, and continue to pursue additional reforms. Specifically, the US government should establish a process for conducting a regular and comprehensive review of the USML to ensure that new technologies and circumstances are accounted for on an ongoing basis.[194] As long as technology continues to evolve, America's export control regime must evolve with it.[195] If industry remains engaged and continues to work with its government partners, we can enjoy the benefits of dynamic new developments in the space world, and succeed in transforming what was once science fiction into an incredible new reality.

---

[193] For example, see the invaluable contributions by Patricia Cooper, formerly of the Satellite Industry Association, interview, 'Talking export controls with Patricia Cooper', available at http://www.sia.org/wp-content/uploads/2013/04/WorldECR-issue-20-201303-PC-Article-p14.pdf; Written Testimony for Patricia A. Cooper, President, Satellite Industry Association, Before the House Foreign Affairs Committee Hearing on Export Controls, Arms Sales, and Reform: Balancing U.S. Interests (Part II), (7 February 2012), available at http://archives.republicans.foreignaffairs.house.gov/112/HHRG-112-FA-WState-PCooper-20120207.pdf; 'SIA Applauds New Export Control Regulations for Satellites' (*Satellite Markets*, 14 May 2014), available at http://www.satellitemarkets.com/news-analysis/sia-applauds-new-export-control-regulations-satellites; and SpaceNews Editor, 'Making a Difference – Patricia Cooper, Former President, Satellite Industry Association' (*SpaceNews*, 8 September 2014), available at http://www.spacenews.com/article/features/41795making-a-difference-patricia-cooper-former-president-satellite-industry.

[194] Gold, *supra* note 6.

[195] Goldstein, *supra* note 75.

PART III

Space governance

9. Regulatory choices for sub-orbital flights carrying humans: elements for consideration

Thierry Herman and Alexander Soucek

1. INTRODUCTION

This chapter provides elements for consideration regarding the regulation of sub-orbital flights carrying humans, in particular those carried out for commercial purposes ('commercial sub-orbital flights'). In that respect, we note dissension over technical typologies and the competence of legal norms. We argue that the issue at stake is not to contemplate trade-offs in a conflict of competence between existing regulatory regimes ('air law versus space law'). The formulation or application of law should be anticipated by clarity over two questions. First, what needs to be achieved? Second, how can this best be achieved? The adjective 'best' here relates to the social functions of law: preventing undesirable behaviour, securing desirable behaviour, providing facilities for private arrangements, settling unregulated disputes.[1] In this chapter we propose elements for discussion rather than conclusive answers. We argue that clarity over the regulatory purposes plays as much a role as the experience gained with existing regimes and precision over definitions and semantics. We recall characteristics of the two systems that are considered apparent 'candidate regimes' for regulating commercial sub-orbital flights. Finally, we present two examples of domestic regulatory approach that deal with different subjects and yet reveal certain commonalities.

[1] J. Raz, *The Authority of Law: Essays on Law and Morality*, Clarendon Press 1979, published to Oxford Scholarship Online, March 2012.

2. IDENTIFYING SUB-ORBITAL FLIGHTS

Sub-orbital flights have a long history. Rocket-powered vehicles moved on parabolic trajectories before the space age started; the flights of the first two US astronauts in space, Alan Shepard and Virgil Grissom, were sub-orbital, reaching a peak altitude of 187 km and 190 km, respectively. Sending sounding rockets on sub-orbital trajectories has been an established practice for decades. It is interesting, although not surprising, that a broader debate as to appropriate legal approaches only started with the advent of one specific sub-orbital activity: the sub-orbital transport of humans, in particular of so-called 'non-professional' astronauts.

While the profiles chosen and technologies employed differ, there is a common characteristic for sub-orbital flights carrying humans: they attempt to profit from the near Earth outer space environment while not entering into an orbit around Earth. The question of possible regulatory approaches to such activities has been discussed for years with different starting points and conclusions, and it is far from being uniformly, and satisfactorily, settled.

From a geographic and physical point of view, sub-orbital flights may be located somewhere between the atmosphere and outer space, or, more accurately, traversing both of them at different times during a flight. In contrast, 'classical' spaceflight activities are considered to consist of placing space objects in orbits around the Earth or injecting them into interplanetary trajectories. Thus, they remain in outer space for longer periods and in relatively stable conditions. We see a similar dichotomy from a technical point of view: sub-orbital flight vehicles designed to carry humans show both features associated with classical aircraft, for example wings supporting them during their passage through denser air and an aerodynamic fuselage profile, and with spacecraft, such as rocket engines or heat shield protection. Finally, the purposes and flight profiles diverge, as do the levels of development maturity, associated national interests or the business case expectations of the commercial players in the field. It therefore comes as no surprise that the question of an appropriate regulatory approach (taking into account flight and passenger safety considerations, liabilities and responsibilities, industry standards, insurances and so on) cannot be answered without attempting to define, attribute and qualify the constituent elements of what is called 'sub-orbital flight'.

3. FOSTERING THE DEVELOPMENT OF COMMERCIAL PUBLIC AIR TRANSPORT IN ITS EARLY DAYS: THE EQUATION TO BE SOLVED

In the early twentieth century the first flights of engine-powered aircraft were soon followed by significant technological developments that opened the door to commercial developments. Among the first commercial endeavours making use of these new technologies was the provision of postal services along established routes directly after the First World War. In parallel, as aeroplanes grew larger and more powerful, the transportation of passengers progressively increased during the 1920s. This being said, access to air transportation was then limited to a very small number of people, that is, only those who could afford the price and were willing to face the risk that flying entailed in those days, much like sub-orbital flights today.

This burgeoning industry was fast to attract the attention of lawyers and lawmakers, both at national and international level, who undertook to establish the conditions needed to foster and support the commercial development of public air transport.[2] As for any legal and regulatory regime, the solutions eventually adopted represent a compromise aimed at striking the right balance between social, economic and political considerations. In the case of commercial public air transportation, the following equation had to be solved:

(1) Attract a growing number of passengers by ensuring an adequate level of safety, reducing the price of flying and offering an acceptable level of compensation should a passenger suffer damage, while
(2) supporting the establishment of a viable industry by limiting the operational and administrative burden stemming from the various

---

[2] The United States Air Commerce Act of 1926 is an example of domestic legislation focused on the safety aspects of commercial aviation with the objective of installing confidence among potential passengers. In order to achieve this objective, the Act established a set of requirements applicable to the airmen's airmanship and the airworthiness of the aircrafts, including the corresponding licensing and certification criteria. It further established requirements applicable to the registration of aircrafts, and laid down the conditions for the setting up of the appropriate navigation infrastructure so as to allow for the provisions of air navigation services. This important piece of legislation is considered as a building block for the development of commercial aviation.

applicable legal and regulatory requirements, and by limiting the exposure to legal and financial risks, and

(3) providing an adequate level of protection with respect to the adverse effect (such as damage to a third-party, environmental damage) stemming from the conduct of the activity.

The international and domestic instruments of air law developed over the years have been geared at solving this equation, taking due account of the status of the industry at the time of adoption of these instruments, while establishing legal certainty through the harmonization of the applicable provisions. These legal instruments and the corresponding lessons learned may be usefully examined and, eventually, applied in support of the development of a commercial sub-orbital flights industry.

4. INGREDIENTS FOR SECURING LEGAL CERTAINTY

Legal certainty will be key in fostering the development of the commercial sub-orbital flights industry: it seems advisable to establish a single set of legal provisions, or clear rules for the application of different sets of provisions, thereby preventing confusion as to the norms applicable to sub-orbital flights. To this effect, the rules should remain the same throughout all phases of flight with respect to, for example, (1) the jurisdiction over the vehicle or the system of vehicles and those on board, (2) the authorization and licensing procedures (a bundled procedure performed by a single administration, even though the applicable technical requirements may vary depending on the type of activity performed, vehicle/s used and flight crew involved) and (3) the applicable liability regimes. This would represent an important element of securing legal certainty.

As a case in point, one can take liability for damages caused to third parties on the ground that, per definition, cannot be tackled through contractual arrangements in relation to the victim. In this instance, social considerations can justify that those who are unwillingly exposed to a risk resulting from the performance of commercial sub-orbital flights deserve to be covered by a special liability regime providing greater protection than under the normal tortious/extra-contractual liability regimes, the cost thereof being borne by those drawing profit from the flight, be it the 'travel agent', the vehicle operator or the carrier. What seems difficult to sustain in terms of social rationale (for example, uncertainty as to the recourse available to the victim) or even commercial rationale (for example, uncertainty in terms of the insurance policy to be contracted by the commercial entity operating the flight), not to mention legal coherence, is the fact that a different liability regime could possibly apply when the damage is caused (a) during take-off or at any point before separation, by a system of vehicles composed of a 'mother vehicle' and a sub-orbital vehicle, (b) by the mother vehicle, at any point after separation, or (c) by the sub-orbital vehicle, after separation, taking into consideration that both the mother vehicle and the sub-orbital vehicle are used with the same objective: to provide the passenger with an 'outer space-like' experience (microgravity, high-altitude view of the Earth, 'astronaut feeling' and so on). In case (c) above, one could even envisage the alternative where, upon separation from the mother vehicle, the sub-orbital vehicle could alternatively be subject to a different regime during the climb and during descent should the vehicle 'derive support in the atmosphere from the reactions of the air', by gliding or otherwise during that latter part of the flight.[3]

Should such uncertainties – arising from the plurality of liability regimes – persist, we would fall short of offering adequate protection to the victims, which could incidentally prove detrimental to the development of the commercial sub-orbital flight industry. A uniform solution should therefore be sought and supported.

5. ELEMENTS FOR CONSIDERATION UNDER AIR LAW

A. International and Domestic Air Law

Air law comprises a comprehensive corpus of legal instruments. It includes instruments of public international air law (such as the Chicago Convention of 1944), private international air law (such as the Montreal Convention of 1999[4] and the Rome Convention of 1952[5]), regional regulations and domestic laws and regulations.

[3] Annex 7 to the Convention on International Civil Aviation (Chicago Convention), 7 December 1944, 61 Stat. 1180, 15 UNTS 295, Ninth Edition ICAO Doc. 7300/9 (2006).

[4] Convention for the Unification of Certain Rules for International Carriage by Air (Montreal Convention), 28 May 1999, S. Treaty Doc. N. 106–45.

[5] Convention on Damage Caused by Foreign Aircraft to Third Parties on the Surface (Rome Convention), 7 October 1952, 310 UNTS 181.

The above instruments of international air law have established uniform sets of rules for international air traffic that apply directly to their addressees, be they States, carriers, passengers or third parties. They provide an adequate legal backdrop offering solutions for solving the equation identified under section 3 above, in particular when sub-orbital flights cease to be limited to domestic traffic within the airspace of a single State. This being said, these international instruments already have an impact on the legal framework applicable to the domestic sub-orbital flights on offer as some of their provisions have been transposed into regional and domestic laws and regulations. This is the case of the Standards and Recommended Practices (SARPS) established under the Annexes to the Chicago Convention that have been transposed pursuant to Article 37 thereof.[6]

Norms of international air law (including bilateral agreements between States) are necessarily complemented by domestic laws and regulations. The development of new, or application of existing legal norms to regulate sub-orbital flights must therefore duly consider the wide body of domestic legal norms applying to aviation and, if not directly relate to it, at least draw inspiration from it.

B. Definition of the Term 'Aircraft'

Discussions on the legal regime applicable to sub-orbital flights involve the question as to whether the vehicle or vehicles forming part of a system of vehicles used to perform the sub-orbital flight qualify as 'aircraft'. Under the Chicago Convention, an 'aircraft' is defined as '[a]ny machine that can derive support in the atmosphere from the reactions of the air other than the reactions of the air against the earth's surface'.[7] In view of this definition, the test to be applied is not whether the machine concerned does actually derive support in the atmosphere from the reactions of the air, but whether its design and technical specifications are such that it can indeed derive such support should the necessary conditions be met in terms of speed, air density and aerodynamics. No one would ever question the fact that an aircraft taxiing on the apron does qualify as an aircraft: the same logic can be applied to an aircraft that, during part of the flight, does not derive support in the atmosphere from the reactions of the air. In line with this reasoning, it

---

[6] See below Section 5C on SARPS.

[7] See Chapter 1 of Annex 1, Annex 7 and Annex 8 to the Chicago Convention. It includes several categories that are classified according to basic characteristics, e.g. aeroplane, helicopter, glider, free balloon.

should also be pointed out that, under air law, the type of propulsion used is not part of the definition of an aircraft.

There is nonetheless reluctance to consider a sub-orbital vehicle (for example, with technical characteristics comparable to those of SpaceShip Two or Lynx) as an aircraft in its full right, and to apply the corresponding provisions of air law throughout its flight. This reluctance likely stems from the approach taken under the US Commercial Space Launch Amendment Act of 2004 (CSLAA)[8] that classifies SpaceShip Two and Lynx as commercial sub-orbital reusable launch vehicles (SRLV) rather than as aircraft; this is a distinct and novel category. The International Civil Aviation Organization (ICAO), in a paper presented to the Legal Subcommittee of the United Nations Committee on the Peaceful Use of Outer Space (UNCOPUOS) in 2010,[9] expresses the opinion that SpaceShip Two does not operate as an aircraft during the ballistic phase of the flight but could be considered to be operating as an aircraft when gliding back to Earth. In this respect, it may be argued that changing the mode of operation does not change the status of the vehicle used. The ICAO concludes on the issue by stating that:

> Vehicles that would affect earth-to-earth connections through sub-orbital space could incorporate the constitutive elements of aircraft and fly as such at least during descending phase while gliding. However, rocket propelled vehicles could be considered as not falling under the classification of aircraft. At this stage, one State seems to prefer to classify such vehicles as rockets.[10]

This cautious conclusion is balanced by the statement that provides that '[s]hould sub-orbital vehicles be considered (primarily) as aircraft, when engaged in international air navigation, consequences would follow under the Chicago Convention, mainly in terms of registration, airworthiness certification, pilot licensing and operational requirements'.[11] The corresponding ICAO SARPS would hence apply to the activity.

---

[8] Initially found under Chapter 701 of Title 49 of the US Code (Transportation), the provisions on Commercial Space Launch Activities have been transferred to Chapter 509 of Title 51 of the US Code (National and Commercial Space Programs).

[9] ICAO secretariat, 'Concept of sub-orbital flights: Information from the Intentional Civil Aviation Organization (ICAO)' (C-WP/12436), 2, presented as a room document at the Forty-ninth session of Legal Subcommittee of the UNCOPUOS, Vienna, Austria, 22 March–1 April 2010, UN Doc. A/AC.105/C.2/2010/CRP.9, 19 March 2010.

[10] *Ibid.*, 5.

[11] *Ibid.*, 3.

It has been suggested that sub-orbital vehicles capable of deriving support in the atmosphere from the reactions of the air be considered as aircrafts throughout the flight, even where rocket propulsion is used during part of the flight and/or where the sub-orbital vehicle transits through outer space for a limited period of time; accordingly, relevant air law provisions would apply.

C. SARPS: Establishing a Uniform Corpus of Air Law

Coming back to the SARPS referred to above, Article 37 Chicago Convention (Adoption of International Standards and Procedures) is central to the development of a uniform legal and regulatory regime for aviation. It provides that '[e]ach contracting State undertakes to collaborate in securing the highest practicable degree of uniformity in regulations, standards, procedures, and organization in relation to aircraft, personnel, airways and auxiliary services in all matters in which uniformity will facilitate and improve air navigation'. The article goes on to say that:

> To this end the International Civil Aviation Organization (ICAO) shall adopt and amend from time to time, as may be necessary, international standards and recommended practices (SARPS) and procedures dealing with (a) Communications systems and air navigation aids including ground marking; (b) Characteristics of airports and landing areas; (c) Rules of the air and air traffic control practices; (d) Licensing of operating and mechanical personnel; (e) Airworthiness of aircraft; (f) Registration and identification of aircraft; (g) Collection and exchange of meteorological information; (h) Log books; (i) Aeronautical maps and charts; (j) Customs and immigration procedures; (k) Aircraft in distress and investigation of accidents; and such other matters concerned with the safety, regularity, and efficiency of air navigation, as may from time to time appear appropriate.

The SARPS have been developed in the form of Annexes to the Chicago Convention, as complemented by ICAO Circulars and Manuals. They have been transposed, and further developed, into the legal order of the contracting States through the adoption of laws and regulations, or the implementation of complying practices. At European level, the European Aviation Safety Agency (EASA) has adapted and transposed ICAO SARPS in the form of so-called 'Essential Requirements'. It is conceivable that, for the benefit of regulatory clarity over sub-orbital flights, specific Essential Requirements for the certification of sub-orbital vehicles could be adopted in the future, it being understood that uniform approaches are preferable to domestic or regional normative fragmentation.

Through the adoption of the SARPS, ICAO has indeed promoted the development of a uniform legal and regulatory regime among contracting States for the purpose of facilitating and improving air navigation at large in all areas identified under Article 37. Provided that one agrees with the above suggestion sub-orbital vehicles qualify as 'aircrafts', the SARPS and the resulting domestic laws and regulations do offer a suitable framework addressing a large number of issues arising in relation to the conduct of (commercial) sub-orbital flights, consistent with the ICAO study referred to under the previous section. Furthermore, Article 37 grants ICAO the power to adopt and amend SARPS in all matters related to 'safety, regularity, and efficiency of air navigation', thereby allowing for the adoption of international standards, recommended practices and procedures meeting the specific needs of commercial sub-orbital flights.

6. ELEMENTS FOR CONSIDERATION UNDER SPACE LAW

A. International and Domestic Space Law and State Responsibility

The corpus of space law is composed of instruments of public international law (among them the Outer Space Treaty of 1967,[12] the Liability Convention of 1972[13] and the Registration Convention[14]) and of national space legislation. The number of non-legally binding instruments complementing 'hard law' instruments is increasing.

The instruments of public international space law apply to States and, partially,[15] to intergovernmental organizations that have declared their acceptance of those rights and obligations; they do not directly grant private legal or natural persons with rights and oblige them. As a central

---

[12] Treaty on the Principles Governing the Activities of States in the Exploration and Use of Outer Space, Including the Moon and Other Celestial Bodies (Outer Space Treaty), opened for signature on 27 January 1967, 18 UST 2410, 610 UNTS 205.

[13] Convention on International Liability for Damage Caused by Space Objects (Liability Convention), opened for signature on 29 March 1972, 24 UST, 961 UNTS 187.

[14] Convention on the Registration of Objects Launched in Outer Space (Registration Convention) opened for signature on 14 January 1975, 28 UST 695, 1023 UNTS 15.

[15] Some of the UN space treaties open the possibility for intergovernmental organizations to declare their acceptance of these instruments; this possibility is not given for the Outer Space Treaty, however.

norm, and innovative in international law, Article VI Outer Space Treaty provides that the States Parties to the Treaty shall bear international responsibility for national activities carried out by 'non-governmental entities' and that the activities carried out by these entities in outer space shall require authorization and continuing supervision by the appropriate State Party. This provision attributes State responsibility for *any* non-governmental activity, and thus goes beyond the classic rule of State attribution in international law. Article VI Outer Space Treaty has resulted in a growing number of domestic laws and regulations that establish licensing schemes for private actors conducting space activities. The schemes adopted at national level usually include certification requirements for space vehicles and a mechanism to cover liabilities arising in connection with the conduct of space activities.

B. 'Classical' Space Law and the 'Sub-orbital' Category

The international *corpus iuris spatialis* developed during the end of the 1950s and throughout the 1960s was largely based on what was considered classical or typical spaceflight at the time: governmental, that is publicly funded and operated, space missions with artificial objects delivered into planetary orbits or interplanetary trajectories using chemical propulsion. Accordingly, the first United Nations General Assembly Resolution on legal principles governing the activities of States in the exploration and use of outer space of 1963[16] laid down a list of principles, which were later incorporated into treaty law and, at least to some extent, have possible customary law character today: the freedom of exploration and use of outer space and celestial bodies; the non-appropriation principle; the principles of international State responsibility and liability for space activities; the principle of registration of space objects and associated State jurisdiction and control over such objects; the applicability of international law to space activities. They are today considered the nucleus of space law.

In contrast to the definition of these partially innovative and globally important principles, early space lawmaking fell short of providing accurate definitions of key terms such as 'outer space', 'space activity' or 'space object', as we will examine later. Moreover, the pace of space lawmaking slowed down considerably by the end of the 1960s, while the nature of 'classical' spaceflight as perceived during the early days of the space age started to change and diversify. Increasingly, private operators

---

[16] Resolution 1962 (XVIII) of 13 December 1963.

commenced space activities, a trend that went hand in hand with growing commercial interests derived from the utilization of outer space. Satellite applications were at the centre of this paradigm shift, which unfolded during the 1970s and 1980s. Already at that time the first considerations were concretized with regard to the possibility of space tourism (i.e. spaceflight for touristic purposes). Although the first private individuals, denominated 'space tourists' or later 'spaceflight participants' in an attempt to distinguish them from professional astronauts, were flying to *orbital* destinations such as space stations, it became clear that a touristic 'outer space-like' experience could be offered more economically without the need of reaching orbital velocity and altitude, that is to say by *sub-orbital* flight profiles.

C. Terminology and Definitions

The discussion of the most adequate regulatory approach to commercial sub-orbital flights and space tourism activities should start from a common understanding of terminology and definitions.

(i) Outer space

The (legal) definition and delimitation of outer space has been a recurring topic of discussion ever since the beginning of international space law development. There is no internationally accepted (legal) boundary between airspace and outer space, and even if there were one, it is doubtful whether it would help to decide on the applicability of the most suitable regulatory regime for sub-orbital flights. As for the latter, two philosophies or approaches can be distinguished: the functional approach, looking at the aim of an activity to determine the applicable material law, and the spatial approach, looking at where an activity takes place, predominantly or at a given time, in order to determine the applicable law. A clear trend in favour of one of the approaches is not yet identifiable.

Some domestic or regional legal systems have made punctual references to a possible boundary, often referring to a qualified yet arbitrary 100 km limit. For example, Australian domestic space law defines the term 'space object' as a launch vehicle or a payload that the launch vehicle is 'to carry into or back from an area beyond the distance of 100 km above mean sea level'.[17] Another, albeit non-domestic, example

---

[17] Space Activities Act 1998, Act No. 123 of 1998 as amended, Part 2, Article 8.

is European Council Regulation 428/2009 which defines the technical term 'space-qualified' as referring 'to products designed, manufactured and tested ... for use in the launch and deployment of satellites or high altitude flight systems operating at altitudes of 100 km or higher'. However, these examples should not hide the fact that neither public international law nor the majority of existing domestic space legislation recognize any formal atmosphere/space boundary. Many States have even repeatedly expressed the view that the definition of such a boundary is not needed.[18]

(ii) Space object

Unlike air law, which contains a comparatively accurate definition of 'aircraft',[19] as discussed above, space law has not yet developed an accurate definition of the term 'space object', or replaced the latter with another concept. Article 1(d) of the Liability Convention provides that the term 'space object' includes component parts of a space object as well as its launch vehicle and parts thereof – in part a tautological definition. What can be deducted from it, however, is that orbital capacity is not a constituent factor in determining what is a 'space object'.

In contrast, some domestic legislators have elected a different philosophy: the recently revised Belgian space law qualifies 'space object' as 'any object launched or meant to be launched in an earth's orbit or beyond',[20] thus clearly setting the orbital characteristic as the discriminating standard. A sub-orbital vehicle would thus not be considered a space object under Belgian law. Alternatively, the French Space Operations Act of 2008 provides that the term 'space operation' means any activity consisting of launching or attempting to launch an object into outer space, or of ensuring the command of a space object during its journey in outer space.[21] No reference is made to orbital capacity. Similarly, State practice differs when it comes to the registration of launch vehicles, which some consider to be space objects by analogy or definition, while others do not.

---

[18] See recent reports of UNCOPUOS Legal Subcommittee sessions, e.g. A/AC.105/C.2/ L.294 (2014).

[19] See for example, Chicago Convention, Annex 8.

[20] Law of 17 September 2005 on the Activities of Launching, Flight Operation or Guidance of Space Objects, consolidated text as revised by the Law of 1 December 2013, Article 3(a).

[21] Loi n° 2008-518 du 3 juin 2008 relative aux opérations spatiales, Article 1 (non-official translation).

(iii) Sub-orbital flight

Neither air law nor space law employ the term 'sub-orbital' in any of their broadly accepted international legal instruments. Space law attaches some legal consequences to being (or being placed) in the Earth's orbit, while other consequences depend on being launched into outer space. An element complicating consistent regulatory approaches is the heterogeneity of activities falling under the broad term 'sub-orbital flight'. The questions of how to legally define the term 'sub-orbital' and which activities to subsume thereunder or to exclude are thus (or should be) a core element of the debate. Several academic contributions focus on the element of height (altitude) to determine whether or not an object is to be considered sub-orbital, overlooking the fact that an orbit is not reached as a function of altitude but of velocity.[22] It has also been pointed out that 'sub-orbital' should not be misunderstood as 'below an orbit', and it has been proposed to employ the term 'non-orbital' instead.[23] Additionally, the mission profile geometry plays an important role. In the following, we provide a list of activities that may be considered to be sub-orbital in principle, that is to say involving an artificial object that enters at least the lower regions of outer space without concluding at least one full revolution (orbit) around Earth:

Sounding rockets: sounding rockets are sub-orbital carriers following a parabolic trajectory using single- or multiple-stage solid propulsion systems and carrying service systems as well as scientific payloads.[24] Depending on the rocket type and mission profile, these objects can reach significant altitude at the angular point of their flight profile. One such rocket, the European MAXUS rocket, reaches an apogee of 750 km above the Earth, almost twice the height of the International Space Station, and thus undoubtedly in outer space. Despite being in outer space and making use of the near-Earth microgravity environment, sounding rockets are not always considered 'space objects' or 'space activities' by domestic legislation. Sweden, operating or authorizing the

---

22 An object could theoretically orbit Earth at an altitude of one metre or lower, if it were to be accelerated to reach the necessary speed to do so.

23 F. von der Dunk, 'International space law' in F. von der Dunk and F. Tronchetti (eds), *Handbook of Space Law*, Edward Elgar Publishing 2015, p. 96 (fn. 249).

24 European Space Agency, 'European user guide to low-gravity platforms' (Noordwijk 2005) available at http://www.esa.int/Our_Activities/Human_Spaceflight/Research/European_user_guide_to_low_gravity_platforms, (last accessed 5 August 2017).

operation of sounding rockets under various programmatic set-ups from its territory, has explicitly excluded this type of activity from the definition of space activity: '[l]aunching of sounding rockets is not designated as space activities'.[25] Sounding rockets are also generally not registered as space objects.

Ballistic missiles: '[a] ballistic missile (BM) is a missile that has a ballistic trajectory over most of its flight path'.[26] Some types of such missiles, such as intercontinental ballistic missiles (ICBMs), cross the boundary regions between airspace and outer space during their flight profile; they do not enter an orbit around Earth and can be considered, from a mathematical point of view, sub-orbital vehicles. Neither international space law nor air law is generally considered applicable to such activities.

Professional sub-orbital flights: some of the early human spaceflight missions were sub-orbital flights. Project Mercury used both sub-orbital and orbital mission profiles; the first two US astronauts in space, Alan Shepard and Gus Grissom (both flights in 1961), flew on parabolic, sub-orbital trajectories and were powered by rockets. Their missions are considered space activities.

High-altitude, high-speed passenger transportation: several concepts for allowing passenger transport by use of sub-orbital trajectories have been developed in the past, aiming at developing ways to cover greater distances in shorter times. Early ideas, although targeted at novel space transportation vehicles, include the HOTOL and Sänger projects. Ongoing study examples are the German concept of the Spaceliner, an 'advanced, visionary concept for a sub-orbital, hypersonic, winged passenger transport … powered by rocket propulsion'[27] or the British Skylon single-stage-to-orbit spaceplane.[28] The main purpose of such vehicles is seen in the transport of passengers or freight from one surface location to another; conceptually, this type of sub-orbital activity can be considered air transport using new technological means.

---

[25] Swedish Act on Space Activities, 1982:963, Section 1.

[26] Definition taken from the federation of American Scientists (FAS), http://fas.org/nuke/intro/missile/basics.htm (last accessed 29 September 2014).

[27] DLR, http://www.dlr.de/irs/desktopdefault.aspx/tabid-11303/#gallery/28471 (last accessed 20 September 2014).

[28] http://www.reactionengines.co.uk/ (last accessed 21 September 2014).

Commercial sub-orbital flights:[29] Generally viewed as an economically viable solution allowing paying passengers to experience 'outer space', sub-orbital tourist travel is a developing niche which will likely see first results in the near- to mid-term future, although it has yet to be implemented successfully. Methods and technologies pursued are diverse. SpaceShip One of Scaled Composites was the vehicle used to win the Ansari X Prize in 2004. Its successor, SpaceShip Two, is envisaged to be used as a tourist transport vehicle by the company Virgin Galactic. Several companies are competing for what is considered a promising market. Rocket-powered aircraft reaching the boundary regions between the upper atmosphere and what may be considered outer space are not a new development: the most prominent historic example is the US X-15 vehicle which was used for some 150 flights during the 1960s, twice exceeding the 100 km altitude mark.

Commercial sub-orbital flights hence constitute one among many types of sub-orbital activities, though they are perhaps the most prominent. It is this category of sub-orbital flights that currently demonstrates the need for suitable legal and regulatory regimes, whether at domestic or international level, or both.

(iv) Space tourism

The term 'space tourism' is not defined in international law. There are several approaches to categorizing touristic space activities. Existing legal regimes such as international space law, air law or environmental regulations may be applicable in part or entirely to such activities; some cases seem clearer than others. Orbital space tourism, that is the carriage of paying spaceflight participants on board orbital vehicles or to space objects in the Earth's orbit and back, is generally viewed as part of a space activity subjected to norms of space law, whereas 'zero gravity flights' on board classical aircraft in airspace are commonly considered subject to certain norms of air law. But it is the broad grey area in between that has given rise to questions of regulatory approaches and, despite several attempts to solve the issue, no uniform approach has yet been seen.

---

29 For reasons of consistency, this paragraph refers to *sub-orbital* space tourism only.

7. TWO EXAMPLES OF REGULATORY CHOICES OFFERING ADDITIONAL ELEMENTS FOR CONSIDERATION

Several domestic legal and regulatory regimes offer further elements for consideration as to the solutions that can be adopted, at either national or international level, in relation to commercial sub-orbital flights. In the United States, lawmakers have opted to promote the development of the commercial sub-orbital flights industry by adopting relaxed requirements and procedures, for example a broad and adaptable definition of a 'sub-orbital rocket'. Solutions adopted in France to govern zero-G flights, though not directly applicable to commercial sub-orbital flights, could provide guidance when devising a legal and regulatory regime for commercial sub-orbital flights.

A. Commercial Sub-orbital Flights under US Domestic Law

In the United States, lawmakers have elected to address all commercial sub-orbital flights under national space law, with a view to fostering this new industry, notably through the adoption of the CSLAA that contains a number of provisions to that effect, as demonstrated through the following examples:

(1) The CSLAA has introduced the possibility for private entities wishing to enter the market to ask the Federal Aviation Administration (FAA) for a so-called 'experimental permit',[30] instead of a licence,[31] for the purpose of testing SRLV. These permits are granted under more relaxed conditions compared to those required to obtain a licence, provided the flights are not performed for compensation or hire[32] (i.e. against payment). Upon completion of the test campaign, the private entity would need to obtain a licence before it could operate commercial sub-orbital flights, either under a launch licence or an operator licence.

(2) The definitions contained in the CSLAA are tailored to ensure that all projects under development in the United States would fall within its scope. In this respect, the CSLAA defines 'sub-orbital trajectory' as 'the intentional flight path of a launch vehicle,

---

30 51 U.S.C.§50906.

31 51 U.S.C.§50905.

32 51 U.S.C.§50906(h).

re-entry vehicle, or any portion thereof, whose vacuum instantaneous impact point does not leave the surface of the Earth'[33] and defines a 'sub-orbital rocket' as a category to which SRLVs belong, as 'a vehicle, rocket propelled in all or in part, intended for flight on a sub-orbital trajectory, and the thrust of which is greater than its lift for the majority of the rocket-powered portion of the ascent'.[34] This definition of a 'sub-orbital rocket' seems to have been inspired by the operational model chosen by Virgin Galactic with SpaceShip One and Two.

As to what concerns the definition of 'sub-orbital rocket', the above impression is confirmed by another provision of the CSLAA that states that 'starting 3 years after the date of enactment of the CSLAA, the Secretary [of transportation] may issue final regulations changing the definition of sub-orbital rocket'.[35] The Secretary of Transportation may do so only if he/she 'has determined that the definition … does not describe, or will not continue to describe, all appropriate vehicles and only those vehicles. In making that determination, the Secretary shall take into account the evolving nature of the commercial space launch industry'.[36] In other words, the definition of a sub-orbital rocket may be amended in order to ensure that all vehicles (and incidentally the private entities developing them) that are deemed to merit protection do benefit from the protective regime established under the CSLAA. Leaving to the Secretary of Transport the possibility of adapting the definition of 'sub-orbital rocket' is consistent with the objective of fostering the development of the sub-orbital flights industry. It may, however, lead to some uncertainties should the same vehicle meet both the definition of an 'aircraft' under air law and that of a 'sub-orbital rocket' under US domestic space law.

(3) The CSLAA provides for a moratorium on safety regulation whereby the Secretary of Transportation is to wait before proposing regulations. This moratorium has been extended a number of times and is now running until 1 October 2023, it being emphasized that, when the time comes, '[a]ny such regulations shall take into consideration the evolving standards of safety in the commercial space flight industry'.[37] The extensions of the moratorium were

---

[33] 51 U.S.C.§50902(20).

[34] 51 U.S.C.§50902(19).

[35] 51 U.S.C.§50922(c)(2).

[36] *Ibid.*

[37] 51 U.S.C.§50905(c)(10).

> considered by some lawmakers as being too protective of industrial interests, to the detriment of public safety, and so the current extension from 2015 until 2023 is accompanied by transitory measures that authorize the Secretary of Transportation to issue limited safety regulations under strict conditions.[38]

Where lawmakers elect to govern commercial sub-orbital flights under space law, air law will nonetheless be found to apply to some important issues. This is the case for Air Traffic Services (ATS) that are subject to Annex 11 to the Chicago Convention. ATS need to be provided when the sub-orbital vehicle crosses flight levels and sectors used by civilian and State aircrafts, or even when the airspace used by the sub-orbital vehicle is closed to other types of traffic, be it permanently or temporarily. In view of the unusual flight profile of sub-orbital flights, ATS should include a coordination mechanism[39] to ensure that the airspace used by a sub-orbital vehicle is clear of other forms of traffic. In this respect, Virgin Galactic announced in 2014 that it had concluded an agreement with the FAA that established the conditions for the use of the US airspace by SpaceShip Two.[40] This agreement reportedly provides that Virgin Galactic shall cooperate with the FAA's Air Traffic Control (ATC) unit in Albuquerque and the New Mexico Spaceport Authority to secure clear airspace in a timely manner.[41]

Coming back to domestic space law in the United States, the Human Space Flight Requirements for Crew and Space Flight Participants[42] issued in December 2006 lay down a series of requirements for 'space flight participants on board a vehicle whose operator is licensed or permitted'.[43] These requirements include the following obligations:

---

38 51 U.S.C.§50905(c)(1) and (2).

39 Where future sub-orbital flight projects involve cross-borders operations, a coordination mechanism between the national Air Navigation Services Providers (ANSPs) of the States concerned or between adjacent Air Traffic Control (ATC) units will also need to be implemented.

40 M. Gannon, 'Virgin Galactic Signs FAA Deal to Clear Airspace for Sub-orbital Flights' (*Space.Com*, 30 May 2014) http://www.space.com/26056-virgin-galactic-faa-airspace-deal.html (last accessed 11 September 2014).

41 See *supra* note 1.

42 Human Space Flight Requirements for Crew and Space Flight Participants, 71 Fed. Reg. 75, 615 (15 December 2006) (to be codified at 14 C.F.R, Parts 401, 415, 431, 435, 440 and 460).

43 14 C.F.R. §460.41.

- 'The licensee or permittee must comply with the insurance requirement for licensed or permitted activities'.[44]
- The operator 'must inform each space flight participant in writing about the risks of the launch and re-entry, including the safety record ... of the vehicle'.[45]
- The operator 'must inform each space flight participant that the US government has not certified the launch vehicle and any re-entry vehicle as safe for carrying crew or space flight participants'.[46]
- The operator 'must provide each space flight participant an opportunity to ask questions orally to acquire a better understanding of the hazards and risks of the mission'.[47]
- The operator 'must train each space flight participant before flight on how to respond to emergency situations'.[48]
- The operator 'must implement security requirements to prevent any space flight participant from jeopardizing the safety of the flight crew or the public'.[49]
- In turn, the flight participant 'must provide consent in writing to participate in a launch or re-entry'[50] and 'must execute a reciprocal waiver of claims with the FAA'[51] in the form of an agreement as referred to above.

These provisions have proven successful in establishing a framework suitable for the development of a commercial sub-orbital flights industry, as demonstrated by the number of commercial sub-orbital ventures in the United States. While using terminology pertaining to the space sector (spaceflight, launch, re-entry, and so on), the issues addressed and the solutions provided bear certain similarities with the regime applied to commercial zero-G flights under French domestic air law.

B. Commercial Zero-G Flights under French Air Law

While clearly not fitting the category of sub-orbital flights, commercial zero-G flights raise, to a certain extent, similar legal and regulatory

---

44 14 C.F.R. §440.9.
45 14 C.F.R. §460.45(a).
46 14 C.F.R. §460.45(c).
47 14 C.F.R. §460.45(f).
48 14 C.F.R. §460.51.
49 14 C.F.R. §460.53.
50 14 C.F.R. §460.45(f).
51 14 C.F.R. §460.49.

issues. It is therefore interesting to examine how these issues have been addressed under a domestic air law regime.

We have shown that air law has proven successful in fostering the development of civil aviation. While it has evolved into a very extensive and far-reaching set of laws and regulations that some may claim to be too cumbersome and stringent[52] (for example, with respect to airworthiness certification) to allow for the development of a sub-orbital flight industry, it is to be emphasized that air law covers a wide array of aerial activities that goes far beyond commercial public transportation of passengers.

So-called 'zero-G flights' are aircraft flights with an unusual flight path including parabolic manoeuvres that offer up to 30 seconds of microgravity experience to passengers on board the aircraft during each parabola. Under French domestic law, such flights are referred to as 'thrill flights'; they are subject to the regulatory regime established under the Ministerial Decree concerning the use of civil aircraft in general aviation (Decree).[53] The zero-G flights operated and commercialized by the private company airZero-G are offered to the public for a price of around €6000 per passenger per flight; a flight typically lasts for 2 hours 30 minutes and includes 12 parabolas. The flights are performed within French airspace, using (since 2014) a modified Airbus A310 and reaching a maximum altitude of 28 000 feet.

The Decree defines the term 'thrill flight' as 'a leisure flight with the same departure point and destination whose aim is to provide thrilling sensations for passengers by performing aerobatics',[54] and lists a number of requirements that apply to the flight operator, the aircraft and the competent administrative services. These requirements include the following[55]:

- The operator must ensure 'that its organization is adapted to the activity performed and that procedures to supervise the operations

---

[52] 'Commercialization of Space: Commercial Space Launch Amendments Act of 2004' (Spring 2004) *Harvard Journal of Law and Technology* 17(2), 626.

[53] Arrêté du 8 février 2012 modifiant l'arrêté du 24 juillet 1991 relatif aux conditions d'utilisation des aéronefs civils en aviation générale et l'arrêté du 28 février 2006 relatif au certificat de navigabilité restreint d'aéronefs de collection, JORF 42 of 18 February 2012, p. 2829, text 11.

[54] *Ibid.*, Article 1.

[55] *Ibid.*, Article 2, which provides for the addition of a Chapter VIII to the arrêté du 24 juillet 1991 relatif aux conditions d'utilisation des aéronefs civils en aviation générale (hereinafter 'Decree').

are in place',[56] and 'that all necessary precautions have been taken to secure a level of safety it deems acceptable' (self assessment).[57]
- The operator must also 'prepare an Operation Manual' to be followed by the flight crew[58] and inform 'the passenger in writing that the flight does not meet the same safety requirements as those established for public air transport'.[59]
- The competent administrative services 'may impose amendments to the Operation Manual where they notice that it does not comply with the applicable regulations or where safety so requires',[60] and 'may suspend the thrill flight activity where it considers that the safety level is not adequate'.[61]
- The aircraft used to perform the flight 'must carry a title of airworthiness recognized as valid in France'[62] and must follow a 'flight profile compatible with the flight domain and the limitations identified in the flight manual of the aircraft used'[63] to perform the 'Thrill flight'.

Based on these requirements, airZero-G has developed contractual provisions in the form of 'General Conditions' to be accepted and adhered to by the passengers. Under these General Conditions, airZero-G informs the passenger that:

- the zero-G flight qualifies as a 'Thrill flight' within the meaning of the Decree;
- the French civil aviation authority[64] has granted the aircraft with a 'temporary permit'[65] that is different from a standard airworthiness certificate granted to aircrafts used for public air transport;
- participating in a zero-G flight carries additional risks compared to a standard commercial flight;
- insurance has been purchased by airZero-G to cover bodily injury to the passenger up to a ceiling amount;

---

56 Decree, chapter VIII, Section 8.1.2.
57 Decree, chapter VIII, Section 8.1.3.
58 Decree, chapter VIII, Section 8.2.
59 Decree, chapter VIII, Section 8.1.6.
60 Decree, chapter VIII, Section 8.2.6.
61 Decree, chapter VIII, Section 8.1.5.
62 Decree, chapter VIII, Section 8.3.
63 Decree, chapter VIII, Section 8.4.
64 Direction Générale de l'Aviation Civile.
65 'Laisser-passer provisoire'.

- should the passenger wish additional insurance coverage, they can elect to do so, airZero-G having negotiated special conditions with an insurance company.

The General Conditions further lay down the applicable regime in terms of delays and cancellation, causes leading to the exclusion of a passenger, medical fitness requirements and applicable laws and jurisdictions. The provisions of the Decree and the General Conditions quoted above address a number of issues and provide solutions that are interesting to consider in the context of the future regulation of commercial sub-orbital flights.

8. CONCLUDING REMARKS

We have shown that the activities discussed under the headline 'sub-orbital flight' differ in nature and objectives, and we notice that the discussion of regulatory choice falls short at times of establishing clarity about the *telos* of required regulation. On the other hand, the issues at stake are often similar, in line with the functions of law: specification of vehicles and infrastructures, certification of these vehicles, contractual and third party liability, the need to have access to navigation infrastructures and services, qualifications of the crew, medical fitness of the passengers and competent jurisdiction.

So, which regulatory regime to choose for commercial sub-orbital flights? The options are not surprising:

(1) apply an existing legal regime;
(2) adapt an existing regime so as to reflect the specific characteristics and needs of sub-orbital flights carrying humans;
(3) apply or adapt two or more existing regimes and blend them together without compromising the functions of law;
(4) develop a novel legal regime.

Regulatory requirements will have to be translated into and further completed by contractual arrangements to be concluded by the different stakeholders.

Whether appropriate legal and regulatory regimes are chosen at domestic level, as has been done already, or at international level depends on the ability of States to jointly agree on the objective to be reached. In this respect, the approach that will eventually be taken by States will depend on their determination to solve the same equation as the one they solved for commercial public air transportation many decades ago.

10. Establishing a regulatory framework for the development and operation of sub-orbital and orbital aircraft in the EU: the role of the EASA

Jean-Bruno Marciacq[1]

1. INTRODUCTION

The Treaty on the Functioning of the European Union[2] allows for the development of common policies for all sectors of transport, including aviation, and its safety. To this end, in 2002 the European legislator established the European Aviation Safety Agency (EASA), located in Cologne, Germany, and gave it responsibility for the regulation of aviation safety, successively encompassing airworthiness, air operations and flight crew licensing (FCL), air traffic management (ATM), air navigation SYSTEMS (ANS), as well as aerodromes (ADR).

Annexes 2, 6 and 8 of the International Civil Aviation Organization (ICAO) of the Chicago Convention[3] define an aircraft as 'any machine that can derive support in the atmosphere from the reactions of the air other than the reactions of the air against the earth's surface'. Aerodynamic lift generated during the atmospheric part of flight is commonly used to sustain and control a vehicle, that is to take off, climb, pull up, perform manoeuvres, fly back to the airport and land. Thus, sub-orbital

[1] The views expressed by the author in this chapter do not represent the views of the European Aviation Safety Agency.

[2] Consolidated version of the Treaty on the Functioning of the European Union annexed to the Final Act of the Intergovernmental Conference which adopted the Treaty of Lisbon, signed on 13 December 2007, *Official Journal C* 326, 26.10.2012, p. 47–390, http://eur-lex.europa.eu/legal-content/EN/TXT/?uri=celex%3A12012E%2FTXT.

[3] Convention on International Civil Aviation, 1944, as amended 2006 (ICAO Doc. 7300/9).

and orbital aircraft are considered to be aircraft as opposed to rockets, which are symmetrical bodies not generating lift and solely sustained by their rocket engine(s).

Consequently, the regulation of sub-orbital and orbital aircraft (SOA) airworthiness, their crew, operations, insertion into the traffic and utilisation of aerodromes would in principle fall under the remit of the EASA, which would have to fulfil its role of protecting European citizens in relation to civil sub-orbital and orbital flights, that is, it must certify SOAs and their operations before they could be used for Commercial Transport in the EU.

Since the EASA was first contacted by potential applicants in 2007, many projects have developed and the situation has evolved. Thus, this chapter intends to update the approach initially proposed at the 3rd International Association for the Advancement of Space Safety (IAASS) in Rome in October 2008 and complemented at the 61st International Astronautical Congress (IAC) in Prague in October 2010. This chapter seeks to incorporate sub-orbital and orbital aircraft into the EU regulatory system and to establish a consistent regulatory framework that allows safe and environmentally controlled operations of SOA in Europe. For further legal and technical details, please refer to the corresponding papers published by the IAASS and IAC.

In addition to this update, a discussion on why SOA are indeed aircraft and not rockets is attached as an appendix to the chapter.

2. THE SOA INDUSTRY

Sub-orbital and orbital aircraft projects are currently under development in the USA and Europe; projected investment is estimated to be around €3 billion.[4]

The function of SOA is to regularly bring passengers and/or payloads to very high altitudes (above 30 km [100K ft or FL1000]) and return them safely back to Earth, without reaching orbital speeds and therefore without being able to stay in space. In order to be financially viable and operationally flexible, most SOA concepts currently under development are aeroplanes (airborne with fixed lifting surfaces) taking off and landing horizontally from/on conventional runways, unlike rocket-launched space vehicles.

---

[4] Estimation: US projects = 1b$ + EU projects = 1b€ + Spaceports = 3b€ in total.

Beyond this functional purpose, the emerging sub-orbital industry intends to use SOA – financing their development by operating SOA for commercial transport (carrying paying passengers and scientific payloads) – as a stepping stone for developing and proving the technology required for the next generation of high altitude/high speed transport (HST) and orbital aircraft (OA) at lower costs and risks. This goal requires achieving regulatory certification of new technological developments such as rocket propulsion and control systems on civil aircraft capable of flying at very high speeds and altitudes, which has not so far been done.

All European designers of SOA[5] have consequently publicly declared their intention to apply to the EASA for certification of their projects, thus guaranteeing their passengers and crew a certain level of safety without jeopardising the reputation of their companies, but also reassuring their investors and insurance providers.

Developing rules is a pre-requisite for a certification process, and has also been requested by the industry, as this allows it to base designs on existing rules. To this end, in 2010 two Rulemaking Tasks were inserted into the EASA Rulemaking Programme. However, as at the time of writing, only preliminary studies have been started due to the lack of resources assigned to the EASA. The purpose of this chapter is therefore to list possible options for establishing a consistent regulatory framework, thereby enabling safe and environmentally controlled operations of SOA in Europe.

3. THE INTERNATIONAL LEGAL CONTEXT

From a legal perspective in relation to the regulation of sub-orbital flights, both the Chicago Convention and the aviation-related legislation in the EU have to be taken into consideration.[6] The European Commission Space Industrial Policy[7] also gives some direction to research on how to handle sub-orbital flights. Finally, as some sub-orbital aeroplanes may also touch, if not enter outer space, international space law issues must also be considered.

---

5 EADS-Astrium, DASSAULT/Swiss Space Systems (S3), Booster Industries, Reaction Engines Limited (REL).

6 K. von Wogau, *Report on Space and Security*, European Parliament 2004–2009 Session Document, 10.6.2008, Ref. A6-0250/2008.

7 COM(2013) 108 final and Memo/13/146.

A. The Chicago Convention

The ICAO was established by the Chicago Convention,[8] which was signed on 7 December 1944 by 52 Contracting States, which today number almost 200. Article 1 of the Chicago Convention recognises the complete and exclusive sovereignty of each Contracting State over the airspace above its territory (including territorial waters). As a consequence, international standards adopted and published by the ICAO are addressed to said States and not directly to natural or legal persons. In other words, ICAO standards do not have direct force of law in the Contracting States. Article 12 of the Chicago Convention clarifies that each Contracting State undertakes to adopt measures to implement and enforce rules of the air in its territory. The same article requires that the regulations in force in the States should conform, to the greatest possible extent, to the standards adopted by the ICAO.

However, Article 44 of the same Convention assigns to the ICAO objectives for the safe and orderly growth of aviation in a larger geographical area (i.e., 'throughout the world') than the territory and territorial waters of Contracting States. In other words, flights over water between Contracting States also come under the scope of the ICAO. Even in this respect, however, the ICAO in practice does not establish law directly applicable to natural or legal persons. Article 17 of the Convention establishes that aircraft have the nationality of the State in which they are registered. Furthermore, the ICAO Council has divided airspace over the entire world into Flight Information Regions (FIRs) also spanning the high seas. Responsibility for establishing rules of the air and providing air navigation services has then been delegated to the contracting State to whom each FIR has been assigned. This is balanced by Article 33, however, which guarantees the mutual recognition of certificates issued by Contracting States.

As a conclusion, ICAO provisions are not directly applicable to aviation personnel or operators, unless transposed into national law by Contracting States. In practice, most States have done so in order to remain part of the international aviation system, which brings them significant economic and social benefits. From the EASA perspective it is amongst the objectives of the Agency to assist Member States in fulfilling their obligations under the Chicago Convention.

---

8 Convention on International Civil Aviation, 1944, as amended 2006 (ICAO Doc. 7300/9).

In 2005 the ICAO published in the form of a Working Paper the results of a Study on the Concept of Sub-orbital Flights.[9] One of its conclusions was that:

> From a spatialist viewpoint, there is no clear indication in international law on the delimitation between airspace and outer space which would permit to conclude on the applicability of either air law or space law to sub-orbital flights. On the other hand, it may be argued from a functionalist viewpoint that air law would prevail since airspace would be the main centre of activities of sub-orbital vehicles in the course of an earth-to-earth transportation, any crossing of outer space being brief and only incidental to the flight.

Furthermore, the Working Paper also stated that: 'Should sub-orbital vehicles be considered (primarily) as aircraft, when engaged in international air navigation, consequences would follow under the Chicago Convention, mainly in terms of registration, airworthiness certification, pilot licensing and operational requirements.'

In a more recent letter dated 17 March 2010 to the UNCOPUOS Legal Subcommittee on the subject:

> the ICAO secretariat informed [the UNCOPUOS] that it is aware of the fact that commercial sub-orbital operations are being planned by various entities and expect that such operations would affect international aviation in some manner. In the meantime, ICAO Secretariat believed that the information contained in its study on the Concept of Sub-orbital flights remains pertinent and therefore the [UNCOPUOS] Legal Subcommittee could consider the study.[10]

B. Aviation Law in the European Union

(i) General

The Rome Treaty of 1957 establishing the European Economic Community, while maintaining independence and sovereignty for the signatory States, established institutions (e.g., the Council and the European Commission) with delegated powers to adopt legally binding measures, which are directly applicable to and effective for natural and legal persons acting in the territory of the Community. In other words, some attributes of the sovereignty had been delegated to a supranational entity without the need to create a larger State.

---

[9] ICAO C-WP/12436 dated 30 May 2005.

[10] Committee on the Peaceful Uses of Outer Space, Legal Subcommittee, Forty-ninth Session 22 March–1 April 2010, dated 19 March 2010.

Transport has historically been an area where the Community has been empowered to establish common policies and common rules, although the initially required unanimity in the Council has greatly delayed the development of binding common rules for aviation in Europe. Technical cooperation via the Joint Aviation Authorities led to the harmonisation of several technical rules, but on a best endeavours principle.

With the introduction of qualified majority voting in 1986 in order to adopt common policies, in particular for air transport, the number and scope of basic EU legal instruments applicable to aviation safety constantly expanded. The major ones are summarised in Table 10.1.[11]

In conclusion, the Member States (28 at the time of writing) of the Community (now the European Union) have progressively discharged their obligation to transpose ICAO standards into law applicable in their territory, not individually but collectively.

Conversely, EU law does not contain 'minimum' requirements, but 'the' requirements. Minimum requirements are in fact necessary to ensure safety, but additional requirements adopted nationally may distort the internal market. EU Member States are therefore not allowed to establish additional requirements in fields where community competence exists.

(ii) The EASA role and procedures

The European Aviation Safety Agency was established by Regulation (EC) No. 1592/2002, later repealed and replaced by Regulation (EC) No. 216/2008 (hereinafter referred to as the 'Basic Regulation'). The same Regulations established the EU's common rules in the field of aviation safety. Initially, those common rules were limited to airworthiness and the environmental compatibility of aeronautical products (i.e., aircraft, engines or propellers). The Basic Regulation does not explicitly define the term 'aircraft'. Therefore, and in line with the Agency's objectives explained above, the ICAO definition contained in Annex 8 to the Chicago Convention applies: 'an aircraft is any machine that can derive support in the atmosphere from the reactions of the air other than the reactions of the air against the earth's surface'. For example, 'hovercraft' are therefore not considered as aircraft and are outside the EASA system. On the other hand, civilian aeroplanes and helicopters, but also sailplanes and aerostats (i.e., either balloons or airships), are subject to the common rules for airworthiness, on the basis of Article 5 of the Basic Regulation.

---

[11] Since their initial issue, numerous regulations listed in this table have been amended or replaced. Therefore, this table only intends to show the initial evolution of the Agency's competences without necessarily providing the latest references to EU legislation currently in force.

Table 10.1 EU main aviation milestones and associated regulations [FTO/ZER/JBM – EASA]

Year	Act	Topic
1991	Directive 670	Mutual recognition of aeronautical licences
1991	Regulation 3922	Harmonisation of technical aeronautical rules
1994	Directive 56	'Independent Investigators'
2002	Regulation 1592	Establishment of EASA
2003	Regulations 1702, 2042	Implementing rules on initial and continuing airworthiness
2003	Directive 42	Safety occurrence reporting
2004	Directive 36	Safety Assessment of Foreign Aircraft (SAFA)
2004	Regulations 549, 550, 551, 552	'Package' of 4 Regulations on the 'Single European Sky' (SES)
2005	Regulation 2111	'Safety list'
2006	Regulation 1899	EU-OPS for commercial air operators
2008	Regulation 216	First extension of EASA system to operations, flight crew licensing and third country operations (TCO)
2009	Regulation 1108	Second extension of EASA system to ATM/ANS and aerodrome safety
2011	Regulations 1178, 1332, 1034, 1035	Implementing rules on air crew licensing, airspace usage requirements, ATM/ANS service provision and oversight, air traffic controllers (ATCO)
2012	Regulations 965, 923	Implementing rules on air operations (EASA-OPS), standardised European rules of the air (SERA)

In this context the EU legislator has decided that:

- Legally binding implementing rules, in compliance with the Basic Regulation, can only be adopted by the European Commission after having received an Opinion from the EASA for that purpose;
- Certification specifications (CS) and applicable means of compliance (AMCs) applicable to specific products can be adopted and published by the EASA, provided an equivalent level of safety (ELoS) can be demonstrated by the applicant to the EASA, but are not legally binding;

- A 'certification basis' has to be defined for each product subject to certification, which is based on the airworthiness codes, but adapting them for each specific case through 'special conditions' (SCs).

Only the EASA is competent to issue type certificates in the EU and – on the basis of treaties concluded between the EU and these States – in Norway, Iceland, Lichtenstein and Switzerland. The type certificate attests that the design of a product complies with the airworthiness requirements set out by the legislator.

The legal framework described above also applies to sub-orbital aeroplanes, since they are considered to be aircraft. Therefore, interested designers consult the EASA, who assist in preparing applications for technical, operational and organisational approvals.

C. EU Space Industrial Policy

On 28 February 2013 the European Commission published its EU-Space Industrial Policy,[12] aiming, amongst other things, to contribute to the global Europe 2020 strategy.[13]

One of the objectives mentioned by this policy is to explore whether commercial spaceflight activities need to be embedded in a legal framework. A study launched by the European Commission in 2012 assessed the market potential of sub-orbital spaceflights, essentially for sensational flights, to then determine whether to develop a European regulatory approach. This study performed by Booz Allen (now Strategy&) interviewed all stakeholders at that time. A second study performed by Strategy& in 2014 extended the conclusions to other more important, yet less visible potentials of the suborbital market, that is LEO cubesat launches, science, defence, and so on.[14]

D. US Regulatory Framework

In 1984 the US Space Launch Act established the basis for licensing and promoting Commercial Space Flights in the USA, which led to the creation of the Office of Commercial Space Transportation under the

---

[12] EC-COM (2013) 108.

[13] EC-COM (2010) 2020.

[14] For more information see http://www.cesmamil.org/wordpress/wp-content/uploads/2014/06/1540-SCATTEIA-CESMA_Scatteia.pdf.

Clinton Administration, later relocated to the US Federal Aviation Administration (FAA/AST).[15]

The US National Space Policy,[16] authorised by George W. Bush in 2006, further aimed to 'encourage an innovative commercial space sector, including the use of prize competitions' such as the successful $10 Million Ansari X-Prize, won by Paul Allen and Burt Rutan of Scaled Composites with the SpaceShipOne/White Knight carrier Two Stage To Space (TSTS) system on 4 October 2004.[17]

Having launched the first commercial sub-orbital spaceflight (Space Ship One), the United States needed to describe the legal basis applicable to those activities. The solution was found in the fact that the same authority (FAA) responsible for issuing certificates/approvals regarding aviation safety could also license launches into outer space. This solution was based on the licence to perform launches into outer space. The Congress adopted the Commercial Space Launch Amendment Act, signed by the President on 23 December 2004. The Commercial Space Launch Act is now the legal basis for the FAA to regulate commercial human spaceflight, with the aim of protecting the safety of the uninvolved public on the ground in terms of launch and re-entry (including air- and spaceworthiness, protection of health and safety of flight crew and flight participants, and training and medical checking of flight crew and participants).

In December 2006, the FAA-AST published its Final Rulemaking, 'Human Space Flight Requirements for Crew and Space Flight Participants', the purpose of which was to establish minimum standards and specific requirements for licensing space launches. The significant difference between licensing and certification must be underlined at this point, since it creates substantial consequences for the approach chosen: in licensing, the operator bears the full responsibility of its operations, whereas in certification, the certificating authority takes a part of the

---

15 K. Wong, 'Developing Commercial Human Space Flight Regulations' Proceedings of the 2nd IAASS Conference, 14–16 May 2007 Chicago, USA (ESA SP-645, July 2007); R. Repcheck, 'FAA's Implementation of the Commercial Space Launch Amendments Act of 2004 – The Experimental Permit', Proceedings of the 1st IAASS Conference, 25–27 October 2005 Nice, France (ESA SP-599, December 2005); US Executive Order 12465; Title of the United States Code, Subtitle IX, Chapter 701.

16 US National Space Policy, 31 August 2006.

17 J.N. Pelton, 'Space Planes and Space Tourism: Developing New Safety Standards and Regulations', Proceedings of the 2nd IAASS Conference, 14–16 May 2007, Chicago (ESA SP-645, July 2007); I.H. Diederiks-Verschoor, *An Introduction to Air Law*, 8th edn, Kluwer Law International 2006, p. 241.

responsibility. There are advantages and drawbacks to both methodologies, which may have to be adapted in order to best fit the existing regulatory framework of each country.

E. Applicability of International Space Law

In addition to the international and European air law context described above, international and national space law has to be considered. Sub-orbital aeroplanes may also enter outer space. Although a clear and commonly accepted legal delimitation does not exist,[18] options are available. For example, every flight which goes beyond 100 km above sea level can be considered as having entered outer space, legally speaking. A 110 km delimitation is used by many outer space lawyers and goes back to a Russian proposal in the United Nations Legal Subcommittee to the Committee on the Peaceful Uses of Outer Space on the agenda item of 'Definition and delimitation of outer space'.[19] So far only one State has seen the necessity to delimit air space and outer space in national legislation and has chosen the 100 km opinion.[20] Those who are not supporting the (arbitrarily set) 100 km (or 110 km) delimitation take a physical approach: air space ends where the air can no longer support the machine and outer space begins where an object can (at least) briefly maintain an orbit. This opinion was first brought up by Theodore von Kármán, which is why the line between air space and outer space is sometimes called the Kármán line. Calculations of that line differ, and this is why some people see it at 53 miles (*c*.84 km) and others at 60 miles (almost 100 km).

The paradox is that although it is not clearly defined, almost all provisions on outer space activities in international law would become applicable whenever the line between air and space law is crossed, wherever it is set. Only a few provisions require a vehicle to enter an orbit.[21] This is why 'sub-orbital' activities may potentially be subject to

---

[18] See e.g. UN Doc. A/AC.105/769 or G. Oduntan, 'The never ending dispute: Legal theories on the spatial demarcation boundary plane between airspace and outer space', in *Hertfordshire Law Journal* 2003, pp. 64 *et seq.*

[19] UN Doc. A/AC.105/769, para. 3.

[20] Section 8 of the Australian Space Activities Act (1998) defines that an object is launched into outer space when it is launched 'in an area beyond the distance of 100 km above sea level'.

[21] E.g., Art. 2(1) of the Convention on Registration of Objects Launched into Outer Space, 1975 (1023 UNTS 15).

most of the space law provisions, should they cross the yet to be legally defined limit of outer space.

If a sub-orbital aircraft passes that line and enters outer space, it is subject to a set of outer space rules, which are very often quite different from the rules provided for in (international) air law. While air law originates from national rules set up under the sovereignty of States concerning their airspace and eventually harmonised on an international level through treaties and agreements,[22] space law originates in international law. No State can claim sovereignty over any part of outer space. Legal provisions can therefore only be set up by international agreements (i.e., at the UN level). Only those issues explicitly mentioned in these agreements as having national concern can be regulated by States.

4. OBJECTIVES/VARIABLE CONFIGURATION

The main objective of the development of a rulemaking framework suitable for SOA operations in the EU is to provide a consistent and harmonised set of regulatory material, in a timely manner, enabling the development and safe operation of sub-orbital aircraft in Europe. To achieve this objective, the SOA regulatory framework should aim at:

- providing efficient means for necessary policy setting and external communication on the EU's SOA-related activities;
- identifying all regulatory material needed to establish a consistent regulatory framework in Europe in a safe, globally interoperable and timely manner;
- developing technical specifications to allow for the certification of sub-orbital aircraft and their operations to/from Europe;
- ensuring that this regulatory material is compatible with existing European and worldwide aviation and space regulations (including FAA licences and certificates) and is compliant with current and future ICAO standards;
- providing efficient means for the necessary internal and external coordination and monitoring/steering on SOA-related issues;
- providing efficient means for the promotion of European views.

Although the objectives may be clear, some areas that may be subject to many variations and combinations are:

---

[22] In the case of the EU, a number of sovereign States collectively exercise their respective sovereignties in fields where common action has been agreed.

- the approach which the regulator will choose (the tool: *who* will actually write the rules); and
- the format which the regulatory framework will take (the product: *what* will the structure of rules look like).

This 'variable geometry' project will largely depend on the option(s) recommended by the EU and the Member States on one hand, and the stakeholders on the other hand. It is a project that will be the subject of a round of consultations and should normally be sanctioned by the publication of a policy on SOA.

Once the regulatory framework is clarified, the establishment of the associated legal framework (top-level rules, binding) and of the technical standards (detailed specifications, non-binding) could be done in sequence, as usual. If resources allow, both tasks could be started in parallel, in order to meet both lawmakers' and industry's respective demands for a clear legal framework, in order to avoid potential legal hurdles and be able to define roles and liabilities on the one hand, and on the other hand detailed technical specifications, in order for designers to develop vehicles according to rules as soon as possible.

Nevertheless, regardless of the entity that will be in charge of regulating and of the form adopted for the legal framework, the EASA system would have to be involved, as evidenced by the flowchart in Figure 10.1 and the following considerations:

- In respect of excluding sub-orbital flights from the scope of the Basic Regulation (i.e., by adding sub-orbital aircraft and flights to Article 1(2) or 4(4) and Annex II), the EASA would have to prepare and propose an Amendment to its Basic Regulation (EC) 216/2008 to specifically exclude these operations from its scope, and would still have to keep an indirect oversight on sub-orbital flights in order to make sure that they do not impact the safety of regular aviation operations (see the arrow from Phase I to Phase III in Figure 10.1).
- In all other cases, at a minimum, the EASA would be involved in drafting technical specifications. If it is tasked with establishing a comprehensive regulatory framework, the EASA would be involved in all phases. This is why, in Figure 10.1, all arrows point to Phase IIb, regardless of the outcomes of Phases I and IIa respectively.

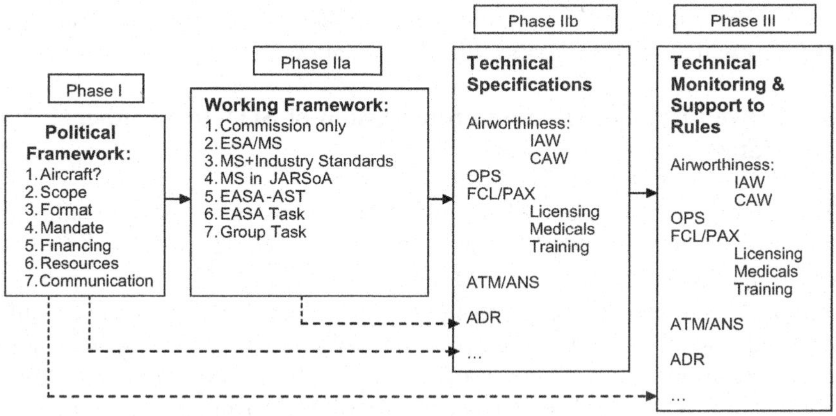

Figure 10.1 SOA rulemaking flowchart

In conclusion, the level of effort required of the EASA, and consequently the corresponding resources, would depend on the legal framework to be established.

5. PROPOSED OPTIONS

Following the publication of the SOA rulemaking activity in the EASA Rulemaking Programme, the initial phase consisting of consultation with most stakeholders and the European Commission started in 2012. As a result, a comprehensive set of options were proposed to the Commission for review and decision, based on economical, technical and political considerations. However, despite the declared need of the European suborbital industry for a regulatory framework – and as an exact contrary to the USA, where all is done by the government and the FAA to support and develop this nascent industry – because SOA test flights were only just starting in the USA, and because of the lack of political interest at both EU and national levels, all regulatory activities in this domain were frozen at that time and the task withdrawn from the EASA Rulemaking Programme.

Since then, all SOA programmes in continental Europe have been abandoned (e.g. Booster) or frozen (Astrium Spaceplane), and some companies such as Swiss Space Systems (S3) even went bankrupt. As stated above, the only remaining 'island' exception is the UK, which has ambitious projects supported by the government, the CAA–UK and the UK Space Agency, such as the Reaction Engines Ltd (REL) 'Skylon'

Orbital Aircraft or its 4-engines High Speed Point-to-Point Transport version, which represents the future of aviation.

A. Leave the Regulation and Supervision of SOA to Member States

This option would essentially maintain the status quo, which would not fulfil the request of most EU stakeholders, as it would result in diverging requirements throughout the Member States and distort competition without offering a level playing field. Furthermore, such approach would pose additional difficulties to potential third country operators, whose commercial operations from, to or within the EU would not be authorised without an ICAO-recognised certificate of airworthiness. Last but not least, in order to be implemented, this option would require SOA to be explicitly removed from the scope of the EASA Basic Regulation, which would require some rulemaking work and 'comitology', but would not remove the obligations on the EASA to monitor the potential impact of SOA operations on the safety of aviation.

B. Participation and Support for National Initiatives

Another possible option is to allow the EASA to technically and legally support national initiatives, such as the one jointly run by the UK-CAA and UK Space Agency in order to establish a plan for allowing SOA operations in the UK, which would currently see, inter alia, a commercial British spaceport be set up in 2018. Despite being interesting and allowing the EASA to participate in what is proposed by individual Member States in the short term, should this type of approach develop, it may result in multiple bilateral agreements, which may not necessarily be endorsed by other Member States, rendering any future harmonisation difficult, if not impossible.

C. Participation in International Groups

Being an extension of the second option to multilateral groups, this option may allow for defining a common framework between all participants, although reaching a consensus may be more complex due to the number of participants. Once a consensus is reached, this option has the advantage of being de facto voluntarily endorsed and implemented by all participants, as was the case prior to the EASA with the Joint Aviation Authorities (JAA), or more recently with unmanned aerial systems by establishing a Joint Authorities Rulemaking for Unmanned Systems

(JARUS). One could envisage establishing an equivalent of that organisation for SOA (i.e., JARSOA). However, establishing such a group and its prerogatives, and then reaching a consensus, might be a lengthy process, which bears the risk of not matching industry target dates, and whose results would still need to be legally endorsed at the national level.

D. EU Policy for SOA

If and when necessary, depending on developments and proposals by industry, the European Commission might consider issuing a 'policy' with the technical help of the EASA, that is a document offering guidance in order to accommodate sub-orbital aeroplanes and their operations in the EU, while waiting for the full regulatory framework to be developed.

Any new policy is developed through the rulemaking procedure of the Agency,[23] which means, among other things, public consultation on the content of the rules, through the Notice of Proposed Amendment (NPA) process. Thus, all stakeholders would have the chance to comment and ask for clarification whenever necessary.

However, despite providing necessary guidance and planning, this policy would not be sufficient for establishing a legal and regulatory framework in the EU.

E. Create a 'Light' Process Similar to FAA-AST Licensing

A possible approach would be to implement a new 'light' regime, similar to the one currently offered by the FAA-AST licensing process in the USA for reusable sub-orbital vehicles. However, since Europe is not a federation of States as in the USA, this would result in the EASA defining high-level essential requirements and leaving implementation to EU Member States, thus resulting in multiple interpretations. This approach would de facto prevent a harmonisation of implementing and technical requirements between Member States, making interoperability impossible. Moreover, the EU and EASA would not bear liability for national implementation of high-level general requirements.

---

23 EASA Management Board Decision 01-2012, available at http://www.easa.europa.eu/management-board/management-board.php.

F. Develop a Full Set of Rules for SOA under EASA Regulatory Framework and Processes

This option takes advantage of the existing aviation regulatory framework and processes. The development of rules would thus follow the classical path of first developing a preliminary Regulatory Impact Assessment (Pre-RIA), followed by Terms of References and the development of a Notice of Proposed Amendment (NPA) to the existing regulations, containing among others an Explanatory Note and a detailed Regulatory Impact Assessment (RIA). This process allows all stakeholders to comment on and amend if necessary the proposed rules, during the consultation period (typically three months). Answers to all comments are provided via the Comments Response Document (CRD). Then, Opinions to the Commission (for amending the Basic Regulation or for creating/amending implementing rules) and Agency Decisions (for non-binding technical specifications) are published. Opinions are further processed by the European Commission, whereas Decisions taken by the Agency can be directly utilised in all Member States. This process, although more comprehensive and lengthy (3 to 5 years in general), bears the advantage of setting regulations for all 28 Member States, ensuring full interoperability. However, compatibility with national space laws would still have to be addressed at national level, unless coordinated by a supranational entity.

G. Phased Approach (Options A to F)

This option has the benefit of progressively combining all the advantages of the above listed options, both in terms of timeline and contents. The process could start by establishing bilateral agreements with the Member States that have made the most advances in developing both equipment and legislation, and then establish a joint group (JARSOA) in order to reach a consensus, publish a policy based on the results of the discussions of the group, and then endorse its conclusions through a full regulatory process. Although more consensual and progressive, this phased approach may be much more lengthy and thus costly than the classical development of rules as is currently the case in the existing aviation rulemaking process. However, it bears the mutual benefit of allowing the industry and operators to develop their respective products and operations, and for authorities to accompany these developments by defining high-level requirements first and then detailing technical requirements along with the technical developments made in the meantime, before publishing a full set of rules.

H. Scope Extended to High Altitude/High Speed Transport (HST) and Orbital Aircraft (OA)

As requested by some EU stakeholders, one could envisage defining a regulatory framework that would embrace the full scope of potential future developments of sub-orbital flights from the outset, that is high altitude aircraft (HAA) and high-speed transport aircraft (HST), as well as orbital aircraft (OA). Although desirable, as it would cover the full spectrum of operations from the ground to orbit, this would necessitate harmonising aviation with orbital and space requirements from the start, what may be a too-ambitious endeavour. For the present time, SOA projects are just about to fly, and high-speed transport and orbital projects are announced only for the next decade, as they require development of more complex technologies, which may not yet exist, thus making the rulemaking exercise hazardous, if not void. What would be desirable and possible, though, is to develop SOA regulations first, keeping in mind future developments and allowing provisions for compatibility with future categories of aircraft.

6. PROPOSED REQUIREMENTS

A. Essential Requirements

Since sub-orbital aeroplanes are very similar to conventional aircraft in their design and operation for the non-rocket propelled and ballistic part of their flight, all essential requirements for 'normal' aircraft can be fully applied for the aerial phase of the flight. For the specific rocket-propelled and ballistic phases of the flight, however, and also due to the very high altitudes and/or speeds reached by SOA, new essential requirements may be introduced.

In particular, Regulations (EC) No. 216/2008 and (EU) 748/2012, are deemed fully applicable to the ground/air phase of the flight, at the exclusion of the rocket-powered and high altitude/ballistic phases of the flight.

In this respect, and in order not to reinvent the wheel, the EASA would be open to considering amongst other things the applicable FAA-AST 14.CFR.400 series requirements as essential requirements for the certification of SOA.

B. Technical Specifications

In order to benefit from both the best engineering practices and the experience developed during a century of aviation and half a century of manned spaceflight design and operations, it is quite obvious that technical specifications should not be established from scratch, but should rather be based on existing and proven safety specifications and practices.

It should be also noted that there is a current evolution of existing aviation technical requirements towards the definition of high-level, objective-based requirements on the one hand, and calling up existing or purposely developed technical specifications on the other hand.

One significant example of this current move is the proposed reformatting of the FAR/CS-23 into essential requirements to be listed in Part-23 and replacing CS-23/AMC/GM with the corresponding ASTM standards. This particular change may affect the referencing of the certification basis of some SOA, as the CS-23 was until now considered to be the reference to be complemented by special conditions. However, it will not affect the requirements as such, as the contents and the objectives of CS-23 will integrally be transferred into the new format.

C. Legal Implications When Entering Outer Space (Spaceworthiness)

While the EASA is competent in certifying sub-orbital aircraft, the jurisdiction of the EU (and all States) ends where outer space begins, wherever this limit lies. States have agreed on a different legal concept for activities carried out in outer space. As part of that concept, the Community of States has established a national responsibility for States for national activities in outer space (Art. VI Outer Space Treaty[24]). States have to authorise and supervise such activities. This responsibility is not transferred to the EU as part of its aviation competences. The EASA cannot deal with this (very short) outer space part of the sub-orbital flight, unless the States responsible agree and it is invested with the authority to act on their behalf.

This responsibility would indirectly include aspects of spaceworthiness as described above (in addition to airworthiness). The spaceworthiness of a sub-orbital aircraft is not directly mentioned in international treaties (or

---

[24] Treaty on Principles Governing the Activities of States in the Exploration and Use of Outer Space, Including the Moon and Other Celestial Bodies, 1967 (610 UNTS 205).

in Community law). However, States can be held internationally liable for damage caused by a space object.[25] Some States[26] have therefore included the approval of the spaceworthiness of a space object[27] within their national legislation on space activities,[28] requiring the operator to carry out the activity in a safe manner and without causing damage to persons or property. Such national legislation would apply in addition to any to be established by the EASA as the certification basis for the air flight part of the operation.

Unlike the approval of airworthiness under the EASA legal framework, only individual objects (not the type of product) are subject to this spaceworthiness approval.

As stated above, the approval of spaceworthiness is not subject to the competences of the European Union, including the EASA. Should a wider scope be required in the future, in order to bring such activities on the crossroads of air and space under one single jurisdiction, this would have to be decided by the (European) legislator. However, appropriate agreements could be established now between the EASA and the responsible State for activity in outer space (typically the State of the operator) in order to avoid double processes, which would be a burden for industry. The EASA is ready to explore this possibility with the appropriate administrations.

7. CONCLUSIONS

If we limit ourselves to the case of sub-orbital aircraft as defined in this chapter, the EASA has a regulatory framework and procedures for certifying them as aircraft. The challenge will be in adapting existing

---

[25] See Art. VII Outer Space Treaty; see also the Convention on International Liability for Damage Caused by Space Objects 1972 (961 UNTS 187).

[26] At the European level, namely: Norway (Act on Launching Objects from Norwegian Territory etc. into Outer Space 1969), Sweden (Act on Space Activities 1982, Decree on Space Activities 1982), UK (Outer Space Act 1986), Belgium (Law on the Activities of Launching, Flight Operations or Guidance of Space Objects 2005), the Netherlands (Rules Concerning Space Activities and the Establishment of a Registry of Space Objects – Space Activities Act 2006) and France (Bill Nr. 2008-518 relating to spatial operations 2008).

[27] A sub-orbital aeroplane or a launched payload thereof may be considered as a space object when entering outer space.

[28] These States have established such national space legislation in order to implement their international obligation to authorise and supervise non-governmental space activities according to Art. VI(2) of the Outer Space Treaty.

technical specifications and the development of special conditions necessary for covering this category of aircraft to ensure an appropriate level of safety.

With the same assumption, operational rules (EASA Part-OPS) and maintenance regulations (Regulation (EC) No. 2042/2003[29]) provide a basis for operations and maintenance, respectively, but would need exemptions to cover the case of sub-orbital aeroplanes. The challenge will be in the identification, development and agreement of such exemptions.

While waiting to develop a policy, and so to follow closely the novel design and techniques used by such aeroplanes, the EASA may also offer to applicants a cooperative research framework, in order to jointly prepare the terrain in the best way for their applications.

Also, in order to pave the way for future long-lasting rulemaking activities, especially concerning operations and licensing, potential applicants are encouraged to route their request for proposed rulemaking as soon as possible via their representatives in EASA consultative forums, such as the Aero Space Defence (ASD) group at the Safety Standards Consultation Committee (SSCC).

Due to the distribution of responsibilities between the different actors in the Community system as well as the distribution of responsibility for the air flight part and the spaceflight part of the operation, a close cooperation between the Agency, the Commission and the EU Member States and their respective national aviation authorities and space agencies would be necessary.

Last but not least, cooperation with ICAO, FAA and ESA is deemed essential, as their expertise and experience would help the Agency considerably in the first phase of its proposed approach, which is to adapt aviation requirements to the novelties introduced by sub-orbital aircraft. In the future, should the scope be extended to Orbital Aircraft, that is beyond the outer limits of the atmosphere, a more global cooperation with all parties would be necessary to explore the technical, operational and legal aspects of this fascinating and challenging endeavour.

---

[29] Commission Regulation (EC) No. 2042/2003 on the continuing airworthiness of aircraft and aeronautical products, parts and appliances, and on the approval of organisations and personnel involved in these tasks (OJ, 315 L 1, 28.11.2003).

APPENDIX: WHY ARE SUB-ORBITAL AIRCRAFT INDEED AIRCRAFT, AND NOT ROCKETS?

1. KEY FACTS

A. ICAO *definition*

'Any machine that can derive support in the atmosphere from the reactions of the air other than the reactions of the air against the earth surface' (ICAO Annexes 2, 6, 7 and 8).

Examples: aeroplanes (fixed wings), lifting bodies, rotorcraft (rotating wings), balloons.

Counterexamples: rockets (pure symmetrical bodies, no lift but thrust), hovercraft.

Corollary 1: 'If it has wings, it is an aircraft'.

Corollary 2: 'If it takes off from and/or lands on a runway, it is an aircraft.'

Corollary 3: 'It is not because you put a rocket engine on an aircraft that it suddenly becomes a rocket; it is still an aircraft'.

B. They *fly* in the atmosphere for 99 per cent of their flight

Even if one of the commercial objectives of SOA operators (besides flying scientific payloads) is to fly passengers above 100 km (to give them an 'astronaut' diploma), the part of the flight above 100 km is very marginal both in time and distance (limited to a few kilometres and seconds) and does not make a fundamental difference in terms of design or exposure.

Moreover, the purpose of sub-orbital flights is not always to reach 100 km (e.g., the XCOR Lynx Mark I's projected ceiling is 64 km; S3's initial project will reach 80 km). The FAA definition of a sub-orbital trajectory is 'having a vacuum instantaneous impact point on earth'; another is 'not orbital', that is, not reaching orbital speeds (and energy).

The limit of outer space is not legally set (the physical atmosphere actually goes up to 400 km) and the scope of ATM is not limited in altitude ('the sky is not the limit!').

It would neither be practical for the EASA and for the applicants, nor would it make sense in terms of engineering, to certify SOA and their operations below 100 km and ignore what happens above, or to license them only for the rocket-propelled or ballistic or whatever phase of their flight.

C. Technically, safety is not managed in the same way as the safety of a rocket is

Rockets follow a launch corridor and are blown out by flight termination devices if deviating too much from their trajectory. This would obviously not be accepted either by flight crews or by paying passengers.

Rockets are *launched through* the atmosphere, aircraft *fly in* the atmosphere. The launch corridors are defined in such a way as to minimise the impact of debris on uninvolved public or property, whereas SOAs have a glide ratio infinitely higher than rockets, thus their trajectory is less predictable and they potentially have a much wider footprint on the ground.

The intersection of the rather vertical rocket launch corridors with the airspace are usually defined as exclusive areas, preventing collisions with other aircraft. This could not be applicable to SOA, which need to be operated from commercial air transport airports for commercial purposes.

Rockets (including manned rockets) have launch statistics of catastrophic failures (loss of rocket) in the order of 1 in 100 flights. This would not be acceptable for paying passengers. Realistic safety objectives closer to aviation ones have to be set, both for the safety of passengers and also for protecting this newly emerging business (*Hindenburg* syndrome).

D. They are the first necessary step for the development and certification at lower risks and costs of high speed/high altitude point-to-point transport aircraft (HST) as well as orbital aircraft (OA)

Thus, it would be more efficient to treat them as aircraft from the beginning (what they are anyway, both from a legal and technical perspective), in order to build on the same regulatory basis and not have to reinvent a regulatory framework and technical requirements when HST and OA projects will only appear from 2020 onwards (e.g., EADS-Astrium Zero Emission High Speed Transport (ZEHST), Reaction Engines Ltd (REL) Skylon and Lapcat, etc.).

E. Otherwise, they would be outside of EU *Common* Competences

As per 216/2008, the EASA shall regulate all aircraft in the EU, whatever form they have. The only exclusions are specified in Basic Regulation Articles 1, 4 and Annex II: governmental, rescue, police and research aircraft. As soon as the operators intend to perform commercial operations (i.e., transporting paying passengers or payloads), aviation rules would apply.

Should a sub-orbital aircraft developer or operator apply for certification from the EASA at the time of writing, the Agency would have to handle it under the current aviation framework by using existing requirements (e.g., CS-23, CS-25, CS-E and EASA-OPS) and adapt them to the specifics introduced by SOA (e.g., very high altitude, meaning, exposure to vacuum and radiation, rocket propulsion and control systems, single pilot operation, etc.) by means of Special Conditions, which would then become public, ensuring an equal treatment for all other stakeholders.

Conversely, there is no such space safety regulatory agency in Europe as NASA in the USA, as each Member State is responsible for rocket launches from its territory. The EASA is not an EU body and it is also not a regulatory entity.

If each EU Member State has to develop its own regulations on sub-orbital flights, not only will this multiply efforts necessarily, it may be difficult or even impossible to harmonise these afterwards. Stakeholders want to be able to operate from all of Europe and the USA, and if possible to operate point-to-point flights.

As a result, SOA would not be able to fly in other EU Member States (unless bilateral agreements are established) and would be confronted with a pre-JAA environment (no harmonisation, approvals in each Member State, etc.), which would be a severe commercial hurdle for stakeholders. Moreover, in the event of accidental overflight of or landing/crash in another Member State, such a regime would also lead to irresolvable legal complications, and most probably to the end of SOA activity.

F. Most EU stakeholders want to be certified by the EASA

On top of not jeopardising the reputation of the stakeholders and of this whole new emerging business, EASA certification gives more confidence to investors and passengers, leading to businesses being better financed and selling more tickets, and reducing the risk of accidents and thus insurance premiums. Therefore, despite the limited additional regulatory and financial burden (which should be limited to the minimum necessary to ensure an acceptable level of safety, as in aviation), EASA certification globally lowers the risks and costs and in the end makes the business more sustainable and more profitable.

G. Conclusions

For engineering, safety, legal and commercial reasons, SOA should be handled as aircraft and not as rockets. There is no harmonised regulatory framework and experience for rockets similar to the one for aviation, nor is there an EU regulatory entity responsible for space safety in Europe. In terms of time and resources, it would not be very efficient to leave it to each EU Member State to regulate SOA on a national basis as 'rockets', since each national regulation may be difficult to harmonise with those of other Member States. It could also be detrimental to the global safety (and environmental protection) of EU citizens (passengers, but mainly uninvolved public), and also to this emerging business, both for EU developers, manufacturers and operators, who want to be able to operate from anywhere in Europe, as well as for operators who want to operate US-developed SOAs from Europe, thus generating a non-negligible commercial spinoff and employment.

2. BACKGROUND INFORMATION

A. Aircraft

ICAO Definitions in Annex 2 (Rules of the Air), Annex 6 (Ops), Annex 7 (Aircraft Nationality and Registration Marks) and Annex 8 (Airworthiness)

Aircraft: 'any machine that can derive support in the atmosphere from the reactions of the air other than the reactions of the air against the earth surface.'

Examples: aeroplanes (fixed wings), lifting bodies, rotorcraft (rotating wings), balloons.

Aeroplane: 'A power-driven heavier-than-air aircraft, deriving its lift in flight chiefly from aerodynamic reactions on surfaces which remain fixed under given conditions of flight.'

Counter examples: rockets, hovercraft, ekranoplans.

ICAO Council 175 Session, 30 May 2005, C-WP/12436 Working Paper 'Concept of Sub-orbital Flights'

'Should sub-orbital vehicles be considered (primarily) as aircraft, when engaged in international air navigation, consequences would follow under the Chicago Convention, mainly in terms of registration, airworthiness certification, pilot licensing and operational requirements (unless they are otherwise classified as State aircraft under Article 3 of the Convention).'

B. Space Object

1. Neither the Convention on International Liability for Damage Caused by Space Objects (29 March 1972, hereinafter the Liability Convention) nor the Convention on Registration of Objects Launched into Outer Space (14 January 1975, hereinafter the Registration Convention) offers a definition of 'space object', but they stipulate that a space object includes its component parts as well as the launch vehicle and parts thereof (Art. I(d) and Art. I(b), respectively).
2. It is specified in the Registration Convention that a space object is to be registered by the launching State, and that the Secretary-General of the United Nations will be duly informed of this with a view to keeping an international register where pertinent information can be recorded (Art. II and Art. III refer). Beyond registration, international space law does not regulate the requirements for the certification of space objects and the licensing of their personnel (even if astronauts are addressed in the Agreement on the Rescue of Astronauts, the Return of Astronauts, and the Return of Objects Launched into Outer Space (22 April 1968)). However, the Treaty on Principles Governing Activities of States in the Exploration and Use of Outer Space Including the Moon and Other Celestial Bodies (27 January 1967, hereinafter the Outer Space Treaty) stipulates, inter alia, in Article VI that '[t]he activities of non-governmental entities in outer space ... shall require authorization and continuing supervision by the appropriate State Party to the Treaty'.
3. SpaceShipOne and Two have been granted a launch licence by the FAA Office of Commercial Space Transportation (FAA-AST) as a reusable launch vehicle (RLV), and classified as a sub-orbital rocket. It nevertheless appears from the UN Register of Space Objects that SpaceShipOne has not been registered as a space object. One of the reasons for this might be that the Registration

Convention applies only to space objects 'launched into earth orbit or beyond' (Art. II), that is, not to objects performing sub-orbital flights per se.

C. Airspace vs Outer Space

1. The United Nations Committee on the Peaceful Uses of Outer Space (UNCOPUOS), which is the UN forum where technical and legal aspects of space activities with global impact are considered, has discussed the issue of the definition and delimitation of outer space from 1962 onward and no definite conclusion has been reached so far.
2. From 1962 onwards, pilots flying the X-15 first sub-orbital aircraft were awarded 'Astronauts Wings' if they had flown over 50 miles (80 km), which is the limit set arbitrarily by the International Aeronautics Federation (FAI) (e.g., a few X-15 pilots became astronauts many years before by sitting in a capsule on top of a rocket).
3. However, the 100 km (62.5 miles) limit known as the Kármán line, beyond which aircraft are no longer capable of generating lift, is commonly accepted as the limit between the airspace and outer space, although the atmosphere goes up to 400 km and beyond.
4. As debated for decades in the framework of UNCOPUOS, it may be questioned whether the vertical limit of airspace would be critical in determining the scope of applicability of air law as opposed to international space law conventions (*spatialist approach*), or whether the type of activities at issue would determine which law should apply (*functionalist approach*). The latter school of thought submits that flights that would be passing merely in transit through (sub)orbital space in the course of an Earth-to-Earth transportation would remain subject to air law.
5. International Navigation
   a. Part I of the Chicago Convention addresses aircraft engaged in international air navigation and Part III deals with international air transport. In this latter respect, Article 96(b) in Part IV defines 'international air service' as 'an air service which passes through the air space over the territory of more than one State'. These elements define the scope of the mandate and competence given to the ICAO.
   b. While certain commercial vehicles might in the future fly from one State to another and transit through sub-orbital

space, current endeavours aim at carrying passengers from and to the same location. However, ascending and descending phases might involve in certain cases the crossing of more than one national airspace. Sub-orbital vehicles considered as civil aircraft crossing foreign airspaces could then be treated as engaging in international air navigation.

c. As regards any applicability of international air law to sub-orbital flights, pertinent Annexes to the Chicago Convention contain associated communication, navigation, surveillance, licensing, operation and airworthiness issues, among others, that would be amenable to their regulation. However, ICAO Annexes currently lack technical requirements in this area. Should it be determined that such flights should be governed by international air law, Assembly Resolution A35-14, Appendix G nevertheless acknowledges that, for certain categories of aircraft or classes of airmen, it may be many years before Standards and Recommended Practices (SARPs) come into force, or that it may be found most practicable not to adopt SARPs. Accordingly, Resolving Clause 2 stipulates that 'certificates and licences issued or rendered valid, under national regulations, by the Contracting State in which the aircraft is registered shall be recognized by the other Contracting States for the purpose of flight over their territories, including landings and take-offs.'

11. The role of UNCOPUOS in the international regulation of non-governmental space activities

Jean-François Mayence

1. THE ROLE OF UNCOPUOS IN THE REGULATION OF SPACE ACTIVITIES FROM THE HISTORICAL PERSPECTIVE

Speaking of the United Nations Committee on the Peaceful Uses of Outer Space (UNCOPUOS) as an actor in the regulation of the space sector may be misleading: UNCOPUOS is no regulating body. It takes part in the international governance of space activities by issuing recommendations, guidelines or technical standards that serve as a common reference for spacefaring nations on a voluntary basis.

The role of UNCOPUOS as a subsidiary body of the United Nations General Assembly has mainly been dedicated, since its establishment in 1958, to the exchange of information among its 18 Member States. At the time of its transformation into a permanent body in 1959, the Committee had 24 Member States. Its terms of reference were focused on reviewing and fostering international cooperation in the area of peaceful uses of outer space, including exchange and dissemination of information and support for national space programmes, as well as studying legal problems related to the exploration of outer space.

This language did not provide for a regulatory mission or authority entrusted to UNCOPUOS, in the sense that it did not vest any authority in the United Nations to manage governmental or non-governmental activities on the basis of treaty provisions. This is no real surprise considering that organizations born between the end of the Second World War and the International Geophysical Year (1957/1958) were not meant to exercise regulating authority, but rather to foster international cooperation in the context of reconciliation and confidence-building among nations.

In that respect, the International Civil Aviation Organization (ICAO) appears to be an exception; according to the provisions of the 1948 Chicago Convention,[1] the organization was designed to establish international technical standards and legal norms in the field of civil aviation and could have served as a model in the later development of international space policy. But this has not been the case. So far, despite the complex issues raised by their development, orbital operations have remained subject to a cautious cooperation between States and to their sovereign regulatory power.

The evanescent character of UNCOPUOS in its early years was enhanced by the fact that, until 1968, it had no ad hoc supporting office. That year, a dedicated service of the United Nations Secretary General, the United Nations Office for Outer Space Affairs, was established. That service helped the development of UNCOPUOS in providing ad hoc expertise in support of UNCOPUOS's work and operational activities in implementing capacity-building programmes.

With the establishment of its two Sub-Committees, particularly its Legal Sub-Committee in 1962, the active involvement of UNCOPUOS in the making of international space law was strengthened. Nevertheless, the mandate of the Legal Sub-Committee was limited to providing expertise, as well as the ad hoc forum, for elaborating treaties and resolutions relating to the exploration and utilization of outer space. Contrary to other fields of activity, sometimes adjacent to space conquest (like aviation), the private sector was left completely out of the loop when reflection upon the principles of international space law occurred.[2] Considering that the International Air Transport Association (IATA), the trade association of world airlines, had existed since 1945 and had had the chance to provide input in the creation of the ICAO and the Chicago Convention system – or that the International Telecommunication Union (ITU) included the private sector in its work through Sector Members and Associates from industry – nothing prevented UNCOPUOS from also establishing a mechanism allowing private actors to consult, either on an ad hoc basic or in a more permanent manner. This does not mean that industries and entrepreneurs had nothing to say in the adoption of

---

1 Convention on International Civil Aviation, 7 December 1944.

2 The exclusion of the private sector did not concern the scientific community as much as it did the industrial community: through the recognition of COSPAR (the Committee on Space Research), established in 1958, scientists were associated with the development of the United Nations' capacity for supporting international space cooperation. See H.L. Van Traa-Engleman, *Commercial Utilization of Outer Space* (Martinus Nijhoff 1993), p. 8.

international norms applicable to their sector, but that their concerns were voiced through the positions adopted by Member States in international bodies, leaving their interest in the hands of more or less powerful lobbies active in the various spacefaring nations.

However, the long debate between the superpowers in the drafting of the 1967 UN Outer Space Treaty, in particular its Article VI, reflected the relative difficulties in arriving at a consensus on the extent to which private enterprises can contribute to space exploration. The concept of 'non-governmental entity' was vague enough to cover all kinds of actors, from scientific institutions to commercial companies.[3] Their place in the legal framework established at that time made them subject to States' omnipotence, setting up a logical connection not always exempt from flaws. For instance, the assimilation of private entities into a State for the purpose of implementing the treaties left open the question of whether the former were bound by provisions such as compliance with international law, including the UN Charter, or whether their activities were susceptible to engaging the liability of their home State.[4]

In recent years the mutation of UNCOPUOS has been driven by the growing strategic character of outer space. From the cooperation between leading space nations and with emerging spacefaring countries, the debate has moved to a more sensitive issue: the rivalry over space resources and their use for national strategic purposes. Being a member of UNCOPUOS has become an objective for States that have not necessarily developed critical technology to access outer space, or have not even considered ratifying the United Nations space treaties. The same goes for observers within a multitude of non-governmental organizations (NGOs) whose applications are sponsored by Member States, sometimes without real representative authority or even direct relation to the space business. One might legitimately wonder whether this trend consists of a phenomenon of direct lobbying to UNCOPUOS or a deeper evolution of the consideration of outer space as an area of global public interest.

---

[3] On this subject, see M. Gerhard, 'Commentary of Article VI', in S. Hobe, B. Schmidt-Tedd and K.-U. Schrogl (eds), *Cologne Commentary on Space Law (Volume 1)*, (Carl Heymann Verlag 2009); M. Couston, *Droit spatial économique* (Sides 1994).

[4] See J.F. Mayence, 'The Concept of State Responsibility for Private Space Activities', in S. Hobe and S. Freeland (eds), *In Heaven as on Earth? The Interaction of Public International Law on the Legal Regulation of Outer Space* (Cologne Institute of Air and Space Law 2013).

2. THE ROLE OF UNCOPUOS IN THE REGULATION OF SPACE ACTIVITIES: THE METHODOLOGICAL PERSPECTIVE

Some States have expressed concerns about the increasing diversity of interests in attendance at the Committee sessions. While it is undeniably suitable to maintain a close link between UNCOPUOS's work and the civil space society, including space technology and applications users, the question remains whether or not a set of criteria for granting membership of or observer status within the Committee should be adopted. Actually, as far as NGOs are concerned, the only condition observers must satisfy is to be granted consultative status with the United Nations Economic and Social Council (ECOSOC).[5] NGOs are generally introduced as organizations pursuing a non-profit goal. It seems to be a tacit condition for becoming observers at UNCOPUOS. As far as commercial organizations, or by extension international associations representing commercial interests, are concerned, it is not certain that their application as observers would be welcomed, even nowadays. The question has remained rather theoretical so far, probably due to the fact that no real assessment of the impact of UNCOPUOS's decisions or work on the commercial space sector has been performed. Whoever discusses UNCOPUOS with the space industry or with space services providers quickly realizes that the Committee and its work have remained an abstraction to them. The unfamiliarity is obvious and, to be clear, mutual. As with the United Nations outer space treaties themselves, the Committee is considered a product of the Cold War, and not in phase with the development of global economic governance. On the other hand, the Committee itself has not shown any proactive interest in incorporating commercial actors in its work. The only recent effort in that direction was the setting up of ad hoc expert groups in the framework of the Working Group of the Scientific and Technical Sub-Committee dedicated to the Long-Term Sustainability of Space Activities. Even in that case, direct participation of the commercial space sector as such was not allowed, but only a representation of their interests through the designation of national experts appointed by Member States was allowed.

The involvement, or at least the consultation, of stakeholders and field actors in the making of intergovernmental regulation is a recurring question for political scientists and lawyers. Substantial literature exists

---

[5] A list of NGOs with consultative status with ECOSOC can be found in document E/2013/INF/6 on the website csonet.org.

on the subject, although studies have often been performed with regard to a particular area of activity.[6] A comprehensive review of the participation of the private sector in the international lawmaking process is still needed to some extent in order to demonstrate whether the involvement of non-governmental entities (and by 'non-governmental' we certainly do not limit ourselves to NGOs) has become, through the time, a *good governance principle.*[7]

Our point is not to discuss whether the effective and direct participation of so-called 'non-State actors'[8] in the intergovernmental regulation-making process is or is not a condition of legitimacy. We leave this discussion to political sciences, although we tend to believe that 'legitimacy' is a fluctuating concept. Even in democratic systems it is not established that the absence of participation of non-State actors in the decision-making process at the international level constitutes per se a breach of legitimacy. The contrary might be true in a monist system, where international law and national law form one single order. But in a dualist system, democratic legitimacy may be achieved according to domestic constitutional law, providing for governmental powers to represent the State and for parliament to assent to or control the government's actions as a condition of the citizens agreeing to be bound by the

---

[6] See notably W. Mattli and T. Buthe, 'Setting International Standards: Technological Rationality or Primacy of Power?', *World Politics* (2003) 56, 1–42; R.B. Mitchell, 'Regime Design Matters: Intentional Oil Pollution and Treaty Compliance', *International Organization* (Summer 1994) 48.3, 425–458; K. Raustiala and D.G. Victor, 'The Regime Complex for Plant Genetic Resources', *International Organization* (Spring 2004) 58, 277–309; N. Hachez and J. Wouters, 'A Glimpse at the Democratic Legitimacy of Private Standards. Assessing the Public Accountability of GLOBALG.A.P.', *Journal of International Economic Law* (2011) 14, 677–710.

[7] See notably the comprehensive study of L. Mosley, 'Private Governance for the Public Good? Exploring Private Sector Participation in Global Financial Regulation', February 2005, available at https://www.princeton.edu/~hmilner/Conference_files/KEOHANE/mosley.pdf, accessed 7 July 2016. This study, although addressing the particular case of global finance regulation, provides a general assessment of how non-State actors are considered in intergovernmental decision policy.

[8] We propose to use this term to designate all kinds of stakeholders and actors, public or private, who carry on operational activities subject to the considered regulation in a dedicated economic area (sectorial and/or geographical), without being formally involved in the decisional process at the governmental level.

regulation.[9] For this reason, we prefer to analyse the participation of non-State actors in the regulation-making process from the legal perspective, and notably the principles of sound governance, as well as with regard to the effectiveness of such regulations.

Major authors[10] have assessed the possibility of more active and systematic participation by non-State actors in the intergovernmental regulation-making process concerning their respective sectors. This is particularly the case for activities dealing with the economic exploitation of *global public goods*, namely resources that can be used by anyone without affecting their availability in substance to others (*non-rivalry*) or the capacity for others to access them (*non-excludability*).[11] Outer space resources qualify, to a large extent, as global public goods, although what sounded true 50 years ago may not do so nowadays: the geostationary orbit, for instance, has been qualified by the ITU as a limited natural resource, while the controversial status of celestial bodies' natural resources as common heritage of mankind is ignored by very recent national laws planning future extra-terrestrial mining activities. But outer space as such, through the previously mentioned topic of the long-term sustainability of space activities, may be the best example of a global public good, with concerns such as space debris illustrating what some are facetiously referring to as a 'global public bad'.

---

9 See, e.g., C. Ryngaert, 'Imposing International Duties on Non-State Actors and the Legitimacy of International Law', Working Paper 42, KU Leuven – Centre for Global Governance Studies, March 2010. On this point, we beg to differ with the author as he considers that: 'in order to ground the binding character of international law for non-State actors, one thus has to prove that either the process of creation of the international legal norm is just, which implies that the non-State actor has at least to some extent participated in the creation of the norm (democratic participation/participatory governance), or that the legal norm or its implementation has in itself an important substantive value'.

10 See J.S. Nye and R.O. Keohane, 'Transnational Relations and World Politics: An Introduction', *International Organizations* (Summer 1971) 25.3, 329–349.

11 See S. Cogolati, L. Hamid and N. Vanstappen, 'Global Public Goods and Democracy: What Role for International Law?', Working Paper 159, KU Leuven – Centre for Global Governance Studies, May 2015; J. Wouters and B. De Meester, 'The Role of International Law in Protecting Public Goods: Regional and Global Challenges', Working Paper 1, KU Leuven – Interdisciplinary Research Group on International Agreements and Development, December 2003; R. Wessel and J. Wouters, 'The Phenomenon of Multilevel Regulation: Interactions between Global, EU and National Regulatory Spheres towards a Research Agenda', Working Paper 8, KU Leuven – Centre for Global Governance Studies, January 2008.

The international institutional management scheme(s) of global public goods has been reviewed by some authors according to a model and a categorization depicting their internal balance of powers. Borrowing from Walzer,[12] Drezner[13] identifies three types of intergovernmental schemes: *clubs*, *neighbourhoods* and *universes*. Clubs are organizations[14] with limited membership, driven by a like-minded philosophy. A symptomatic example is the Group of 7 (aka G7) gathering of the seven Western nations considered to be the most powerful (economically and politically). Neighbourhoods are constituted on the premise of regional cooperation (e.g., the European Union), while universes are organizations open to all countries (provided they abide by common basic principles). This distinction, although using a terminology that rather belongs to video games, is relevant for UNCOPUOS, particularly with regard to its historical evolution. The Committee, established at the time where only a couple of nations had the capacity to explore outer space, was constituted by the two cold war superpowers – the USA and USSR – plus a number of founding members, mainly representatives of the 'two blocks'.[15] Progressively, the composition of UNCOPUOS has moved towards a much broader sample of States (see above), until it developed a practice of accepting any candidate member or observer on the sole basis of objective conditions. By slightly caricaturing, we could say, that, from a club, UNCOPUOS has apparently turned into a universe, on the assumption that outer space, as a global public good, should be managed by the entire world community. We say 'apparently' only because the current trend within UNCOPUOS reminds us that the original club has never ceased to exist. As soon as space issues become strategic, the real decisions are not made within the Committee, or even at multilateral level.

---

12 M. Walzer, *Spheres of Justice: A Defense of Pluralism and Equity* (Basic Books 1983).

13 D. Drezner, 'Who Rules? The Regulation of Globalization', August 2002, available at http://danieldrezner.com/research/whorules.pdf, accessed 7 July 2016. See, notably, his assessment of the US policy with regard to the role of ITU in the management of Internet domains assignation (p. 23).

14 By 'organizations' we do not necessarily mean institutional organizations established as (international) legal entities, but rather any kind of intergovernmental cooperative mechanisms.

15 Albania, Argentina, Australia, Austria, Belgium, Brazil, Bulgaria, Canada, Czechoslovakia, France, Hungary, India, Iran, Italy, Japan, Lebanon, Mexico, Poland, Romania, Sweden, the Union of Soviet Socialist Republics, the United Arab Republic, the United Kingdom of Great Britain and Northern Ireland and the United States of America.

Apart from very specific and limited provisions here and there,[16] international treaties generally do not feature a systematic role for non-State actors in the making of their implementing regulation. Therefore, it seems hazardous to establish a general principle of law in that regard. We must also recognize that, even in their domestic legal order, States are reluctant to establish a general right for non-State actors to be included in the regulation-making process. Consultation from the targeted sector is increasingly included as an element in the procedure of decision-making by the government, but it remains on a case-by-case basis, without clear and precise directives on how and whom to consult.

At the international level, this case-by-case approach seems even more random and improvised. Some sectors do not have international representative associations able to voice the interests and expectations of their members from various countries or regions. The rigidity of intergovernmental bodies often impedes participation of non-State actors or confines them to the role of passive observers. Those who oppose direct involvement of non-State actors in the intergovernmental process decry the non-representative character of non-State actors and argue that those actors should state their claim through their respective national governments. Therefore, there is no need for them to reiterate their concerns at the international level.

This last argument must be opposed by several considerations:

(a) Participation of non-State actors means that they are 'represented by themselves'. Governments, although they might take into account concerns or interests expressed by their citizens, are expected to state their own position and nothing more;
(b) The transnational representation of a dedicated sector of activity allows participants to develop common positions and to affirm them with more weight than through national governments. This prevents a certain dilutive effect that may occur because of elements 'lost in translation';[17]

---

[16] See, notably, the example of the Aarhus Convention mentioned by Cogolati, Hamid and Vanstappen, *supra* note 10, at 17–18, which demonstrates the need to intelligently combine top-down and bottom-up approaches in modern decision-making processes on public outcomes.

[17] This is due either to the fact that the same sector has more or less impact on political decisions from one country to another, or to the fact that national stakeholders might realize that some of their concerns are shared by counterparts in other countries, or even to the language used to express the concerns by sometimes non-technical experts.

(c) Most of the time intergovernmental work is organized in successive phases. A first technical phase involves governmental experts and is often dedicated to the primary drafting of the regulatory text. The draft may be coming and going between working groups and plenary sessions. Achieving consensus among the States involved quite often necessitates long negotiations and drafting efforts until the text is 'shaped' in a more political form, sometimes modifying its original content. In this context, consultation with non-State actors through national channels – if deemed feasible – could not guarantee faithful reporting of their respective positions. The reality of the intergovernmental decision-making process requires non-State actors to be directly involved in negotiations in order that governments benefit from their expertise. This should not prevent governments from making the final decision on the text of the regulations.

As previously mentioned, we would now like to consider the benefits of the participation of non-State actors in the regulation-making process from the point of view of the effectiveness of a regulation. In the case of space activities in particular, industry and operators play a fundamental role in establishing standards and safety/quality procedures. Self-regulation has been the starting point for most governmental initiatives in the field of nuclear power use or space traffic management. The same phenomenon can be observed right now in the sub-orbital flight sector, where operators are expected to develop the generic rules applicable to commercial activities they intend to carry on. It might seem obvious, but regulation is better applied and complied with when it has been conceived in accordance with technical facts. And in the space business, technical facts belong to the industry. Experience shows that the space industry is willing to abide by international regulations as long as such regulations are based on relevant technical facts, as well as on a reasonable assessment of their impact on technical and financial feasibility. International regulation may allow a smoother interface between national standards and thereby facilitate better globalization of the sector and more effective international cooperation or competition. Involving non-State actors may be seen as a way to alleviate bureaucracy, avoiding 'regulation for the sake of regulation', as well as a way to make private actors responsible for the implementation of the norms. This requires not only the necessary decision scheme at intergovernmental level, but also a sufficient level of *mutual confidence* between States and their economic communities, and their respective representatives.

This mutual confidence might be the missing element at UNCOPUOS, and there may be two reasons for this:

(a) First, the heavy and rather inflexible working procedures of UNCOPUOS. It seems that the Committee is lacking forums and subordinate bodies dedicated to informal consultations. Any working group or sub-level expert groups have to be approved by the Member States by consensus. Some States have shown a clear and constant reluctance to allow the participation of commercial stakeholders in their own capacity. At the beginning of the 2000s, however, a special solution was adopted to handle the item of the draft UNIDROIT Protocol for international guarantees on space assets.[18] An informal consultation mechanism was set up and opened to Member States on a voluntary basis to deal with some sensitive issues related to financial institutions and their connection with the United Nations outer space treaties. A few meetings were held in 2000 and 2001 and allowed some progress on those issues, but they also highlighted the difficulty of obtaining recognition and acceptance of the results of those informal consultations by the Committee. Held '*in the margin of*' the formal sessions, the findings and conclusions reached by the participating States were not considered (sometimes by the very same States) as '*opposable*' to the Committee.

    The attitude of the Committee toward the commercial space sector is regrettable to the extent that it prevents synergies between political governance and field actors. Prior consultations with and feedback from non-State actors would certainly be two elements enriching the work of UNCOPUOS. That being said, it is not contested that consultation with the commercial sector should not turn into hard lobbying in favour of particular interests. It is also obvious that States should remain in charge of the final decision. But the principle of including operational stakeholders in the governance process is today widely accepted. It is a shame that UNCOPUOS's practice goes in the opposite direction.

(b) Second, the space sector itself has not always developed clever and adaptive strategies defining its own role in the framework of the United Nations. Private enterprises still lack the ability to speak the

---

[18] Protocol to the UNIDROIT Convention on International Interests in Mobile Equipment on Matters Specific to Space Assets, done in Berlin, 9 March 2012.

UN language and the imagination to conceive their functional contribution to the achievement of humanist purposes, such as the benefit of outer space to mankind. Actually, nothing in the United Nations outer space treaties prohibits commercial activities in outer space. The important role reserved for such activities in allowing a global sharing of the welfare derived from space technology and applications is left wide open. For instance, the 1979 United Nations Moon Agreement envisages the exploitation of natural resources in a framework of principles that is not meant to impede private enterprise. However, industry has developed an a priori opposition to that Agreement, allegedly for the reason that it would nip in the bud any business opportunity before it can even arise. This illustrates the poor imagination from the commercial sector in proposing solutions that would allow a consolidated business plan as a means to implement the requirements under the concept of Common Heritage of Mankind.

Anyone with even limited experience in the space business is fully aware of the role of the industry and the service providers in the definition, the development and the implementation of space programmes and, by extension, of the principles of international space law. It is up to the commercial space sector to demonstrate its responsibility to do business in compliance with the international rules and, thereby, become a player in the field of international governance of space.

3. THE ROLE OF UNCOPUOS IN THE REGULATION OF SPACE ACTIVITIES: CONCLUSION AND PROPOSALS FOR A WAY FORWARD

The ability of UNCOPUOS to respond to critical issues in the utilization of space resources is the main condition of its sustainability (to the extent that the sustainability of an institution is considered a goal in itself). Space debris, orbital traffic management, small satellites policy, access to satellite data, sub-orbital flights, asteroid mining[19] – all of these are

[19] On that latter subject, UNCOPUOS's reaction, in the coming years, to the national initiatives setting legal frameworks for mineral resource exploitation on celestial bodies and asteroids (i.e., Luxembourg governmental statement – see http://www.spaceresources.lu (accessed 8 August 2017) – and the US Asteroid Act of 2015 – see Chapter 51 US Code as amended 25 November 2015) will provide a significant indication of this international body's relevance for the

topics on which a consensual solution by States is expected from the Committee. Other bodies and agencies have already started to occupy the field: ICAO (with sub-orbital flights), UNIDROIT (with financial mechanisms), ITU (with orbital resources management), or recently the Permanent Court of Arbitration with its rules for space dispute settlement. If UNCOPUOS wants to remain *the* international body in charge of space governance, it must expand its sphere of influence to all actors and stakeholders, that is to the commercial space sector. On the other hand, private entrepreneurs must understand that the legal regime of the United Nations outer space treaties will not evolve dramatically in the coming years. The reason for that is not so much the attachment of States to the old-fashioned principles inherited from the Cold War, but rather the fear that, considering the growing rivalry resulting from the status of space resources as a core element of global economic and strategic dominance today, no consensus could be reached on alternative and updated provisions governing space activities. Besides, it is not the values of the United Nations outer space treaties that should not be questioned, but rather the language by which those values are expressed and the respective role of States, international organizations and private enterprises in making them a reality.

With the development of national space laws, governments have seemed to regain control over the regulation of the space sector. Those laws often go beyond the implementation of the United Nations outer space treaties: they feature elements of policies and regulations in connection with other national priorities. This is where the private space sector may play an active role in consulting the authorities and as those actors primarily concerned in any regulation. The perimeter of the law will then be defined not only by reference to the instruments resulting from UNCOPUOS's work, but also by the influence of the views expressed by those who will have to comply with the law's provisions.

Reconciliation of politics and society is not only a national issue, but also a challenging task for international lawyers and policy makers.

---

application of the United Nations outer space treaties. The interpretation of Article II of the 1967 UN Outer Space Treaty made by certain States may constitute a challenge to the international space community and, although UNCOPUOS is not the competent body for dealing with authoritative interpretation of any treaty, its capacity to maintain a cooperative debate on their status and application must be considered as an indicator of its credibility.

12. National and international regulatory aspects of commercial space activities: self-regulation as the way forward?

Katrin Nyman-Metcalf

1. INTRODUCTION

Space law went from nothing to a rather sophisticated legal regime in just a few years: when the first satellite was launched into orbit (the Soviet Sputnik, 4 October 1957) there was basically no space law. About a decade later there were several international conventions, resolutions from the United Nations (UN), agreements, protocols of different organizations and a vast amount of academic and other literature. After this flying start, developments slowed down or at least changed. There have been various instruments adopted in more recent times, but the pace has been slower and instead of generally applicable conventions, there are mainly special agreements, contracts or memoranda of understanding targeted at specific issues and between concerned parties only.[1] With space use growing constantly, the need for a legal regime has not diminished, but rather new issues have been added and new players have entered the field. Regulation needs to fit a technologically complex and international area of activity, not falling under the jurisdiction of any one State but to be used for the benefit of mankind, in which technological developments can be very fast. Another area that to some extent fits this description is cyberspace (the Internet), for which self-regulation has become an important means of governance while still maintaining the

[1] E. Walter, 'The privatisation and commercialisation of outer space' (Chapter 4.1) in C. Brünner and A. Soucek (eds) *Outer Space in Society, Politics and Law* (Springer 2011) 493–518, at 502.

maximum freedom of action. Consequently, it is interesting to examine whether self-regulation may have something to offer for outer space as well.

The basic principles of space law are well known: outer space is the province of mankind, it cannot be appropriated and it may be used by all countries. These principles were set out in the 1963 Space Declaration[2] and repeated in the 1967 Outer Space Treaty.[3] It may be discussed to what extent such principles have become customary law.[4] The International Court of Justice (ICJ), in the *North Sea Continental Shelf* case,[5] has stated that even if the number of ratifications of a convention is not high, non-ratification may be due to reasons other than active disapproval of the content. Matters can become customary law quickly, as stated in Judge Sorensen's dissenting opinion in the *North Sea Continental Shelf* case,[6] where he mentioned the extremely dynamic process of evolution in modern times. The combined process of codification and progressive development of international law may be the best way to achieve new international law.[7] Indeed, the expression 'instant customary law' originates from space law[8] and although not universally accepted, has gained wide recognition.

The discussion on the extent and manner in which space law principles have become customary law does not, however, change the fact that the principles are known and accepted by the States active in space. Most such States have also ratified the 1967 Outer Space Treaty, the main legally binding space law document. What is also well known is that when it comes to more detailed rules on space use, there is an absence of the kind of common understanding that may exist for basic principles. Legal development has moved from a more inclusive process under the auspices of the UN to rule-making through agreement by concerned parties. The most famous example is in connection with the International

---

2 Declaration of Legal Principles Governing the Activities of States in the Exploration and Use of Outer Space, GA Res. 1662/XVIII, 13 December 1963.

3 Treaty on Principles Governing the Activities of States in the Exploration and Use of Outer Space, including the Moon and Other Celestial Bodies, London, Moscow, Washington 27 January 1967, 610 UNTS 205. In force 10 October 1967.

4 F. Lyall and P.B. Larsen, *Space Law. A Treatise* (Ashgate 2009) 77–80.

5 *North Sea Continental Shelf* case [1969] ICJ Reports, 43.

6 *North Sea Continental Shelf* case, Diss. Op. Judge Sorensen [1969] ICJ Reports, 242.

7 Lyall and Larsen, *supra* note 4, at 77.

8 B. Cheng 'United Nations Resolutions on Outer Space: "Instant" International Customary Law?' (1965) IJIL 5, 23–48.

Space Station, which is governed by an intergovernmental agreement that mainly stipulates that the space treaties apply to the Space Station, with additional Memoranda of Understanding and Implementing Agreements with relevant details.[9] Such treaty-making can be seen as an illustration of the fact that space activities have become more commonplace and considerations around them are less concerned with the benefit for mankind than with practical considerations. Lyall and Larsen compare the modern law of outer space with the law relating to the sea, in the sense that there is on the one hand the law of the sea (public international law concerning the relations between States) and on the other hand maritime law (regulating commercial activities), a distinction that is becoming relevant also for space law.[10] Requirements on space legislation change when the nature of the activities changes.[11] More specific and concrete rules are needed and it can be more difficult to reach agreement on these, while at the same time States are more concerned about the need for detailed regulations when the questions are no longer only about research and experiments, but actually about activities that may make money. In such a situation, self-regulation may also find its place.

2. WHAT IS SELF-REGULATION?

There is no single accepted definition of self-regulation, although the general concept as such is quite easy to understand simply from the term itself: regulation by those same subjects that are the ones being regulated, or at least regulation by their peers, by creating some form of a regulatory body such as an industry association or similar. This body will develop rules, such as codes of conduct, and a system to enforce them, with its mandate deriving from the consent of its members.[12] This short description gives a good introduction to the attraction of self-regulation: if those being regulated make the regulations, the chances are that they

[9] This is well described at the European Space Agency (ESA) website: www.esa.int/Our_Activities/Human_Spaceflight/International_Space_Station/International_Space_Station_legal_framework (accessed 5 August 2017).

[10] Lyall and Larsen, *supra* note 4, at 559.

[11] A. Soucek, 'International law' (Chapter 3.4.1) in C. Brünner and A. Soucek (eds) *Outer Space in Society, Politics and Law* (Springer 2011) 294–405, at 360.

[12] B. Morgan and K. Yeung, *An Introduction to Law and Regulation. Texts and Materials.* (Cambridge University Press 2007, reprint 2009) 95.

will understand better than a third party the dangers of regulation, how it may threaten to overly restrict the activity in question and how respect may be obtained for necessary regulation. From the viewpoint of the State, the reasons to accept self-regulation are that it can provide a useful addition to formal regulation and if it works well, it may be possible for the State to refrain from regulating an activity, which can be beneficial as it saves costs, avoids legal cases and in some instances promotes values, such as freedom of the media in the case of Internet or press self-regulation. In an international environment such as that of outer space activities, the fact that self-regulation can transcend national borders is also attractive.

For new media, Keller explains the obvious advantages for the State of it being more effective to rely directly on the expertise of companies most closely involved with rapid technological and commercial changes; the State is able to transfer enforcement costs to the private sector and it can distance itself from tasks that may otherwise look like censorship.[13] Morgan and Yeung point out that there are also sceptics of self-regulation, who find that members of certain professions exaggerate the need for regulating themselves.[14] Nevertheless, the interest for self-regulation is growing. Morgan and Yeung place self-regulation in the context of the facilitative role of law, as this role is seen especially in the context of consensus and cooperation, which is needed for self-regulation. They highlight how many different instruments fall within a broad category of regulatory instruments based on the consent of the participants of the system; 'pure' self-regulation and many mixed systems. The consensual basis can come from legal support, from social consensus or from other reasons. What is essential is that the system is created not through direct State coercion (through law), but finds its origins in the will of the parties, even if this desire may be reached through a variety of means, including State encouragement.[15]

Self-regulation exists in many different contexts, and due to this, also in many different forms. For a system to be seen as self-regulation, it should be set up and managed by the sector it is regulating rather than by the State. Different forms of co-regulation are becoming increasingly important, with a self-regulatory system linked to the official State system in various ways. This may include the possibility for sanctions issued by the self-regulatory system to be realized through the State

---

13 P. Keller, *European and International Media Law* (Oxford University Press 2011) 31.

14 Morgan and Yeung, *supra* note 12, at 93.

15 *Ibid.*, 92.

system, or for a formal role for the self-regulatory system through legislation. The degree of interlinkage and interaction determines whether a system can still be seen as a self-regulatory one.

The term self-regulation can also be used for regulation by an individual subject of its own activities. This is also called consumer choice or responsibility, and it means that the person using a certain item or conducting an activity decides whether to protect himself against risks and how, to what extent and whether to interact with others, and in that case on what premises. Such self-regulation, without a unified system and relying only on the individual, is often not possible in modern society because of the high level of interdependence between actors, the risk of effects on others of any activity, potential environmental hazards and so on. It is simply not enough to rely solely on the judgement and goodwill of an individual in respect of any potentially risky activity. It may be possible to rely on such personal 'self-regulation' in cases where there are technical solutions to avoid risks, in the sense that machinery or other equipment can be designed to prevent risks, for example an automatic turn-off function when a piece of equipment overheats, or an inability to drive a vehicle faster than a pre-set speed. The use of such technology may be obligatory under the law instead of regulating the activity of individual users.[16] As technology develops so rapidly, there may be new ways to use technology to make choices that we cannot imagine today; the motor industry with its development of self-driving cars serves as a good illustration of the kind of potential that exists.

3. FORMS OF REGULATION OF OUTER SPACE ACTIVITIES

As mentioned above, space law started mainly through UN treaties, often anticipating issues that were thought to become relevant for outer space use. In the more recent decades, space law has been more reactive and related to concrete activities. If traditionally law comes after the facts, in fast-moving situations this may not always be possible and several authors have, according to Pop, used space as an example of an area

---

[16] For some examples and a discussion, see R. Brownsword and M. Goodwin, *Law and the Technologies of the Twenty-First Century. Texts and Materials* (Cambridge University Press 2012) 441–442.

where law cannot wait for facts.[17] Such a view is not universal among space lawyers, however, as some have also stressed that early lawmaking creates solutions looking for problems. However, the opposite may be that law lags too much behind facts.[18] In many space areas the level of predictability of the law is not satisfactory, as Kayser states.[19] This is despite the high costs and complexity of space activities, both of which contribute to parties having a high degree of interest in legal certainty. In such a situation a need arises for other forms of instruments to supplement legally binding ones, to clarify the situation, interpret notions and create obligations in different ways than only through national legislation.

The idea of codes of conduct, known in respect of outer space, is similar to self-regulation (and the terms are sometimes even used interchangeably).[20] The similarity is found in the legally non-binding nature of the instruments, which does not however mean that the parties drafting the instruments do not presume that they will be followed. The rules are drafted by those who will also observe them, which should mean that the parties are generally willing to implement the rules as they have been instrumental in their drafting. Space debris mitigation is an essential issue for outer space that until now has been handled by various codes of conduct or guidelines.[21] In the space context, the codes of conduct that exist (or are in the process of being drafted) are generally made by States, while self-regulation could involve private parties. There is nothing preventing hybrids between different categories or a mixture of

---

[17] V. Pop, *Who Owns the Moon? Extraterrestrial Aspects of Land and Mineral Resources Ownership* (Space Regulations Library series 4, Springer 2008) 27–28.

[18] Pop, *supra* note 17, at 27–28.

[19] V. Kayser, *Launching Space Objects: Issues of Liability and Future Prospects* (Space Regulations Library series 1, Kluwer 2001) 258.

[20] The term soft law is used as a common term for various types of non-binding instruments. It can be used for self-regulation, codes of conduct, etc., as the term itself is not strictly defined. Various soft law instruments are relevant for outer space, to fill gaps in binding rules and to deal with new emerging issues. On this see F. Tronchetti, 'Soft law' (Chapter 4.5) in C. Brünner and A. Soucek (eds) *Outer Space in Society, Politics and Law* (Springer 2011) 619–637.

[21] K.-U. Schrogl 'Space and its sustainable uses' (Chapter 4.4) in C. Brünner and A. Soucek (eds) *Outer Space in Society, Politics and Law* (Springer 2011) 604–618, at 605. Planetary protection is another area where very brief and general legal rules are complemented only by non-binding technical specifications. *Ibid.*, at 614.

parties. The International Telecommunication Union (ITU), for example, has already had private firms as members parallel to States for some time, even if the type of membership and associated rights differ between the categories.[22]

The European Union (EU) has been involved in the drafting and promotion of a Code of Conduct for Outer Space Activities,[23] focused to a large extent on space debris mitigation but relevant generally for the various uses of outer space. The code is not a piece of EU legislation; the EU External Action Service facilitates the negotiations that encompass any interested parties (UN members), and it is not intended to be a binding document.[24] Whether it is a first step toward later making a binding convention is an open question.[25] By early 2016 the work on the Code had largely stopped due to disagreements between States on its nature as well as on the process for creating such a document. Space debris mitigation as a topic for self-regulation would not be so far from the current system with guidelines and codes of conduct. The difference here would be more in the parties making the rules than in their content.

For a long time after the start of the space age, the number of countries that had space legislation was very low. Although, as in any area regulated by international conventions, there should be appropriate national rules to ensure that any international obligations are transformed into legal obligations at the national level, space was seen by many legislators as exotic and unlikely to be of practical interest. Some countries, such as Norway and Sweden, passed short space acts quite early to ensure systems of licensing, whereas other countries relied on various pieces of legislation (for example, on communications, administrative practices, State responsibility for dangerous activities and similar).

---

22 Codes of conduct may be seen as the first step, indeed as a stepping stone, towards binding agreements; but they do not have to be seen in this way. They can also have a value in their own right. See N.-L. Remuss, 'Space and security' (Chapter 4.2) in C. Brünner and A. Soucek (eds) *Outer Space in Society, Politics and Law* (Springer 2011) 519–568, at 539. Self-regulation is normally not seen as a step toward binding regulation but rather as an alternative to it.

23 http://www.consilium.europa.eu/uedocs/cmsUpload/st14455.en10.pdf (accessed 5 August 2017).

24 http://eeas.europa.eu/non-proliferation-and-disarmament/outer-space-activities/index_en.htm (accessed 5 August 2017).

25 On the Code of Conduct negotiations, see Remuss, *supra* note 22, at 540–545.

France, for example, passed a special space law only in 2008.[26] Many States may have felt (and still feel) that the amount of actual or potential space activities is so low that specialized legislation is not needed.

However, as Lyall and Larsen point out, if there is space legislation, the chances for effective and uniform control of space launches and other space activities improve.[27] National law does not lose its importance even if there is functioning self-regulation. Such regulation supplements the national legislation, fills the gaps that national law may leave, and for that reason in some instances may be shaped by national law. This raises a challenge in an international arena like outer space regarding what effect self-regulation carried out by an international body may have, if it does not fit with national laws. The voluntary nature of self-regulation is a benefit in this context as it may avoid questions about jurisdiction, if those involved in the self-regulatory system accept it through one or other mechanism. Such a mechanism and the extent of possible partners can be decided to fit the system in question. If there is an issue of enforcement, the jurisdiction issue – concerning whether the self-regulatory system should be linked to an official State enforcement system (like some press regulatory systems) – may return.

Even with the gradual privatization of space, this is still a sector where States play an important role. There is a legal reason for this: States are liable for all activities for which they are seen to be the launching State in accordance with the Outer Space Treaty (Article VII). It is not likely that the Treaty would be changed in this respect. The important role for the State is partly a consequence of the early State dominance that still remains to some extent, but there are also other reasons, like the importance of space for defence and the dual use nature of many space activities. Dual use gives rise to export control issues. This is an area where the USA and Europe have had differing views, with the USA being more restrictive but finding its policy challenged as the market for space components grows and becomes more international.[28] Previously, the high cost and high risk of space activities were reasons for State involvement, but with public budget cutbacks in almost all countries, private investment is needed instead to deal with such high costs and risks. Public–private partnerships for space have been investigated, for

---

[26] I. Marboe 'National space legislation' (Chapter 3.4.3) in C. Brünner and A. Soucek (eds) *Outer Space in Society, Politics and Law* (Springer 2011) 439–463, at 448.

[27] Lyall and Larsen, *supra* note 4, at 471.

[28] M.C. Mineiro, *Space Technology Export Controls and International Cooperation in Outer Space* (Springer 2012) 63.

example in satellite communications.[29] These different considerations in favour of State involvement in the sector to some extent speak against self-regulation, as the field for such regulation is not open; but at the same time this does not mean that no issues exist for which self-regulation may indeed be relevant. As mentioned, self-regulation can also encompass mixtures between the private and the public sector.

4. SELF-REGULATION: FROM CYBERSPACE TO OUTER SPACE?

The question of the need for or usefulness of a self-regulatory system is linked to liberalization of the sector in question, since in a heavily regulated and/or State-run sector there is little room for self-regulation. This is why the question may become relevant for the space sector in a not too distant future. The same background led to the interest for self-regulation in the communications law sphere, for media, telecommunications and in particular anything related to modern online media or various services using the Internet.

Self-regulation has proven to be useful in areas where there is rapid change, for example through the introduction of new technologies. Regulation produced by the affected parties can more quickly and easily take into consideration all sorts of new factors that the legislator may struggle with, given the time usually involved in the regular legislative process; technological developments may occur so quickly that once a draft law has reached the stage of adoption, some of its content is obsolete. Modern media (primarily the Internet) is one such area. Keller discusses how the contemporary liberal democratic State, which is porous and interlinked with non-State actors, has broadened its methods of law and administration with co- and self-regulation broadly used to maintain effective and responsive control over new media.[30] Considerations of media regulation are quite different from those related to space uses (for example, due to the sensitive issue of freedom of expression and how to ensure this while also balancing it with other rights such as privacy). However, some of the considerations for regulation of modern media and cyberspace are similar to issues that can arise for outer space uses. This includes not only the mentioned issue of rapid technological change (and the need to understand the technical side of what is regulated) but also

---

29 C. Venet, 'The economic dimension' (Chapter 2.1.3) in C. Brünner and A. Soucek (eds), *Outer Space in Society, Politics and Law* (Springer 2011) 61.

30 Keller, *supra* note 13, at 5 and 30.

the international character of the area, where problems stemming from a lack of jurisdiction for any one State may mean that co- and self-regulation could be a necessary alternative.[31] These similarities mean that the forces that have driven both States and sector participants to embrace self-regulation in cyberspace may be relevant also for outer space, even if the regulation emanating from the efforts would likely look rather different as far as substantive provisions as well as enforcement systems are concerned.

For cyberspace, there may be an overly optimistic attitude to self-regulation as it is the only acceptable form of regulation for many of the sector participants. The Internet developed so quickly that it has remained largely unregulated, as legislators and regulators were faced with a *fait accompli* that was difficult to address once they had caught up with developments. Authoritarian States make various attempts to restrict the Internet, while societies with freedom of expression and free media mainly lack specific regulation and merely rely on regular criminal law and similar.[32] This is not without problems, when matters such as copyright prove difficult to regulate in the cyber environment and when cybercriminals can so easily move from one jurisdiction to another, as the physical location of the server is irrelevant for their activities. There is also a tension between data protection (privacy) and the free flow of information. This creates a system in which self-regulation is appropriate, but whether that also means that sector participants are interested in it is another matter. Rather, in the situation where the alternative to self-regulation appears to be not official regulation but rather no regulation at all, the incentive for the actors to regulate must look different. This said, many authors agree that cyberspace has necessitated taking a new look at regulation, which definitely includes an important place for self-regulation.[33]

---

31 *Ibid.*, 32–33.

32 The United States in its framework policy for global electronic commerce from 1997 expressed that the policy for the new medium should be minimal State intervention with an emphasis on market forces. Similar pronouncements were made by the EU in their Green Paper on regulatory implications of communications convergence, also in 1997. Keller, *supra* note 13, at 94–95.

33 Brownsword and Goodwin, *supra* note 16, at 348. For data protection in the EU, an emphasis on self-regulation to enable less strict official regulation is an important element in the debate. See, for example, Keller, *supra* note 13, at 340.

5. ISSUES FOR SELF-REGULATION OF OUTER SPACE ACTIVITIES

Initially, space exploration and use was to a large extent handled by States through State organs. States were the main actors and clients for the most space activities, and these activities were seen to be of scientific and technological interest rather than of economic benefit.[34] The exception was communications satellites, where especially (and initially exclusively) in the USA the private sector was active from the start. Even in this context, there was a different kind of State involvement than in other totally private areas of the economy, as State involvement was still needed for launching activities. Gradually, space activities have become privatized to an increasing extent.

There was a general trend of liberalization of various previously State-owned and State-run sectors starting in the 1990s (or to some extent the late 1980s), such as telecommunications, transport and energy. This coincided with the fall of the Soviet Union and the end of communism in Europe, so this development was chronologically not so different in the former Eastern bloc as it was in the Western European States. The developments affected space primarily through communications liberalization; Venet mentions the World Trade Organization (WTO) Agreement on Basic Telecommunications from 1997 as well as the privatization of INMARSAT and INTELSAT.[35] INTELSAT was privatized in 2001 and INMARSAT in 1998. This means that communication satellite capability nowadays rests mainly with private firms.[36] The private companies are linked with special organizations (the International Telecommunications Satellite Organization (ITSO)[37] for INTELSAT) that manage the aspects that private companies do not handle, as a legacy of the creation of the organization as an international governmental body. However, by and large the sector is dealt with as any commercial sector.

Not only communications activity but also related activities like launching are becoming privatized, even if not yet to the same extent. The passing of the 1984 Commercial Space Launch Act of the United States (amendment Act 2004) meant that launching was opened up for private entities. The Act establishes a licensing system and various

---

[34] Venet, *supra* note 29, at 55.

[35] *Ibid.*, at 56.

[36] K. Nyman-Metcalf, 'Space for the benefit of mankind? New developments and old problems' in *Annals of Air and Space Law Vol. XXXIV* (Institute of Air and Space Law, McGill University 2009) 621–644, at 639–641.

[37] http://www.itso.int (accessed 1 September 2013).

requirements for licensees, including demands for insurance or other means to meet any claims.[38] The situation regarding requirements for insurance or more generally the handling of risks present at different times in the space activities vary in different national legislations. This is still an emerging issue and there are no detailed international rules or international harmonization.[39] With more States participating in space activities and a growing market for satellites, the market for launch services also became interesting and international trade increased. This coincided with the United States changing its policy to favour commercialization of space (after the 1986 Challenger space shuttle disaster), as well as with the fall of the Soviet Union; therefore, the preconditions for increased private activity in space were quite rapidly created.[40] In space use language, there are new diverse, private payloads waiting for launch. Launching is still a predominantly State-run activity (in Europe, through ESA), but commercial launching capabilities are being developed. The possible potential for self-regulation may arise in relation to both payloads and launching.

Traffic rules for space or space traffic management is increasingly discussed as space use intensifies.[41] This is an area where self-regulation could give the parties the chance to influence regulation so that it meets their needs and is not overly restrictive. It is not too early to think about such regulation, given how fast developments may occur with new technologies. Cyberspace may be used as an analogy here: developments were much faster than most people could have predicted, so any discussions on if and how to regulate this area tended to be bypassed by developments, thus rendering regulation attempts obsolete. The kind of rules for space traffic management for which Schrogl sees a need[42] (with information requirements including definitions of what information is needed and how it should be procured, safety provisions and management and implementation rules) can fit into the self-regulatory concept.

The question of space debris has until now been regulated by various instruments such as codes of conduct, which are a form of self-regulation

---

38 Marboe, *supra* note 26, at 445.

39 C. Gaubert, 'Insurance in the context of national authorisation' (Chapter 5) in F. von der Dunk (ed.) *National Space Legislation in Europe* (Martinus Nijhoff 2011) 163–177.

40 Mineiro, *supra* note 28, at 55. The Space Law Committee of the International Law Association has developed a draft Model Law on National Space Legislation.

41 Schrogl, *supra* note 21, at 609.

42 *Ibid.*, 610.

although they do not fit perfectly with the most accepted definition of this type of regulation, being produced to a large extent by States. But as States are still important actors in space and as other actors are able to have some input in the discussion of such rules (even if so far these actors are mainly international organizations rather than private firms), there are many similarities with the main principles of self-regulation. Complementing the codes of conduct between States (or States and other parties) with codes of conduct for private parties can be a very useful way to manage this important space use issue. Codes of conduct drawn up by States in many instances include provisions on how standards can also be imposed on private parties.

Other new issues like the question of space assets have led to some new initiatives, though not yet to any binding new rules. New areas of law like private international law have become relevant. In 2001 the International Institute for the Unification of Private International Law (UNIDROIT) completed the Space Assets Protocol to the Convention on International Interests in Mobile Equipment, intending to provide a registry for space assets, primarily for financing purposes.[43] The Protocol has not yet been adopted by the institute but is in draft form. Such rules may find support in self-regulatory rules.

Space tourism is a new use of outer space that attracts a lot of attention as it tickles the imagination and creates a lot of media attention, even if it is still a marginal activity. The possibility of space tourists raises new legal issues, among other reasons due to the way people going into space have been defined previously, that is to say as astronauts: envoys of mankind. The reason for such language and for provisions related to such a view on persons going to space is linked to the understanding that they go there in order to undertake experiments, scientific work and generally carry out activities that are of benefit for mankind. Persons travelling to space purely for their own pleasure do not fit well into this category and into existing definitions.[44] The 2004 US Commercial Space Launch Amendments Act is one of the very few legal reflections of the possibility of space tourists, introducing requirements for operators on explicit informed consent of customers.[45] However, the Federal Aviation Administration (FAA) in the USA has refrained from regulating the sector in order to let it get started. The Commercial Space Launch

---

[43] V. Kopal, 'Origins of space law and the role of the United Nations' (Chapter 3.1) in C. Brünner and A. Soucek (eds) *Outer Space in Society, Politics and Law* (Springer 2011) 221–233, at 230. Also Walter, *supra* note 1, at 508.

[44] Soucek, *supra* note 11, at 330. See also Walter, *supra* note 1, at 500–501.

[45] Marboe, *supra* note 26, at 446.

Competitiveness Act of November 2015 extended the period during which the FAA will not regulate until October 2023.[46] In Russia, which has experience of space tourists, the legal issues have been handled on a case-by-case basis with no specific legal requirements.[47] The nature of the issue is such that both national and international rules will be needed, if and when the phenomenon becomes more widespread.[48] As far as Europe is concerned, there may also be scope for EU law, given that air law to a large extent is affected by EU law, not least as far as passenger rights are concerned.[49] The definition of space tourists vis-à-vis existing provisions on space travellers as well as other rules for such persons could be a possible subject for self-regulation. Another possibility would be that the definition is created in a legal instrument but the rules of behaviour for the tourists, or perhaps rather for the firms taking them into space, would be suitable for self-regulation.

A further topic that could profit from self-regulation is access to data from remote sensing. Here there exists a mixture of rules from different instruments, including agreements and voluntary decisions by parties possessing the data. As there is likely to be more and more emphasis on data and its availability, some self-regulatory principles and rules could add important clarity.[50]

Self-regulation may be handled by an existing organization, that is to say an organization that is not established for this purpose but to which the parties assign some powers through consent. Indeed, this may be a good way to achieve some credibility for the system even from the start, provided of course that the organization is well chosen. If self-regulation becomes relevant for space activities, there are various organizations to choose from. There are global as well as regional organizations and although none of them appear automatically suitable and relevant, this could be reconsidered when the extent and type of regulation required is known.

---

[46] http://spacenews.com/industry-committee-to-start-work-on-human-space flight-safety-standards/ (accessed 7 August 2017).

[47] Marboe, *supra* note 26, at 448.

[48] Lyall and Larsen, *supra* note 4, at 565.

[49] M. Gerhard, 'Space tourism – the authorisation of sub-orbital space transportation' (Chapter 9) in F. von der Dunk (ed.) *National Space Legislation in Europe* (Martinus Nijhoff 2011) 263–296, especially at 287 and 294–296.

[50] On access to data, see A. Soucek *et al.*, 'Data policies: a survey of principles and trends' (Chapter 4.7) in C. Brünner and A. Soucek (eds), *Outer Space in Society, Politics and Law* (Springer 2011) 665–689.

6. SOME CONCLUDING REMARKS

Self-regulation does presume a desire to cooperate, at least to some extent. It is not necessary to have perfectly functioning cooperation between various parties for self-regulation to work, but there is a need for a minimum level of mutual interest for any system to get the required amount of adherence. Outer space has shown both examples of a 'space race' between rival countries and (at the same time!) a surprising willingness to cooperate with UN involvement as well as other joint efforts even in the early days of space use, in the very cold period of the Cold War. The international political situation today might appear more suitable for international cooperation since the world has become increasingly globalized and most States have close relations with other States, even those with whom they do not have entirely friendly relations. However, in the context of this chapter, one new parameter is of interest, namely the privatization of space activities. The motivation for private firms to cooperate is different to that of States. The reasons to want or not to want to cooperate are different: these are questions of commercial interests and competition rather than State prestige and international law initiatives. When discussing the motivation for self-regulation, there is an understanding that there exists a kind of trade-off: if there is functioning self-regulation, there is less need for State regulation. This can be a strong incentive to self-regulate. If a viable system is created, the temptation for the State to regulate lessens as it sees that the sector is in fact handled in a competent manner.

Space use has become much more commonplace. Launches are no longer the headline news that they were in the first decades of space use. At the same time, use of outer space is still a high-tech and in many ways high-risk environment. The perception of it from those not engaged themselves may be that it is even more complex and risky than it actually is. Whatever the case, it is an area where much of the activity is alien to those not directly involved, who are therefore not best placed to consider the most suitable type of regulation. Perhaps those involved are willing to take certain risks that the outside observer deems too high. Perhaps some activities are not as risky as they appear to be, but the objective description of them still is.

Self-regulation has practical benefits in that it can take into account technological changes quickly and react flexibly. It is regulation made by experts: those who are to be regulated make the rules and they know what is needed to conduct an activity. But in addition to this, a moral case can also be made for self-regulation: it may help to ensure that those involved in an activity take responsibility and act ethically. From this viewpoint, it is not too early to think about self-regulation for any activity, in outer space or elsewhere.

Index